T0183612

Lecture Notes in Computer Science 12337

Founding Editors

Gerhard Goos
  Karlsruhe Institute of Technology, Karlsruhe, Germany
Juris Hartmanis
  Cornell University, Ithaca, NY, USA

Editorial Board Members

Elisa Bertino
  Purdue University, West Lafayette, IN, USA
Wen Gao
  Peking University, Beijing, China
Bernhard Steffen
  TU Dortmund University, Dortmund, Germany
Gerhard Woeginger
  RWTH Aachen, Aachen, Germany
Moti Yung
  Columbia University, New York, NY, USA

More information about this series at http://www.springer.com/series/7407

Jianer Chen · Qilong Feng ·
Jinhui Xu (Eds.)

Theory and Applications
of Models of Computation

16th International Conference, TAMC 2020
Changsha, China, October 18–20, 2020
Proceedings

 Springer

Editors
Jianer Chen
Department of Computer Science
Texas A&M University
College Station, TX, USA

Qilong Feng
School of Computer Science
and Engineering
Central South University
Changsha, China

Jinhui Xu
Department of Computer Science
and Engineering
State University of New York at Buffalo
Buffalo, NY, USA

ISSN 0302-9743   ISSN 1611-3349   (electronic)
Lecture Notes in Computer Science
ISBN 978-3-030-59266-0   ISBN 978-3-030-59267-7   (eBook)
https://doi.org/10.1007/978-3-030-59267-7

LNCS Sublibrary: SL1 – Theoretical Computer Science and General Issues

© Springer Nature Switzerland AG 2020
This work is subject to copyright. All rights are reserved by the Publisher, whether the whole or part of the material is concerned, specifically the rights of translation, reprinting, reuse of illustrations, recitation, broadcasting, reproduction on microfilms or in any other physical way, and transmission or information storage and retrieval, electronic adaptation, computer software, or by similar or dissimilar methodology now known or hereafter developed.
The use of general descriptive names, registered names, trademarks, service marks, etc. in this publication does not imply, even in the absence of a specific statement, that such names are exempt from the relevant protective laws and regulations and therefore free for general use.
The publisher, the authors and the editors are safe to assume that the advice and information in this book are believed to be true and accurate at the date of publication. Neither the publisher nor the authors or the editors give a warranty, expressed or implied, with respect to the material contained herein or for any errors or omissions that may have been made. The publisher remains neutral with regard to jurisdictional claims in published maps and institutional affiliations.

This Springer imprint is published by the registered company Springer Nature Switzerland AG
The registered company address is: Gewerbestrasse 11, 6330 Cham, Switzerland

Preface

The 16th Annual Conference on Theory and Applications of Models of Computation (TAMC 2020) was held during October 18–20, 2020, in Changsha, China. The workshop brings together researchers working on all aspects of computer science for the exchange of ideas and results.

TAMC 2020 was the 16th conference in the series. The main themes of TAMC 2020 were computability, complexity, algorithms, information theory and their extensions to machine learning theory, and foundations of artificial intelligence. 83 submissions were received from more than 13 countries and regions. The TAMC 2020 Program Committee selected 37 papers for presentation at the conference. In addition, we had two plenary speakers, Gregory Gutin (Royal Holloway, University of London, UK) and Xianfeng David Gu (State University of New York at Stony Brook, USA). Thanks for their contributions to the conference and proceedings.

We would like to thank the Program Committee members and external reviewers for their hard work in reviewing and selecting papers. We are also very grateful to all the authors who submitted their work to TAMC 2020. We thank the members of the Editorial Board who agreed to publish this volume in the *Lecture Notes in Computer Science* series and the editors at Springer for their encouragement, cooperation, and hard work throughout the preparation of these proceedings.

May 2020

Jianer Chen
Qilong Feng
Jinhui Xu

Organization

Program Committee

Anthony Bonato	Ryerson University, Canada
Yixin Cao	The Hong Kong Polytechnic University, Hong Kong, China
Jianer Chen	Texas A&M University, USA
Hu Ding	University of Science and Technology of China, China
Thomas Erlebach	University of Leicester, UK
Qilong Feng	Central South University, China
Seok-Hee Hong	The University of Sydney, Australia
Ziyun Huang	Penn State Erie, The Behrend College, USA
Aaron D. Jaggard	U.S. Naval Research Laboratory, USA
Steffen Lempp	University of Wisconsin-Madison, USA
Jian Li	Tsinghua University, China
Shi Li	State University of New York at Buffalo, USA
Mia Minnes	University of California, San Diego, USA
Evanthia Papadopoulou	University of Lugano (USI), Switzerland
A. Pavan	Iowa State University, USA
Anil Seth	IIT Kanpur, India
Xiaoming Sun	Institute of Computing Technology, Chinese Academy of Sciences, China
Shin-Ichi Tanigawa	The University of Tokyo, Japan
Takeshi Tokuyama	Tohoku University, Japan
Haitao Wang	Utah State University, USA
Lusheng Wang	City University of Hong Kong, Hong Kong, China
Ge Xia	Lafayette College, USA
Jinhui Xu	State University of New York at Buffalo, USA
Boting Yang	University of Regina, Canada
Christos Zaroliagis	CTI, University of Patras, Greece
Guochuan Zhang	Zhejiang University, China
Huaming Zhang	University of Alabama in Huntsville, China
Martin Ziegler	KAIST, China

Additional Reviewers

Abam, Mohammad Ali
Ackerman, Eyal
Coiteux-Roy, Xavier
Crole, Roy
Della Vedova, Gianluca
Deng, Shichuan
Eiben, Eduard
Fujii, Kaito
Guan, Chaowen
Higashikawa, Yuya
Hitchcock, John M.
Horn, Paul
Huang, Jiawei
Huang, Lingxiao
Huang, Zengfeng
Iwamasa, Yuni
Jiang, Zhihao
Jung, Achim
Kanellopoulos, Panagiotis
Kawachi, Akinori
Kontogiannis, Spyros
Krishnaswamy, Ravishankar
Kulkarni, Janardhan
LeGall, Francois
Li, Wenjun

Loff, Bruno
Lukovszki, Tamas
Martin, Keye
Niewerth, Matthias
Nomikos, Christos
Qin, Ruizhe
Quanrud, Kent
Rajagopal Padmanabhan, Madhavan
Shi, Feng
Shioura, Akiyoshi
Soma, Tasuku
Tsichlas, Kostas
Variyam, Vinodchandran
lachos, Evangelos
Voudouris, Alexandros
Wang, Di
Wang, Minghua
Wang, Zixiu
Ward, Justin
Yang, Fan
Yu, Haikuo
Yu, Wei
Zhang, Peng
Zhang, Tianyi
Zhou, Yufan

Contents

Semilattices of Punctual Numberings

Nikolay Bazhenov$^{1,2(\boxtimes)}$ ⓘ, Manat Mustafa^3 ⓘ, and Sergei Ospichev^{1,2} ⓘ

^{1} Sobolev Institute of Mathematics, 4 Acad. Koptyug Avenue,
Novosibirsk 630090, Russia
^{2} Novosibirsk State University, 2 Pirogova St., Novosibirsk 630090, Russia
{bazhenov,ospichev}@math.nsc.ru
^{3} Department of Mathematics, School of Sciences and Humanities,
Nazarbayev University, 53 Qabanbaybatyr Avenue, Nur-Sultan 010000, Kazakhstan
manat.mustafa@nu.edu.kz

Abstract. The theory of numberings studies uniform computations for classes of mathematical objects. A large body of literature is devoted to investigations of computable numberings, i.e. uniform enumerations for families of computably enumerable sets, and the reducibility $\leq$ among these numberings. This reducibility, induced by Turing computable functions, aims to classify the algorithmic complexity of numberings.

The paper is inspired by the recent advances in the area of punctual algebraic structures. We recast the classical studies of numberings in the punctual setting—we study *punctual numberings*, i.e. uniform computations for families of primitive recursive functions. The reducibility $\leq_{pr}$ between punctual numberings is induced by primitive recursive functions. This approach gives rise to upper semilattices of degrees, which are called *Rogers pr-semilattices*. We prove that any infinite Rogers pr-semilattice is dense and does not have minimal elements. Furthermore, we give an example of infinite Rogers pr-semilattice, which is a lattice. These results exhibit interesting phenomena, which do not occur in the classical case of computable numberings and their semilattices.

Keywords: Numbering · Upper semilattice · Rogers semilattice · Primitive recursion · Friedberg numbering · Online computation · Punctual structure

1 Introduction

The theory of numberings gives a formal approach to studying uniform computations for classes of mathematical objects. One of the first important applications of this theory is provided by Gödel [15], who employed an effective numbering of

---

The work was supported by Nazarbayev University Faculty Development Competitive Research Grants N090118FD5342. The first author was partially supported by the grant of the President of the Russian Federation (No. MK-1214.2019.1). The third author was partially supported by the program of fundamental scientific researches of the SB RAS No. I.1.1, project No. 0314-2019-0002.

© Springer Nature Switzerland AG 2020
J. Chen et al. (Eds.): TAMC 2020, LNCS 12337, pp. 1–12, 2020.
https://doi.org/10.1007/978-3-030-59267-7_1

first-order formulae in the proof of his seminal incompleteness theorems. Kleene's results [24], in a sense, created numberings as a separate object of study: in particular, he constructed a universal partial computable function. After that, the foundations of the modern theory of numberings were developed by Kolmogorov and Uspenskii [25,35] and, independently, by Rogers [33].

Let S be a countable set. A *numbering* of S is a surjective map ν from the set of natural numbers ω onto S. A standard tool for measuring the algorithmic complexity of numberings is provided by the notion of *reducibility* between numberings: A numbering ν is *reducible* to another numbering μ (denoted by $\nu \leq \mu$) if there is total computable function $f(x)$ such that $\nu(n) = \mu(f(n))$ for all $n \in \omega$. In other words, there is an effective procedure which, given a ν-index of an object from S, computes a μ-index for the same object.

Since 1960s, the investigations of *computable numberings* have become a fruitful area of research. Let S be a countable family of computably enumerable (c.e.) sets. A numbering ν of the family S is *computable* if the set

$$\{\langle n, x \rangle : n \in \omega, \ x \in \nu(n)\}$$

is c.e. The family S is *computable* if it has a computable numbering. Informally speaking, the computability of S means that there is a procedure, which provides a *uniform enumeration* of the family S.

In a standard recursion-theoretic way, the notion of reducibility between numberings give rise to the *Rogers upper semilattice* (or *Rogers semilattice* for short) of a computable family S: This semilattice contains the degrees of all computable numberings of S. Here two numberings have the same degree if they are reducible to each other. Rogers semilattices allow one to measure computations of a given family and also used as a tool to classify properties of computable numberings for different families.

To name only a few, computable numberings and the corresponding Rogers semilattices were studied by Badaev [3,4], Ershov [11,12], Friedberg [14], Goncharov [16,17], Lachlan [26,27], Mal'tsev [28], Pour-El [32], Selivanov [34], and many other researchers. Note that computable numberings are closely connected to algorithmic learning theory (see, e.g., the recent papers [1,9,21]). For a survey of results and bibliographical references on computable numberings, the reader is referred to the monograph [12] and the articles [2,5,13].

Goncharov and Sorbi [19] started developing the theory of generalized computable numberings: roughly speaking, these numberings provide uniform descriptions for families of sets belonging to the levels of various computability-theoretic hierarchies. In this direction, much work has been done for the hyperarithmetical hierarchy [2,7,31] and the Ershov hierarchy [6,18,20,30].

The prior investigations in the theory of numberings are mainly motivated by the general area of *computable* or *effective mathematics*. This area aims to understand and calibrate the algorithmic content of mathematical objects. The roots of this direction go back to the introduction of non-recursive mathematical methods at the beginning of the 20th century, as discussed in [29]. Following the agenda of computable mathematics, the theory of numberings generally employs the Turing computability framework.

Our paper is inspired by the recent developments in computable structure theory: Kalimullin, Melnikov, and Ng [22] introduced the notion of a *punctual* (or *fully primitive recursive*) *structure*. An infinite structure S in a finite signature is *punctual* if the domain of S is equal to ω, and the basic functions and relations of S are primitive recursive.

The notion of punctuality essentially *eliminates* all instances of unbounded search in Turing computable algorithms. This feature allows one to mimic any reasonable "online" algorithm, i.e. an algorithm, which has to make decisions on the fly. A typical example of such an algorithm is online colouring, say, of a tree: given the nth vertex of an input tree, you have to decide its colour right at the moment (you cannot wait for the $(n+1)$-th vertex to appear). The reader is referred to the survey [8] for a detailed discussion of the motivation behind punctuality, the known results in this area, and further bibliographical references.

The aim of this paper is to introduce online computational models into the study of numberings. Our approach is based on the punctuality paradigm of [8, 22].

Definition 1.1. *Let S be a family of total functions acting from ω into ω. We say that a numbering ν of the family S is* punctual *if the function*

$$g_\nu(n, x) := (\nu(n))(x)$$

is primitive recursive. A family S is punctual *if it has a punctual numbering.*

Informally speaking, the punctuality of S means that one can promptly (and uniformly) compute every function from S. The punctuality paradigm requires that we have to *modify* the notion of the reduction between numberings accordingly:

Definition 1.2. *Let ν and μ be numberings. We say that ν is* punctually reducible *to μ, denoted by $\nu \leq_{pr} \mu$, if there is a primitive recursive function $f : \omega \to \omega$ such that*

$$\nu(n) = \mu(f(n)), \text{ for all } n \in \omega.$$

In this case, we say that f punctually reduces ν to μ.

In a natural way, Definitions 1.1 and 1.2 give rise to the notion of *Rogers pr-semilattice* of a punctual family S, see Sect. 2 for a formal definition. Most of the problems on numberings in terms of Rogers semilattices, in the general setting, can be formulated as follows:

- Find global algebraic properties of Rogers semilattices (such as cardinality, type of the algebraic structure, ideals, segments, covers, etc.)
- Describe invariants, and among them the number of maximal and minimal elements, to distinguish different Rogers semilattices.
- Classify numberings which generate special elements in Rogers semilattices (extremal elements, limit points, split elements, etc.).

Based on the motivation above, in our paper we show that the theory of punctual numberings exhibits striking differences with the classical numbering theory—The main results of the article can be summarized as follows. Let $\mathcal{S}$ be an infinite punctual family.

- Below an arbitrary degree from $\mathcal{R}_{pr}(\mathcal{S})$, one can build an infinite descending chain and an infinite antichain (Theorems 3.1 and 4.1). In contrast, in the classical setting, the Rogers semilattice of, say, the family $\mathcal{T}_0 := \{\emptyset\} \cup \{\{i\} : i \in \omega\}$ has the *least* element.
- The semilattice $\mathcal{R}_{pr}(\mathcal{S})$ is dense (Theorem 6.1), i.e. given two degrees $\mathbf{a} < \mathbf{b}$, there is always a degree $\mathbf{c}$ with $\mathbf{a} < \mathbf{c} < \mathbf{b}$. On the other hand, standard recursion-theoretic methods show that the classical Rogers semilattice of $\mathcal{T}_0$ has an initial segment containing precisely two degrees.
- There is a simple example of a family $\mathcal{S}_1$ such that the structure $\mathcal{R}_{pr}(\mathcal{S}_1)$ is an infinite lattice (Proposition 5.1). Note that Selivanov [34] proved that any infinite classical Rogers semilattice *cannot* be a lattice.

The proofs of these results employ some techniques developed for finitely generated punctual structures in [22] and Section 7 of [8].

2 Preliminaries and General Facts

Given a total function $f \colon \omega \to \omega$ and a non-zero natural number m, by $f \upharpoonright m$ we denote the following tuple:

$$f \upharpoonright m := (f(0), f(1), \ldots, f(m-1)).$$

For a set $A \subseteq \omega$, by χ_A we denote the characteristic function of A. For a pair of natural numbers (k, ℓ), the value $\langle k, \ell \rangle$ is its standard Cantor index, i.e.

$$\langle k, \ell \rangle = \frac{(k+\ell)(k+\ell+1)}{2} + k.$$

Suppose that ν is a numbering of a family $\mathcal{S}_0$, and μ is a numbering of a family $\mathcal{S}_1$. Note that the condition $\nu \leq \mu$ always implies that $\mathcal{S}_0 \subseteq \mathcal{S}_1$. Clearly, if $\nu \leq_{pr} \mu$, then $\nu \leq \mu$.

Numberings ν and μ are *equivalent* (denoted by $\nu \equiv \mu$) if $\nu \leq \mu$ and $\mu \leq \nu$. The *punctual equivalence* $\equiv_{pr}$ is defined in a similar way. The numbering $\nu \oplus \mu$ of the family $\mathcal{S}_0 \cup \mathcal{S}_1$ is defined as follows:

$$(\nu \oplus \mu)(2x) = \nu(x), \quad (\nu \oplus \mu)(2x+1) = \mu(x).$$

It is not hard to establish the following fact (see, e.g., Proposition 3 in [12, p. 36]). If $\trianglelefteq \in \{\leq, \leq_{pr}\}$ and ξ is a numbering of a family $\mathcal{S}_2$, then

$$(\nu \trianglelefteq \xi \,\&\, \mu \trianglelefteq \xi) \iff (\nu \oplus \mu \trianglelefteq \xi).$$

Let $\mathcal{S}$ be a punctual family of functions. By $Com_{pr}(\mathcal{S})$ we denote the set of all punctual numberings of $\mathcal{S}$. Since the relation $\equiv_{pr}$ is a congruence on the structure

$(Com_{pr}(\mathcal{S}); \leq_{pr}, \oplus)$, we use the same symbols $\leq_{pr}$ and $\oplus$ on numberings of $\mathcal{S}$ and on $\equiv_{pr}$-equivalence classes of these numberings.

The quotient structure $\mathcal{R}_{pr}(\mathcal{S}) := (Com_{pr}(\mathcal{S})/\equiv_{pr}; \leq_{pr}, \oplus)$ is an upper semi-lattice. We call the structure $\mathcal{R}_{pr}(\mathcal{S})$ the *Rogers pr-semilattice* of the punctual family $\mathcal{S}$.

Let $\mathcal{T}$ be a family of (Turing) computable functions acting from ω into ω. A numbering ν of the family $\mathcal{T}$ is *computable* if the function g_ν from Definition 1.1 is computable. Note that this definition is consistent with the notion of computable numbering from the introduction: If we identify functions from $\mathcal{T}$ with their graphs, then we will get precisely the same notions.

We say that a family $\mathcal{T}$ is *Turing computable* if it has a computable number-ing. The definition of *Rogers semilattice* $\mathcal{R}_c(\mathcal{T})$ is obtained in a similar way to the semilattice $\mathcal{R}_{pr}(\mathcal{S})$, modulo the following modification: one needs to consider all computable numberings of $\mathcal{T}$ and the standard reducibility $\leq$ between them.

If ν is a numbering, then by η_ν we denote the corresponding equivalence relation on ω:

$$m \ \eta_\nu \ n \ \Leftrightarrow \ \nu(m) = \nu(n).$$

A numbering ν is *negative* if the relation η_ν is co-c.e., i.e. η_ν is the complement of a c.e. set. A numbering ν is *decidable* if the relation η_ν is (Turing) computable. Numbering ν is *Friedberg* if η_ν is the identity relation.

2.1 Basics of Punctuality

The *restricted Church–Turing thesis* for primitive recursive functions says the following: A function is primitive recursive if and only if it can be described by an algorithm that uses only bounded loops. Informally speaking, one needs to eliminate all instances of `while ... do`, `repeat ... until`, and `goto` in a Pascal-like programming language.

Our proofs will exploit the restricted Church–Turing thesis without an explicit reference.

We fix a computable list $(p_e)_{e \in \omega}$ of all unary primitive recursive functions. We emphasize that the list is computable, but it *cannot be* primitive recursive. Nevertheless, the following function *can be* treated as a punctual object:

$$p_e[t](x) := \begin{cases} p_e(x), & \text{if the value } p_e(x) \text{ is computed in} \\ & \text{at most } t \text{ computational steps,} \\ \text{undefined}, & \text{otherwise.} \end{cases}$$

Without loss of generality, one may also assume the following: if $p_e(x)$ is equal to N, then for any $t \leq \max(e, x, N)$, the value $p_e[t](x)$ is undefined. The formal details can be recovered from Section 10 of [8].

2.2 First Facts About Punctual Numberings

Proposition 2.1. *Let $\mathcal{S}$ be a finite punctual family. Then the semilattice $\mathcal{R}_{pr}(\mathcal{S})$ contains precisely one element.*

Proof. Suppose that $\mathcal{S} = \{f_0, f_1, \ldots, f_m\}$. We fix a natural number N such that the strings $f_i \upharpoonright N$, $i \leq m$, are pairwise different.

It is sufficient to establish the following fact: For arbitrary punctual numberings ν and μ of $\mathcal{S}$, one can build a function g, which punctually reduces ν to μ.

The desired g is constructed as follows. We fix indices a_i, $i \leq m$, such that $\mu(a_i) = f_i$. For an arbitrary index $k \in \omega$, we promptly find the number $j \leq m$ such that $\nu(k) \upharpoonright N = f_j \upharpoonright N$. Then we define $g(k) := a_j$.   □

Let $\mathcal{T}$ be an infinite, Turing computable family of functions. Mal'tsev [28] showed that every computable numbering of $\mathcal{T}$ is negative. Ershov [10] proved that for any computable numbering ν of $\mathcal{T}$, there is a computable Friedberg numbering μ of $\mathcal{T}$ such that $\mu \leq \nu$.

Here we adapt these facts to the punctual setting.

Definition 2.1. *Let $\mathcal{S}$ be a punctual family, and let ν be a punctual numbering of $\mathcal{S}$. We say that ν is* strongly punctually decidable *(or spd for short) if it satisfies the following:*

1. *the equivalence relation η_ν is primitive recursive, and*
2. *for any η_ν-equivalence class C, either C contains only one element, or $C = [0]_{\eta_\nu}$.*

Proposition 2.2. *Suppose that $\mathcal{S}$ is a punctual family, and ν is a punctual numbering of $\mathcal{S}$. Then:*

(a) ν is negative, and
(b) there is a strongly punctually decidable numbering $\mu \in Com_{pr}(\mathcal{S})$ such that $\mu \leq_{pr} \nu$.

Proof. The item (a) is an easy corollary of the Mal'tsev's result mentioned above. We sketch the proof of the item (b).

Since the numbering ν is negative, the set

$$I_\nu := \{n \in \omega : (\forall i < n)(\nu(i) \neq \nu(n))\}$$

is computably enumerable. Choose a primitive recursive function h such that $range(h) = I_\nu$. We define a new primitive recursive function

$$\widehat{h}(x) := \begin{cases} h(x), & \text{if } (\forall i < x)(h(i) \neq h(x)), \\ h(0), & \text{otherwise.} \end{cases}$$

The desired numbering is defined as $\mu := \nu \circ \widehat{h}$. It is not hard to establish that μ satisfies the conditions of the proposition.   □

3   Warming Up: Absence of Minimal Elements

The section contains an introduction to punctual constructions: we give a detailed proof of the result below, which serves as a good starting point.

Theorem 3.1. *Let S be an infinite punctual family. Then the semilattice $\mathcal{R}_{pr}(S)$ does not contain minimal elements. Consequently, the structure $\mathcal{R}_{pr}(S)$ is infinite.*

Proof. Let α be a punctual numbering of S. By Proposition 2.2, there is a spd numbering $\nu \in Com_{pr}(S)$ such that $\nu \leq_{pr} \alpha$. In order to prove the theorem, we build a numbering $\mu \in Com_{pr}(S)$ such that $\mu \leq_{pr} \nu$ and $\nu \not\leq_{pr} \mu$.

Our construction will satisfy the following series of requirements:

P_e: The function p_e does not punctually reduce ν to μ.

The key difference between our construction and a typical injury argument (of recursion theory) is the following: Our requirements *do not injure* each other, and we will satisfy only one requirement P_e at a time.

Strategy for P_e. Suppose that the P_e-strategy starts working at a stage s^e of the construction, and N^e is the least index such that the object $\mu(N^e)$ is still undefined at the beginning of the stage s^e.

We wait until the first stage $t > s^e$ with the following properties:

(a) There is a (least) number $w^e \leq t$ such that $w^e \geq e$, $\nu(w^e) \neq \nu(0)$, and we have not used the object $\nu(w^e)$ in our definition of μ before.
(b) For this particular w^e, the value $p_e[t](w^e)$ is already defined.

Note that checking whether $\nu(w^e)$ is equal to $\nu(0)$ is a punctual procedure, since the numbering ν is spd.

While waiting for this t to appear, we should not delay the definition of the numbering μ, so, one by one, we put:

$$\mu(N^e) := \nu(0),\ \mu(N^e + 1) := \nu(0),\ \mu(N^e + 2) := \nu(0),\dots.$$

When the desired stage t is achieved, we proceed as follows:

1. For each $k \leq p_e(w^e)$, if the object $\mu(k)$ is still undefined, then put $\mu(k) := \nu(0)$.
2. Let m be the least index such that at this moment, $\mu(m)$ is still undefined. Set $\mu(m + \ell) := \nu(\ell)$, for every $\ell \leq w^e$.

It is clear that the described actions ensure that the requirement P_e is forever satisfied: Our choice of the witness w^e guarantees that we have $\nu(w^e) \neq \mu(p_e(w^e))$.

The construction is arranged as follows: We start the P_0-strategy and wait until it is satisfied. When P_0 is satisfied, we immediately start the P_1-strategy. After P_1 is satisfied, we start P_2, etc.

Verification. Since the family S is infinite, each strategy P_e will eventually find its witness w^e, and after that, P_e will eventually become satisfied. Therefore, we deduce $\nu \not\leq_{pr} \mu$.

The constructed numbering μ is punctual: Indeed, for an index $k \in \omega$, one can just look at the stage $k+1$ of the described construction. At this stage $k+1$, we can promptly find an index $r(k)$ such that $\mu(k)$ is equal to $\nu(r(k))$. This shows the punctuality of μ, and furthermore, the function r punctually reduces μ to ν.

Informally speaking, the punctuality of μ is ensured by elimination of unbounded searches: Surely, the P_e-strategy wants to "catch" a particularly good stage t, but this quest for t does not delay the construction at all—while doing the t-search, our definition of μ just executes a straightforward filler action (copying $\nu(0)$ for appropriate μ-indices).

Now it is enough to show that the numbering μ has an index for every element of S. This is ensured by the assignment $\mu(m + \ell) := \nu(\ell)$ given above—after satisfying P_e, μ copies the long initial segment $\nu(0), \nu(1), \ldots, \nu(w^e)$. $\square$

The classical result of Khutoretskii [23] shows that for any computable family T, its semilattice $\mathcal{R}_c(T)$ is either one-element, or infinite. Proposition 2.1 and Theorem 3.1 together imply that the punctual setting exhibits a similar behavior:

Corollary 3.1. *For an arbitrary punctual family S, its Rogers pr-semilattice is either one-element or infinite.*

4   Infinite Antichain

Theorem 4.1. *Let S be an infinite punctual family, and β be a punctual numbering of S. Then the semilattice $\mathcal{R}_{pr}(S)$ contains an infinite antichain under the degree of β.*

Proof. Here we give a construction, which builds *two* $\leq_{pr}$-incomparable punctual numberings ν and μ of S. This construction admits a straightforward generalization to the case of countably many incomparable elements.

We apply Proposition 2.2 and fix a spd numbering α of the family S such that $\alpha \leq_{pr} \beta$. Our numberings ν and μ will copy different pieces of α. We satisfy the following series of requirements:

P_e: p_e: $\nu \not\leq_{pr} \mu$, i.e. p_e does not punctually reduce ν to μ.
Q_i: p_i: $\mu \not\leq_{pr} \nu$.

We fix a (punctual) ordering of the requirements: $P_0 < Q_0 < P_1 < Q_1 < \ldots$. This means that we will satisfy P_0, then Q_0, then P_1, etc.

The P_e- and Q_i-requirements are very similar, so we give a description only for a P_e-strategy. Essentially, this is a slightly modified version of the strategy from Theorem 3.1.

The P_e-strategy. By the *background action* of the P_e-strategy, we mean the following: Whenever we are waiting for some object to be found, we do not delay our construction, and we just put

$$\nu(N) = \mu(N) := \alpha(0), \ \nu(N+1) = \mu(N+1) := \alpha(0), \ \ldots,$$

starting with an appropriate index N. This N is typically clear from the context (recall the N^e from Theorem 3.1).

The strategy P_e waits until the first (large enough) stage t with the following property: There is an index $w \leq t$ such that $\alpha(w) \neq \alpha(0)$, and the object $\alpha(w)$ has not been employed in the construction before.

When this w is found, we choose the least m^e such that $\nu(m^e)$ is still undefined, and we set $\nu(m^e) := \alpha(w)$. We wait for the first stage t_1 such that the value $p_e[t_1](m^e)$ is defined. After that, proceed as follows:

- For every $k \leq p_e(m^e)$, if $\mu(k)$ is still undefined, then put $\mu(k) := \alpha(0)$.
- Ensure that both μ and ν copy pieces of α: set $\mu(N_0 + \ell) = \nu(N_1 + \ell) := \alpha(\ell)$, where $\ell \leq w$ and the indices N_0, N_1 are chosen in an appropriate way.

These actions guarantee that $\nu(m^e) = \alpha(w) \neq \mu(p_e(m^e))$, and the P_e-requirement is satisfied.

The Q_i-strategy is essentially the same as that of P_e, modulo the following modification: ν and μ need to switch places in the strategy description.

The construction is arranged similarly to Theorem 3.1: our requirements are satisified one by one, according to their priority ordering.

Verification mimics that of Theorem 3.1. Clearly, the numberings ν and μ are $\leq_{pr}$-incomparable. Moreover, they are punctual, and both of them can be punctually reduced to α. The copying of large α-pieces implies that both ν and μ index the whole family S.   □

5   Lattices

Selivanov [34] obtained the following result: For any computable family T, if the semilattice $\mathcal{R}_c(T)$ is infinite, then it *cannot* be a lattice.

The propositions of this section show that the result of Selivanov *cannot be transferred* to the punctual setting: Sometimes, an infinite $\mathcal{R}_{pr}(S)$ is a lattice, and sometimes it is not.

Proposition 5.1. *Let S be a punctual family containing the following functions: for $i \in \omega$,*

$$g_i(x) := i, \quad \text{for all } x.$$

Then the structure $\mathcal{R}_{pr}(S)$ is an infinite lattice with the greatest element.

For reasons of space, the proof of Proposition 5.1 is omitted.

Proposition 5.2. *There exists an infinite punctual family S such that the semilattice $\mathcal{R}_{pr}(S)$ contains a minimal pair. Consequently, $\mathcal{R}_{pr}(S)$ is not a lattice.*

The desired family S of Proposition 5.2 is defined via its punctual Friedberg numbering: for $k, \ell \in \omega$, set $\nu\langle k, 0 \rangle := \chi_{\{2k\}}$ and $\nu\langle k, \ell+1 \rangle := \chi_{\{2k, 2\ell+1\}}$. For reasons of space, further proof is omitted.

6   Density

Recall that Theorem 3.1 proves that every infinite Rogers pr-semilattice is downwards dense. Proposition 5.1 gives an example of infinite $\mathcal{R}_{pr}(\mathcal{S})$ having the greatest element. Thus, in general, pr-semilattices are not upwards dense.

This section contains two results. First, we provide an example of upwards dense pr-semilattice. After that, we establish a general result (Theorem 6.1) which proves density for an arbitrary infinite $\mathcal{R}_{pr}(\mathcal{S})$.

Proposition 6.1. *Let $\mathcal{S}$ be a punctual family containing the following functions: $g_i := \chi_{\{i\}}$, for $i \in \omega$. Then the structure $\mathcal{R}_{pr}(\mathcal{S})$ does not have maximal elements.*

Proof (sketch). Let ν be an arbitrary punctual numbering of the family $\mathcal{S}$. In order to prove the proposition, we want to build a punctual numbering μ of *some subfamily* $\mathcal{S}_0 \subseteq \mathcal{S}$ such that $\mu \not\leq_{pr} \nu$. Indeed, the existence of such μ is sufficient for us, since this implies that $\nu \leq_{pr} \nu \oplus \mu$ and $\nu \oplus \mu \not\leq_{pr} \nu$.

We satisfy the series of requirements

$$P_e \colon p_e \colon \mu \not\leq_{pr} \nu.$$

The P_e-strategy. Choose the least w such that we have not talked about the object $\mu(w)$ before. Wait for the least stage s such that the value $p_e[s](w)$ is defined, and there is a number $N \leq s$ with $(\nu(p_e(w)))(N) = 1$.

While waiting for this s, just propagate

$$\mu(w)(0) := 0, \ \mu(w)(1) := 0, \ \mu(w)(2) := 0, \ \ldots.$$

When the stage s and the corresponding N is found, set $\mu(w) := \chi_{\{N+s+1\}}$.

The construction is arranged in a straightforward way. We note that in order to ensure the punctuality of μ, the P_e-strategy also has to implement some simple background actions—e.g., one by one, we set $\mu(w+1) := \chi_{\{0\}}$, $\mu(w+2) := \chi_{\{0\}}$, etc.

Clearly, μ indexes a subfamily of $\mathcal{S}$. Since every P_e is eventually satisfied, we have $\mu \not\leq_{pr} \nu$.   □

Theorem 6.1. *Let $\mathcal{S}$ be an infinite punctual family. Suppose that ν and μ are punctual numberings of $\mathcal{S}$ such that $\nu <_{pr} \mu$, i.e. $\nu \leq_{pr} \mu$ and $\nu \not\equiv_{pr} \mu$. Then there is a numbering $\xi \in Com_{pr}(\mathcal{S})$ such that $\nu <_{pr} \xi <_{pr} \mu$.*

For reasons of space, the proof of Theorem 6.1 is omitted.

Acknowledgements. Part of the research contained in this paper was carried out while the first and the last authors were visiting the Department of Mathematics of Nazarbayev University, Nur-Sultan. The authors wish to thank Nazarbayev University for its hospitality.

References

1. Ambos-Spies, K., Badaev, S., Goncharov, S.: Inductive inference and computable numberings. Theor. Comput. Sci. **412**(18), 1652–1668 (2011). https://doi.org/10.1016/j.tcs.2010.12.041
2. Badaev, S., Goncharov, S.: Computability and numberings. In: Cooper, S.B., Löwe, B., Sorbi, A. (eds.) New Computational Paradigms, pp. 19–34. Springer, New York (2008). https://doi.org/10.1007/978-0-387-68546-5_2
3. Badaev, S.A.: Computable enumerations of families of general recursive functions. Algebra Log. **16**(2), 83–98 (1977). https://doi.org/10.1007/BF01668593
4. Badaev, S.A.: Minimal numerations of positively computable families. Algebra Log. **33**(3), 131–141 (1994). https://doi.org/10.1007/BF00750228
5. Badaev, S.A., Goncharov, S.S.: Theory of numberings: open problems. In: Cholak, P., Lempp, S., Lerman, M., Shore, R. (eds.) Computability Theory and Its Applications, Contemporary Mathematics, vol. 257, pp. 23–38. American Mathematical Society, Providence (2000). https://doi.org/10.1090/conm/257/04025
6. Badaev, S.A., Lempp, S.: A decomposition of the Rogers semilattice of a family of d.c.e. sets. J. Symb. Logic **74**(2), 618–640 (2009). https://doi.org/10.2178/jsl/1243948330
7. Bazhenov, N., Mustafa, M., Yamaleev, M.: Elementary theories and hereditary undecidability for semilattices of numberings. Arch. Math. Log. **58**(3–4), 485–500 (2019). https://doi.org/10.1007/s00153-018-0647-y
8. Bazhenov, N., Downey, R., Kalimullin, I., Melnikov, A.: Foundations of online structure theory. Bull. Symb. Log. **25**(2), 141–181 (2019). https://doi.org/10.1017/bsl.2019.20
9. Case, J., Jain, S., Stephan, F.: Effectivity questions for Kleene's recursion theorem. Theor. Comput. Sci. **733**, 55–70 (2018). https://doi.org/10.1016/j.tcs.2018.04.036
10. Ershov, Y.L.: Enumeration of families of general recursive functions. Sib. Math. J. **8**(5), 771–778 (1967). https://doi.org/10.1007/BF01040653
11. Ershov, Y.L.: On computable enumerations. Algebra Log. **7**(5), 330–346 (1968). https://doi.org/10.1007/BF02219286
12. Ershov, Y.L.: Theory of Numberings. Nauka, Moscow (1977). (in Russian)
13. Ershov, Y.L.: Theory of numberings. In: Griffor, E.R. (ed.) Handbook of Computability Theory. Studies in Logic and the Foundations of Mathematics, vol. 140, pp. 473–503. North-Holland, Amsterdam (1999). https://doi.org/10.1016/S0049-237X(99)80030-5
14. Friedberg, R.M.: Three theorems on recursive enumeration. I. Decomposition. II. Maximal set. III. Enumeration without duplication. J. Symb. Log. **23**(3), 309–316 (1958). https://doi.org/10.2307/2964290
15. Gödel, K.: Über formal unentscheidbare Sätze der Principia Mathematica und verwandter Systeme I. Monatshefte für Mathematik und Physik **38**(1), 173–198 (1931). https://doi.org/10.1007/BF01700692
16. Goncharov, S.S.: Computable single-valued numerations. Algebra Log. **19**(5), 325–356 (1980). https://doi.org/10.1007/BF01669607
17. Goncharov, S.S.: Positive numerations of families with one-valued numerations. Algebra Log. **22**(5), 345–350 (1983). https://doi.org/10.1007/BF01982111
18. Goncharov, S.S., Lempp, S., Solomon, D.R.: Friedberg numberings of families of n-computably enumerable sets. Algebra Log. **41**(2), 81–86 (2002). https://doi.org/10.1023/A:1015352513117

19. Goncharov, S.S., Sorbi, A.: Generalized computable numerations and nontrivial Rogers semilattices. Algebra Log. **36**(6), 359–369 (1997). https://doi.org/10.1007/BF02671553
20. Herbert, I., Jain, S., Lempp, S., Mustafa, M., Stephan, F.: Reductions between types of numberings. Ann. Pure Appl. Logic **170**(12), 102716 (2019). https://doi.org/10.1016/j.apal.2019.102716
21. Jain, S., Stephan, F.: Numberings optimal for learning. J. Comput. Syst. Sci. **76**(3–4), 233–250 (2010). https://doi.org/10.1016/j.jcss.2009.08.001
22. Kalimullin, I., Melnikov, A., Ng, K.M.: Algebraic structures computable without delay. Theor. Comput. Sci. **674**, 73–98 (2017). https://doi.org/10.1016/j.tcs.2017.01.029
23. Khutoretskii, A.B.: On the cardinality of the upper semilattice of computable enumerations. Algebra Log. **10**(5), 348–352 (1971). https://doi.org/10.1007/BF02219842
24. Kleene, S.C.: Introduction to Metamathematics. Van Nostrand, New York (1952)
25. Kolmogorov, A.N., Uspenskii, V.A.: On the definition of an algorithm. Uspehi Mat. Nauk. **13**(4), 3–28 (1958). (in Russian)
26. Lachlan, A.H.: Standard classes of recursively enumerable sets. Z. Math. Logik Grundlagen Math. **10**(2–3), 23–42 (1964). https://doi.org/10.1002/malq.19640100203
27. Lachlan, A.H.: On recursive enumeration without repetition. Z. Math. Logik Grundlagen Math. **11**(3), 209–220 (1965). https://doi.org/10.1002/malq.19650110305
28. Mal'cev, A.I.: Positive and negative numerations. Sov. Math. Dokl. **6**, 75–77 (1965)
29. Metakides, G., Nerode, A.: The introduction of nonrecursive methods into mathematics. In: The L. E. J. Brouwer Centenary Symposium (Noordwijkerhout, 1981). Studies in Logic and the Foundations of Mathematics, vol. 110, pp. 319–335. North-Holland, Amsterdam (1982). https://doi.org/10.1016/S0049-237X(09)70135-1
30. Ospichev, S.S.: Friedberg numberings in the Ershov hierarchy. Algebra Log. **54**(4), 283–295 (2015). https://doi.org/10.1007/s10469-015-9349-2
31. Podzorov, S.Y.: Arithmetical D-degrees. Sib. Math. J. **49**(6), 1109–1123 (2008). https://doi.org/10.1007/s11202-008-0107-8
32. Pour-El, M.B.: Gödel numberings versus Friedberg numberings. Proc. Am. Math. Soc. **15**(2), 252–256 (1964). https://doi.org/10.2307/2034045
33. Rogers, H.: Gödel numberings of partial recursive functions. J. Symb. Log. **23**(3), 331–341 (1958). https://doi.org/10.2307/2964292
34. Selivanov, V.L.: Two theorems on computable numberings. Algebra Log. **15**(4), 297–306 (1976). https://doi.org/10.1007/BF01875946
35. Uspenskii, V.A.: Systems of denumerable sets and their enumeration. Dokl. Akad. Nauk SSSR **105**, 1155–1158 (1958). (in Russian)

Partial Sums on the Ultra-Wide Word RAM

Philip Bille$^{(\boxtimes)}$ [ID], Inge Li Gørtz [ID], and Frederik Rye Skjoldjensen

DTU Compute, Technical University of Denmark, Kgs. Lyngby, Denmark
{phbi,inge}@dtu.dk

Abstract. We consider the classic partial sums problem on the ultra-wide word RAM model of computation. This model extends the classic w-bit word RAM model with special ultrawords of length w^2 bits that support standard arithmetic and boolean operation and scattered memory access operations that can access w (non-contiguous) locations in memory. The ultra-wide word RAM model captures (and idealizes) modern vector processor architectures.

Our main result is a new in-place data structure for the partial sum problem that only stores a constant number of ultrawords in addition to the input and supports operations in doubly logarithmic time. This matches the best known time bounds for the problem (among polynomial space data structures) while improving the space from superlinear to a constant number of ultrawords. Our results are based on a simple and elegant in-place word RAM data structure, known as the Fenwick tree. Our main technical contribution is a new efficient parallel ultra-wide word RAM implementation of the Fenwick tree, which is likely of independent interest.

Keywords: Ultra-wide word RAM model · Partial sums · Fenwick tree

1 Introduction

Let $A[1, \ldots, n]$ be an array of integers of length n. The *partial sums problem* is to maintain a data structure for A under the following operations:

- sum(i): return $\sum_{k=1}^{i} A[k]$.
- update(i, Δ): set $A[i] \leftarrow A[i] + \Delta$.

The partial sums problem is a classic and well-studied data structure problem [1–4,9,12,14,16–19,21–24,31,32,38]. Partial sums is a natural range query problem with applications in areas such as list indexing and dynamic ranking [12], dynamic arrays [3,32], and arithmetic coding [14,34]. From a lower bound perspective, the problem has been central in the development of new techniques for proving lower bounds [29]. In classic models of computation the complexity of the partial sums problem is well-understood with tight logarithmic upper and lower bounds on the operations [31]. Hence, a natural question is if

© Springer Nature Switzerland AG 2020
J. Chen et al. (Eds.): TAMC 2020, LNCS 12337, pp. 13–24, 2020.
https://doi.org/10.1007/978-3-030-59267-7_2

practical models of computation capturing modern hardware advances will allow us the overcome the logarithmic barrier.

One such model is the *RAM with byte overlap* (RAMBO) model of computation [6,7,17]. The RAMBO model extends the standard w-bit word RAM model [20] with special words where individual bits are shared among other words, i.e., changing a bit in a word will also change the bit in the words that share that bit. The precise model depends on the layout of shared bits. This memory architecture is feasible to design in hardware and prototypes have been built [27]. In the RAMBO model Brodnik et al. [8] gave a time-space trade-off for partial sums that uses $O(n^{w/2^\tau} + n)$ space and supports operations in $O(\tau)$ time and for a parameter τ, $1 \le \tau \le \log\log n$. Here, the n term in the space bound is for the special words with shared bits (organized in a tree layout) and the $O(n^{w/2^\tau})$ term is for standard words. Plugging in constant τ, this gives an $O(n^{\epsilon w} + n)$ space and constant time solution, for any $\epsilon > 0$. At the other extreme, with $\tau = \log\log n$, this gives an $O(n)$ space and $O(\log\log n)$ time solution.

More recently, Farzan et al. [13] introduced the *ultra-wide word RAM* (UWRAM) model of computation. The UWRAM model also extends the word RAM model, but with special *ultrawords* of length w^2 bits. The model supports standard arithmetic and boolean operations on ultrawords and *scattered* memory access operations that access w locations in memory specified by an ultraword in parallel. The UWRAM captures modern vector processor architectures [11,28,33,36]. We present the details of the UWRAM model in Sect. 2. Farzan et al. [13] showed how to simulate algorithms on RAMBO model on the UWRAM model at the cost of slightly increasing space. Simulating the above solution for partial sums they gave a time-space trade-off for partial sums that uses $O(n^{w/2^\tau} + nw\log n)$ space and supports operations in $O(\tau)$ time and for a parameter τ, $1 \le \tau \le \log\log n$. For constant τ, this is $O(n^{\epsilon w} + nw\log n)$ space and constant time, for any $\epsilon > 0$, and for $\tau = \log\log n$ this is $O(nw\log\log n)$ space and $O(\log\log n)$ time.

1.1 Setup and Results

We revisit the partial sums problem on the UWRAM and present a simple new algorithm that significantly improves the space overhead of the previous solutions. Let A be an array of n w-bit integers. An *in-place data structure* for the partial sums problem is a data structure that modifies the input array A, e.g., by replacing some of the entries in A, to efficiently support operations. In addition to the modified array the data structure is only allowed to store $O(1)$ of ultrawords. This definition extends the standard in-place/implicit data structure concept [10,15,30,35,37] to the UWRAM, by allowing a constant number of ultrawords to be stored instead of (standard) words. Clearly, without this modification computation on ultrawords is impossible. As in Farzan et al. [13] we distinguish between the *restricted UWRAM* that supports a minimal set of instructions on ultrawords consisting of addition, subtraction, shifts, and bitwise boolean operations and the *multiplication UWRAM* that extends the instruction set of the restricted UWRAM with a multiplication operation on ultrawords. We show the following main result:

Theorem 1. *Given an array A of n w-bit integers, we can construct in-place partial sums data structures for A that support* sum *and* update *operations in* $O(\log \log n)$ *time on a restricted UWRAM.*

Compared to the previous result, Theorem 1 matches the $O(\log \log n)$ time bound of Farzan et al. [13] (with parameter $\tau = \Theta(\log \log n)$) while improving the space overhead from $O(nw \log n)$ to a constant number of ultrawords. This is important in practical applications since modern vector processors have a very limited number of ultrawords available.

Technically, our solution is based on a simple and elegant in-place word RAM data structure, called the *Fenwick tree* (see Sect. 3 for a detailed description). The Fenwick tree support operations in $O(\log n)$ by sequentially traversing an implicit tree structure. We show how to efficiently compute the access pattern on the tree structure in parallel using prefix sum computations on ultrawords. Then, given the locations to access we use scattered memory operations to access them all in parallel. In total, this leads to the exponential improvement of Fenwick trees. The main bottleneck in our algorithm is the prefix sum computation. Interestingly, if we allow multiplication we can compute prefix sums in constant time leading to the following Corollary for the multiplication UWRAM:

Corollary 1. *Given an array A of n w-bit integers, we can construct in-place partial sums data structures for A that support* sum *and* update *operations in constant time on a multiplication UWRAM.*

Multiplication (or prefix sum computation) is not an AC^0 operation (it cannot be implemented by a constant depth, polynomial size circuit) and therefore likely not practical to implement on ultraword. However, Corollary 1 shows that we can achieve significant improvements on the UWRAM with special operations. Since UWRAM capture modern processors, we believe it is worth investigating further, and that our work is a first step in this direction.

1.2   Outline

The paper is organized as follows. In Sect. 2 and 3 we review the UWRAM model of computation and the Fenwick tree. In Sect. 4 we present our UWRAM implementation of the Fenwick tree. Finally, in Sect. 4.4 we discuss extensions of the result and open problems.

2   The Ultra-Wide Word RAM Model

The *word RAM* model of computation [20] consists of an infinite memory of w-bit words and an instruction set of arithmetic, boolean, and memory access instructions such as the ones available in standard programming languages such

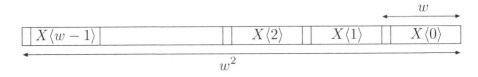

Fig. 1. The layout of an ultraword of w^2 divided into w words each of w bits. The leftmost bit of each word is reserved to be a test bit.

as C. We assume that we can store a pointer into the input in a single word and hence $w \geq \log n$, where n is the size of the input. The time complexity of a word RAM algorithm is the number of instructions and the space complexity is the number of words used by the algorithm.

The *ultra-wide word RAM* (UWRAM) model of computation [13] extends the word RAM model with special ultrawords of w^2 bits. We distinguish between the *restricted UWRAM* that supports a minimal set of instructions on ultrawords consisting of addition, subtraction, shifts, and bitwise boolean operations and the *multiplication UWRAM* that additionally supports multiplication. The time complexity is the number of instruction (on standard words or ultrawords) and the space complexity is the number of (standard) words used by the algorithm. The restricted UWRAM captures modern vector processor architectures [11,28, 33,36]. For instance, the Intel AVX-512 vector extension [33] support similar operations on 512-bit wide words (i.e., a factor of 8 compared to $64^2 = 4096$).

2.1 Word-Level Parallelism

Due to their similarities, we can adopt many word-level parallelism techniques from the word RAM to the UWRAM. We briefly review the key primitives and techniques that we will use.

Let X be an ultraword of w^2 bits. We often view X as divided into w words of w consecutive bits each. See Fig. 1. We number the words in X from right-to-left starting from 0 and use the notation $X\langle j \rangle$ to denote the jth word in X. Similarly, the bits of each word $X\langle j \rangle$ are numbered from right-to-left starting from 0. If only the rightmost $\ell \leq w$ words in X are non-zero, we say that X has *length* ℓ. For simplicity in the presentation, we reserve the leftmost bit of each word to be a *test bit* for word-level parallelism operations. One may always remove this assumption at no asymptotic cost, e.g., by using two words in an ultraword to simulate each single word.

We now show how to implement common operations on ultrawords that we will use later. Most of these are already available in hardware on modern vector processor architectures. Componentwise arithmetic and bitwise operation are straightforward to implement using standard word-level parallelism techniques from the word RAM. For instance, given ultrawords X and Y, we can compute the componentwise addition, i.e., the ultraword Z such that $Z\langle j \rangle = X\langle j \rangle + Y\langle j \rangle$ for $j = 0, \ldots, w - 1$ by adding X and Y and & 'ing with the mask $(01^{w-1})^w$ to clear any test bits (we use exponentiation to denote bit repetition, i.e., $0^3 1 =$ 0001). We can also compare X and Y componentwise by |'ing in the test bits of X, subtracting Y, and masking out the test bits by &'ing with $(10^{w-1})^w$. The jth test bit of the result contains a 1 iff $X\langle j \rangle \geq Y\langle j \rangle$. Given X and another ultraword T containing only test bits, we can extract the words in X according to the test bits, i.e., the ultraword E such that $E\langle j \rangle = X\langle j \rangle$ if the jth test bit of T is 1 and $E\langle j \rangle = 0$ otherwise. To do so we copy the test bits by a subtracting $(0^{w-1}1)^w$ from T and &'ing the result with X. All of the above mentioned operation take constant time on a restricted UWRAM. Given an ultraword X of length ℓ, the *prefix sum* of X is the ultraword P of length ℓ, such that $P\langle j \rangle = \sum_{k \leq j} X\langle k \rangle$. We assume here that the integers computed in the prefix sum never exceed the maximum size available in a word such that $P\langle j \rangle$ is always well-defined. We need the following result.

Lemma 1. *Given an ultraword X of length ℓ we can compute the prefix sum of X in $O(\log \ell)$ time on a restricted UWRAM and in $O(1)$ time on a multiplication UWRAM.*

Proof. First consider the restricted UWRAM. We implement a standard parallel prefix-sum algorithm [25] (see also the survey by Blelloch [5]). For simplicity, we assume that ℓ is a power of two. The algorithm consists of two phases that conceptually construct and traverse a perfectly balanced binary tree T of height $\log \ell$ whose leaves are the ℓ words of X.

Given an internal node v in T, let v_{left} and v_{right} denote the left and right child of v, respectively. The first phase performs a bottom-up traversal of T and computes for each node v an integer $b(v)$. If v is a leaf, $b(v)$ is the corresponding integer in X and if v is an internal node $b(v) = b(v_{\text{left}}) + b(v_{\text{right}})$. The second phase performs a top-down traversal of T and computes an integer $t(v)$. If v is the root then $t(v) = 0$ and if v is an internal node then $t(v_{\text{left}}) = t(v)$ and $t(v_{\text{right}}) = t(v_{\text{left}}) + b(v_{\text{right}})$. After the second phase the integers at the leaves is the prefix sum shifted by a single element and missing the last element. We shift and add the last element to produce the final prefix sum. Since T is perfectly balanced we can implement each level of a phase in constant time using shifting and addition. The final shift and addition of the last element takes constant time. It follow that the total time is $O(\log \ell)$. During the computation we only need to maintain all of the values in a constant number of ultrawords.

Next consider the multiplication instruction set. We can then simply multiply X with the constant $(0^{w-1}1)^w$ and mask out the ℓ rightmost words of the result to produce the prefix sum. See Hagerup [20] for a detailed description of why this is correct. In total this uses $O(1)$ time.

2.2 Memory Access

The UWRAM supports standard memory access operation to read or write a single word or a sequence of w contiguous words. More interestingly, the UWRAM also supports *scattered* access operations that access w memory locations (not necessarily contiguous) in parallel. Given an ultraword A containing w memory addresses, a *scattered read* loads the contents of the addresses into an ultraword X, such that $X\langle j\rangle$ contains the contents of memory location $A\langle j\rangle$. Given two ultrawords A and X *scattered write* sets the contents memory location $A\langle j\rangle$ to be $X\langle j\rangle$. Scattered memory accesses captures the memory model used by IBM's *Cell* architecture [11]. Scattered memory access operations were also proposed by Larsen and Pagh [26] in the context of the I/O model of computation.

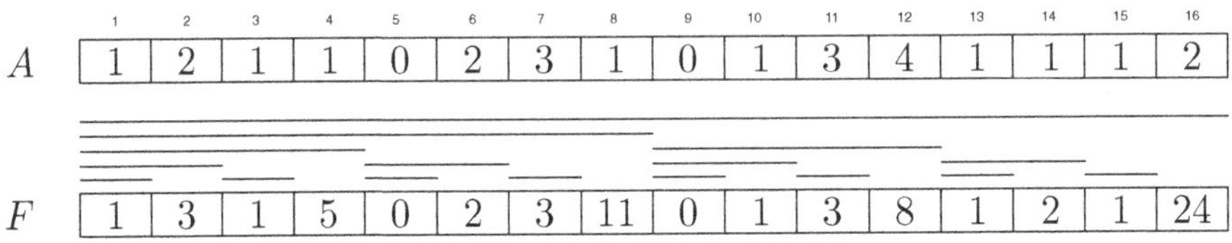

	1	2	3	4	5	6	7	8	9	10	11	12	13	14	15	16
A	1	2	1	1	0	2	3	1	0	1	3	4	1	1	1	2

F	1	3	1	5	0	2	3	11	0	1	3	8	1	2	1	24

Fig. 2. A array A and the Fenwick tree F. The lines above F indicate the partial sum of A stored at the rightmost endpoint of the line. For instance, the $F[12] = A[9] + A[10] + A[11] + A[12] = 0 + 1 + 3 + 4 = 8$.

3   Fenwick Trees

Let A be an array of n w-bit integers and assume for simplicity that n is a power of two. The Fenwick tree [14,34] is an in-place data structure that replaces the array A as follows. If $n = 1$, then leave A unchanged. Otherwise, replace all values at even entries $A[2i]$ by the sum $A[2i-1] + A[2i]$. Then, recurse on the subarray $A[2, 4, \ldots, n]$. The resulting array F stores a subset of the partial sums of A organized in a tree layout (see Fig. 1).

To answer $\mathsf{sum}(i)$ query, we compute a sequence of indices in F and add the values in F at these indices together. Let $\mathsf{rmb}(x)$ denote the position of the rightmost bit in an integer x. Define the *sum sequence* $i^s_1, \ldots, i^s_r$ given by $i^s_1 = i$ and $i^s_j = i^s_{j-1} - 2^{\mathsf{rmb}(i^s_{j-1})}$, for $j = 2, \ldots, r$. The final element i^s_r is 0. We compute and return $F[i^s_1] + F[i^s_2] + \cdots + F[i^s_{r-1}]$. For instance, for $i = 13 = (1101)_2$ the sum sequence is $13, 12, 8, 0 = (1101)_2, (1100)_2, (1000)_2, (0000)_2$. Hence, $\mathsf{sum}(13) = F[13] + F[12] + F[8] = 1 + 8 + 11 = 20 = A[1] + \cdots + A[13]$. We access at most $O(\log n)$ entries in F and hence the total time for sum is $O(\log n)$. Note that we can always recover the original array A using the sum operation, since $A[i] = \mathsf{sum}(i) - \mathsf{sum}(i-1)$.

To compute $\mathsf{update}(i, \Delta)$, we compute a sequence of indices in F and add Δ to the values in F at each of these indices. Define the *update sequence* $i^u_1, \ldots, i^u_t$ given by $i^u_1 = i$ and $i^u_j = i^u_{j-1} + 2^{\mathsf{rmb}(i^u_{j-1})}$, for $j = 2, \ldots, t$. The final element i^u_t is $2n$. We set $F[i^u_1] = F[i^u_1] + \Delta, \ldots, F[i^u_t] = F[i^u_{t-1}] + \Delta$. For instance, for $i = 13$ the update sequence is $13, 14, 16, 32$. Hence, $\mathsf{update}(13, 5)$ adds 5 to $F[13]$, $F[14]$, and $F[16]$. Similar to the sum operation, the total running time for update is $O(\log n)$.

4   Partial Sums on the Ultra-Wide Word RAM

We now present an efficient implementation of Fenwick trees on the UWRAM model of computation. We only store the Fenwick tree, as the array F described in Sect. 3 and a constant number of ultraword constants that we use for computation. We first show some basic properties of the sum and update sequences in Sect. 4.1, before presenting our UWRAM implementation of the operations in Sects. 4.2 and 4.3.

4.1   Computing Sum and Update Sequences

To compute the sum and update sequences we cannot directly apply the recursive definitions, since this would need $\Omega(\log n)$ steps. Instead, we show how to express the sequences as a prefix sum that we can efficiently derive from the input integer i. Then, using Lemma 1 we will show how to compute it in on the UWRAM in the following sections.

Let $i_1^s, \ldots, i_r^s$ and $i_1^u, \ldots, i_t^u$ be the sum sequence and update sequences, respectively, for i as defined in Sect. 3. Define the *offset sum sequence* $o_1^s, \ldots, o_{r-1}^s$ and *offset update sequence* $o_1^u, \ldots, o_{t-1}^u$ for i to be the sequences of differences of the sum and update sequences, respectively, that is, $o_j^s = i_{j+1}^s - i_j^s$, for $j = 1, \ldots, r-1$ and $o_j^u = i_{j+1}^u - i_j^u$, for $j = 1, \ldots, t-1$. By definition, we have that

$$i_j^s = i + \left(\sum_{k<j} o_k^s \right) \qquad\qquad i_j^u = i + \left(\sum_{k<j} o_k^u \right) \qquad (1)$$

We also have that $o_j^s = -2^{\mathsf{rmb}(i_j^s)}$ and hence each sum offset is a power of 2 corresponding to the rightmost 1 bit in i_j^s. Thus, o_1^s corresponds to the rightmost 1 in $i_1^s = i$. Adding $o_1^s = -2^{\mathsf{rmb}(i)}$ (i.e., subtracting $2^{\mathsf{rmb}(i)}$) "clears" the rightmost 1 bit in i. Thus, o_2^s corresponds to the 1 bit in i immediately to left of the rightmost 1 bit. In general, we have that $o_j^s = -2^b$, where b is the position of the jth rightmost bit in i, for $j = 1, \ldots, r-1$. For instance, for $i = 13 = (1101)_2$ the offset sum sequence is $-1, -4, -8$ corresponding to the three 1 bits in the binary representation of i.

Similarly, for the update offsets, we have that $o_j^u = 2^{\mathsf{rmb}(i_j^u)}$. Hence, o_1^u corresponds to rightmost 1 in i. Adding $o_1^u = 2^{\mathsf{rmb}(i_1^u)}$ clears the rightmost consecutive group of 1 bits in i and flips the following 0 bit to 1. In general, we have that $o_j^u = 2^b$, where b is the position of the jth rightmost 0 to the left of $\mathsf{rmb}(i)$, for $j = 2, \ldots, t-1$. For instance, for $i = 13 = (01101)_2$ the offset update sequence is $1, 2, 16$.

4.2   Sum

To compute the $\mathsf{sum}(i)$, the main idea is to first construct the sum sequence in an ultraword, then use a scattered read to retrieve the entries from F in parallel into another ultraword, and finally sum the entries of this ultraword to compute the final result. We do this in 3 steps as follows. See Fig. 3 for an example of the computed ultrawords during the algorithm.

I	13	13	13	13

M	8	4	2	1

O	8	4		1

P	13	5		1

P'	5		1	

S	8		12	13

Fig. 3. Computing the sum sequence for $i = 13 = (1011)_2$. Words with 0 are left blank. I contains duplicates of i. M is a precomputed mask. O is the bitwise & of I and M. P is the prefix sum of the non-zero words in O. P' is P shifted left by one word. S is the sum sequence obtained by componentwise subtraction of P' from I.

Step 1: Compute Offsets Compute the ultraword O such that $O\langle j \rangle = 2^j$ if -2^j is an offset for i and 0 otherwise, i.e., the non-zero entries of O is the offset sequence for i. To do so we first construct the ultraword I consisting of $\log n$ duplicates of i, i.e., $I\langle j \rangle = i$ for $j = 1, \ldots, \log n$. We then compute the bitwise & of I and a mask M, such that $M\langle j \rangle = 2^j$ for $j = 1, \ldots, \log n$, i.e., bit j of $M\langle j \rangle = 1$ and the other bits of $M\langle j \rangle$ are 0. By the discussion in Sect. 4.1 the resulting ultraword is O.

On the multiplication UWRAM we can construct I in constant time by multiplying i with $(0^{w-1}1)^w$. On the restricted UWRAM we can construct I in $O(\log \log n)$ time by repeatedly doubling using shifts and bitwise |. The rest of the computation takes constant time in both models.

Step 2: Compute Sum Sequence Compute an ultraword S of length $\log n$ whose non-zero entries is the sum sequence $i_1^s, \ldots, i_{r-1}^s$. To do so we first compute the prefix sum P of the non-zero words of O, i.e., we compute the prefix sum of O and then extract the words corresponding to non-zero words in O. Then we shift P by 1 word to the left to produce an ultraword P' and finally subtract P' from I to produce an ultraword S. By (1) the non-zero words in S is the sum sequence for i.

By Lemma 1 the prefix sum computation takes constant time on a multiplication UWRAM and $O(\log \log n)$ time on a restricted UWRAM. The remaining steps take constant time.

Step 3: Compute Sum Finally, we compute $F[i_1^s] + F[i_2^s] + \cdots + F[i_{r-1}^s]$. To do so we do a scattered read on S to retrieve $F[i_1^u], \ldots, F[i_{s-1}^u]$ into a single ultraword F' and compute a prefix sum on F'. The sum is then the last word in the result. The scattered read takes constant time. The prefix sum computation takes constant time on a multiplication UWRAM and $O(\log \log n)$ time on a restricted UWRAM. We assume here that $F[0] = 0$. If not we may simply temporarily set $F[0] = 0$ during the computation. Also note that it suffices to perform the first phase of the prefix sum computation as discussed in the proof of Lemma 1 since we only need the sum of all of the retrieved entries.

In total, the sum operation takes constant time on a multiplication UWRAM and $O(\log \log n)$ time on a restricted UWRAM.

4.3   Update

We compute update(i, Δ) similar to our algorithm for sum. We describe how to modify each step of sum.

In step 1, we modify the computation of the ultraword O such that it now contains the update offsets, that is, $O\langle j \rangle = 2^j$ if 2^j is an update offset for i and 0 otherwise. To do so we now construct a mask M such that $M\langle j \rangle$ contains a 0 in bit j if j is to the left of $\mathsf{rmb}(i)$ and 1 elsewhere. We then compute a bitwise $|$ of M and I and negate the result. Finally, we set word $\mathsf{rmb}(i)$ of the result to be $2^{\mathsf{rmb}(i)}$. By the discussion in Sect. 4.1 the resulting ultraword is O.

In step 2, since O now contains the offsets and not the negative offsets, we change the final subtraction to an addition to produce the update sequence stored in a single ultraword U.

In step 3, we do a scattered read on U to retrieve $F[i_1^u], \ldots, F[i_{s-1}^u]$ into a single ultraword F'. We then duplicate Δ to all words in an ultraword D and add D to F' to produce an ultraword F''. Finally, we do a scattered write on U and F'' to update F.

The changes are straightforward to implement in the same time as above. Hence, the update operation takes constant time on a multiplication UWRAM and $O(\log \log n)$ time on a restricted UWRAM.

In summary, we use $O(\log \log n)$ time on a restricted UWRAM and $O(1)$ time on a multiplication UWRAM for both operation. We only store the Fenwick tree in the array F and a constant number of ultrawords. This completes the proof of Theorem 1 and Corollary 1.

4.4   Extensions and Open Problems

We sometimes also consider the following operations in the context of partial sums:

- access(i): return $A[i]$.
- select(j): return the smallest i such that sum$(i) \geq j$

As mentioned access is trivial to support since $A(i) = \mathsf{sum}(i) - \mathsf{sum}(i - 1)$. In contrast, the select operation do not seem to easily lend itself to an efficient parallel implementation on the UWRAM. While it is straightforward to implement in $O(\log n)$ time by "top-down" traversal of the Fenwick tree our techniques do not appear be useful to speed up this solution on the UWRAM. We leave it as an open problem to investigate the complexity of the select operation on the UWRAM.

Our results leave the precise relation between UWRAM and RAMBO model of computation open. While Farzan et al. [13] show how to simulate RAMBO algorithms with a small overhead in space our results show that a direct approach to designing UWRAM algorithms can produce significantly better results. We wonder what the precise relation between the models are and if stronger simulation results are possible.

References

1. Ben-Amram, A.M., Galil, Z.: A generalization of a lower bound technique due to Fredman and Saks. Algorithmica **30**(1), 34–66 (2001)
2. Ben-Amram, A.M., Galil, Z.: Lower bounds for dynamic data structures on algebraic RAMs. Algorithmica **32**(3), 364–395 (2002)
3. Bille, P., et al.: Dynamic relative compression, dynamic partial sums, and substring concatenation. Algorithmica **80**(11), 3207–3224 (2018). Announced at ISAAC 2016
4. Bille, P., Christiansen, A.R., Prezza, N., Skjoldjensen, F.R.: Succinct partial sums and Fenwick trees. In: Fici, G., Sciortino, M., Venturini, R. (eds.) SPIRE 2017. LNCS, vol. 10508, pp. 91–96. Springer, Cham (2017). https://doi.org/10.1007/978-3-319-67428-5_8
5. Blelloch, G.E.: Prefix sums and their applications. In: Synthesis of Parallel Algorithms (1990)
6. Brodnik, A.: Searching in constant time and minimum space (Minimae res magni momenti sunt). Ph.D. thesis, University of Waterloo (1995)
7. Brodnik, A., Carlsson, S., Fredman, M.L., Karlsson, J., Munro, J.I.: Worst case constant time priority queue. J. Syst. Softw. **78**(3), 249–256 (2005)
8. Brodnik, A., Karlsson, J., Munro, J.I., Nilsson, A.: An $O(1)$ solution to the prefix sum problem on a specialized memory architecture. In: Navarro, G., Bertossi, L., Kohayakawa, Y. (eds.) TCS 2006. IIFIP, vol. 209, pp. 103–114. Springer, Boston, MA (2006). https://doi.org/10.1007/978-0-387-34735-6_12
9. Burkhard, W.A., Fredman, M.L., Kleitman, D.J.: Inherent complexity trade-offs for range query problems. Theor. Comput. Sci. **16**(3), 279–290 (1981)
10. Chan, T.M., Chen, E.Y.: Optimal in-place algorithms for 3-D convex hulls and 2-D segment intersection. In: Proceedings of the 25th SOCG, pp. 80–87 (2009)
11. Chen, T., Raghavan, R., Dale, J.N., Iwata, E.: Cell broadband engine architecture and its first implementation—a performance view. IBM J. Res. Dev. **51**(5), 559–572 (2007)
12. Dietz, P.F.: Optimal algorithms for list indexing and subset rank. In: Dehne, F., Sack, J.-R., Santoro, N. (eds.) WADS 1989. LNCS, vol. 382, pp. 39–46. Springer, Heidelberg (1989). https://doi.org/10.1007/3-540-51542-9_5
13. Farzan, A., López-Ortiz, A., Nicholson, P.K., Salinger, A.: Algorithms in the ultra-wide word model. In: Jain, R., Jain, S., Stephan, F. (eds.) TAMC 2015. LNCS, vol. 9076, pp. 335–346. Springer, Cham (2015). https://doi.org/10.1007/978-3-319-17142-5_29
14. Fenwick, P.M.: A new data structure for cumulative frequency tables. Softw. Pract. Exp. **24**(3), 327–336 (1994)

15. Franceschini, G., Muthukrishnan, S., Pătraşcu, M.: Radix sorting with no extra space. In: Arge, L., Hoffmann, M., Welzl, E. (eds.) ESA 2007. LNCS, vol. 4698, pp. 194–205. Springer, Heidelberg (2007). https://doi.org/10.1007/978-3-540-75520-3_19

16. Frandsen, G.S., Miltersen, P.B., Skyum, S.: Dynamic word problems. J. ACM **44**(2), 257–271 (1997)

17. Fredman, M., Saks, M.: The cell probe complexity of dynamic data structures. In: Proceedings of the 21st STOC, pp. 345–354 (1989)

18. Fredman, M.L.: A lower bound on the complexity of orthogonal range queries. J. ACM **28**(4), 696–705 (1981)

19. Fredman, M.L.: The complexity of maintaining an array and computing its partial sums. J. ACM **29**(1), 250–260 (1982)

20. Hagerup, T.: Sorting and searching on the word RAM. In: Morvan, M., Meinel, C., Krob, D. (eds.) STACS 1998. LNCS, vol. 1373, pp. 366–398. Springer, Heidelberg (1998). https://doi.org/10.1007/BFb0028575

21. Hampapuram, H., Fredman, M.L.: Optimal biweighted binary trees and the complexity of maintaining partial sums. SIAM J. Comput. **28**(1), 1–9 (1998)

22. Hon, W.K., Sadakane, K., Sung, W.K.: Succinct data structures for searchable partial sums with optimal worst-case performance. Theor. Comput. Sci. **412**(39), 5176–5186 (2011)

23. Husfeldt, T., Rauhe, T.: New lower bound techniques for dynamic partial sums and related problems. SIAM J. Comput. **32**(3), 736–753 (2003)

24. Husfeldt, T., Rauhe, T., Skyum, S.: Lower bounds for dynamic transitive closure, planar point location, and parentheses matching. In: Karlsson, R., Lingas, A. (eds.) SWAT 1996. LNCS, vol. 1097, pp. 198–211. Springer, Heidelberg (1996). https://doi.org/10.1007/3-540-61422-2_132

25. Ladner, R.E., Fischer, M.J.: Parallel prefix computation. J. ACM **27**(4), 831–838 (1980)

26. Larsen, K.G., Pagh, R.: I/O-efficient data structures for colored range and prefix reporting. In: Proceedings of the 23rd SODA, pp. 583–592 (2012)

27. Leben, R., Miletic, M., Špegel, M., Trost, A., Brodnik, A., Karlsson, J.: Design of high performance memory module on PC100. In: Proceedings of the ECSC, pp. 75–78 (1999)

28. Lindholm, E., Nickolls, J., Oberman, S., Montrym, J.: NVIDIA Tesla: a unified graphics and computing architecture. IEEE Micro **28**(2), 39–55 (2008)

29. Miltersen, P.B.: Cell probe complexity-a survey. In: Proceedings of the 19th FSTTCS, p. 2 (1999)

30. Munro, J.I., Suwanda, H.: Implicit data structures for fast search and update. J. Comput. Syst. Sci. **21**(2), 236–250 (1980)

31. Pătraşcu, M., Demaine, E.D.: Logarithmic lower bounds in the cell-probe model. SIAM J. Comput. **35**(4), 932–963 (2006). Announced at SODA 2004

32. Raman, R., Raman, V., Rao, S.S.: Succinct dynamic data structures. In: Dehne, F., Sack, J.-R., Tamassia, R. (eds.) WADS 2001. LNCS, vol. 2125, pp. 426–437. Springer, Heidelberg (2001). https://doi.org/10.1007/3-540-44634-6_39

33. Reinders, J.: AVX-512 Instructions. Intel Corporation, Santa Clara (2013)

34. Ryabko, B.Y.: A fast on-line adaptive code. IEEE Trans. Inf. Theory **38**(4), 1400–1404 (1992)

35. Salowe, J., Steiger, W.: Simplified stable merging tasks. J. Algorithms **8**(4), 557–571 (1987)

36. Stephens, N., et al.: The ARM scalable vector extension. IEEE Micro **37**(2), 26–39 (2017)
37. Williams, J.W.J.: Algorithm 232: heapsort. Commun. ACM **7**, 347–348 (1964)
38. Yao, A.C.: On the complexity of maintaining partial sums. SIAM J. Comput. **14**(2), 277–288 (1985)

Securely Computing the n-Variable Equality Function with $2n$ Cards

Suthee Ruangwises$^{(\boxtimes)}$ ⓘ and Toshiya Itoh ⓘ

Department of Mathematical and Computing Science,
Tokyo Institute of Technology, Tokyo, Japan
`ruangwises.s.aa@m.titech.ac.jp, titoh@c.titech.ac.jp`

Abstract. Research on the area of secure multi-party computation using a deck of playing cards, often called card-based cryptography, started from the introduction of the "five-card trick" to compute the logical AND function by den Boar in 1989. Since then, many protocols to compute various functions have been developed. In this paper, we propose a new card-based protocol that securely computes the n-variable *equality function* using $2n$ cards. We also show that the same technique can be applied to compute any *doubly symmetric function* $f : \{0,1\}^n \to \mathbb{Z}$ using $2n$ cards, and any symmetric function $f : \{0,1\}^n \to \mathbb{Z}$ using $2n + 2$ cards.

Keywords: Card-based cryptography · Secure multi-party computation · Equality function · Symmetric function · Doubly symmetric function

1 Introduction

During a two-candidate election, a group of n friends decides that they should discuss about the election only if everyone in the group supports the same candidate. However, each person does not know other people's preferences and wants to hide his/her own preference from the others unless they all support the same candidate in order to avoid awkwardness in the conversation. How can they know whether their preferences all coincide without leaking any other information?

In terms of secure multi-party computation, this situation can be viewed as a group of n players where the ith player has a bit a_i of either 0 or 1. Define the *equality function* $E(a_1, ..., a_n) = 1$ if $a_1 = ... = a_n$ and $E(a_1, ..., a_n) = 0$ otherwise. Our goal is to design a protocol that announces only the value of $E(a_1, ..., a_n)$ without leaking any other information, such as the preference of any player or the number of players who support each candidate (not even probabilistic information).

Secure multi-party computation is one of the most actively studied research areas in cryptography. It involves situations where multiple parties want to compare their private information without revealing it. In particular, this paper focuses on secure multi-party computation using a deck of playing cards, often

© Springer Nature Switzerland AG 2020
J. Chen et al. (Eds.): TAMC 2020, LNCS 12337, pp. 25–36, 2020.
https://doi.org/10.1007/978-3-030-59267-7_3

called card-based cryptography. The benefit of card-based protocols is that they provide solutions to real-world situations using only a small deck of cards, which is portable and can be found in everyday life, and do not require computers. Moreover, these straightforward protocols are easy to understand and verify the correctness and security, even for non-experts.

1.1 Related Work

The first research on card-based cryptography started in 1989 with the "five-card trick" introduced by den Boer [3] to compute the logical AND function on two players' bits a and b. This protocol uses three identical ♣ cards and two identical ♡ cards.

Throughout this paper, a bit 0 is encoded by a *commitment* ♣♡ and a bit 1 by a commitment ♡♣. We give each player one ♣ card and one ♡ card, and put another ♣ card face-down on a table. The first player then places his commitment of a face-down to the left of the ♣ card, while the second player places his commitment of b face-down to the right of it. Then, we swap the fourth and the fifth cards from the left, resulting in the following four possible sequences.

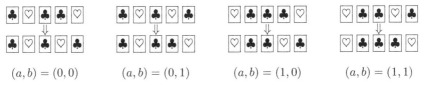

$(a,b) = (0,0)$ $(a,b) = (0,1)$ $(a,b) = (1,0)$ $(a,b) = (1,1)$

Observe that there are only two possible sequences in a cyclic rotation of the deck, ♡♣♡♣♣ and ♡♡♣♣♣, with the latter showing up if and only if $a = b = 1$. We can obscure the initial position of the cards by making a *random cut* to shuffle the deck into a uniformly random cyclic permutation, i.e. a permutation uniformly chosen at random from $\{\text{id}, \pi, \pi^2, \pi^3, \pi^4\}$ where $\pi = (1\ 2\ 3\ 4\ 5)$, before turning all cards face-up. Hence, we can determine whether $a \wedge b = 1$ from the cycle.

Since the introduction of the five-card trick, several other protocols to compute the AND function have been developed. These subsequent results [1,2,4,5,7,8,10,13,16] aimed to either reduce the number of required cards or improve properties of the protocol involving output format, running time, type of shuffles, etc.

Apart from the AND function protocol, various kinds of protocols have been developed as well, such as the XOR function protocol [2,8,9], the copy protocol [8] (creating multiple copies of the commitment), the *majority function* protocol [12] (deciding whether there are more 0s or 1s in the inputs), and the adder protocol [6] (adding bits and storing the sum in binary representation). Nishida et al. [11] proved that any n-variable Boolean function can be computed with $2n + 6$ cards, and any such function that is symmetric can be computed with $2n + 2$ cards.

1.2   The Six-Card Trick

For the equality function, the case $n = 2$ is a negation of the XOR function, which can be easily computed with four cards. For the case $n = 3$, Shinagawa and Mizuki [14] developed the following protocol called the "six-card trick" to compute the function $E(a, b, c)$ on three players' bits a, b, and c using six cards.

First, the players put the commitments of a, b, and c face-down on a table in this order from left to right. Then, we rearrange the cards into a (2 4 6) permutation, i.e. move the second leftmost card to the fourth leftmost position, the fourth card to the sixth position, and the sixth card to the second position, resulting in the following eight possible sequences.

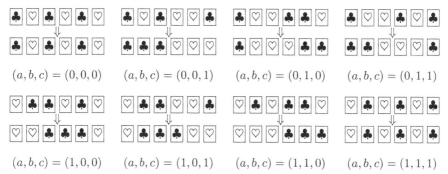

Observe that there are only two possible sequences in a cyclic rotation of the deck, ♣♣♣♡♡♡ and ♣♡♣♡♣♡, with the latter showing up if and only if $a = b = c$, i.e. $E(a, b, c) = 1$. Again, we can obscure the initial position of the cards by making a random cut before turning all cards face-up, hence we can determine the value of $E(a, b, c)$ from the cycle.

The six-card trick has a benefit that it uses only one random cut. However, the technique used in this protocol heavily relies on the symmetric nature of the special case $n = 3$, suggesting that there might not be an equivalent protocol using $2n$ cards for a general n. In fact, in [14] they found by using a computer that in the case $n = 4$, an eight-card protocol that uses only one random cut does not exist.

1.3   Our Contribution

In this paper, we develop a card-based protocol that securely computes the n-variable equality function using $2n$ cards. We also show that the same technique can be applied to compute any *doubly symmetric function* (see the definition in Sect. 4.1) $f : \{0, 1\}^n \to \mathbb{Z}$ using $2n$ cards, and any symmetric function $f : \{0, 1\}^n \to \mathbb{Z}$ using $2n + 2$ cards.

2   Basic Operations

First, we will introduce basic operations on a deck of cards that will be used in our protocols.

2.1 Random Cut

Suppose we have a sequence of cards $(x_0, x_1, ..., x_{k-1})$. A random cut is an operation to shuffle the deck into a uniformly random cyclic permutation, shifting the sequence into $(x_r, x_{r+1}, ..., x_{r+k-1})$, where r is a uniformly random integer from $\{0, 1, ..., k-1\}$ and the indices are taken in mod k.

$$\boxed{?}\,\boxed{?}\,...\,\boxed{?} \;\;\Rightarrow\;\; \boxed{?}\,\boxed{?}\,...\,\boxed{?}$$
$$\;\;x_0\;x_1\quad\;\; x_{k-1}\qquad\quad x_r\;\, x_{r+1}\quad x_{r+k-1}$$

In real world, a random cut can be performed by applying a *Hindu cut*, which is a basic shuffling operation commonly used in card games [17].

2.2 Random k-Section Cut

A *random k-section cut* is a generalization of a *random bisection cut* introduced by Mizuki and Sone [8]. Suppose we have a sequence of km cards $(x_0, x_1, ..., x_{km-1})$. We divide the cards into k blocks $B_0, ..., B_{k-1}$, with each block B_i consisting of m consecutive cards $x_{im}, x_{im+1}, ..., x_{(i+1)m-1}$.

Then, we shuffle the blocks into a uniformly random cyclic permutation, shifting the order of them into $(B_r, B_{r+1}, ..., B_{r+k-1})$, where r is a uniformly random integer from $\{0, 1, ..., k-1\}$ and the indices are taken in mod k. This operation shifts the sequence of cards into $(x_{rm}, x_{rm+1}, ..., x_{(r+k)m-1})$, where the indices are taken in mod km.

In real world, a random k-section cut can be performed by putting each block of cards into an envelope and applying a random cut on the pile of envelopes before taking the cards out.

2.3 XOR with a Random Bit

Recall that we encode 0 and 1 by commitments ♣♡ and ♡♣, respectively. Suppose we have a sequence of k bits $(a_1, a_2, ..., a_k)$ as an input, with each a_i encoded by a commitment (x_i, y_i). We want to securely perform the XOR operation with the same random bit on every input bit, i.e. output the sequence $(a_1 \oplus r, a_2 \oplus r, ..., a_k \oplus r)$ where $r \in \{0, 1\}$ is a uniformly random bit.

We can achieve this by applying a random 2-section cut in a way similar to the copy protocol of Mizuki and Sone [8]. First, arrange the cards as $X = (x_1, x_2, ..., x_k, y_1, y_2, ..., y_k)$ and apply a random 2-section cut on X. Then, for each $i = 1, 2, ..., k$, take the ith and the $(i+k)$-th cards from X in this order as the commitment of the ith output bit.

Observe that after applying the random 2-section cut, the sequence X will become either $(x_1, x_2, ..., x_k, y_1, y_2, ..., y_k)$ or $(y_1, y_2, ..., y_k, x_1, x_2, ..., x_k)$ with equal probability. In the former case, the commitment of every ith output bit will be (x_i, y_i), which is $a_i \oplus 0$; in the latter case, the commitment of every ith output bit will be (y_i, x_i), which is $a_i \oplus 1$. Therefore, the correctness of the operation is verified.

2.4   Adding Two Integers in $\mathbb{Z}/k\mathbb{Z}$

For $k \geq 3$, we first introduce two schemes of encoding integers in $\mathbb{Z}/k\mathbb{Z}$, the ♣-scheme and the ♡-scheme. The ♣-scheme uses one ♣ card and $k-1$ ♡ cards arranged in a row. An integer i corresponds to an arrangement where the ♣ card is the $(i+1)$-th card from the left, e.g.. ♡♣♡ encodes 1 in $\mathbb{Z}/3\mathbb{Z}$. Conversely, the ♡-scheme uses one♡ card and $k-1$ ♣ cards arranged in a row. An integer i corresponds to an arrangement where the ♡ card is the $(i+1)$-th card from the left, e.g. ♣♣♡♣ encodes 2 in $\mathbb{Z}/4\mathbb{Z}$.

Suppose we have integers a and b in $\mathbb{Z}/k\mathbb{Z}$, with a encoded in ♡-scheme by a sequence of face-down cards $X = (x_0, x_1, ..., x_{k-1})$, and b encoded in ♣-scheme by a sequence of face-down cards $Y = (y_0, y_1, ..., y_{k-1})$. We want to securely compute the sum $a + b \pmod{k}$ and have it encoded in ♡-scheme without using any additional card.

The intuition of this protocol is that we transform a and b into $a - r$ and $b + r$ for a random $r \in \mathbb{Z}/k\mathbb{Z}$, reveal $b + r$, and then shift the cards encoding $a - r$ to the right by $b + r$ positions to make them encode $(a - r) + (b + r) = a + b$. This technique was first used by Shinagawa et al. [15] in the context of using regular k-gon cards to encode integers in $\mathbb{Z}/k\mathbb{Z}$.

First, take the cards from X and Y in the following order and place them on a single row from left to right: the leftmost card of X, the rightmost card of Y, the second leftmost card of X, the second rightmost card of Y, and so on. The cards now form a new sequence $Z = (x_0, y_{k-1}, x_1, y_{k-2}, ..., x_{k-1}, y_0)$.

$$X: \boxed{?}\,\boxed{?}\,...\,\boxed{?} \quad Y: \boxed{?}\,\boxed{?}\,...\,\boxed{?} \quad \Rightarrow \quad Z: \boxed{?}\,\boxed{?}\,\boxed{?}\,\boxed{?}\,...\,\boxed{?}\,\boxed{?}$$

with labels $x_0\ x_1$ ... x_{k-1} under X, $y_0\ y_1$ ... y_{k-1} under Y, and x_0, x_1, x_{k-1} above Z with y_{k-1}, y_{k-2}, y_0 below.

Apply a random k-section cut on Z, transforming the sequence into $(x_r, y_{-r+k-1}, x_{r+1}, y_{-r+k-2}, ..., x_{r+k-1}, y_{-r})$ for a uniformly random $r \in \mathbb{Z}/k\mathbb{Z}$, where the indices are taken in mod k.

$$Z: \boxed{?}\,\boxed{?}\,\boxed{?}\,\boxed{?}\,...\,\boxed{?}\,\boxed{?} \quad \Rightarrow \quad Z: \boxed{?}\ \boxed{?}\ \boxed{?}\ \boxed{?}\ ...\ \boxed{?}\ \boxed{?}$$

with x_0, x_1, x_{k-1} above the left Z and y_{k-1}, y_{k-2}, y_0 below; x_r, x_{r+1}, x_{r+k-1} above the right Z and y_{-r+k-1}, y_{-r+k-2}, y_{-r} below.

Take the cards in Z from left to right and place them at these positions in X and Y in the following order: the leftmost position of X, the rightmost position of Y, the second leftmost position of X, the second rightmost position of Y, and so on. We now have sequences $X = (x_r, x_{r+1}, ..., x_{r+k-1})$ and $Y = (y_{-r}, y_{-r+1}, ..., y_{-r+k-1})$, which encode $a - r$ and $b + r$, respectively.

$$Z: \boxed{?}\ \boxed{?}\ \boxed{?}\ \boxed{?}\ ...\ \boxed{?}\ \boxed{?} \quad \Rightarrow \quad X: \boxed{?}\ \boxed{?}\ ...\ \boxed{?} \quad Y: \boxed{?}\ \boxed{?}\ ...\ \boxed{?}$$

with x_r, x_{r+1}, x_{r+k-1} above Z and y_{-r+k-1}, y_{-r+k-2}, y_{-r} below; $x_r\ x_{r+1}\ x_{r+k-1}$ under X and $y_{-r}\ y_{-r+1}\ y_{-r+k-1}$ under Y.

Turn all cards in Y face-up to reveal $s = b+r$. Note that this revelation does not leak any information of b because $b + r$ has an equal probability to be any integer in $\mathbb{Z}/k\mathbb{Z}$ no matter what b is. Then, we shift the cards in X to the right by s positions, transforming X into $(x_{r-s}, x_{r-s+1}, ..., x_{r-s+k-1})$.

$$X: \boxed{?}\ \boxed{?}\ ...\ \boxed{?} \quad \Rightarrow \quad X: \boxed{?}\ \boxed{?}\ ...\ \boxed{?}$$

with $x_r\ x_{r+1}\ x_{r+k-1}$ under the left X and $x_{r-s}\ x_{r-s+1}\ x_{r-s+k-1}$ under the right X.

Therefore, we now have a sequence X encoding $a - r + s = (a-r) + (b+r) = a + b$ in $\heartsuit$-scheme as desired.

3   Our Main Protocol

We get back to our main problem. Observe that if we treat each input a_i as an integer, the value of $E(a_1, ..., a_n)$ depends only on the sum $s_n = \sum_{i=1}^{n} a_i$. Therefore, we will first develop a protocol to compute that sum. The intuition of this protocol is that for each $k = 2, 3, ..., n$, we inductively compute the sum $s_k = \sum_{i=1}^{k} a_i$ in $\mathbb{Z}/(k+1)\mathbb{Z}$. Note that since s_k is at most k, its value in $\mathbb{Z}/(k+1)\mathbb{Z}$ does not change from its actual value.

3.1 Summation of the First k Bits

We will show that if we have two additional cards, one ♣ and one ♡, we can compute the sum s_k for every $k = 2, 3, ..., n$ by the following procedure.

First, swap the two cards in the commitment of a_1 and place an additional ♣ card face-down to the right of them. The resulting sequence, called C_1, encodes a_1 in $\mathbb{Z}/3\mathbb{Z}$ in ♡-scheme.

$$
\begin{array}{ll}
\text{Case } a_1 = 0: & \clubsuit \ \heartsuit \qquad\qquad \heartsuit \ \clubsuit \qquad\qquad \heartsuit \ \clubsuit \ \clubsuit \\
\text{Case } a_1 = 1: & \heartsuit \ \clubsuit \qquad\qquad \clubsuit \ \heartsuit \qquad\qquad \clubsuit \ \heartsuit \ \clubsuit
\end{array}
$$

$$a_1: \boxed{?}\,\boxed{?} \ \Rightarrow \ \boxed{?}\,\boxed{?} \ \Rightarrow \ C_1: \boxed{?}\,\boxed{?}\,\boxed{\clubsuit}$$

Then, put an additional ♡ card face-down to the right of the commitment of a_2. The resulting sequence, called C_2, encodes a_2 in $\mathbb{Z}/3\mathbb{Z}$ in ♣-scheme.

$$
\begin{array}{ll}
\text{Case } a_2 = 0: & \clubsuit \ \heartsuit \qquad\qquad \clubsuit \ \heartsuit \ \heartsuit \\
\text{Case } a_2 = 1: & \heartsuit \ \clubsuit \qquad\qquad \heartsuit \ \clubsuit \ \heartsuit
\end{array}
$$

$$a_2: \boxed{?}\,\boxed{?} \ \Rightarrow \ C_2: \boxed{?}\,\boxed{?}\,\boxed{\heartsuit}$$

We then apply the addition protocol introduced in Sect. 2.4 to store the sum $s_2 = a_1 + a_2$ in $\mathbb{Z}/3\mathbb{Z}$ encoded in ♡-scheme in C_1. We also now have two ♡ cards and one ♣ card from C_2 after we turned them face-up. These cards are called *free cards* and are available to be used later in the protocol.

$$C_1 \text{ encoding } s_2: \boxed{?}\,\boxed{?}\,\boxed{?}$$

$$\text{free cards from } C_2: \boxed{\clubsuit}\,\boxed{\heartsuit}\,\boxed{\heartsuit}$$

Inductively, for each $k \geq 3$, after we finish computing s_{k-1}, we now have a sequence C_1 of k face-down cards encoding s_{k-1} in $\mathbb{Z}/k\mathbb{Z}$ in ♡-scheme. We also have $k - 1$ free ♡ cards and one free ♣ card from C_{k-1} after we turned them face-up. Append the free ♣ card face-down to the right of C_1, making the sequence now encode s_{k-1} in $\mathbb{Z}/(k+1)\mathbb{Z}$ in ♡-scheme. Also, place the $k - 1$ free ♡ cards face-down to the right of the commitment of a_k. The resulting sequence, called C_k, encodes a_k in $\mathbb{Z}/(k+1)\mathbb{Z}$ in ♣-scheme.

$$C_1 \text{ encoding } s_{k-1}: \boxed{?}\,\boxed{?}\,...\,\boxed{?} \quad \Rightarrow \quad C_1 \text{ encoding } s_{k-1}: \boxed{?}\,\boxed{?}\,...\,\boxed{?}\,\boxed{\clubsuit}$$

$$\text{commitment of } a_k: \boxed{?}\,\boxed{?} \quad \Rightarrow \quad C_k \text{ encoding } a_k: \boxed{?}\,\boxed{?}\,\boxed{\heartsuit}\,...\,\boxed{\heartsuit}$$

$$\text{free cards from } C_{k-1}: 1 \times \boxed{\clubsuit}, (k-1) \times \boxed{\heartsuit}$$

Then, apply the addition protocol to compute the sum $s_{k-1} + a_k \pmod{k+1}$ $= s_k \pmod{k+1} = s_k$ and have it encoded in ♡-scheme by C_1 as desired.

$$C_1 \text{ encoding } s_k: \boxed{?}\boxed{?} \cdots \boxed{?}$$

$$\text{free cards from } C_k: 1 \times \boxed{\clubsuit} \,,\, k \times \boxed{\heartsuit}$$

Therefore, starting with one additional $\clubsuit$ card and one additional $\heartsuit$ card, we can compute the sum $s_k = \sum_{i=1}^{k} a_i$ for every $k = 2, 3, ..., n$.

3.2 Putting Together

The summation protocol introduced in Sect. 3.1 requires two additional cards to compute s_k. However, we can compute the equality function without using any additional card by the following procedure.

First, apply the random bit XOR protocol in Sect. 2.3 to transform the input into $(a_1 \oplus r, a_2 \oplus r, ..., a_n \oplus r)$ for a random bit $r \in \{0, 1\}$. Then, turn the two cards encoding the nth bit face-up to reveal $a_n \oplus r$. Note that this revelation does not leak any information of a_n because seeing $\clubsuit\heartsuit$ and $\heartsuit\clubsuit$ each has probability $1/2$ no matter whether a_n is 0 or 1.

If the cards are $\clubsuit\heartsuit$, i.e. $a_n \oplus r = 0$, the equality function outputs 1 if and only if $a_i \oplus r = 0$ for every $i = 1, ..., n-1$, which is equivalent to $\sum_{i=1}^{n-1}(a_i \oplus r) = 0$. Note that we now have one free $\clubsuit$ card and one free $\heartsuit$ card from the cards we just turned face-up. With these two additional cards, we can apply the summation protocol to compute $\sum_{i=1}^{n-1}(a_i \oplus r)$ as desired. On the other hand, if the two rightmost cards are $\heartsuit\clubsuit$, i.e. $a_n \oplus r = 1$, the equality function outputs 1 if and only if $a_i \oplus r = 1$ for every $i = 1, ..., n-1$, which is equivalent to $\sum_{i=1}^{n-1}(a_i \oplus r \oplus 1) = 0$. Therefore, we can swap the two cards encoding every bit so that each ith bit becomes $a_i \oplus r \oplus 1$ and then apply the same protocol.

Note that the final sum is encoded in $\heartsuit$-scheme by a row of n cards, where the equality function outputs 1 if and only if the sum is zero, i.e. the $\heartsuit$ card is at the leftmost position. However, we do not want to reveal any information about the actual value of the sum except whether it is zero or not. Therefore, we apply a final random cut on the sequence of $n-1$ rightmost cards (all cards in the row except the leftmost one) to make all the cases where the sum is not zero indistinguishable. Finally, we turn all cards face-up and locate the position of the $\heartsuit$ card. If it is the leftmost card in the row, then output 1; otherwise output 0.

We use one random 2-section cut in the random bit XOR operation, $n-2$ random k-section cuts for computing the sum of $n-1$ bits, and one random cut in the final shuffle. Therefore, the total number of shuffles used in the whole protocol is n.

4 Applications

4.1 Computing Other Symmetric Functions

A function $f : \{0, 1\}^n \to \mathbb{Z}$ is called symmetric if

$$f(a_1, ..., a_n) = f(a_{\sigma_1}, ..., a_{\sigma_n})$$

for any $a_1, ..., a_n$ and any permutation $(\sigma_1, ..., \sigma_n)$ of $(1, ..., n)$. A symmetric function f is called doubly symmetric if

$$f(a_1, ..., a_n) = f(1 - a_1, ..., 1 - a_n)$$

for any $a_1, ..., a_n$. For example, the equality function is doubly symmetric, while the majority function is symmetric but not doubly symmetric. Another example of a doubly symmetric function is $f(a_1, ..., a_n) = a_1 \oplus ... \oplus a_n$ for an even n.

Observe that for any symmetric function $f : \{0, 1\}^n \to \mathbb{Z}$, the value of $f(a_1, ..., a_n)$ depends only on the sum $\sum_{i=1}^{n} a_i$, hence f can be written as

$$f(a_1, ..., a_n) = g\left(\sum_{i=1}^{n} a_i\right)$$

for some function $g : \{0, ..., n\} \to \mathbb{Z}$. Also, if f is doubly symmetric, we have $g(a) = g(n - a)$ for any $a \in \{0, ..., n\}$.

Our protocol can also be applied to compute any doubly symmetric function. Let $f : \{0, 1\}^n \to \mathbb{Z}$ be any doubly symmetric function and let $g : \{0, ..., n\} \to \mathbb{Z}$ be a function such that

$$f(a_1, ..., a_n) = g\left(\sum_{i=1}^{n} a_i\right).$$

First, we apply the random bit XOR protocol with a random bit $r \in \{0, 1\}$ to every input a_i and then reveal $a_n \oplus r$ (without leaking any information of a_n since $a_n \oplus r$ has an equal probability to be 0 and 1 no matter whether a_n is 0 or 1).

Since f is doubly symmetric, if $a_n \oplus r = 0$, we have

$$f(a_1, ..., a_n) = f(a_1 \oplus r, ..., a_n \oplus r)$$

$$= g\left(\sum_{i=1}^{n}(a_i \oplus r)\right)$$

$$= g\left(\sum_{i=1}^{n-1}(a_i \oplus r)\right),$$

so we can apply the summation protocol to compute $\sum_{i=1}^{n-1}(a_i \oplus r)$. On the other hand, if $a_n \oplus r = 1$, we have $a_n \oplus r \oplus 1 = 0$, so

$$f(a_1, ..., a_n) = f(a_1 \oplus r \oplus 1, ..., a_n \oplus r \oplus 1)$$

$$= g\left(\sum_{i=1}^{n}(a_i \oplus r \oplus 1)\right)$$

$$= g\left(\sum_{i=1}^{n-1}(a_i \oplus r \oplus 1)\right),$$

hence we can swap the two cards encoding every bit and apply the same protocol to compute $\sum_{i=1}^{n-1}(a_i \oplus r \oplus 1)$.

For each $b \in \operatorname{Im} f = \operatorname{Im} g$, let $P_b = \{a \in \{0,1,...,n\}|g(a) = b\}$. We now have a row of n cards encoding the sum in $\heartsuit$-scheme. Recall that in $\heartsuit$-scheme, an integer i corresponds to an arrangement where the $(i+1)$-th card from the left being $\heartsuit$. Therefore, we can take from the row all the cards corresponding to integers in P_b, i.e. the $(i+1)$-th card from the left for every $i \in P_b$, apply a random cut on them, and put them back into the row at their original positions in order to make all the cases where the sum is in P_b indistinguishable. We need to separately apply such random cut for every $b \in \operatorname{Im} f$ such that $|P_b| > 1$. These random cuts ensure that turning the cards face-up does not reveal any information about the sum except the output value of g. Finally, we turn all cards face-up to reveal an integer s and output $g(s)$. The number of required cards is $2n$, and the total number of shuffles is at most $n - 1 + |\operatorname{Im} f|$.

For a function that is symmetric but not doubly symmetric, we can directly apply the summation protocol to compute the sum $s_n = \sum_{i=1}^{n} a_i$, apply the above random cut for every $b \in \operatorname{Im} f$ such that $|P_b| > 1$, and output $g(s_n)$, although it requires two additional cards at the beginning. Therefore, the number of required cards is $2n + 2$, and the total number of shuffles is at most $n - 1 + |\operatorname{Im} f|$.

4.2   Optimality

There is a protocol developed by Mizuki et al. [6] that can compute the sum of n input bits using only $O(\log n)$ cards, but their protocol restricts the order of submission of the inputs so that the cards can be reused. Any protocol that the inputs are submitted simultaneously requires at least $2n$ cards as we need two cards for a commitment of each person's bit, hence our protocol is the optimal one for computing any doubly symmetric function.

For computing symmetric functions that are not doubly symmetric, the protocol of Nishida et al. [11] also uses $2n + 2$ cards to compute any symmetric function $f : \{0,1\}^n \to \{0,1\}$. Their protocol has a benefit that the output is in *committed-format*, i.e. encoded in the same format as the input ($\clubsuit\heartsuit$ for 0 and $\heartsuit\clubsuit$ for 1), so the output can be securely used as an input of another function. However, our protocol uses fewer number of shuffles and also has a benefit that the output is not restricted to be binary, hence supporting functions with more than two possible outputs (an example of such function is the majority function that supports the case of a tie for an even n, which has three possible outputs).

5   Future Work

For computing the equality function or any doubly symmetric function, our protocol is optimal in terms of number of cards as it matches the trivial lower bound of $2n$. However, there is still an open problem to find a committed-format protocol that uses $2n$ cards, or a non-committed-format one with the same number of cards but uses a fewer number of shuffles. For symmetric functions that are not doubly symmetric, an open problem is to find a protocol that computes such functions with less than $2n + 2$ cards.

Another interesting future work is to prove the lower bound of the number of cards or the number of shuffles required to compute such functions, either for a committed-format protocol or for any protocol.

References

1. Abe, Y., Hayashi, Y., Mizuki, T., Sone, H.: Five-card AND protocol in committed format using only practical shuffles. In: Proceedings of the 5th ACM on ASIA Public-Key Cryptography Workshop (APKC 2018), pp. 3–8 (2018)
2. Crépeau, C., Kilian, J.: Discreet solitary games. In: Stinson, D.R. (ed.) CRYPTO 1993. LNCS, vol. 773, pp. 319–330. Springer, Heidelberg (1994). https://doi.org/10.1007/3-540-48329-2_27
3. Boer, B.: More efficient match-making and satisfiability *the five card trick*. In: Quisquater, J.-J., Vandewalle, J. (eds.) EUROCRYPT 1989. LNCS, vol. 434, pp. 208–217. Springer, Heidelberg (1990). https://doi.org/10.1007/3-540-46885-4_23
4. Koch, A.: The Landscape of Optimal Card-based Protocols. Cryptology ePrint Archive https://eprint.iacr.org/2018/951/20181009:160322 (2018)
5. Koch, A., Walzer, S., Härtel, K.: Card-based cryptographic protocols using a minimal number of cards. In: Iwata, T., Cheon, J.H. (eds.) ASIACRYPT 2015. LNCS, vol. 9452, pp. 783–807. Springer, Heidelberg (2015). https://doi.org/10.1007/978-3-662-48797-6_32
6. Mizuki, T., Asiedu, I.K., Sone, H.: Voting with a logarithmic number of cards. In: Mauri, G., Dennunzio, A., Manzoni, L., Porreca, A.E. (eds.) UCNC 2013. LNCS, vol. 7956, pp. 162–173. Springer, Heidelberg (2013). https://doi.org/10.1007/978-3-642-39074-6_16
7. Mizuki, T., Kumamoto, M., Sone, H.: The five-card trick can be done with four cards. In: Wang, X., Sako, K. (eds.) ASIACRYPT 2012. LNCS, vol. 7658, pp. 598–606. Springer, Heidelberg (2012). https://doi.org/10.1007/978-3-642-34961-4_36
8. Mizuki, T., Sone, H.: Six-card secure AND and four-card secure XOR. In: Deng, X., Hopcroft, J.E., Xue, J. (eds.) FAW 2009. LNCS, vol. 5598, pp. 358–369. Springer, Heidelberg (2009). https://doi.org/10.1007/978-3-642-02270-8_36
9. Mizuki, T., Uchiike, F., Sone, H.: Securely computing XOR with 10 cards. Australas. J. Comb. **36**, 279–293 (2006)
10. Niemi, V., Renvall, A.: Secure multiparty computations without computers. Theor. Comput. Sci. **191**, 173–183 (1998)
11. Nishida, T., Hayashi, Y., Mizuki, T., Sone, H.: Card-based protocols for any boolean function. In: Jain, R., Jain, S., Stephan, F. (eds.) TAMC 2015. LNCS, vol. 9076, pp. 110–121. Springer, Cham (2015). https://doi.org/10.1007/978-3-319-17142-5_11
12. Nishida, T., Mizuki, T., Sone, H.: Securely computing the three-input majority function with eight cards. In: Dediu, A.-H., Martín-Vide, C., Truthe, B., Vega-Rodríguez, M.A. (eds.) TPNC 2013. LNCS, vol. 8273, pp. 193–204. Springer, Heidelberg (2013). https://doi.org/10.1007/978-3-642-45008-2_16
13. Ruangwises, S., Itoh, T.: AND protocols using only uniform shuffles. In: van Bevern, R., Kucherov, G. (eds.) CSR 2019. LNCS, vol. 11532, pp. 349–358. Springer, Cham (2019). https://doi.org/10.1007/978-3-030-19955-5_30

14. Shinagawa, K., Mizuki, T.: The six-card trick: secure computation of three-input equality. In: Lee, K. (ed.) ICISC 2018. LNCS, vol. 11396, pp. 123–131. Springer, Cham (2019). https://doi.org/10.1007/978-3-030-12146-4_8

15. Shinagawa, K., et al.: Multi-party computation with small shuffle complexity using regular polygon cards. In: Au, M.-H., Miyaji, A. (eds.) ProvSec 2015. LNCS, vol. 9451, pp. 127–146. Springer, Cham (2015). https://doi.org/10.1007/978-3-319-26059-4_7

16. Stiglic, A.: Computations with a deck of cards. Theor. Comput. Sci. **259**, 671–678 (2001)

17. Ueda, I., Nishimura, A., Hayashi, Y., Mizuki, T., Sone, H.: How to implement a random bisection cut. In: Martín-Vide, C., Mizuki, T., Vega-Rodríguez, M.A. (eds.) TPNC 2016. LNCS, vol. 10071, pp. 58–69. Springer, Cham (2016). https://doi.org/10.1007/978-3-319-49001-4_5

Polynomial Kernels for Paw-Free Edge Modification Problems

Yixin Cao[1]([⊠]) [ID], Yuping Ke[1], and Hanchun Yuan[2]

[1] Department of Computing, Hong Kong Polytechnic University,
Hong Kong, China
yixin.cao@polyu.edu.hk
[2] School of Computer Science and Engineering, Central South University,
Changsha, China

Abstract. Let H be a fixed graph. Given a graph G and an integer k, the H-free edge modification problem asks whether it is possible to modify at most k edges in G to make it H-free. Sandeep and Sivadasan (IPEC 2015) asks whether the paw-free completion problem and the paw-free edge deletion problem admit polynomial kernels. We answer both questions affirmatively by presenting, respectively, $O(k)$-vertex and $O(k^4)$-vertex kernels for them. This is part of an ongoing program that aims at understanding compressibility of H-free edge modification problems.

Keywords: Kernelization · Paw-free graph · Graph modification

1 Introduction

A graph modification problem asks whether one can apply at most k modifications to a graph to make it satisfy certain properties. By modifications we usually mean additions and/or deletions, and they can be applied to vertices or edges. Although other modifications are also considered, most results in literature are on vertex deletion and the following three edge modifications: edge deletion, edge addition, and edge editing (addition/deletion).

Compared to the general dichotomy results on vertex deletion problems [1,5], the picture for edge modification problems is far murkier. Embarrassingly, this remains true for the simplest case, namely, H-free graphs for fixed graphs H. This paper is a sequel to [2], and we are aiming at understanding for which H, the H-free edge modification problems admitting polynomial kernels. Our current focus is on the four-vertex graphs; see Fig. 1 (some four-vertex graphs are omitted because they are complement of ones presented here) and Table 1.[1] We refer the reader to [2] for background of this research and related work.

---

[1] Disclaimer: Independent of our work, Eiben et al. [3] obtain similar results for edge modification problems to paw-free graphs. They are also able to develop a polynomial kernel for the editing problem.

Supported by RGC grants 15201317 and 15226116, and NSFC grant 61972330.

© Springer Nature Switzerland AG 2020
J. Chen et al. (Eds.): TAMC 2020, LNCS 12337, pp. 37–49, 2020.
https://doi.org/10.1007/978-3-030-59267-7_4

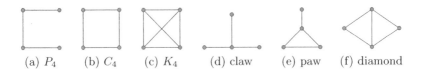

(a) P_4 (b) C_4 (c) K_4 (d) claw   (e) paw   (f) diamond

Fig. 1. Graphs on four vertices (their complements are omitted).

Table 1. The compressibility results of H-free edge modification problems for H being four-vertex graphs. Note that every result holds for the complement of H; e.g., the answers are also no when H is $2K_2$ (the complement of C_4).

H	Completion	Deletion	Editing
K_4	Trivial	$O(k^3)$	$O(k^3)$ [8]
P_4	$O(k^3)$	$O(k^3)$	$O(k^3)$ [4]
Diamond	Trivial	$O(k^3)$	$O(k^8)$ [2]
Paw	$O(k)$ [this paper]	$O(k^4)$ [this paper]	$O(k^6)$ [3]
Claw	Unknown	Unknown	Unknown
C_4	No	No	No [4]

In this paper, we show polynomial kernels for both the completion and edge deletion problems when H is the paw (Fig. 1(e)). They answer open problems posed by Sandeep and Sivadasan [7].

Theorem 1. *The paw-free completion problem has a $38k$-vertex kernel.*

Theorem 2. *The paw-free edge deletion problem has an $O(k^4)$-vertex kernel.*

It is easy to see that each component of a paw-free graph is either triangle-free or a complete multipartite graph with at least three parts [6]. This simple observation motivates us to take the modulator approach. Here by modulator we mean a set of vertices that intersect every paw of the input graph by at least two vertices. Note that the deletion of all the vertices in the modulator leaves the graph paw-free. We then study the interaction between the modulator M and the components of $G - M$, which are triangle-free or complete multipartite. We use slightly different modulators for the two problems under study.

2   Paw-Free Graphs

All graphs discussed in this paper are undirected and simple. A graph G is given by its vertex set $V(G)$ and edge set $E(G)$. For a set $U \subseteq V(G)$ of vertices, we denote by $G[U]$ the subgraph induced by U, whose vertex set is U and whose edge set comprises all edges of G with both ends in U. We use $G - X$, where $X \subseteq V(G)$, as a shorthand for $G[V(G) \setminus X]$, which is further shortened as $G - v$ when $X = \{v\}$. For a set E_+ of edges, we denote by $G + E_+$ the graph obtained by adding edges in E_+ to G,—its vertex set is still $V(G)$ and its edge set becomes $E(G) \cup E_+$. The graph $G - E_-$ is defined analogously. A paw is shown in Fig. 1(e).

For the paw-free completion (resp., edge deletion) problem, a *solution* of an instance (G, k) consists of a set E_+ (resp., E_-) of at most k edges such that $G + E_+$ (resp., $G - E_-$) is paw-free. For a positive integer k, a *k-partite* graph is a graph whose vertices can be partitioned into k different independent sets, called parts, and a k-partite graph is *complete* if all the possible edges are present, i.e., there is an edge between every pair of vertices from different parts. A *complete multipartite graph* is a graph that is complete k-partite for some $k \geq 3$. Note that here we exclude complete bipartite graphs for convenience.

Proposition 1 ([6]). *A graph G is paw-free if and only if every component of G is triangle-free or complete multipartite.*

In other words, if a connected paw-free graph contains a triangle, then it is necessarily a complete multipartite graph. Another simple fact is on the adjacency between a vertex and a (maximal) clique in a paw-free graph.

Proposition 2 ($\star^2$). *Let K be a clique in a paw-free graph. If a vertex v is adjacent to K, then $|K \setminus N[v]| \leq 1$.*

A set $M \subseteq V(G)$ of vertices is a *modulator* of a graph G if every paw in G intersects M by at least two vertices. Note that $G - M$ is paw-free. The following three propositions characterize the interaction between the modulator M and the components of $G - M$.

Proposition 3 ($\star$). *Let M be a modulator of G. If $v \in M$ forms a triangle with some component C of $G - M$, then all the neighbors of v are in M and C.*

In other words, if a vertex v in M forms a triangle with a component of $G - M$, then v is a "private" neighbor of this component. As we will see, these components (forming triangles with a single vertex from M) are the focus of our algorithms.

Proposition 4 ($\star$). *Let G be a graph and M a modulator of G. If a vertex $v \in M$ forms a triangle with an edge in a triangle-free component C of $G - M$, then (i) v is adjacent to all the vertices of C; and (ii) C is complete bipartite.*

We say that a triangle-free component of $G - M$ is of type I if it forms a triangle with some vertex in M, or type II otherwise. By Proposition 4, for each type-I triangle-free component, all its vertices have a common neighbor in M. A component is trivial if it consists of a single vertex. Note that all trivial components of $G - M$ are type-II triangle-free components.

Proposition 5. *Let G be a graph and M a modulator of G. For any complete multipartite component C of $G - M$ and vertex $v \in M$ adjacent to C, the set of vertices in C that are nonadjacent to v is either empty or precisely one part of C.*

---

[2] The proof of a proposition marked with a $\star$ is deferred to the full version.

Proof. Suppose that the parts of C are $U_1, \ldots, U_p$. We have nothing to prove if all the vertices in C are adjacent to v. In the following we assume that, without loss of generality, v is adjacent to $u \in U_1$ and nonadjacent to $w \in U_p$. We need to argue that v is adjacent to all vertices in the first $p-1$ parts but none in the last part. Any vertex $x \in U_i$ with $1 < i < p$ makes a clique with u and w. It is adjacent to v by the definition of the modulator ($\{u, v, w, x\}$ cannot induce a paw) and Proposition 2. Now that v is adjacent to some vertex from another part ($p \geq 3$), the same argument implies $U_1 \subseteq N(v)$. To see $U_p \cap N(v) = \emptyset$, note that a vertex $w' \in U_p \cap N(v)$ would form a paw together with u, v, w, contradicting the definition of the modulator.   □

A *false twin class* of a graph G is a vertex set in which every vertex has the same open neighborhood. It is necessarily independent. The following is immediate from Proposition 5.

Corollary 1. *Let M be a modulator of G, and C a complete multipartite component of $G - M$. Each part of C is a false twin class of G.*

The preservation of false twins by all minimum paw-free completions may be of independent interest.

Lemma 1 ($\star$). *Let G be a graph and E_+ a minimum set of edges such that $G + E_+$ is paw-free. A false twin class of G remains a false twin class of $G + E_+$.*

3   Paw-Free Completion

The safeness of our first rule is straightforward.

Rule 1. *If a component of G contains no paw, delete it.*

Behind our kernelization algorithm for the paw-free completion problem is the following simple and crucial observation. After Rule 1 is applied, each remaining component of G contains a paw, hence a triangle, and by Proposition 1, we need to make it complete multipartite. We say that a vertex v and an edge xy *dominate* each other if at least one of x and y is adjacent to v. Note that an edge dominates, and is dominated by, both endpoints of this edge. Every edge in a complete multipartite graph dominates all its vertices, and hence in a yes-instance, every edge "almost" dominates vertices in the component.

Lemma 2 ($\star$). *Let G be a connected graph containing a paw and uv an edge in G. We need to add at least $|V(G) \setminus N[\{u, v\}]|$ edges incident to u or v to G to make it paw-free.*

For the paw-free completion problem, we build the modulator using the procedure in Fig. 2, whose correctness is proved in Lemma 3. Starting from an empty set of paws, we greedily add paws: If a paw does not intersect any previously chosen paw with two or more vertices, then add it. All the vertices of the selected paws already satisfy the definition of the modulator. After that, we have two more steps, taking all the degree-one vertices of all paws in G, and deleting a vertex from $M \cap G'$ for certain component G' of G. Their purposes are to simplify the disposal of triangle-free components of $G - M$: In particular, (iii) and (iv) of Lemma 3 are instrumental for dealing with, respectively, type-I and type-II triangle-free components of $G - M$.

---

0. $\mathcal{F} \leftarrow \emptyset$; $M \leftarrow \emptyset$;
1.   **for each** paw F of G **do**
1.1.   **if** $|F \cap F'| \leq 1$ for each paw F' in $\mathcal{F}$ **then**
             $\mathcal{F} \leftarrow \mathcal{F} \cup \{F\}$;
             add the vertices of F to M;
1.2.   **else** add the degree-one vertex of F to M;
2.   **for each** component G' of G **do**
2.1.   **if** an isolated vertex v of $G' - M$ dominates all the edges in G' **then**
             find an edge uw in $G[N(v)]$;
             remove u from M;
3.   **return** M.

Fig. 2. The construction of the modulator for G.

Lemma 3. *Let (G, k) be an instance of the paw-free completion problem. The vertex set M constructed in Fig. 2 has the following properties.*

(i) The construction is correct and its result is a modulator of G.

(ii) For each component G' of G, we need to add at least $|M \cap G'|/4$ edges to G' to make it paw-free.

(iii) Let C be a triangle-free component of $G - M$. If C is nontrivial and any vertex in C is contained in a triangle, then C is of type I.

(iv) For each isolated vertex v in $G - M$, there is an edge in $G_v - N[v]$, where G_v is the component of G containing v.

Proof. We may assume without loss of generality that G is connected and contains a paw; otherwise it suffices to work on its components that contain paws one by one, because both the construction and all the statements are component-wise.

We denote by M' the set of vertices added to M in step 1. Note that it is a modulator of G because vertices added in step 1.1 already satisfy the definition. Let X be the set of isolated vertices in $G - M'$ each of which dominates all the edges in G. If X is empty, then step 2 is not run, $M = M'$ and we are done. In the rest, $X \neq \emptyset$. We argue first that X is a false twin class. Vertices in X are pairwise nonadjacent by definition. Suppose for contradiction that $N(x_1) \neq N(x_2)$ for $x_1, x_2 \in X$, then there is a vertex v in $N(x_1) \setminus N(x_2)$ or in $N(x_2) \setminus N(x_1)$. But then x_2 does not dominate edge vx_1, or x_1 does not dominate edge vx_2, contradicting the definition of X. We then argue that any vertex $x \in X$ is in a triangle. By assumption, G contains a triangle uvw. If $x \in \{u, v, w\}$, then we are done. Otherwise, x must be adjacent to at least two of $\{u, v, w\}$ to dominate all the three edges in this triangle. Note that $N(x) \in M'$ because x is isolated in $G - M'$. This justifies step 2.1 of the construction of M. Note that it removes only one vertex from M'.

Now we prove by contradiction that M is a modulator of G. Suppose that there is a paw F with $|F \cap M| \leq 1$. By construction, $|F \cap M'| \geq 2$, which means $|F \cap M| = 1$ and the only vertex in $M' \setminus M$ is in F. Let $\{v\} = M' \setminus M$ and $\{u\} = F \cap M$; note that the other two vertices of F are in $V(G) \setminus M'$. Since any vertex in X is isolated in $G - M'$ and dominates all the edges of G, every component of $G - M'$ is trivial, which means that the two vertices in $F \setminus \{u, v\}$ are not adjacent. Therefore, one of u and v must be the degree-three vertex of F, and the other is a degree-two vertex of F. But the degree-one vertex of F has been added to M' in step 1.1 or 1.2, a contradiction. This justifies (i).

Let U_1 and U_2 be the sets of vertices added to M' in steps 1.1 and 1.2 respectively; $U_1 \cup U_2 = M'$. For each paw F added in step 1.1, at least one of its missing edges needs to be added to G to make it paw-free. This edge is not in any previous selected paw F', because we add F only when $|F \cap F'| \leq 1$. Therefore, we need to add at least $|U_1|/4$ edges to $G[U_1]$ to make it paw-free. On the other hand, each vertex v in U_2 is the degree-one vertex of some paw F, (it is possible that all other three vertices of F are in U_1,) we need to add at least one edge incident to v. Therefore, we need to add at least $|U_2|/2$ edges incident to vertices in U_2 to G to make it paw-free. Note that these two sets of edges we need to add are disjoint. The total number of edges we need to add to G to make it paw-free is at least $|U_1|/4 + |U_2|/2 \geq |U_1 \cup U_2|/4 = |M'|/4 \geq |M|/4$. This concludes assertion (ii).

Assertion (iii) follows from Proposition 4 if the triangle has two vertices from C: Note that the other vertex must be from M because C itself is triangle-free. Let the vertices in the triangle be $u, v \in M$ and $w \in C$. If C contains the vertex in $M' \setminus M$, then $X \subseteq C$ because it is a false twin class, and there is a vertex in M making a triangle with C, and it follows from Proposition 4. Now that C is a nontrivial component of $G - M'$, we can find a neighbor x of w in C. Note that it is adjacent to at least one of u and v; otherwise, x is the degree-one vertex of the paw induced by $\{x, u, v, w\}$ and should be in M'. As a result, x is adjacent to at least one of u and v, and then we can use Proposition 4.

Assertion (iv) follows from the construction of M and the fact that X is a false twin class we proved above.   □

Corollary 2. *If (G, k) is a yes-instance, then M contains at most $4k$ vertices.*

We proceed only when $|M| \leq 4k$. A consequence of this modulator is a simple upper bound on the number of vertices in all the type-II triangle-free components of $G - M$. Note that all trivial components of $G - M$ are considered here.

Lemma 4 ($\star$). *Let (G, k) be a yes-instance to the paw-free completion problem on which Rule 1 is not applicable, and M the modulator of G. The total number of vertices in all the type-II triangle-free components of $G - M$ is at most $2k$.*

Hereafter we consider the components G' of G one by one; let $M' = M \cap V(G')$. If all components of $G' - M'$ are type-II triangle-free components, then a bound of the size of $V(G') \setminus M'$ is given in Lemma 4. In the rest, at least one component of $G' - M'$ is a type-I triangle-free component or a complete multipartite component. The way we bound $|V(G') \setminus M'|$ for such a component is to show, after applying some reductions, the minimum number of edges we need to add to G' to make it paw-free is linear on $|V(G') \setminus M'|$. The first one is very straightforward.

Lemma 5 $(\star)$. *If two components in $G' - M'$ are not type-II triangle-free components, then we need to add at least $|V(G') \setminus M'|/2$ edges to G' to make it paw-free.*

Henceforth, $G' - M'$ has precisely one type-I triangle-free component or one complete multipartite component, but not both. Each part of such a component is an independent set (recall that a type-I triangle-free component is complete bipartite by Proposition 4). The next two propositions are on independent sets I of G. The first is about the cost of separating vertices in I into more than one part; it also means that a sufficiently large independent set cannot be separated. The second states that if each of the vertices in I is adjacent to all the other vertices, then we can remove all but one vertex in I from the graph.

Proposition 6 $(\star)$. *Let G' be a connected graph containing a paw, and I an independent set of G'. If we do not add all the missing edges between I and $N(I)$, then we need to add at least $|I| - 1$ edges among I to G' to make it paw-free.*

Proposition 7 $(\star)$. *Let I be an independent set in a component G' of a graph G. If every vertex in I is adjacent to every vertex in $V(G') \setminus I$, then (G, k) is a yes-instance if and only if $(G - (I \setminus \{v\}), k)$ is a yes-instance for any $v \in I$. Moreover, if $G - I$ is connected, then (G, k) is a yes-instance if and only if $(G - I, k)$ is a yes-instance.*

We are now ready to consider type-I triangle-free components.

Lemma 6. *Let C be a type-I triangle-free component of $G' - M'$ and let $L \uplus R$ be the bipartition of C with $|L| \geq |R|$. If any of the following conditions is satisfied, then we need to add at least $|C|/32$ edges to G' to make it paw-free.*

(i) $|L| \leq 4|M'|$;
(ii) there is an edge in $G' - N[L]$;
(iii) $V(G') \neq N[C]$ and $|L| \leq 2|R|$;
(iv) there are $|L|/2$ or more missing edges between L and $N(L)$;
(v) $|L| \leq |R| + |M'|$ and $G - N[R]$ has an edge; or
(vi) $|L| \leq |R| + |M'|$ and there are $|R|/2$ or more missing edges between R and $N(R)$.

Proof. (i) If $|L| \leq 4|M'|$, then $|C| = |L| + |R| \leq 2|L| \leq 8|M'|$, and it follows from Lemma 3(ii). (ii) By Lemma 2, we need to add at least $|L| \geq |C|/2$ edges.

(iii) Since C is complete bipartite and $|L| \geq |R|$, we can find a matching of size $|R|$ between L and R. By Lemma 2, for each vertex $v \in V(G') \setminus N[C]$, the number of edges between v and C we need to add is at least $|R| = (2|R|+|R|)/3 \geq (|L|+|R|)/3 = |C|/3$. (iv) By Proposition 6, we need to add at least $|L|/2 \geq |C|/4$ edges.

In the rest, (v) and (vi), $|L| \leq |R|+|M'|$. We may assume none of the previous conditions is satisfied. Therefore, $|L| > 4|M'|$, which means $|L| \leq 2|R|$. Also note that the proofs for these two conditions are almost the same as conditions (ii) and (iv) respectively. (v) By Lemma 2, we need to add at least $|R| \geq |C|/3$ edges. (vi) By Proposition 6, we need to add at least $|R|/2 \geq |C|/6$ edges.   □

We say that a type-I triangle-free component C of $G' - M'$ is *reducible* if none of the conditions in Lemma 6 holds true.

Rule 2 ($\star$). *Let C be a type-I triangle-free component of $G' - M'$ and let $L \uplus R$ be the bipartition of C with $|L| \geq |R|$. If C is reducible, then add all the missing edges between L and $N(L)$ and all the missing edges between $V(G') \setminus N[L]$ and $N(L)$; decrease k accordingly; and remove all but one vertex from $(V(G') \setminus N[L]) \cup L$.*

In the last we consider the complete multipartite components of $G' - M'$.

Lemma 7 ($\star$). *Let C be a complete multipartite component of $G' - M'$, and let P^* be a largest part of C. If any of the following conditions is satisfied, then we need to add at least $|C|/12$ edges to G' to make it paw-free.*

(i) $|C| \leq 3|M'|$;
(ii) there is an edge in $G' - N[C]$;
(iii) $|P^*| > 2|C|/3$ and $G' - N[P^*]$ has an edge;
(iv) $|P^*| \leq 2|C|/3$ and $V(G') \neq N[C]$; or
(v) $|P^*| \leq 2|C|/3$ and $V(G') = N[C]$, and for every part P of C,
   – $G' - N[P]$ contains an edge, or
   – there are at least $|P|$ missing edges between $V(G') \setminus N[P]$ and $N(P)$.

We say that a complete multipartite component C of $G' - M'$ is *reducible* if none of the conditions in Lemma 7 holds true.

Rule 3 ($\star$). *Let C be a reducible complete multipartite component of $G' - M'$ and P^* a largest part of C.*

(1) If $|P^*| > 2|C|/3$, then add all the missing edges between $V(G') \setminus N[P^*]$ and $N(P^*)$; decrease k accordingly; and remove $(V(G') \setminus N[P^*]) \cup P^*$ from G.
(2) Otherwise, find a part P such that $V(G') \setminus N[P]$ is an independent set and there are less than $|P|$ missing edges between $V(G') \setminus N[P]$ and $N(P)$. Add all the missing edges between $V(G') \setminus N[P]$ and $N(P)$; decrease k accordingly; and remove $P \cup (V(G') \setminus N[P])$ from G.

We summarize our kernelization algorithm for the paw-free completion problem in Fig. 3 and use it to prove our main result of this section.

procedure **reduce**(G, k)

0.   **if** $k < 0$ **then return** a trivial no-instance;
1.   remove all paw-free components from G;
2.   construct modulator M;
3.   **if** $|M| > 4k$ **then return** a trivial no-instance;
4.   **if** $> 2k$ vertices in type-II triangle-free components of $G - M$ **then**
4.1.   **return** a trivial no-instance;
5.   **for each** component G' of G **do**
5.1. $M' \leftarrow V(G') \cap M$;
5.2.   **if** 2 components in $G' - M'$ are not type-II triangle-free components **then**
              goto 5;
5.3.   **if** $G' - M'$ has a type-I triangle-free component C **then**
              if C is reducible **then** apply Rule 2 and **return reduce**(G, k);
5.4.   **if** $G' - M'$ has a complete multipartite component C **then**
              if C is reducible **then** apply Rule 3 and **return reduce**(G, k);
6.   **if** $|V(G)| \leq 38k$ **then return** (G, k);
7.   **else return** a trivial no-instance.

Fig. 3. The kernelization algorithm for the paw-free completion problem.

Proof (of Theorem 1). We use the algorithm described in Fig. 3. The correctness of steps 0 and 1 follows from the definition of the problem and Rule 1 respectively. Steps 2 and 3 are justified by Lemma 3 and Corollary 2. Step 4 is correct because of Lemma 4, and after that we only need to consider the components of $G - M$ that are not type-II triangle-free components, which are dealt with in step 5. The cost of a component of G is the minimum number of edges we need to add to it to make it paw-free.

If two components of $G' - M'$ are not type-II triangle-free components, then by Lemma 5, the cost of G' is at least $|V(G')\backslash M|/2$. Therefore, there is nothing to do for step 5.2. Henceforth, $G' - M'$ has precisely one type-I triangle-free component or one complete multipartite component, but not both. The algorithm enters step 5.3 if there is a type-I triangle-free component C in $G' - M'$. If C is reducible, we rely on the correctness of Rule 2; otherwise, the cost of G' is at least $|C|/32$ by Lemma 6. The algorithm enters step 5.4 if there is a complete multipartite component C in $G' - M'$. If C is reducible, we rely on the correctness of Rule 3; otherwise, the cost of G' is at least $|C|/12$ by Lemma 7.

When the algorithm reaches step 6, neither of Rules 2 and 3 is applicable. There are at most $4k$ vertices in M, at most $2k$ vertices in all the type-II triangle-free components of $G - M$. On the other hand, for each other vertex, there is an amortized cost of at least $1/32$. Therefore, if (G, k) is a yes-instance, then the number of vertices is at most $38k$, and this justifies steps 6 and 7.

We now analyze the running time of this algorithm. When each time the algorithm calls itself in step 5.3 or 5.4, it removes at least one vertex from the graph. Therefore, the recursive calls can be made at most n times. On the other hand, each step clearly takes polynomial time. Therefore, the algorithm returns in polynomial time.   □

4   Paw-Free Edge Deletion

For this problem, we construct the modulator in the standard way. We greedily find a maximal packing of edge-disjoint paws. We can terminate by returning "no-instance" if there are more than k of them. Let M denote the set of vertices in all the paws found; we have $|M| \leq 4k$. It is a modulator because every paw not included shares at least an edge with some chosen one, hence at least 2 vertices.

The safeness of the following rule is straightforward: If we do not delete this edge, we have to delete a distinct one from each of the paws, hence $k + 1$.

Rule 4. *Let uv be an edge of G. If there exist $k + 1$ paws such that for any pair of them, the only common edge is uv, then delete uv from G and decrease k by 1.*

We first deal with complete multipartite components of $G - M$.

Rule 5 ($\star$). *Let C be a complete multipartite component of $G - M$. From each part of C, delete all but $k + 1$ vertices.*

Rule 6 ($\star$). *Let C be a complete multipartite component of $G - M$. Delete all but $k + 4$ parts of C that are adjacent to all vertices in $N(C)$.*

Lemma 8 ($\star$). *After Rules 5 and 6 are applied, there are at most $O(k^3)$ vertices in the complete multipartite components of $G - M$.*

In the following, we assume that Rule 4 is not applicable. We mark some vertices from each of the triangle-free components that should be preserved, and then remove all the unmarked vertices. Recall that a triangle-free component of $G - M$ is of type I or type II depending on whether it forms a triangle with some vertex in M.

The following simple observation is a consequence of Proposition 3 and the definition of type-I triangle-free components.

Corollary 3. *If a vertex in M is adjacent to the triangle-free components of $G - M$, then either it is adjacent to precisely one type-I triangle-free component, or it is adjacent to only type-II triangle-free components.*

By Proposition 4, a type-I triangle-free component C of $G - M$ is complete bipartite.

Rule 7. *Let $\mathcal{C}$ be all the type-I triangle-free components of $G - M$, and let $U = \bigcup_{C \in \mathcal{C}} V(C)$.*

(i) For each $S \subseteq M$ with $|S| = 3$ and each $S' \subseteq S$, mark $k + 1$ vertices from $\{x \in U \mid N(x) \cap S = S'\}$.

(ii) *For each $C \in \mathcal{C}$ with bipartition $L \uplus R$ do the following. For each $S \subseteq M$ with $|S| = 2$ and each $S' \subseteq S$, mark $k + 3$ vertices from $\{x \in L \mid N(x) \cap S = S'\}$ and $k + 3$ vertices from $\{x \in R \mid N(x) \cap S = S'\}$.*

Delete all the unmarked vertices from U.

Lemma 9. *Rule 7 is safe.*

Proof. Let G' be the graph obtained after applying Rule 7. If (G, k) is a yes-instance, then (G', k) is a yes-instance. For the other direction, suppose that (G', k) is a yes-instance, with a solution E_-. We prove by contradiction that $G - E_-$ is paw-free as well. A paw F in $G - E_-$ contains at least one deleted vertex, because $G' - E_-$ is paw-free, and at most two deleted vertices, because otherwise F is a paw of G and should be in the modulator.

Consider first that F contains only one deleted vertex x. Let C be the triangle-free component of $G - M$ containing it. If all the other three vertices in F are from M, then in step (i) we have marked $k + 1$ vertices in C that have the same adjacency to $F \setminus \{x\}$ as x in G. Since $|E_-| \leq k$, the adjacency between $F \setminus \{x\}$ and at least one of these marked vertex is unchanged. This vertex forms a paw with $F \setminus \{x\}$ in $G' - E_-$, a contradiction. Now at most two vertices of F are from M. We may assume without loss of generality that $x \in L$, where $L \uplus R$ is the bipartition of C. In step (ii) we have marked $k + 3$ vertices in L that have the same adjacency to $F \cap M$ as x; let them be Q. By Proposition 4, every vertex in $Q \cup \{x\}$ is adjacent to all vertices in R; on the other hand, no vertex in $Q \cup \{x\}$ is adjacent to any vertex in another component of $G - M$ different from C. Therefore, all vertices in $Q \cup \{x\}$ have the same adjacency to $F \setminus L$ in G. Since $|E_-| \leq k$, the adjacency between $F \setminus \{x\}$ and at least one vertex in Q is unchanged (noting that $|Q \cap F| \leq 2$). This vertex forms a paw with $F \setminus \{x\}$ in $G' - E_-$, a contradiction.

In the rest, F contains two deleted vertices x and y. If x and y are adjacent, then they are from the different parts of some component $C = L \uplus R$. Without loss of generality, we assume that $x \in L$ and $y \in R$. Since $|F \cap M| \leq 2$, by step (ii), we can find two set $Q_1 \subseteq L$ and $Q_2 \subseteq R$ that have the same adjacency to $F \cap M$ as x and y respectively. Note that $|Q_1| \geq k + 3$ and $|Q_2| \geq k + 3$. Each vertex in Q_1 has the same adjacency to $F \setminus \{x\}$. The situation is similar for Q_2 and $F \setminus \{y\}$. For $i = 1, 2$, since $|E_-| \leq k$ and $|Q_i \cap F| \leq 2$, the adjacency between $F \setminus \{x\}$ and at least one vertex in Q_i is unchanged. These two vertices form a paw with $F \setminus \{x, y\}$ in $G' - E_-$, a contradiction (because $Q_1 \uplus Q_2$ is complete bipartite). Now that x and y are not adjacent, then they are in the same part or in different components. Then one of x and y is the degree-one vertex of F and the other is a degree-two vertex of F, and we can get that the adjacency of x and y to $F \cap M$ are different. By Proposition 4, the component(s) containing x and y is complete bipartite, then x and y are adjacent to all vertices in the part that does not contain them in corresponding component. By step (ii), we can find two set Q_1 in the part containing x and Q_2 in the part containing y that have the same adjacency to $F \cap M$ as x and y respectively. Then $Q_1 \neq Q_2$. Since $|E_-| \leq k$, $|Q_1 \cap F| \leq 2$ and $|Q_2 \cap F| \leq 2$, at least one vertex in Q_1 and at least one vertex in Q_2 are unchanged. These two vertices form a paw with $F \backslash \{x, y\}$ in $G' - E_-$, a contradiction. □

Lemma 10 ($\star$). *After Rule 7 is applied, there are at most $O(k^4)$ vertices in all the type-I triangle-free components of $G - M$.*

Finally, we deal with type-II triangle-free components of $G - M$.

Rule 8. *Let $\mathcal{C}$ be all the type-II triangle-free components of $G - M$, and let $U = \bigcup_{C \in \mathcal{C}} V(C)$.*

(i) For each $S \subseteq M$ with $|S| = 3$ and each $S' \subseteq S$, mark $k + 1$ vertices from $\{x \in U \mid N(x) \cap S = S'\}$.

(ii) Mark all the vertices in non-trivial components that form a triangle with M, and for each of them, mark $k + 1$ of its neighbors in C.

Delete all the unmarked vertices from C.

Lemma 11. *Rule 8 is safe.*

Proof. Let G' be the graph obtained after applying Rule 8. If (G, k) is a yes-instance, then (G', k) is a yes-instance. For the other direction, suppose that (G', k) is a yes-instance, with a solution E_-. We prove by contradiction that $G - E_-$ is paw-free as well. A paw F in $G - E_-$ contains at least one deleted vertex since $G' - E_-$ is paw-free.

By the definition of type-II triangle-free components, no triangle contains an edge in $\mathcal{C}$, implying that the triangle t in F contains no edge in $\mathcal{C}$. Note that if F contains three vertices in $\mathcal{C}$, then t must contain an edge in $\mathcal{C}$, a contradiction. If F contains precisely one vertex v in $\mathcal{C}$, then by step (i), we can find a vertex v' in $G' - E_-$ such that v' has the same adjacency to $F \cap M$ as v, implying that $F \setminus \{v\} \cup \{v'\}$ in $G' - E_-$ forms a paw. If F contains two vertices x and y in $\mathcal{C}$, then either x or y is in a triangle t of F. Without loss of generality, we assume that x is in t, implying that x is marked in step (ii). If y is adjacent to x, then by step (ii), there are $k + 1$ marked vertices adjacent to x; let them be Q. The vertices in Q are not adjacent to any vertex in $F \cap M$ since no triangle in G contains an edge in $\mathcal{C}$. Then, each vertex in Q forms a paw with $F \setminus \{y\}$ in G'. Since $|E_-| \leq k$, there is a vertex v' in Q forms a paw with $F \setminus \{y\}$ in $G' - E_-$, a contradiction. If x is not adjacent to y, by step (i), there are $k + 1$ marked vertices Q' having the same adjacency to $F \cap M$ as y such that each vertex in Q' is not adjacent to x since no triangle in G contains an edge in $\mathcal{C}$. Then, each vertex in Q' forms a paw with $F \setminus \{y\}$ in G'. Since $|E_-| \leq k$, there is a vertex v' in Q' forms a paw with $F \setminus \{y\}$ in $G' - E_-$, a contradiction. □

Lemma 12 ($\star$). *After Rule 8 is applied, there are at most $O(k^4)$ vertices in all the type-II triangle-free components of $G - M$.*

References

1. Cai, L.: Fixed-parameter tractability of graph modification problems for hereditary properties. Inf. Process. Lett. **58**(4), 171–176 (1996)
2. Cao, Y., Rai, A., Sandeep, R.B., Ye, J.: A polynomial kernel for diamond-free editing. In: ESA 2018, pp. 10:1–10:13 (2018)
3. Eiben, E., Lochet, W., Saurabh, S.: A polynomial kernel for paw-free editing (2019). arXiv:1911.03683
4. Guillemot, S., Havet, F., Paul, C., Perez, A.: On the (non-)existence of polynomial kernels for P_l-free edge modification problems. Algorithmica **65**(4), 900–926 (2013)
5. Lewis, J.M., Yannakakis, M.: The node-deletion problem for hereditary properties is NP-complete. J. Comput. Syst. Sci. **20**(2), 219–230 (1980)
6. Olariu, S.: Paw-free graphs. Inf. Process. Lett. **28**(1), 53–54 (1988)
7. Sandeep, R.B., Sivadasan, N.: Parameterized lower bound and improved kernel for diamond-free edge deletion. In: IPEC 365–376 (2015)
8. Tsur, D.: Kernel for K_t-free edge deletion (2019). arXiv:1908.03600

Floorplans with Walls

Katsuhisa Yamanaka[1] and Shin-ichi Nakano[2(✉)]

[1] Iwate University, Morioka, Japan
[2] Gunma University, Kiryu, Japan
nakano@cs.gunma-u.ac.jp

Abstract. Let P be a set of n points in the proper inside of an axis-aligned rectangle R, and each point in P is either h-type, v-type or f-type. We wish to partition R into a set S of $n+1$ rectangles by n line segments so that each point in P is on the common boundary line segment between two rectangles in S, and also each h-type point in P is on a horizontal line segment and each v-type point in P is on a vertical line segment. (Each f-type point in P is on a line segment. f-type menas free type). Such a partition of R is called a feasible floorplan of R with respect to P. Each point in P corresponds to the location of a structurally necessary horizontal or vertical wall, or a column (pillar) to support upper part, and a feasible floorplan is a floorplan achieving suitable earthquake resistance. An algorithm to enumerate all feasible floorplans of R with respect to P is known when P consists of only f-type points.

In this paper when P consists of the three type points we give an efficient algorithm to enumerate all feasible floorplans of R with respect to P. The algorithm is based on the reverse search method, and enumerates all feasible floorplans in $O(|S_P|n)$ time using $O(n)$ space, after $O(n \log n)$ time preprocessing, where S_P is the set of the feasible floorplans of R with respect to P.

Keywords: Enumeration · Floorplan · Algorithm

1 Introduction

Let P be a set of n points in the proper inside of an axis-aligned rectangle R, and each point in P is either h-type, v-type or f-type. Those are shortened forms of horizontal type, vertical type and free type. We wish to partition R into a set S of $n + 1$ rectangles by n line segments so that each point in P is on the common boundary line segment between two rectangles in S, and also each h-type point in P is on a horizontal line segment and each v-type point in P is on a vertical line segment. Each f-type point in P is on either a horizontal or vertical line segment. We call such a partition of R a feasible floorplan of R with respect to P. Figure 1(b) illustrates the 8 feasible floorplans of R with respect to the point set P in Fig. 1(a). For simplicity we assume no two points have the same x-coordinate, and no two points have the same y-coordinate (Otherwise one can slightly modify the locations). Intuitively each point in P is the location

© Springer Nature Switzerland AG 2020
J. Chen et al. (Eds.): TAMC 2020, LNCS 12337, pp. 50–59, 2020.
https://doi.org/10.1007/978-3-030-59267-7_5

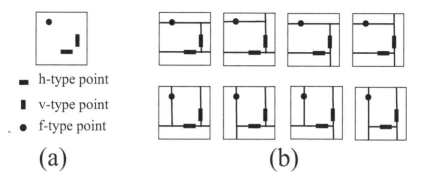

■ h-type point

▮ v-type point

• f-type point

(a)   (b)

Fig. 1. (a) An example of a rectangle R and a set P of three points in R, (b) all feasible floorplans of R with respect to P.

of a structurally necessary horizontal or vertical wall, or a column to support upper part, and a feasible floorplan is a floorplan achieving suitable earthquake resistance.

When P consists of only f-type points, several results are known. Ackerman et al. [1,2] gave an algorithm to enumerate all feasible floorplans with respect to P. The algorithm is based on the reverse search method [3,4] and enumerates all feasible floorplans in either $O(|S_P|n)$ time using $O(n)$ space or $O(|S_P|\log n)$ time using $O(n^3)$ space, where S_P is the set of feasible floorplans with respect to P. Yamanaka et al. [8] designed a faster algorithm, which is also based on the reverse search method. The algorithm is simple and uses only $O(n)$ space, and enumerates all feasible floorplans in $O(|S_P|)$ time. Some efficient algorithms using a similar technique are designed [5–7].

In this paper we consider a more general problem in which P consists of the three type points. We give an efficient algorithm to enumerate all feasible floorplans of R with respect to P. The algorithm is based on the reverse search method, and enumerates all feasible floorplans in $O(|S_P|n)$ time using $O(n)$ space, after $O(n\log n)$ time preprocessing, where S_P is the set of all feasible floorplans of R with respect to P.

The rest of the paper is organized as follows. Section 2 gives some definitions. Section 3 defines a tree structure among the feasible floorplans. Section 4 gives our enumeration algorithm. Finally Sect. 5 is a conclusion.

2   Preliminaries

In this section we give some definitions.

Let P be a set of n points in the proper inside of an axis-aligned rectangle R, and each point in P is either h-type, v-type or f-type. We assume that no two points in P have the same x-coordinate, and no two points have the same y-coordinate. A *feasible floorplan* is a partition of R into a set S of $n+1$ rectangles by n line segments so that each point in P is on the common boundary line segment between two rectangles in S, and also each h-type point in P is on a horizontal line segment and each v-type point in P is on a vertical line segment, and each f-type point in P is on either a horizontal or vertical line segment. We assume that no four rectangles in S share a common corner point in a floorplan. One can observe that every maximal line segment not on R contains exactly one point in P in any feasible floorplan of R with respect to P. Let S_P be the set of all feasible floorplans with respect to P. For R and P in Fig. 1(a), all 8 feasible floorplans in S_P are illustrated in Fig. 1(b).

Let Q be a feasible floorplan of R with respect to P. A line segment containing no end point of other line segment except at its endpoint is called *a basic line segment*. Each maximal line segment consists of one or more basic line segments. A maximal vertical line segment containing a point $p \in P$ is $type(u,d)$ if it contains u end points of maximal horizontal line segments above p and d end points of maximal horizontal line segment below p. Thus if a basic line segment is also a maximal vertical line segment then it is $type(0,0)$. A basic vertical line segment with $type(0,0)$ is *a fixed v-wall* if it contains a v-type point in P.

3 Family Tree

In this section we define a tree structure among the feasible floorplans of R with respect to P.

Let Q_r be the feasible floorplan of R with respect to P such that each h-type or f-type point is on a horizontal line segment having its left end on the left vertical line segment of R and its right end on the right vertical line segment of R, and each v-type point in P is on a fixed v-wall. We will show that Q_r corresponds to the root of the tree structure. See examples in Fig. 4(d) and Fig. 5.

Given a feasible floorplans $Q \neq Q_r$ of R with respect to P, we define the parent feasible floorplan $P(Q)$ of R with respect to P, as follows. Let s be the leftmost maximal vertical line segment in Q which is neither the left vertical line segment of R nor a fixed v-wall. Since $Q \neq Q_r$ such s always exists. Let $p \in P$ be the point on s. Since s is a vertical line segment, p is either v-type or f-type. We have the following two cases to consider.

Case 1. s is $type(0,0)$. See Fig. 2(a).
In this case s consists of exactly one basic vertical line segment, and p is f-type. If it were v-type then it would contradict the choice of s. We (1) remove s from Q, then (2) append a horizontal line segment s' containing p as a basic horizontal line segment, then extend s' to left so that s' has the left end on the left vertical line segment of R and shrink crossing fixed v-walls so that they remain fixed v-walls in the resulting floorplan, as illustrated in Fig. 2. Intuitively this is rotation of s. Note that s' has exactly one basic horizontal line segment on the right of p.

Case 2. Otherwise. See Fig. 3(a).
In this case s consists of two or more basic vertical line segments. We (1) remove s from Q, then (2) extend each maximal horizontal line segment t having left

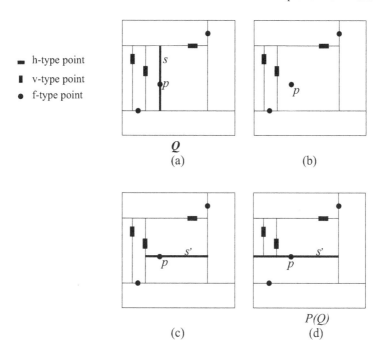

Fig. 2. The parent floorplan $P(Q)$ of Q in Case 1.

end on s to left so that t has the left end on the left vertical line segment of R, and shrink the fixed v-walls crossing with some t so that they remain fixed v-walls, (Note that each t crosses with only fixed v-walls by the definition of s. See Fig. 3(c)) then (3) extend each maximal horizontal line segment t having right end on s to the right so that t has exactly one basic horizontal line segment on the right of p, then (4) finally execute one of the following two subcases.

Case 2(a). p is f-type.
Append a horizontal line segment s' containing p as a basic horizontal line segment then extend to left so that s' has the left end on the left vertical line segment of R and shrink crossing fixed v-walls so that they remain fixed v-walls. See an example in Fig. 3(d). Intuitively this is shrink and rotation of s. Note that s' has exactly one basic horizontal line segment on the right of p.

Case 2(b). p is v-type.
Append a vertical line segment s' containing p as a fixed v-wall. Intuitively this is shrink of s.

We have the following fact.

Fact 1. If $P(Q)$ is derived from Q by rotation of a vertical line segment s to a horizontal line segment s' (Case 1 or Case 2(a)), then s' has exactly one basic horizontal line segment on the right of p in $P(Q)$.

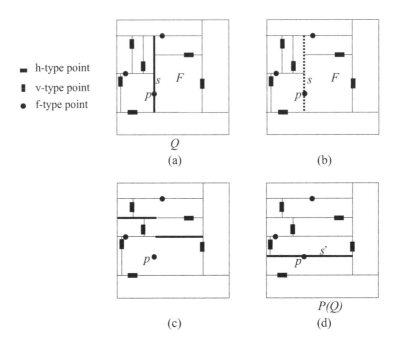

Fig. 3. The parent floorplan $P(Q)$ of Q in Case 2.

Note that the number of maximal vertical line segments of $P(Q)$ is one less or equal to that of Q. We have defined $P(Q)$ for each feasible floorplan Q except Q_r. We say $P(Q)$ is the *parent* of Q and Q is a *child* of $P(Q)$.

Given a feasible floorplan Q in S_P, which is the set of all feasible floorplans of R with respect to P, by repeatedly computing its parent, we can have the unique sequence Q, $P(Q)$, $P(P(Q))$, $\cdots$ of feasible floorplans with respect to P which eventually ends with Q_r. Note that the total length of the vertical line segments in those feasible floorplans is decreasing in the sequence, and Q_r has the minimum such length. See an example of such sequence in Fig. 4.

By merging those sequences we define the family tree T_P of S_P in which the vertices of T_P correspond to feasible floorplans of R with respect to P, and each edge corresponds to each relation between some Q and $P(Q)$. See Fig. 5.

4   Algorithm

In this section we design an algorithm to enumerate all feasible floorplans of R with respect to P. The algorithm is based on the reverse search method [3,4].

If we have an algorithm to compute all child floorplans of a given feasible floorplan of R with respect to P, then by recursively executing the algorithm from Q_r, we can compute all feasible floorplans of R with respect to P. This is the outline of the reverse search method [3,4]. We are now going to design such a all-children-enumeration algorithm.

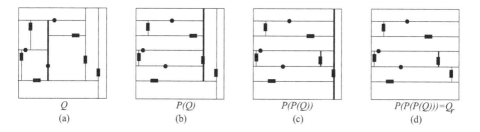

Fig. 4. The removing sequence.

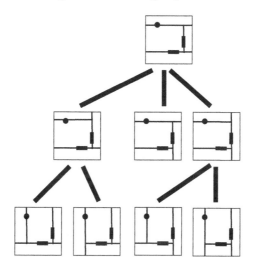

Fig. 5. The family tree.

Let s' be the maximal leftmost vertical line segment in Q which is neither the left vertical line segment of R nor a fixed v-wall, and $p' \in P$ be the point on s' (If $Q = Q_r$ we hypothetically regard s' the right vertical line segment of R, and we regard p' any point on s'). One can observe that each possible child floorplan of Q is one of the following three types for some maximal horizontal line segment s. See Fig. 6.

Type 1: $C(s, 0, 0)$
In this type s is a maximal horizontal line segment containing a f-type point $p \in P$ locating on the left of p' and s has no upper or lower end point of other vertical line segment on the right of p, so s has exactly one basic horizontal line segment on the right of p. Thus the horizontal line segment between p and the right end point of s is (a part of) basic, but s may have an upper or lower end point of some fixed v-wall on the left of p.

$C(s, 0, 0)$ is the floorplan constructed from Q by (1) removing s from Q, then (2) appending a vertical line segment s'' containing p as a basic line segment, then (3) extend the fixed v-walls having end points on s (on the left of p) so that they remain fixed v-walls.

Now $C(s, 0, 0)$ is also a feasible floorplan with respect to P. Intuitively this is the child floorplan derived from Q by rotation of s.

Note that if s is a horizontal line segment containing a point in P locating on the "right" of p' then resulting floorplan $C(s, 0, 0)$ is not a child of Q, since the leftmost vertical line segment of the resulting floorplan is not s'' but s', so the parent of the resulting floorplan is not Q. Thus we do not need check $C(s, 0, 0)$ with such s. Also note that if s has an upper or lower end point of other vertical line segment on the right of p, then s has two or more basic horizontal line segment on the right of p in Q, so $C(s, 0, 0)$ is not a child of Q by **Fact 1**. We can observe otherwise $C(s, 0, 0)$ is a child of Q.

Type 2(a): $C(s, u, d)$

In this type s is a maximal horizontal line segment containing a f-type point $p \in P$ locating on the left of p' and s has no upper or lower end point of other vertical line segment on the right of p, so s has exactly one basic horizontal line segment on the right of p. Thus, as in Type 1, the horizontal line segment between p and the right end point of s is basic, but s may have an upper or lower end point of some fixed v-wall on the left of p.

Let u' be the number of maximal horizontal line segments above p and d' the number of maximal horizontal line segments below p in Q. For example for the floorplan Q in Fig. 6, $u' = 3$ and $d' = 2$. For two integers $u < u'$ and $d < d'$, $C(s, u, d)$ is the floorplan constructed from Q by (1) removing s from Q, then (2) appending a vertical line segment s'' containing p as a basic line segment, then (3) extending s'' upward and downward so that it becomes $type(u, d)$, then (4) shrinking each maximal horizontal line segment t properly intersecting s'' so that it has an end point on s'', then (5) extend the fixed v-walls locating on the left of p and having end points on some shrinked horizontal line segment so that they remain fixed v-walls. See examples in Fig. 6. Intuitively this is a child floorplan derived from Q by rotation and extension of s. If $u = d = 0$ then this is just $C(s, 0, 0)$.

Note that if s is a horizontal line segment containing a point in P locating on the "right" of p' then resulting floorplan $C(s, u, d)$ is not a child of Q, since the leftmost vertical line segment of the resulting floorplan is not s'' but s'. Thus we do not need check $C(s, u, d)$ with such s. Also note that, as in Type 1, if s has an upper or lower end point of other vertical line segment on the right of p, then s has two or more basic horizontal line segment on the right of p in Q, so $C(s, u, d)$ is not a child of Q by **Fact 1**. Similarly if some t in (4) has an upper or lower end point of other vertical line segment on the "right" of p in Q (See Q in Fig. 7) then t has two or more basic horizontal line segment on the right of p in Q then $C(s, u, d)$ is not a child of Q, since the basic line segments in t locating right of p except the leftmost one cannot exist in $C(s, u, d)$ (See the dashed line of $C(s, 1, 0)$ in Fig. 7) so $P(C(s, u, d))$ is not Q. We can observe otherwise $C(s, u, d)$ is a child of Q.

Type 2(b): $C(s, u, d)$

In this type s is a fixed v-wall containing a v-type point $p \in P$ locating left of p'. Let u' be the number of maximal horizontal line segments above p and d' the number of maximal horizontal line segments below p in Q. For two integers $u < u'$ and $d < d'$, $C(s, u, d)$ is the floorplan constructed from Q by (1) extending s upward and downward so that it becomes $type(u, d)$, then (2) shrinking each maximal horizontal line segment t properly intersecting s so that it has an end point on s, then (3) extend the fixed v-walls locating on the left of p and having end points on some shrinked horizontal line segment so that they remain fixed v-walls. Intuitively this is the child floorplan derived from Q by extension of s.

Note that if s is locating on the "right" of p' then resulting floorplan $C(s, u, d)$ is not a child of Q, since the leftmost vertical line segment of the resulting floorplan is not s but s'. Thus we do not need check $C(s, u, d)$ with such s. Similarly if some t in (2) has an upper or lower end point of other vertical line segment on the "right" of p, then t has two or more basic horizontal line segment on the right of p, then, similar to Type 2(a) above, $C(s, u, d)$ is not a child of Q. We can observe otherwise $C(s, u, d)$ is a child of Q.

We have the following lemma.

Lemma 1. *Based on the analysis above one can enumerate all child floorplans of given Q.*

We now explain a data structure required for our child enumeration algorithm. We regard each corner of a rectangle as a vertex, each basic line segment as an edge and a floorplan as a graph. We store and maintain the current floorplan using some standard data structure for plane graphs during the execution of our enumeration algorithm. This part needs $O(n)$ space. We can efficiently trace the basic segments on the boundary of each rectangle. Also given a vertex and a direction (up/down/left/right) we can find the neighbour vertex in constant time.

We also maintain the list of the maximal horizontal line segments having a point in P on the left of the current leftmost vertical line segment. We assume that those horizontal line segments are sorted in the list by the x-coordinates of the points in P on the horizontal line segments. For Q_r such list can be constructed in $O(n \log n)$ time. Thus we need $O(n)$ space for the list and can update it efficiently.

For each recursive call we need $O(n)$ memory and the depth of the call is at most n so this part needs $O(n^2)$ space in total.

We have the following lemma.

Lemma 2. *Given a child floorplan $C(s, u, d)$ of Q one can check if $C(s, u+1, d)$ is a child floorplan of Q or not, and if it is a child floorplan of Q one can generate $C(s, u + 1, d)$ in $O(n)$ time.*

Proof. Let t be the maximal horizontal line segment containing the upper end point of s'' in $C(s, u, d)$. We have the following three cases.

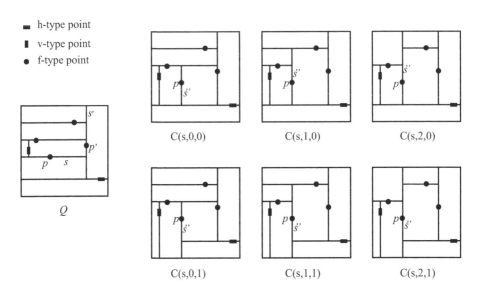

- ■ h-type point
- ▮ v-type point
- ● f-type point

Fig. 6. The child floorplans with respect to s.

If t has a point in P on the left of $p \in P$ on s'', and t has exactly one basic line segments on the right of p in $C(s, u, d)$, then shrinking t from $C(s, u, d)$ so that the right end point of t is on s'' then extending s'' upward so that it has one more basic line segment results in $C(s, u + 1, d)$ and it is a child of Q. The number of different segments between them is clearly $O(n)$, so one can generates $C(s, u + 1, d)$ from $C(s, u, d)$ in $O(n)$ time.

If t has a point in P on the left of $p \in P$ on s'', and t has two or more basic line segments on the right of p in $C(s, u, d)$ (See $C(s, 0, 0)$ in Fig. 7) then $C(s, u + 1, d)$ is not a child of Q, since the basic line segments of t locating right of p except the leftmost one cannot exist in $C(s, u + 1, d)$ (See the dashed line of $C(s, 1, 0)$ in Fig. 7) so $P(C(s, u + 1, d))$ is not Q.

If t has a point in P on the right of $p \in P$ on s'' then removing the part of t locating left of p from $C(s, u, d)$ then extending s'' upward so that it has one more basic line segment and modifying the fixed v-walls having end points on t

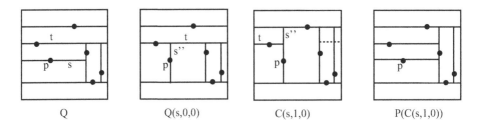

Fig. 7. $C(s, 1, 0)$ is not a child of Q.

so that they remain fixed v-walls results in $C(s, u + 1, d)$, and it is a child of Q. The number of different segments between them is also $O(n)$.

Thus in $O(n)$ time we can check if $C(s, u + 1, d)$ is a child of Q or not, and if it is a child we can generate $C(s, u + 1, d)$ from $C(s, u, d)$. □

Similarly given $C(s, u, d)$ one can check if $C(s, u, d + 1)$ is a child floorplan of Q or not, and if it is a child floorplan of Q one can generate $C(s, u, d + 1)$ from $C(s, u, d)$ in $O(n)$ time. Thus we have the following lemma.

Lemma 3. *One can enumerate all child floorplans of a given feasible floorplan Q with respect to P in $O(kn)$ time, where k is the number of child floorplans of Q.*

Since we need $O(kn)$ time for each vertex of the family tree, where k is the number of child floorplans of the floorplan corresponding to the vertex, the algorithm above runs in $O(|S_P|n)$ time in total, where S_P is the set of feasible floorplans with respect to P.

We have the following theorem.

Theorem 1. *After $O(n \log n)$ time preprocessing one can enumerate all feasible floorplans with respect to P in $O(|S_P|n)$ time and $O(n^2)$ space.*

5   Conclusion

In this paper we have designed a simple and efficient algorithm to enumerate all feasible floorplans with respect to a given set P of points. Our algorithm enumerate all such floorplans in $O(|S_P|n)$ time after $O(n \log n)$ time preprocessing, where S_P is the set of floorplans with respect to P, and $|P| = n$.

Can we enumerate all feasible floorplans with respect to P in $O(|S_P|)$ time?

References

1. Ackerman, E., Barequet, G., Pinter, R.Y.: On the number of rectangulations. In: Proceedings of the SODA, pp. 729–738 (2004)
2. Ackerman, E., Barequet, G., Pinter, R.Y.: On the number of rectangulations of a planar point set. J. Comb. Theory Ser. A **113**, 1072–1091 (2006)
3. Avis, D.: Generating rooted triangulations without repetitions. Algorithmica **16**, 618–632 (1996)
4. Avis, D., Fukuda, K.: Reverse search for enumeration. Discrete Appl. Math. **65**, 21–46 (1996)
5. Nakano, S.: Enumerating floorplans with n rooms. In: Eades, P., Takaoka, T. (eds.) ISAAC 2001. LNCS, vol. 2223, pp. 107–115. Springer, Heidelberg (2001). https://doi.org/10.1007/3-540-45678-3_10
6. Nakano, S.: Efficient generation of plane trees. Inf. Process. Lett. **84**, 167–172 (2002)
7. Li, Z., Nakano, S.: Efficient generation of plane triangulations without repetitions. In: Orejas, F., Spirakis, P.G., van Leeuwen, J. (eds.) ICALP 2001. LNCS, vol. 2076, pp. 433–443. Springer, Heidelberg (2001). https://doi.org/10.1007/3-540-48224-5_36
8. Yamanaka, K., Rahman, M.S., Nakano, S.-I.: Floorplans with columns. In: Gao, X., Du, H., Han, M. (eds.) COCOA 2017. LNCS, vol. 10627, pp. 33–40. Springer, Cham (2017). https://doi.org/10.1007/978-3-319-71150-8_3

A Primal-Dual Randomized Algorithm for the Online Weighted Set Multi-cover Problem

Wenbin Chen[1,2(✉)], Fufang Li[1], Ke Qi[1], Miao Liu[1], and Maobin Tang[1]

[1] School of Computer Science and Cyber Engineering, Guangzhou University, Guangzhou, People's Republic of China
cwb2011@gzhu.edu.cn
[2] Guangxi Key Laboratory of Cryptography and Information Security, Guilin 541004, Guangxi, China

Abstract. Given a ground set $\mathcal{U}$ of n elements and a family of m subsets $\mathcal{S} = \{S_i : S_i \subseteq \mathcal{U}\}$. Each subset $S \in \mathcal{S}$ has a positive cost $c(S)$ and every element $e \in \mathcal{U}$ is associated with an integer coverage requirement $r_e > 0$, which means that e has to be covered at least r_e times. The *weighted set multi-cover problem* asks for the minimum cost subcollection which covers all of the elements such that each element e is covered at least r_e times.

In this paper, we study the online version of the weighted set multi-cover problem. We give a randomized algorithm with competitive ratio $8(1+\ln m)\ln n$ for this problem based on the primal-dual method, which improve previous competitive ratio $12 \log m \log n$ for the online set multi-cover problem that is the special version where each cost $c(S)$ is 1 for every subset S.

1 Introduction

The *weighted set multi-cover* problem is the generalization of the set cover problem, which is defined as follows. Given a ground set $\mathcal{U} = \{1, \ldots, n\}$ of n elements and a family of m subsets $\mathcal{S} = \{S_i : 1 \leq i \leq m\}$, where $S_i \subseteq \mathcal{U}$ for all i. Each subset $S \in \mathcal{S}$ has a positive cost $c(S)$ and every element $e \in \mathcal{U}$ is associated with an integer coverage requirement $r_e > 0$, which means that e has to be covered at least r_e times. The goal is to find a minimum cost subcollection that covers all of the elements such that each element e is covered at least specified times r_e. When all $r_e = 1$, the set multi-cover problem becomes the set cover problem. Let $R = \max_{e \in \mathcal{U}} r_e$. We assume that $R = O(n)$.

Similarly, the *online weighted set multi-cover* problem is the generalization of the online set cover problem, which is described as follows. An adversary gives elements and their coverage requirement to the algorithm from $\mathcal{U}$ one-by-one. When a new element e and its coverage requirement r_e are given, the algorithm has to cover it at least r_e times by choosing some sets of $\mathcal{S}$ containing it. We assume that the elements of $\mathcal{U}$ and the coverage requirement of elements and

© Springer Nature Switzerland AG 2020
J. Chen et al. (Eds.): TAMC 2020, LNCS 12337, pp. 60–68, 2020.
https://doi.org/10.1007/978-3-030-59267-7_6

the members of $\mathcal{S}$ are known in advance to the algorithm, however, the set of elements given by the adversary is not known in advance to the algorithm. The objective is to minimize the total cost of the sets chosen by the algorithm.

The performance of an online algorithm is measured by the competitive ratio, which is defined as follows. Given an instance I of a minimization optimization problem M. Let $OPT(I)$ denote the optimum cost of off-line algorithms for instance I. If for each instance I of M, an online algorithm OA outputs a solution with cost at most $c \cdot OPT(I) + \alpha$, where α is a constant independent of the input sequence, then the competitive ratio of OA is c. If for each instance I of M, a randomized online algorithm ROA outputs a solution with expected cost at most $c \cdot OPT(I) + \alpha$, where α is independent of the input sequence, then the competitive ratio of ROA is c.

The set cover problem has wide application and is a well-known problem in algorithms and complexity. In [11], Karp shows that the set cover problem is NP-compete. Johnson [10] and Lovasz [13] give the greedy approximation algorithm for the unweighted set cover problem. Chvatal [7] proposes the greedy approximation algorithm for the weighted set cover problem. These greedy algorithms are of approximation ratio H_n, where $H_n = 1 + 1/2 + \ldots + 1/n$. Lund and Yannakakis show that the approximation ratio $O(\log n)$ for the set cover problem is essentially tight [14]. Later, Feige proves that it is impossible to have an approximation algorithm for the set cover problem with approximation ratio better than $O(\log n)$ [8]. Rajagopalan and Vazirani propose primal-dual RNC approximation algorithms for the set mullti-cover and covering integer programs problems [15]. Noga Alon et al. study the online set cover problem. Based on the techniques from computational learning theory, Noga Alon et al. propose a deterministic algorithm for this problem with competitive ratio $O(\log m \log n)$ [1]. The set cover problem is related to the budgeted maximum coverage problem, which is a flexible model for many applications [16–20].

In the areas of exact and approximation algorithms, the primal-dual method is one of powerful design methods. To our best of knowledge, the first time that the primal-dual method is used to the design of online algorithms is in Young' work about weighted paging [21], where he design an k-competitive online algorithm. In recent several years, Buchbinder and Naor have shown that the primal-dual method can be widely used to the design and analysis of online algorithms for many problems such as ski-rental, ad-auctions, routing and network optimization problems and so on [2–6].

In [12], Kuhnle et al. introduce the online set multi-cover problem and design randomized algorithms with $12 \log m \log n$-competitive ratio. In this paper, we study the online weighted set multi-cover problem. We present an $8(1 + \ln m) \ln n$ competitive randomized algorithm for this problem based on the primal-dual method. Specially, when each cost $c(S)$ is 1 for every subset S, the online weighted set multi-cover problem become the online set multi-cover problem. Thus, our algorithm improve Kuhnle et al.'s competitive ratio for the online set multi-cover problem.

2 A Fractional Primal-Dual Algorithm For the Online Weighted Set Multi-cover Problem

In this section, we design a fractional algorithm for the online weighted set multi-cover problem via the primal-dual method. A fractional algorithm allows an element e is fractionally covered its f_S part by a set S such that $\sum_{e \in S} f_S = 1$.

First, the weighted set multi-cover problem can be formulated as a 0–1 integer program as follows.

$$
\begin{array}{ll}
\text{Minimize} & \sum_{S \in \mathcal{S}} c(S) x_S \\
\text{Subject to} & \sum_{S: e \in S} x_S \geq r_e,\ e \in \mathcal{U} \\
& x_S \in \{0, 1\},\ S \in \mathcal{S}
\end{array}
$$

Its Linear Programs relaxation is as follows.

$$
\begin{array}{ll}
\text{Minimize} & \sum_{S \in \mathcal{S}} c(S) x_S \\
\text{Subject to} & \sum_{S: e \in S} x_S \geq r_e,\ e \in \mathcal{U} \\
& -x_S \geq -1,\ S \in \mathcal{S} \\
& x_S \geq 0,\ S \in \mathcal{S}
\end{array}
$$

Its Dual Programs is as follows

$$
\begin{array}{ll}
\text{Maximize} & \sum_{e \in U} r_e y_e - \sum_{S \in \mathcal{S}} z_S \\
\text{Subject to} & \sum_{e \in S} y_e - z_S \leq c(S),\ S \in \mathcal{S} \\
& y_e \geq 0,\ e \in \mathcal{U} \\
& z_S \geq 0,\ S \in \mathcal{S}
\end{array}
$$

In the following, we design the online fractional algorithm for the weighted set multi-cover problem via the primal-dual design method developed in recent years [2–6] (see Algorithm 2.1).

1: At time t, when an element e with coverage requirement r_e arrives:
2: If the primal constraints $\sum_{S: e \in S} x_S \geq r_e$ corresponding to e is satisfied, then do
   nothing.
3: Otherwise, do the following:
4: While $\sum_{S: e \in S} x_S < r_e$:
5:      Continuously increase y_e .
6:      If $x_S = 0$ and $(\sum_{e \in S} y_e) - z_S = c(S)$, then set $x_S \leftarrow \frac{1}{m}$.
7:      If $\frac{1}{m} \leq x_S < 1$, then x_S increase by the following function:
       $x_S \leftarrow \frac{1}{m} \cdot \exp(\frac{1}{c(S)}[(\sum_{e \in S} y_e) - z_S - c(S)])$.
8:      If $x_S = 1$, then z_S is increased at the same ratio as y_e.

Algorithm 2.1: The online fractional algorithm for the weighted set multi-cover problem.

Theorem 1. *The fractional online algorithm for the weighted set multi-cover problem is of competitive ratio $2(1 + \ln m)$.*

Proof. Let P denote the value of the objective function of the primal solution and D denote the value of the objective function of the dual solution. Initially, let $P = 0$ and $D = 0$. In the following, we prove three claims:

(1) The primal solution produced by the fractional algorithm is feasible.
(2) Every dual constraint in the dual program is violated by a factor of at most $(1 + \ln m)$.
(3) $P \leq 2D$.

By three claims and weak duality of linear programs, the theorem follows immediately.

First, we prove the claim (1) as follows. Consider a primal constraint $\sum_{S:e \in S} x_S \geq r_e$. In each *While* iteration (From line 5 to line 8 in the fractional algorithm), when this new primal constraint $\sum_{S:e \in S} x_S \geq r_e$ becomes be satisfied, the variable x_S stop increasing its value and its value is not greater than 1. Upon x_S become 1, the fractional algorithm begin to increase z_S and y_e at the same ratio. After that, the increases of z_S and y_e cannot result in infeasibility.

Second, we prove the claim (2) as follows. Consider any dual constraint $\sum_{e \in S} y_e - z_S \leq c(S)$. Since its corresponding variable x_S is not greater than 1, we get that:

$$x_S = \frac{1}{m} \cdot \exp(\frac{1}{c(S)}[(\sum_{e \in S} y_e) - z_S - c(S)]) \leq 1.$$

So $\exp(\frac{1}{c(S)}[(\sum_{e \in S} y_e) - z_S - c(S)]) \leq m.$

Then, $(\sum_{e \in S} y_e) - z_S - c(S) \leq c(S) \ln m.$

Thus, we get that: $(\sum_{e \in S} y_e) - z_S \leq c(S)(1 + \ln m).$

Third, we prove claim (3) as follows. The contribution to the primal cost consists of two parts. Let C_1 denote the contribution part which is from (6) of the fractional algorithm, where variables x_S are increased from $0 \rightarrow \frac{1}{m}$. Let C_2 denote the other contribution part which is from (7) of the fractional algorithm, where variables x_S are increased from $\frac{1}{m}$ up to at most 1 by the exponential function.

Bounding C_1: Let $\tilde{x}_S = \min(x_S, \frac{1}{m})$. We bound the term $\sum_{S \in \mathcal{S}} c(S)\tilde{x}_S$. To do this, we need the following several facts.

First, from the fractional algorithm, we get that if $x_S > 0$, and therefore $\tilde{x}_S > 0$, then:

$$\sum_{e \in S} y_e - z_S \geq c(S). \tag{1}$$

We call (1) as the primal complementary slackness condition.

At the time t, let $B'(S) = \{S | x_S = 1, e \in S\}$. Then $|B'(S)| \leq r_e$ since otherwise the constraint at time t has been already satisfied and the fractional algorithm stops increasing the variable y_e. Thus, $(m-1)|B'(S)| \leq (m-1)r_e$. So $\frac{m-|B'(S)|}{m} \leq r_e - |B'(S)|$. Since $\tilde{x}_S \leq \frac{1}{m}$, $\sum_{S \in \mathcal{S} \backslash B'(S)} \tilde{x}_S \leq \frac{m-B'(S)}{m}$. Hence

$$\sum_{S \in \mathcal{S} \backslash B'(S)} \tilde{x}_S \leq r_e - |B'(S)| \tag{2}$$

Also, it follows from the algorithm that if $z_S > 0$, then:

$$x_S \geq 1. \tag{3}$$

We call (2) as the dual complementary slackness and (3) as the second dual complementary slackness condition.

Using the primal and dual complementary slackness conditions, we show the following conclusions:

$$\sum_{S \in \mathcal{S}} c(S) \tilde{x}_S$$

$$\leq \sum_{S \in \mathcal{S}} (\sum_{e \in S} y_e - z_S) \tilde{x}_S \tag{4}$$

$$= \sum_{S \in \mathcal{S}} (\sum_{e \in S} y_e \tilde{x}_S) - \sum_{S \in \mathcal{S}} z_S \tilde{x}_S \tag{5}$$

$$= \sum_e (\sum_{S: e \in S} \tilde{x}_S) y_e) - \sum_{S \in \mathcal{S}} z_S \tilde{x}_S \tag{6}$$

$$\leq \sum_e r_e y_e - \sum_{S \in \mathcal{S}} z_S \tag{7}$$

Where inequality (4) follows from inequality (1) and equality (6) follows by changing the order of summation. As for the reason why inequality (7) holds, we consider some time t. At the time t when e with coverage requirement r_e arrive. From the fractional algorithm, we know that z_S is increased at the same ratio as y_e only when $x_S = 1$. Thus, $\frac{dy_e}{dt} = \frac{dz_S}{dt}$ only when $S \in B'(S)$. Hence, the increasing ratio of the left-hand side of (7) at the time t is $(\sum_{S \in \mathcal{S} \backslash B'(S)} \tilde{x}_S)\frac{dy_e}{dt}$. But, at the time t, the increasing ratio of the right-hand side of (7) is $(r_e - |B'(S)|)\frac{dy_e}{dt}$. By inequality (2), we get $(\sum_{S \in \mathcal{S} \backslash B'(S)} \tilde{x}_S)\frac{dy_e}{dt} \leq (r_e - |B'(S)|)\frac{dy_e}{dt}$. So inequality (7) holds

Thus, C_1 is at most D.

Bounding C_2 : At some time t, we show that the increase ΔC_2 is most ΔD in the same round.

$$\Delta C_2 = \sum_{S \in \mathcal{S}, \frac{1}{m} \leq x_S < 1} c(S) \cdot \Delta x_S \tag{8}$$

From the line 7 of the fractional algorithm, we get that $\frac{dx_S}{dy_e} = \frac{1}{c(S)} \cdot x_S$. So, $\Delta x_S = \frac{1}{c(S)} \cdot x_S \cdot \Delta y_e$. Thus, we get that:

$$\Delta C_2 = (\sum_{S \in \mathcal{S}, \frac{1}{m} \leq x_S < 1} x_S) \cdot \Delta y_e \qquad (9)$$

At the time t, the new primal constraints are not yet satisfied, so we get that: $\sum_{S \in \mathcal{S}, \frac{1}{m} \leq x_S < 1} x_S + \sum_{x_S=1} 1 < r_e$. Thus, $\sum_{S \in \mathcal{S}, \frac{1}{m} \leq x_S < 1} x_S < r_e - \sum_{x_S=1} 1$. Hence,

$$\Delta C_2 \leq (r_e - \sum_{x_S=1} 1) \cdot \Delta y_e \qquad (10)$$

From the line 8 of the fractional algorithm, $\Delta y_e = \Delta z_S$ when $x_S = 1$ in the same sound at the time t. So,

$$\Delta C_2 \leq r_e \cdot \Delta y_e - \sum_{x_S=1} \Delta z_S = \Delta D \qquad (11)$$

Thus, $C_2 \leq D$.

Hence, we get that $P = C_1 + C_2 \leq 2D$. So, claim (3) holds. Furthermore, the theorem holds. □

3   Randomized Algorithm for the Online Weighted Set Multi-cover Problem

In this section, we design a randomized algorithm for the online weighted set multi-cover problem with competitive ratio $8(1 + \ln m) \ln n$.

1: For each set $S \in \mathcal{S}$, $4 \ln n$ independently random variables $V(S, i)$ are uniformly chosen from $[0, 1]$ at random.

2: For every set $S \in \mathcal{S}$, let $\varepsilon(S) = \min_{i=1}^{4 \ln n} V(S, i)$.

3: At time t, a new element e and its cover requirement r_e arrives. Let c_e is the times that e has been covered at time t and let $u_e = r_e - c_e$. If $c_e \geq r_e$, then do nothing.

4: Otherwise, we use Algorithm 2.1 to compute the values of x_S in the unsatisfied primal constraint that corresponds to e, and let $\mathcal{C}$ denote the cover set, then do the following:

5: **for** $j = 1$ to u_e **do**

6:     For all unchosen sets $S \in \mathcal{S} \backslash \mathcal{C}$ that appears in the unsatisfied primal constraint that corresponds to e, when $x_S \geq \varepsilon(S)$, take one of these sets to the cover $\mathcal{C}$.

7:     $S \leftarrow \mathcal{S} \backslash \{S\}$; $\mathcal{C} \leftarrow \mathcal{C} \cup S$.

8: **end for**

Algorithm 3.1: The randomized online algorithm for the weighted set multi-cover problem.

Theorem 2. *The randomized algorithm is of competitive ratio $8(1 + \ln m) \ln n$.*

Proof. First, we show that the randomized algorithm produces a feasible solution with high probability $1 - O(\frac{1}{n^2}) > \frac{1}{2}$.

Consider any an element e, assume that it appears at time t. let A_i denote the event that e isn't covered in the i-th round from 5-th to 7-th line in the randomized algorithm. Let $\mathcal{S}_{t_i}$ denote the unchosen sets of $\mathcal{S}$ and $\mathcal{C}_{t_i}$ denote the chosen sets at the beginning of in the i-th round. $c(e, t_i)$ denote the number of e has been covered at the beginning of in the i-th round.

Then, we compute the probability that A_i occurs. Consider any $j(1 \leq j \leq 4 \ln n)$, let D_j denote the event that e is not covered due to j, which means that for all unchosen sets $S \in \mathcal{S}_{t_i}$ and $e \in S$, none of the value of $V(S, j)$ is less than x_S. Thus, $Pr(A_i = 1) = \bigcap\limits_{1 \leq j \leq 4 \ln n} Pr(D_j = 1)$

The probability $Pr(V(S, i) \leq x_S)$ is x_S. So $Pr(D_j = 1) = \prod\limits_{S \in \mathcal{S}_{t_i} | e \in S} (1 - x_S)$. Since $1 - x \leq exp(-x)$, we get that: $Pr(D_j = 1) \leq exp(- \sum\limits_{S \in \mathcal{S}_{t_i} | e \in S} x_S)$.

Since all x_S consist of a fractional solution after the fractional algorithm, we get that $\sum\limits_{S \in \mathcal{S} : e \in S} x_S \geq r_e$. Thus, $\sum\limits_{S \in \mathcal{S}_{t_i} | e \in S} x_S + \sum\limits_{S \in \mathcal{C}_{t_i} | e \in S} x_S \geq r_e$. So $\sum\limits_{S \in \mathcal{S}_{t_i} | e \in S} x_S \geq r_e - \sum\limits_{S \in \mathcal{C}_{t_i} | e \in S} x_S = r_e - c(e, t_i)$. Hence, $Pr(D_j = 1) \leq exp(- \sum\limits_{S \in \mathcal{S}_{t_i} | e \in S} x_S) \leq exp(-n_i)$, where $n_i = r_e - c(e, t_i)$. So, $Pr(D_j = 1) \leq exp(-1)$. Hence, $Pr(A_i = 1) \leq (exp(-1))^{4 \ln n} = exp(-4 \ln n) = \frac{1}{n^4}$.

So, the probability that e is not covered r_e times is $Pr(A_1 = 1 \vee \ldots \vee A_{u_e} = 1) \leq \sum_{i=1}^{u_e} Pr(A_i = 1) \leq \sum_{i=1}^{u_e} \frac{1}{n^4} = \frac{n_e}{n^4} \leq \frac{r_e}{n^4} \leq \frac{R}{n^4} \leq \frac{O(n)}{n^4} = O(\frac{1}{n^3})$.

By the union bound that the probability of union events is at most the sum of the probability of each event, the probability that there is an element e which is not covered r_e times is at most $n \times O(\frac{1}{n^3}) = O(\frac{1}{n^2})$ since there are at most n elements.

Hence, the randomized algorithm produces a feasible solution with high probability $1 - O(\frac{1}{n^2}) > \frac{1}{2}$.

Second, we show that the expected cost of the solution of randomized algorithms is $O(\log n)$ times the fractional solution.

Let B_i denote the event that $V(S, i) \leq x_S$. Then, $Pr(B_i = 1) = x_S$. The probability that the set S is chosen to the solution is at most the probability that there exists an $i, 1 \leq i \leq 4 \ln n$, such that $V(S, i) \leq x_S$.

Thus, the probability that S is chosen to the solution is at most the probability of $\bigcup_{i=1}^{4 \ln n} B_i$. By the union bound this probability is at most the sum of the probabilities of the different events, which is $4x_S \ln n$. Therefore, using the linearity of expectation, the expected cost of the solution is at most $4 \ln n$ times the cost of the fractional solution.

By Theorem 1, the cost of the fractional solution is $2(1 + \ln m)$ times the optimal solution. So the competitive ratio of the randomized algorithm is $8(1 + \ln m) \ln n$. $\qquad\square$

4 Conclusion

In this paper, we have studied the online version of the weighted set multi-cover problem. We have proposed a $8(1 + \ln m) \ln n$-competitive randomized algorithm for this problem based on the primal-dual method. An interesting open problem is to design deterministic algorithms for the online weighted set multi-cover problem.

Acknowledgments. We would like to thank the anonymous referees for their careful readings of the manuscripts and many useful suggestions.

Wenbin Chen's research has been supported by the National Science Foundation of China (NSFC) under Grant No. 11271097, and by the Project of Ordinary University Innovation Team Construction of Guangdong Province Under No. 2015KCXTD014 and No. 2016KCXTD017. This work has been also supported by the Natural Science Foundation of China (U1936116), the Guangxi Key Laboratory of Cryptography and Information Security (GCIS201807). FuFang Li's work had been co-financed by: Natural Science Foundation of China under Grant No. 61472092; Guangdong Provincial Science and Technology Plan Project under Grant No. 2013B010401037; and GuangZhou Municipal High School Science Research Fund under grant No. 1201421317. Ke Qi's research has been supported by the Guangzhou Science and Technology Plan Project under Grant No. 201707010283 and the National Science Foundation of Guangdong Province under Grant No. 2017A030313374. Miao Liu's research has been supported by the Guangzhou Municipal Universities project 1201620342.

References

1. Alon, N., Awerbuch, B., Azar, Y., Buchbinder, N., Naor, J.: The online set cover problem. In: STOC 2003, pp. 100–105 (2003)
2. Bansal, N., Buchbinder, N., Naor, J.: A primal-dual randomized algorithm for weighted paging. In: FOCS 2007, pp. 507–517 (2007)
3. Buchbinder, N., Jain, K., Naor, J.S.: Online primal-dual algorithms for maximizing ad-auctions revenue. In: Arge, L., Hoffmann, M., Welzl, E. (eds.) ESA 2007. LNCS, vol. 4698, pp. 253–264. Springer, Heidelberg (2007). https://doi.org/10.1007/978-3-540-75520-3_24
4. Buchbinder, N., Naor, J.: Online primal-dual algorithms for covering and packing problems. In: Brodal, G.S., Leonardi, S. (eds.) ESA 2005. LNCS, vol. 3669, pp. 689–701. Springer, Heidelberg (2005). https://doi.org/10.1007/11561071_61
5. Buchbinder, N., Naor, J.: Improved bounds for online routing and packing via a primal-dual approach. In: Proceedings of the 47th Symposium on Foundations of Computer Science (FOCS), pp. 293–304 (2006)
6. Buchbinder, N., Naor, J.: The design of competitive online algorithms via a primal-dual approach. Found. Trends Theor. Comput. Sci. **3**(2–3), 93–263 (2009)
7. Chvatal, V.: A greedy heuristic for the set covering problem. Math. Oper. Res. **4**, 233–235 (1979)
8. Feige, U.: A threshold of ln n for approximating set cover. In: Proceedings of the 28th ACM Symposium on the Theory of Computing, pp. 312–318 (1996)
9. Feige, U.: A threshold of $\ln n$ for approximating set cover. J. ACM **45**(4), 634–652 (1998)

10. Johnson, D.S.: Approximation algorithms for combinatorial problems. J. Comput. Syst. Sci. **9**, 256–278 (1974)
11. Karp, R.M.: Reducibility among combinatorial problems. In: Miller, R.E., Thatcher, J.W. (eds.) Complexity of Computer Computations, pp. 85–103. Plenum Press, New York (1972)
12. Kuhnle, A., Li, X., Smith, J.D., Thai, M.T.: Online set multicover algorithms for dynamic D2D communications. J. Comb. Optim. **34**(4), 1237–1264 (2017). https://doi.org/10.1007/s10878-017-0144-y
13. Lovász, L.: On the ratio of optimal integral and fractional covers. Discrete Math. **13**, 383–390 (1975)
14. Lund, C., Yannakakis, M.: On the hardness of approximating minimization problems. In: Proceedings of the 25th ACM Symposium on Theory of Computing, pp. 286–293 (1993)
15. Rajagopalan, S., Vazirani, V.V.: Primal-dual RNC approximation algorithms for set cover and covering integer programs. SIAM J. Comput. **109**(28), 525–540 (1998)
16. Sun, Z., Li, L., Li, X., Xing, X., Li, Y.: Optimization coverage conserving protocol with authentication in wireless sensor networks. Int. J. Distrib. Sens. Netw. **13**(3), 1–16 (2017)
17. Sun, Z., Li, C., Xing, X., Wang, H., Yan, B., Li, X.: K-degree coverage algorithm based on optimization nodes deployment in wireless sensor networks. Int. J. Distrib. Sens. Netw. **13**(2), 1–16 (2017)
18. Sun, Z., Shu, Y., Xing, X., et al.: LPOCS: a novel linear programming optimization coverage scheme in wireless sensor networks. J. Ad Hoc Sens. Wirel. Netw. **33**(1/4), 173–197 (2016)
19. Sun, Z., Zhang, Y., Xing, X., et al.: EBKCCA: a novel energy balanced k-coverage control algorithm based on probability model in wireless sensor networks. KSII Trans. Internet Inf. Syst. **10**(8), 3621–3640 (2016)
20. Sun, Z., Wang, H., Wu, W., Xing, X.: ECAPM: an enhanced coverage algorithm in wireless sensor network based on probability model. Int. J. Distrib. Sens. Netw. **2015**Article ID 203502, 11 pages (2015)
21. Young, N.E.: The k-server dual and loose competitiveness for paging. Algorithmica **11**, 525–541 (1994). Preliminary version appeared in SODA'91 titled "On-Line Caching as Cache Size Varies"

Sumcheck-Based Delegation of Quantum Computing to Rational Server

Yuki Takeuchi[1]([✉]), Tomoyuki Morimae[2,3], and Seiichiro Tani[1]

[1] NTT Communication Science Laboratories, NTT Corporation,
3-1 Morinosato Wakamiya, Atsugi, Kanagawa 243-0198, Japan
{yuki.takeuchi.yt,seiichiro.tani.cs}@hco.ntt.co.jp
[2] Yukawa Institute for Theoretical Physics, Kyoto University,
Kitashirakawa Oiwakecho, Sakyoku, Kyoto 606-8502, Japan
tomoyuki.morimae@yukawa.kyoto-u.ac.jp
[3] JST, PRESTO, 4-1-8 Honcho, Kawaguchi, Saitama 332-0012, Japan

Abstract. Recently, a new model of delegated quantum computing has been proposed, namely, rational delegated quantum computing. In this model, after a client delegates quantum computing to a server, the client pays a reward to the server. In this paper, we propose novel one-round rational delegated quantum computing protocols. The construction of the previous rational protocols depends on gate sets, while our sumcheck technique can be easily realized with any local gate set. We also show that a constant reward gap can be achieved if two non-communicating but entangled rational servers are allowed. Furthermore, we show, under a certain condition, the equivalence between *rational* and *ordinary* delegated quantum computing protocols.

Keywords: Quantum computing · Rational interactive proof · Game theory

1 Introduction

1.1 Background

Delegated quantum computing enables a client with weak computational power to delegate quantum computing to a remote (potentially malicious) server in such a way that the client can efficiently verify whether the server faithfully computes the delegated problem (i.e., can verify the server's integrity). Due to the size of a universal quantum computer and the difficulty of maintaining it, it is expected that first generation full-fledged quantum computers will be used in the delegated-quantum-computing style. Furthermore, since quantum operations

This work is partially supported by MEXT Quantum Leap Flagship Program (MEXT Q-LEAP) Grant Number JPMXS0118067394, JST PRESTO No. JPMJPR176A, and JSPS Grant-in-Aid for Young Scientists (B) No. JP17K12637.

© Springer Nature Switzerland AG 2020
J. Chen et al. (Eds.): TAMC 2020, LNCS 12337, pp. 69–81, 2020.
https://doi.org/10.1007/978-3-030-59267-7_7

and communication are too demanding (for current technologies), the client's operations and their communication should be made classical.

One of the most important open problems in the field of quantum computing is whether a classical client can efficiently delegate universal quantum computing to a quantum server while efficiently verifying the server's integrity. In delegated quantum computing, the honest server's computational power should be bounded by polynomial-time quantum computing, because delegated quantum computing with a server having unbounded computational power is unrealistic. This limitation is the large difference between delegated quantum computing and interactive proof systems for BQP. In interactive proof systems, the computational power of the prover (i.e., the server) is unbounded. Therefore, this open problem cannot be straightforwardly solved from the well-known containment BQP $\subseteq$ PSPACE$=$IP [1].

In this paper, we take a different approach to construct protocols for classical-client delegated quantum computing. We consider delegating quantum computing to a rational server. This model was first proposed by Morimae and Nishimura [2] based on the concept of rational interactive proof systems [3]. We note again that the computational power of the server is bounded by BQP[1] in rational delegated quantum computing, while it is unbounded in the rational interactive proof systems. In rational delegated quantum computing, after the client interacts with the server, the client pays a reward to the server depending on the server's messages and the client's random bits. In *ordinary* delegated quantum computing, the server may be malicious. On the other hand, in *rational* one, the server is always rational, i.e., he/she tries to maximize the expected value of the reward. In the real world, there are several situations where service providers want to maximize their profits. Since rational delegated quantum computing reflects such situations, this model can be considered as another possible situation for delegated quantum computing. In Ref. [2], it was shown that the classical client can delegate universal quantum computing to the rational quantum server in one round.

1.2 Our Contribution

As our main contribution, we propose a novel one-round delegated quantum computing protocol with a classical client and a rational quantum server. More precisely, we construct protocols where the classical client can efficiently delegate to the rational quantum server the estimation of output probabilities of n-qubit quantum circuits. Their estimation has many applications such as estimating the expected values of observables, which are quantities interested especially by physicists, and solving decision problems in BQP. Specifically, we consider any n-qubit polynomial-size quantum circuit with k-qubit output measurements, where $k = O(\log n)$. Since the goal of our rational protocol is to delegate the estimation of the output probabilities, we, for clarity, refer to our protocol as delegated

---

[1] For simplicity, we sometimes use complexity classes to represent computational powers. For example, we say that a server (a client) is a BQP server (a BPP client) when he/she performs polynomial-time quantum (probabilistic classical) computing.

quantum estimating protocol. As shown in the full paper [4], our argument can also be used to construct a one-round rational delegated quantum computing protocol for any BQP problem. Intuitively, using a certain BQP-complete problem [5], any BQP problem can be reduced to the estimation of the probability of the first qubit being projected onto $|1\rangle$. Therefore, our argument works. Furthermore, if a delegated quantum circuit is approximately sparse, our result can be generalized to the estimation of output probabilities with n-qubit output measurements. For general quantum circuits, such generalization is still open.

Our protocols can be applied to a broader class of universal gate sets than the previous protocols [2]. They work for any universal gate set each of whose elementary gates acts on at most $O(\log n)$ qubits, while the previous protocols are tailored for Clifford gates plus $T \equiv |0\rangle\langle 0| + e^{i\pi/4}|1\rangle\langle 1|$ or classical gates plus the Hadamard gate. Note that we only consider gate sets whose elementary gates can be specified with a polynomial number of bits.

Four conditions should be satisfied by practical rational delegated quantum computing protocols:

1. The reward is upper-bounded by a constant.
2. The reward is always non-negative if the BQP server takes an optimal strategy that maximizes its expected value.[2]
3. The maximum of the expected value of the reward is lower-bounded by a constant.
4. The reward gap [6] is larger than a constant. Here, simply speaking, the reward gap is a minimum loss on the expected value of the server's reward incurred by the server's behavior that makes the client accept an incorrect answer. Note that such behavior may require computational power beyond BQP, while we limit the optimal strategy maximizing the expected value to one that can be executed in quantum polynomial time.

The protocols of Ref. [2] and our protocol satisfy only conditions 1–3. Whether the above four conditions can be satisfied simultaneously is an open problem. In Ref. [2], it is shown that if the reward gap is larger than $1/f(n)$ with a polynomial $f(n)$, a super-polynomial increase of the reward (i.e., the violation of the first condition) is unavoidable in one-round protocols with a single server unless $\mathsf{BQP} \subseteq \Sigma_3^\mathsf{P}$. Since this inclusion is considered unlikely given the oracle separation between BQP and PH [7], this implies that it may be impossible to satisfy the above four conditions simultaneously in one-round protocols with a single server.

As the second contribution, for BQP problems, we construct a multi-rational-server delegated quantum computing protocol that satisfies all four conditions simultaneously. In the full paper [4], we also discuss whether a single server is sufficient under the (widely believed) assumption that the learning with errors (LWE) problem is hard for polynomial-time quantum computation.

---

[2] More precisely, the server takes an optimal strategy that can be executed in quantum polynomial time, because we assume that the computational power of the server is bounded by BQP. Throughout this paper, the server's optimization is limited to one that can be performed in quantum polynomial time unless explicitly noted otherwise.

Finally, apart from these results, we show that under the certain condition introduced in Ref. [12], *rational* and *ordinary* delegated quantum computing protocols can be converted from one to the other and vice versa. This equivalence may provide a new approach to tackle the open problem of whether a classical client can efficiently delegate universal quantum computing to a (non-rational) quantum server while efficiently verifying the server's integrity. Based on this equivalence, we give an amplification method for the reward gap.

2   Preliminaries

2.1   Rational Delegated Quantum Computing

In this subsection, we define rational delegated quantum computing. Following the original definition of rational interactive proof systems [3], we first define the transcript $\mathcal{T}$, the server's view $\mathcal{S}$, and the client's view $\mathcal{C}$ as follows:

Definition 1. *We assume that k is odd. Given an instance x and a round i, we define the ith transcript $\mathcal{T}_i$, the ith server's view $\mathcal{S}_i$, and the ith client's view $\mathcal{C}_i$ as follows ($0 \le i \le k$):*

- *$\mathcal{T}_0 = \mathcal{S}_0 = \mathcal{C}_0 = \{x\}$.*
- *When i is odd, $\mathcal{T}_i = \{\mathcal{T}_{i-1}, a_i\}$, where a_i is the ith server's message. On the other hand, when $i(> 0)$ is even, $\mathcal{T}_i = \{\mathcal{T}_{i-1}, b_i\}$, where b_i is the ith client's message.*
- *For odd i, $\mathcal{S}_i = \{\mathcal{S}_{i-2}, \mathcal{T}_{i-1}, V_i\}$, where V_i is a quantum circuit used to compute a_i. Note that $\mathcal{S}_i$ and V_i are not defined for even i because the even-numbered round is a communication from the client to the server.*
- *For even i, $\mathcal{C}_i = \{\mathcal{C}_{i-2}, \mathcal{T}_{i-1}, r_i\}$, where r_i is a random bit string used to compute b_i. Note that $\mathcal{C}_i$ is not defined for odd i because the odd-numbered round is a communication from the server to the client.*

For all i, messages a_i and b_i are polynomial lengths. Particularly, b_i is generated from $\mathcal{C}_i$ in classical polynomial time. The quantum circuit V_i is decided from $\mathcal{S}_{i-2}$.

Based on Definition 1, we define the following k-round interaction between a BPP client and a server:

Definition 2. *Let k be odd. This means that the protocol begins with the server's step. When k is even, the following definition can be adopted by adding a communication from the server to the client at the beginning of the protocol. Let us consider the following k-round interaction:*

1. *A BPP client interacts with a server k times. In the ith round for odd i, the server sends a_i to the client. In the ith round for even i, the client sends b_i to the server.*
2. *The client efficiently calculates a predicate on the instance x and the kth transcript $\mathcal{T}_k$. If the predicate evaluates to $o = 1$, the client answers YES. On the other hand, if $o = 0$, the client answers NO.*

3. *The client efficiently calculates the reward $R \in [0, c]$ and pays it to the server, where c is a positive constant. Note that it is not necessary for the client and server to know the value of c. The reward function $R : \{0,1\}^* \times \{0,1\}^{poly(|x|)} \times \{0,1\}^{poly(|x|)} \to \mathbb{R}_{\geq 0}$ depends on the instance $x \in \{0,1\}^*$, the kth transcript $\mathcal{T}_k \in \{0,1\}^{poly(|x|)}$, and the client's random bits $r_{k+1} \in \{0,1\}^{poly(|x|)}$.*

Rational delegated quantum computing for decision problems is defined as follows:

Definition 3. *The k-round interaction defined in Definition 2 is called a k-round rational delegated quantum computing protocol for decision problems if and only if the following conditions hold: let $\mathbb{E}[f]$ denote the expectation value of a function f. Let $\mathcal{D}_k$ be a distribution that the kth transcript follows. For a language $L \subseteq \{0,1\}^*$ in BQP, if $x \in L$, there exists a classical polynomial-time predicate and a distribution $\mathcal{D}_{YES}$ that can be generated in quantum polynomial time, such that*

$$\Pr[o = 1 \mid \mathcal{D}_k = \mathcal{D}_{YES}] \geq \frac{2}{3} \tag{1}$$

and

$$\mathbb{E}_{\mathcal{T}_k \sim \mathcal{D}_{YES}, r_{k+1}}[R(x, \mathcal{T}_k, r_{k+1})] \geq c_{YES} \tag{2}$$

with some positive constant $c_{YES} \leq c$.

On the other hand, if $x \notin L$, there exists a classical polynomial-time predicate and a distribution $\mathcal{D}_{NO}$ that can be generated in quantum polynomial time, such that

$$\Pr[o = 0 \mid \mathcal{D}_k = \mathcal{D}_{NO}] \geq \frac{2}{3} \tag{3}$$

and

$$\mathbb{E}_{\mathcal{T}_k \sim \mathcal{D}_{NO}, r_{k+1}}[R(x, \mathcal{T}_k, r_{k+1})] \geq c_{NO} \tag{4}$$

with some positive constant $c_{NO} \leq c$.

To generate distributions $\mathcal{D}_{YES}$ and $\mathcal{D}_{NO}$, the server decides the ith message a_i following a distribution $\mathcal{D}_i$ that can be generated in quantum polynomial time and satisfies

$$\mathcal{D}_i = \text{argmax}_{\mathcal{D}_i} \mathbb{E}_{\mathcal{D}_k, \mathcal{T}_k \sim \mathcal{D}_k, r_{k+1}}[R(x, \mathcal{T}_k, r_{k+1}) | \mathcal{D}_i, \mathcal{S}_i], \tag{5}$$

where the expectation is taken over all possible distributions $\mathcal{D}_k$ that are compatible with the current server's view $\mathcal{S}_i$. Here, we consider only the maximizations that can be performed in quantum polynomial time.

Since the server's computational power is bounded by BQP, it is in general hard for the server to select an optimal message that satisfies Eqs. (1) and (2). Therefore, the server's message a_i should be probabilistically generated. That is why we consider the distribution $\mathcal{D}_{YES}$. The same argument holds for the NO case.

The value 2/3 in Eqs. (1) and (3) can be amplified to $1 - 2^{-f(|x|)}$, where $f(|x|)$ is any polynomial in $|x|$, using the standard amplification method (i.e., by repeating steps 1 and 2, and then taking the majority vote among outputs in step 2). We here mention that the above rational delegated quantum computing protocol satisfies conditions 1–3 in Sect. 1. This is straightforward from $R \in [0, c]$ and Eqs. (2) and (4).

The server would like to generate the ith message a_i following a distribution that maximizes the expected value of the finally obtained reward. However, at that time, the server cannot predict the future distribution $\mathcal{D}_k$. Therefore, the server also takes the expectation over all possible distributions $\mathcal{D}_k$. The distribution $\mathcal{D}_i$ in Eq. (5) is a distribution that maximizes such expected reward.

All of our rational protocols except for one in Sect. 3 are in accordance with Definition 3. Our rational protocol in Sect. 3 is a rational delegated quantum computing protocol for function problems, which can be defined in a similar way.

2.2   Reward Gap

Guo *et al.* have introduced the reward gap [6]. For convenience, we define a strategy s as a set $\{a_i\}_i$ of the server's messages, which may be adaptively decided according to the previous client's messages. When we focus on the dependence on the server's messages, we write $\mathbb{E}_{\mathcal{T}_k \sim \mathcal{D}, r_{k+1}}[R(x, \mathcal{T}_k, r_{k+1})]$ by $\mathbb{E}_{s \sim \mathcal{D}'}[R(x, s)]$ for short. For decision problems, the reward gap is defined as follows:

Definition 4. *Let $\mathcal{D}'$ be a distribution that the server's strategy s follows. Let $\mathcal{D}'_{\max}$ be the distribution $\mathcal{D}'$, where each message a_i follows the distribution in Eq. (5). We say that a rational delegated quantum computing protocol has a $1/\gamma(|x|)$-reward gap if for any input x,*

$$\mathbb{E}_{s \sim \mathcal{D}'_{\max}}[R(x, s)] - \max_{s \in S_{\text{incorrect}}} \mathbb{E}[R(x, s)] \geq \frac{1}{\gamma(|x|)}, \tag{6}$$

where $\gamma(|x|)$ is any function of $|x|$, and $S_{\text{incorrect}}$ is the set of the server's strategies that make the client output an incorrect answer. Here, the expectation is also taken over the client's random bits. Note that $S_{\text{incorrect}}$ may include strategies that cannot be executed in quantum polynomial time.

From Definition 3, if the server's strategy s follows the distribution $\mathcal{D}'_{\max}$, the client outputs a correct answer with high probability. $\mathbb{E}_{s \sim \mathcal{D}'_{\max}}[R(x, s)]$ is the maximum expected value of the reward paid to the rational BQP server. On the other hand, $\max_{s \in S_{\text{incorrect}}} \mathbb{E}[R(x, s)]$ is the maximum expected value of the reward paid to the *malicious* computationally-unbounded server if the server wants to maximize the expected value as much as possible while deceiving the client. This is because the client outputs an incorrect answer when the server takes the strategy $s \in S_{\text{incorrect}}$. As a result, the reward gap represents how much benefit the rational server can obtain compared with the malicious one. For function problems, we can define the reward gap in a similar way.

3   Sumcheck-Based Rational Delegated Quantum Computing

In this section, we construct a rational delegated quantum computing protocol for estimating output probabilities of n-qubit quantum circuits, which we call the rational delegated quantum estimating protocol. Particularly, we consider any n-qubit polynomial-size quantum circuit with $O(\log n)$-qubit output measurements. We also show that our protocol satisfies conditions 1–3 mentioned in Sect. 1.

Let $\{q_z\}_{z\in\{0,1\}^k}$ be the output probability distribution of the quantum circuit U, where $q_z \equiv \langle 0^n|U^\dagger(|z\rangle\langle z|\otimes I^{\otimes n-k})U|0^n\rangle$ and I is the two-dimensional identity operator. We show that if the quantum server is rational, the classical client can efficiently obtain the estimated values $\{p_z\}_{z\in\{0,1\}^k}$ with high probability such that $|p_z - q_z| \leq 1/f(n)$ for any z and any polynomial $f(n)$. Therefore, for example, the classical client can approximately sample with high probability in polynomial time from the output probability distribution $\{q_z\}_{z\in\{0,1\}^k}$ of the quantum circuit U. Before proposing our rational delegated quantum estimating protocol, we calculate q_z using the Feynman path integral. Let $U = u_L \ldots u_2 u_1 \equiv \prod_{i=L}^{1} u_i$, where u_i is an elementary gate in a universal gate set for all i, and L is a polynomial in n. The probability q_z is calculated as follows:

$$q_z = \sum_{s\in\{0,1\}^{(2L-1)n-k}} g(z,s), \tag{7}$$

where

$$g(z,s) \equiv \langle 0^n|u_1^\dagger \left(\prod_{j=L}^{2} u_j|s^{(j-1)}\rangle\langle s^{(j-1)}|\right)^\dagger |zs^{(L)}\rangle\langle zs^{(L)}| \tag{8}$$

$$\left(\prod_{i=L}^{2} u_i|s^{(L+i-1)}\rangle\langle s^{(L+i-1)}|\right) u_1|0^n\rangle,$$

and s is a shorthand notation of the $(2L-1)n-k$ bit string $s^{(1)}s^{(2)}\ldots s^{(2L-1)}$. As an important point, given z and s, the function $g(z,s)$ can be calculated in classical polynomial time. This is because each elementary gate acts on at most $O(\log n)$ qubits. Furthermore, from Eq. (8), $0 \leq (1 + \text{Re}[g(z,s)])/2 \leq 1$, where $\text{Re}[g(z,s)]$ is the real part of $g(z,s)$.

To construct our rational delegated quantum estimating protocol, we use the rational sumcheck protocol [8]. The rational sumcheck protocol enables the client to efficiently delegate to the rational server the calculation (or approximation) of $\sum_{i=1}^{l} x_i$, where x_i is an integer for any i. To fit the rational sumcheck protocol to our case, we generalize it for the case of the complex number x_i. As a result, we can set $x_i = g(z,s)$ and z to be a certain fixed value. Our protocol runs as follows:

[**Protocol 1**]

1. For all $z \in \{0,1\}^k$, the rational server and the client perform the following steps:
   (a) The rational server sends to the client a real non-negative number y_z, which is explained later. (Note that y_z is represented by a bit string with logarithmic length; therefore, the message size from the server to the client is logarithmic.)
   (b) The client samples s uniformly at random from $\{0,1\}^{(2L-1)n-k}$.
   (c) The client flips a coin that lands heads with probability $(1+\mathrm{Re}[g(z,s)])/2$. If the coin lands heads, the client sets $b_z = 1$; otherwise, $b_z = 0$.
   (d) Let $Y_z \equiv [y_z + 2^{(2L-1)n-(k+1)}]/2^{(2L-1)n-k}$. The client calculates the reward

$$R(y_z, b_z) \equiv \frac{1}{2^k}\left[2Y_z b_z + 2(1-Y_z)(1-b_z) - Y_z^2 - (1-Y_z)^2 + 1\right], \quad (9)$$

which is the (slightly modified) Brier's scoring rule [9]. This scoring rule guarantees that the expected value of the reward is maximized when y_z is equal to the probability of $b_z = 1$ up to additive and multiplicative factors. Then, the client pays the reward $R(y_z, b_z)$ to the rational server.

2. The client calculates

$$p_z \equiv \frac{y_z}{\sum_{z\in\{0,1\}^k} y_z} \quad (10)$$

   for all z.

Since the sampling in step (c) can be approximately performed in classical polynomial time, what the client has to do is simply efficient classical computing. Furthermore, since the repetitions in step 1 can be performed in parallel, this is a one-round protocol. Note that except for the communication required to pay the reward to the server, Protocol 1 only requires one-way communication from the server to the client.

We show that p_z satisfies $\sum_{z\in\{0,1\}^k} |p_z - q_z| \le 1/f(n)$ for any fixed polynomial $f(n)$ with high probability. This means that p_z is an approximated value of q_z for each z with high probability. More precisely, we show the following theorem:

Theorem 1. *Let $f(n)$ and $h(n)$ be any polynomials in n. Let $q_z = \langle 0^n | U^\dagger(|z\rangle\langle z| \otimes I^{\otimes n-k})U|0^n\rangle$, and p_z be the probability given in Eq. (10). Then, for any $f(n)$ and $h(n)$, there exists Protocol 1 such that $\sum_{z\in\{0,1\}^k} |p_z - q_z| \le 1/f(n)$ with probability of at least $1 - e^{-h(n)}$.*

The proof is given in the full paper [4]. The intuitive idea is that the expected value of our reward function increases as y_z becomes to be close to $q_z/2$ for all z. Therefore, the rational server essentially sends approximated values of $\{q_z\}_{z\in\{0,1\}^k}$ to the client.

From Theorem 1, by approximately sampling from $\{p_z\}_{z\in\{0,1\}^k}$, the client can approximately sample from $\{q_z\}_{z\in\{0,1\}^k}$ with high probability. Given the values of $\{p_z\}_{z\in\{0,1\}^k}$, the approximate sampling from $\{p_z\}_{z\in\{0,1\}^k}$ can be classically performed in polynomial time.

In Protocol 1, we assume that $(1 + \mathrm{Re}[g(z,s)])/2$ can be exactly represented using a polynomial number of bits. If this is not the case, the classical client has to approximate $(1 + \mathrm{Re}[g(z,s)])/2$. As a result, as shown in the full paper [4], the expected value of the reward is maximized when $y_z = q_z/2 + \delta$, where the real number δ satisfies $|\delta| \leq 2^{-f'(n)}$ for a polynomial $f'(n)$. Therefore, even in the approximation case, the classical client can efficiently obtain the estimated values of the output probabilities of quantum circuits.

Next, we show the following theorem:

Theorem 2. *In Protocol 1, the total reward $\sum_{z\in\{0,1\}^k} R(y_z, b_z)$ is between $3/2 - O(1/2^{(2L-1)n-k})$ and $3/2 + O(1/2^{(2L-1)n-k})$ for $b_z \in \{0,1\}$ and any real values $y_z \in [0,1/2]$. Furthermore, the maximum expected value of the total reward is lower-bounded by $3/2 + O\left(1/2^{2(2L-1)n-k}\right)$.*

The proof is given in the full paper [4]. From this theorem, Protocol 1 satisfies conditions 1–3 in Sect. 1.

4   Multi-Rational-Server Delegated Quantum Computing with a Constant Reward Gap

In this section, we consider the reward gap. Although a large reward gap is desirable to incentivize the server to behave optimally, our sumcheck-based protocol has only an exponentially small gap as in the existing rational delegated quantum computing protocols [2]. It is open as to whether a constant reward gap is possible. However, in this subsection, we show that if non-communicating but entangled multiservers are allowed, we can construct a rational delegated quantum computing protocol with a constant reward gap for BQP problems while keeping three conditions 1–3 in Sect. 1. To this end, we utilize multiprover interactive proof systems for BQP. In some multiprover interactive proof systems proposed for BQP, the computational ability of the honest provers is bounded by BQP but that of the malicious provers is unbounded (e.g., Refs. [10,11]). Simply speaking, these multiprover interactive proof systems satisfy the following: for any language $L \in$ BQP, there exists a $poly(|x|)$-time classical verifier V interacting with a constant number of non-communicating but entangled provers, such that for instances x, if $x \in L$, then there exists a $poly(|x|)$-time quantum provers' strategy in which V accepts with probability of at least $2/3$, and if $x \notin L$, then for any (computationally-unbounded) provers' strategy, V accepts with probability of at most $1/3$. We denote the above interaction between V and provers as π_L for the language $L \in$ BQP.

Using the above multiprover interactive proof systems and the construction used in Ref. [3], we construct the following rational delegated quantum computing protocol:

[Protocol 2]

1. For a given BQP language L and an instance x, one of M rational servers sends $b \in \{0,1\}$ to the client. As shown in Theorem 3, if the server is rational, $b = 1(0)$ when x is in L (x is not in L).
2. If $b = 1$, the client and M servers simulate π_L for the language L and instance x; otherwise, the client and M servers simulate $\pi_{\bar{L}}$ for the complement $\bar{L}$ and the instance x.
3. The client pays reward $R = 1/M$ to each of the M servers if the simulated verifier accepts. On other hand, if the simulated verifier rejects, the client pays $R = 0$.
4. The client concludes $x \in L$ if $b = 1$; otherwise, the client concludes $x \notin L$.

Note that since BQP is closed under complement, $\pi_{\bar{L}}$ exists for the complement $\bar{L}$. Here, we notice that even if the simulated verifier accepts, each server can obtain only $1/M$ as the reward. However, since the number M of the servers is two in the multiprover interactive proof systems in Refs. [10,11], the reward $1/M$ paid to each server can be made $1/2$. Furthermore, when we use the results in Refs. [10,11], the number of rounds in Protocol 2 becomes a constant.

We clarify the meaning of "rational" in multi-rational-server delegated quantum computing. We can consider at least two possible definitions of "rational". One is that each server wants to maximize each reward, and the other is that all servers want to collaboratively maximize their total reward. Fortunately, in Protocol 2, these two definitions are equivalent. In other words, the total reward is maximized if and only if the reward paid to each server is maximized. Hereafter, we therefore do not distinguish between these two definitions.

Before we show that Protocol 2 has a constant reward gap, we show that if the servers are rational, the client's answer is correct. More formally, we prove the following theorem:

Theorem 3. *In Protocol 2, if the servers are rational, i.e., take the strategy that maximizes the expectation value of the reward, then $b = 1$ if and only if $x \in L$.*

Proof. First, we consider the YES case, i.e., the case where x is in L. If $b = 1$, the client and the servers perform π_L for the language L and the instance x. Therefore, when the servers simulate the honest provers in π_L, the client accepts with probability of at least $2/3$. On the other hand, if $b = 0$, the client accepts with probability less than or equal to $1/3$. This is because x is a NO instance for the complement $\bar{L}$, i.e., $x \notin \bar{L}$. In $\pi_{\bar{L}}$, when the answer is NO, the acceptance probability is at most $1/3$ for any provers' strategy. Since the completeness-soundness gap $1/3$ is a positive constant, one of the rational servers sends $b = 1$ if $x \in L$. By following the same argument, one of them sends $b = 0$ when $x \notin L$.

From this proof, we notice that the reward gap has the same value as the completeness-soundness gap.[3] Protocol 2 has a $1/3$ reward gap, which is constant. Furthermore, it can be straightforwardly shown that Protocol 2 also satisfies conditions 1–3 mentioned in Sect. 1 as follows. Since the total reward $M \times R$

---

[3] Precisely speaking, since the computational power of the server is bounded by BQP, the server sends $b = 0(1)$ with an exponentially small probability when the correct

paid to M servers is 0 or 1, the first and second conditions are satisfied. When the servers behave rationally, the client accepts with probability at least $2/3$. Therefore, the expected value of the total reward paid to the rational servers is at least $2/3$, which satisfies the third condition.

5   Relation Between Rational and Ordinary Delegated Quantum Computing Protocols

In Sect. 4, by incorporating *ordinary* delegated quantum computing into *rational* delegated quantum computing, we have shown that the four conditions can be simultaneously satisfied. In this section, we consider the reverse direction, i.e., constructing *ordinary* delegated quantum computing protocols from *rational* delegated quantum computing protocols. By combining this construction with the idea in Sect. 4, we obtain an equivalence (under a certain condition) between these two types of delegated quantum computing. Note that in ordinary ones, the server's ability is unbounded in NO cases (i.e., when $x \notin L$).

To construct ordinary delegated quantum computing protocols from rational ones, we consider the general $poly(|x|)$-round rational delegated quantum computing protocol defined in Definition 3, which we call RDQC for short. By adding two conditions for RDQC, we define constrained RDQC as follows:

Definition 5. *The constrained RDQC protocol is an RDQC protocol defined in Definition 3 such that*

1. There exists a classically efficiently computable polynomial $f(|x|)$ such that

$$c_{\mathrm{YES}} - \max_{s \in S_{\mathrm{incorrect}}, x \notin L} \mathbb{E}[R(s, x)] \geq \frac{1}{f(|x|)}, \tag{11}$$

2. The upper-bound c of the reward is classically efficiently computable.

The first condition was introduced in Ref. [12]. It is worth mentioning that the second condition is satisfied in our sumcheck-based protocol, while the first condition is not. Note that the left-hand side of Eq. (11) is not the reward gap.

We show that an *ordinary* delegated quantum computing protocol with a single BQP server and a single BPP client can be constructed from any constrained RDQC protocol. To this end, we show the following theorem:

Theorem 4. *If a language L in BQP has a k-round constrained RDQC protocol, then L has a k-round interactive proof system with the completeness-soundness gap $1/(cf(|x|))$ between an honest BQP prover and a BPP verifier.*

---

answer is YES (NO). Therefore, the finally obtained reward gap is decreased by the inverse of an exponential from the original completeness-soundness gap. However, this is negligible because the original completeness-soundness gap is a constant.

The proof is essentially the same as that of Theorem 4 in Ref. [12].

As shown in the full paper [4], from Theorem 4, we can show that if there exists a constant-round constrained RDQC protocol for BQP, then $\mathsf{BQP} \subseteq \prod_2^p$, which seems to be unlikely due to the oracle separation between BQP and PH [7].

We show that the reverse conversion is also possible using the idea in Sect. 4.

Theorem 5. *If a language L in* BQP *has an interactive proof system with an honest* BQP *prover and a* BPP *verifier, then L has a constrained RDQC protocol.*

The detail is given in the full paper [4].

Finally, by applying Theorems 4 and 5, we give the following amplification method for the reward gap:

Corollary 1. *The reward gap of the constrained RDQC can be amplified to a constant.*

The proof is given in the full paper [4]. Here, we explain the basic idea of the proof. Using the conversion between rational and ordinary delegated quantum computing protocols, we show that the amplification of the reward gap can be replaced with that of the soundness-completeness gap. This means that the traditional amplification method for the soundness-completeness gap can be used to amplify the reward gap. Remarkably, this amplification method works even if the original constrained RDQC protocol has only an exponentially small reward gap. This is because the original constrained RDQC protocol satisfies Eq. (11).

References

1. Shamir, A.: IP=PSPACE. In: Proceedings of the 31st Annual Symposium on Foundations of Computer Science, pp. 11–15. IEEE, St. Louis (1990)
2. Morimae, T., Nishimura, H.: Rational proofs for quantum computing. arXiv:1804.08868
3. Azar, P.D., Micali, S.: Rational proofs. In: Proceedings of the 44th Symposium on Theory of Computing, pp. 1017–1028. ACM, New York (2012)
4. Takeuchi, Y., Morimae, T., Tani, S.: Sumcheck-based delegation of quantum computing to rational server. arXiv:1911.04734
5. Aharonov, D., Ben-Or, M., Eban, E.: Interactive proofs for quantum computations. In: Proceedings of Innovations in Computer Science 2010, pp. 453–469. Tsinghua Univ. Press, Beijing (2010)
6. Guo, S., Hubáček, P., Rosen, A., Vald, M.: Rational arguments: single round delegation with sublinear verification. In: Proceedings of the 5th Conference on Innovations in Theoretical Computer Science, pp. 523–540. ACM, New Jersey (2014)
7. Raz, R., Tal, A.: Oracle separation of BQP and PH. In: Proceedings of the 51st Annual Symposium on Theory of Computing, pp. 13–23. ACM, New York (2019)
8. Guo, S., Hubáček, P., Rosen, A., Vald, M.: Rational sumchecks. In: Kushilevitz, E., Malkin, T. (eds.) TCC 2016. LNCS, vol. 9563, pp. 319–351. Springer, Heidelberg (2016). https://doi.org/10.1007/978-3-662-49099-0_12
9. Brier, G.W.: Verification of forecasts expressed in terms of probability. Mon. Weather. Rev. **78**, 1–3 (1950). https://journals.ametsoc.org/mwr/article/78/1/1/96424/VERIFICATION-OF-FORECASTS-EXPRESSED-IN-TERMS-OF

10. Coladangelo, A., Grilo, A.B., Jeffery, S., Vidick, T.: Verifier-on-a-leash: new schemes for verifiable delegated quantum computation, with quasilinear resources. In: Ishai, Y., Rijmen, V. (eds.) EUROCRYPT 2019. LNCS, vol. 11478, pp. 247–277. Springer, Cham (2019). https://doi.org/10.1007/978-3-030-17659-4_9
11. Grilo, A.B.: A simple protocol for verifiable delegation of quantum computation in one round. In: Proceedings of the 46th International Colloquium on Automata, Languages, and Programming, pp. 28:1–28:13. EATCS, Patras (2019)
12. Chen, J., McCauley, S., Singh, S.: Efficient rational proofs with strong utility-gap guarantees. In: Deng, X. (ed.) SAGT 2018. LNCS, vol. 11059, pp. 150–162. Springer, Cham (2018). https://doi.org/10.1007/978-3-319-99660-8_14

Online Removable Knapsack Problems for Integer-Sized Items

Kanaho Hanji[✉], Hiroshi Fujiwara, and Hiroaki Yamamoto

Shinshu University, Nagano, Japan
19w2100j@shinshu-u.ac.jp, {fujiwara,yamamoto}@cs.shinshu-u.ac.jp

Abstract. In the online removable knapsack problem, a sequence of items, each labeled with its profit and its size, is given in one by one. At each arrival of an item, a player has to decide whether to put it into a knapsack or not. The player is also allowed to discard some of the items that are already in the knapsack. The objective is to maximize the total profit of the knapsack. Iwama and Taketomi gave an optimal algorithm for the case where the profit of each item is equal to its size. In this paper we consider a case with an additional constraint that the size of the knapsack is a positive integer N and the sizes of the items are all integral. For each of the cases $N = 1, 2, \ldots$, we design an algorithm and prove its optimality. It is revealed that the competitive ratio is not monotonic with respect to N.

1 Introduction

In the knapsack problem, a player receives a set of items, each with a profit and a size, and packs some of the items into a knapsack so that the total profit is maximized and the total size does not exceed the knapsack size [5]. Iwama and Taketomi first proposed an online version called the *online removable knapsack problem* [4], which has the following additional settings: (i) The player receives items one by one. Each time an item arrives, the player has to decide whether to put it into the knapsack or throw it away. The player does not know when the sequence of items ends. (ii) When an item arrives, the player can discard some of items that are already in the knapsack.

The *competitive ratio* is often used as an evaluation measure of algorithms for online problems [1]. The competitive ratio is the maximum ratio of the online profit to the optimal offline profit. The online profit is the total profit obtained by an algorithm for items arriving one by one. The optimal offline profit is the optimal profit of a clairvoyant player who knows the whole item sequence beforehand. The smaller the competitive ratio is, the better the algorithm is.

The paper [4] gave an optimal online algorithm with competitive ratio $\frac{\sqrt{5}+1}{2}(\approx 1.618)$ for the online removable knapsack problem under a setting that for each item, its profit is equal to its size. In this paper, we impose a further additional setting that the knapsack size is a positive integer N and the size of each item is also an integer between 1 and N. For each N, we design an

© Springer Nature Switzerland AG 2020
J. Chen et al. (Eds.): TAMC 2020, LNCS 12337, pp. 82–93, 2020.
https://doi.org/10.1007/978-3-030-59267-7_8

optimal algorithm for the problem. The motivation of our study comes from application to practical problems: Although rational numbers are mostly handled in computation, we deal with them as a problem over integral numbers by reduction.

Table 1. The competitive ratio for integer-sized items with profit = size.

Knapsack size	1 to 3	4	5	6	7	8	9
Competitive ratio	1	$\frac{4}{3}(\approx 1.33)$	$\frac{5}{4}(\approx 1.25)$	$\frac{4}{3}(\approx 1.33)$	$\frac{7}{5}(\approx 1.4)$	$\frac{4}{3}(\approx 1.33)$	$\frac{3}{2}(\approx 1.5)$

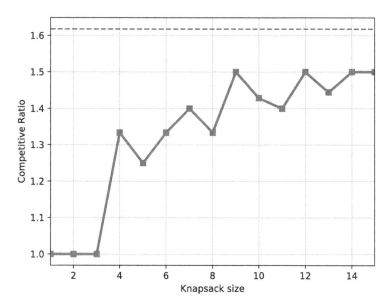

Fig. 1. The competitive ratio for integer-sized items with profit = size. The dashed line stands for the competitive ratio $(= \frac{\sqrt{5}+1}{2})$ for arbitrary-sized items with profit = size [4].

1.1 Our Contribution

Our algorithm achieves a competitive ratio of

$$\max\left\{\frac{N}{\lceil Nt\rceil + 1}, \min\left\{\frac{N}{\lceil Nt\rceil}, \frac{\lceil Nt\rceil}{N - \lceil Nt\rceil + 1}\right\}\right\} \qquad (1)$$

for each N, where $t = \frac{\sqrt{5}-1}{2}$. This is optimal since we show that the competitive ratio of any algorithm is at least that value. The lower bound is derived in a similar way as in the paper [4]. In contrast, the design of our algorithm involves a more careful analysis for individual N. See Table 1 and Fig. 1 for the values of the competitive ratios.

It is observed that each competitive ratio of our algorithm is lower than $\frac{\sqrt{5}+1}{2}$, which is a natural consequence as our problem is more restricted than the original problem in [4]. Nevertheless, our result reveals a non-intuitive relation between N and the competitive ratio: The competitive ratio does not increase monotonically as N grows. For example, the competitive ratio for $N = 4$ is $\frac{4}{3}$, while the competitive ratio for $N = 5$ is $\frac{5}{4} < \frac{4}{3}$.

1.2   Previous Research

The knapsack problem is one of many famous combinatorial optimization problems [5]. The task is to choose a subset of items so that the total profit is maximized and the total size does not exceed the knapsack size. To consider an online version of optimization problems is a natural interest of researchers [1,2,6]. Iwama and Taketomi first proposed the online removable knapsack problem [4]. For the case where the profit of each item is equal to its size, they gave an optimal online algorithm with a competitive ratio of $\frac{\sqrt{5}+1}{2}$.

Han et al. studied the online removable knapsack problem in which the profit of an item is a convex function of its size [3]. They provided a $\frac{5}{3}$ (≈ 1.666)-competitive algorithm, as well as a lower bound of $\frac{\sqrt{5}+1}{2}$ (≈ 1.618) on the competitive ratio.

2   Online Removable Knapsack Problem

2.1   Problem Statement

In the online removable knapsack problem, a player receives a sequence of items $u_1, u_2, \ldots$ one by one. When item u arrives, the player has to determine (i) whether to put u into the knapsack or not, and (ii) a subset of items in the knapsack to be discarded, so that the total size of items in the knapsack does not exceed the knapsack size. Let $|u|$ stand for the size of item u. A sequence of items is displayed as $\sigma = (|u_1|, |u_2|, |u_3|, \ldots)$. We do not define the symbol for profit since we will assume that the profit of an item always equals its size of it. Let $|B|$ denote the total profit ($=$ total size) of items in the knapsack. The objective of this problem is to maximize $|B|$.

2.2   Evaluation of Algorithms

The competitive ratio is most commonly used as a performance measure of an algorithm for an online optimization problem [1]. For the online removable knapsack problem, the competitive ratio of algorithm ALG is defined as the maximum value over the ratio of the total profit by ALG to the optimal total profit in the setting that the entire sequence is known in advance. That is,

$$R_{ALG} = \max_{\sigma} \frac{OPT(\sigma)}{ALG(\sigma)}.$$

It can be said that the smaller the competitive ratio is, the better the algorithm is.

2.3 Additional Settings

We mention two additional settings to the online removable knapsack problem. The first setting (I) is to assume that for every item, the profit equals the size. In the paper [4], an optimal algorithm for the online removable knapsack problem with the setting (I), which we present as Theorem 1 below. Note that with the setting (I), each item u_i satisfies $0 < |u_i| \leq 1$ and it always holds that $0 < |B| \leq 1$.

Theorem 1 ([4]). *For the online removable knapsack problem with the setting (I), an algorithm whose competitive ratio is $\frac{\sqrt{5}+1}{2} (\approx 1.618)$ exists and is optimal.*

The second setting (II) is to let the size of the knapsack be a positive integer N and to let the size of each item be in the set $\{1, 2, 3, \ldots, N\}$. Hereinafter, such an item is called an *integer-sized* item. Under the two settings (I) and (II), the range of the total profit in the knapsack is $0 < |B| \leq N$. Below is our main theorem.

Theorem 2. *For the online removable knapsack problem with the setting (I) and (II), an algorithm whose competitive ratio is $\frac{1}{t'(N)}$ exists and is optimal, where $t'(N) = \dfrac{1}{\max\left\{\frac{N}{\lceil Nt \rceil + 1}, \min\left\{\frac{N}{\lceil Nt \rceil}, \frac{\lceil Nt \rceil}{N - \lceil Nt \rceil + 1}\right\}\right\}}$ and $t = \frac{\sqrt{5}-1}{2}$.*

Theorem 2 is proven by Lemmas 1, 3, 4, 5, and 6 in the following sections.

3 Lower Bounds

We give a lower bound for the problem with the settings (I) and (II) by a similar way to the proof in the paper [4], which is summarized as the following: An arbitrary algorithm receives an input item sequence with two or possibly three items. The first two items of the sequence cannot be packed into the knapsack together. Depending on the choice of which item to pack, the adversary decides whether the third item follows or not.

Such a derivation may look too simple and possible to improve. However, the analysis in Sect. 4 clarifies that the resulting bound is tight.

Lemma 1. *Let the size of knapsack be $N \geq 3$ and $t = \frac{\sqrt{5}-1}{2}$. For any algorithm A, it follows that*

$$R_A \geq \max\left\{\frac{N}{\lceil Nt \rceil + 1}, \min\left\{\frac{N}{\lceil Nt \rceil}, \frac{\lceil Nt \rceil}{N - \lceil Nt \rceil + 1}\right\}\right\}.$$

Proof. We consider two scenarios: One is $\sigma_c = (\lceil Nt \rceil + 1, N - \lceil Nt \rceil, \lceil Nt \rceil)$ or a prefix of σ_c with two items. The other is $\sigma_f = (\lceil Nt \rceil, N - \lceil Nt \rceil + 1, \lceil Nt \rceil - 1)$ or a prefix of σ_f with two items. For each of these sequences, the first item and the second item cannot be put together into the knapsack since $\lceil Nt \rceil + 1 + N - \lceil Nt \rceil > N$ and $\lceil Nt \rceil + N - \lceil Nt \rceil + 1 > N$.

The adversary, observing the behavior of algorithm A, determines whether the player receives the third item or not. For σ_c, if algorithm A puts the second item into the knapsack, then the third item is not given. Thus, we have $R_A \geq \frac{\lceil Nt \rceil + 1}{N - \lceil Nt \rceil}$, since the optimal choice from $(\lceil Nt \rceil + 1, N - \lceil Nt \rceil)$ is just the first one. If algorithm A puts the first item of σ_c into the knapsack, then the third item arrives. Thus, we have $R_A \geq \frac{N}{\lceil Nt \rceil + 1}$, since the optimal choice from $(\lceil Nt \rceil + 1, N - \lceil Nt \rceil, \lceil Nt \rceil)$ is the second and third one. For σ_f likewise, if algorithm A puts the second item into the knapsack, then the third item is not given. Thus, we have $R_A \geq \frac{\lceil Nt \rceil}{N - \lceil Nt \rceil + 1}$, since the optimal choice from $(\lceil Nt \rceil, N - \lceil Nt \rceil + 1)$ is just the first one. If algorithm A puts the first item of σ_f into the knapsack, then the third item arrives. Thus, we have $R_A \geq \frac{N}{\lceil Nt \rceil}$, since the optimal choice from $(\lceil Nt \rceil, N - \lceil Nt \rceil + 1, \lceil Nt \rceil - 1)$ is the second and third one.

From above, we obtain a lower bound as

$$R_A \geq \max \left\{ \min \left\{ \frac{N}{\lceil Nt \rceil + 1}, \frac{\lceil Nt \rceil + 1}{N - \lceil Nt \rceil} \right\}, \min \left\{ \frac{N}{\lceil Nt \rceil}, \frac{\lceil Nt \rceil}{N - \lceil Nt \rceil + 1} \right\} \right\}.$$

Here, a simple derivation leads to the inequality

$$\frac{N}{\lceil Nt \rceil + 1} < \frac{\lceil Nt \rceil + 1}{N - \lceil Nt \rceil}$$

for all N. Using this, the lower bound is rewritten in a simpler form

$$R_A \geq \max \left\{ \frac{N}{\lceil Nt \rceil + 1}, \min \left\{ \frac{N}{\lceil Nt \rceil}, \frac{\lceil Nt \rceil}{N - \lceil Nt \rceil + 1} \right\} \right\}.$$

$\square$

Lemma 1 does not tell anything about the cases of $N = 1$ and 2. In Sect. 4, it is revealed that a trivial lower bound of 1 is tight for both these cases.

4   Upper Bounds

4.1   Algorithm of Iwama and Taketomi Revisited

In this section we show that for each N, there exists an algorithm which has a competitive ratio that is equal to the lower bound in Sect. 3. The algorithm of Iwama and Taketomi [4] with a little modification satisfies our purpose. The algorithm classifies each item by its size and then decides how to pack it.

We follow the notation of Iwama and Taketomi for the item size classification. The classes are S, M, L, and X, which are ordered so that the size increases.

We sometimes identify classes with disjoint sets of item sizes. Later, some classes may be set to be empty sets. We write the maximum (or minimum) size of class S as S_{max} (or S_{min}, respectively). For M, L, and X, we also use the same notation.

We first describe the algorithm of Iwama and Taketomi [4] for generalized classes of items S, M, L, and X.

Algorithm IT [4]: For each item u, do the following operations. Here, $|B|$ is the total profit of items that are in the knapsack immediately before the arrival of u.

- If $|B| \in X$, discard u.
- Otherwise, if $|u| + |B| \leq N$, put u into the knapsack.
- Otherwise, if $|u| \in X$, put only u into the knapsack.
- Otherwise, if only items of S or M size exists in the knapsack, put u into the knapsack and discard the items of S or M size in the knapsack until $|B|$ becomes less than N.
- Otherwise, for the item l which is the item of L size in the knapsack, if $|u| + |l| \leq N$, put u and l only,
  otherwise, put the smaller item of u and l.

The reader may think that even if we apply the original item classification to algorithm IT, that is to say, not modifying the work of [4] at all, the algorithm can achieve tight bounds. However, it is revealed that such a simple application fails. For example, when $N = 40$, our lower bound on the competitive ratio is $\frac{25}{16}$. For $N = 40$, the item classification of Iwama and Taketomi is, after a straightforward rounding, that $S_{min} = 1$, $S_{max} = 11$, $M_{min} = 12$, $M_{max} = 14$, $L_{min} = 15$, $L_{max} = 25$, and $X_{min} = 26$. When the item sequence is $\sigma = (1, 25, 15)$, the algorithm of Iwama and Taketomi [4] yields a knapsack consisting of items $B = \{1, 15\}$ and therefore $|B| = 16$. However, at this time, the competitive ratio becomes more than $1.625 = \frac{26}{16} (> \frac{25}{16})$ since the total profit of the optimal knapsack is $|OPT| = |\{1, 25\}| = 26$, which means that the algorithm cannot achieve a tight bound.

We pick up the proof of competitiveness from the paper [4] and interpret it for a generalized item size classification as the following lemma. .

Lemma 2 ([4]). *Let c be a positive real number. Suppose that a family of classes X, L, M, and S satisfies the following condition: Class X, L, and M satisfy*

$$X_{max} = N, \tag{2}$$

$$X_{min} \geq \left\lceil \frac{N}{c} \right\rceil, \tag{3}$$

$$L_{max} = X_{min} - 1, \tag{4}$$

$$\frac{L_{max}}{L_{min}} \leq c, \tag{5}$$

$$2L_{min} \geq X_{min}, \tag{6}$$

$$M_{max} \leq N - L_{max}, \tag{7}$$

$$and \ M_{min} \geq X_{min} - L_{min}. \tag{8}$$

Class S is empty or satisfies

$$S_{max} = M_{min} - 1 \tag{9}$$
$$\text{and } S_{min} = 1. \tag{10}$$

Then, it holds that

$$R_{IT} \leq c.$$

4.2 A Tight Upper Bound for Each Knapsack Size

In the paper [4], the authors set a target parameter $\frac{1}{t} = \frac{\sqrt{5}+1}{2} = 1.618...$ and gave an algorithm whose competitive ratio less than $\frac{1}{t}$. Our analysis involves a similar parameter. For given N, we let $\frac{1}{t'(N)}$ be the value of the lower bound of the competitive ratio, which is derived in Sect. 3. That is to say, we define

$$\frac{1}{t'(N)} = \max\left\{\frac{N}{\lceil Nt \rceil + 1}, \min\left\{\frac{N}{\lceil Nt \rceil}, \frac{\lceil Nt \rceil}{N - \lceil Nt \rceil + 1}\right\}\right\}. \tag{11}$$

For the case of $N = 1$, a trivial algorithm achieves a competitive ratio of 1. For each of the cases of $N = 2$, 3, and 4, we individually show by Lemma 2 that algorithm IT achieves a competitive ratio of $\frac{1}{t'(N)}$.

Lemma 3. *For each of the cases of $N = 2$, 3, and 4, let $c = \frac{1}{t'(N)}$. Then, there exists an item classification which satisfies the conditions of Lemma 2, for $N = 2$, the classes M and S are empty, the class $X = 2$ and the class $L = 1$. Then, the competitive ratio is 1 since this classification satisfies the condition of Lemma 2. For $N = 3$, the class S is empty, the class $X = \{3\}$, the class $L = \{2\}$, and the class $M = \{1\}$. Then, the competitive ratio is 1 since this classification satisfies the condition of Lemma 2. For $N = 4$, the class M is empty, the class $X = \{4, 3\}$, the class $L = \{2\}$, and the class $S = \{1\}$. Then, the competitive ratio is $\frac{4}{3}$ since this classification satisfies the condition of Lemma 2.*

As for general $N \geq 5$, we classify N into three subsets and discuss competitiveness separately along the classification. Note that $\frac{N}{\lceil Nt \rceil + 1} < \frac{N}{\lceil Nt \rceil}$. We divide the set of all integers ≥ 5 into the following three subsets with respect to the magnitude relation between $\frac{\lceil Nt \rceil}{N - \lceil Nt \rceil + 1}$, $\frac{N}{\lceil Nt \rceil + 1}$, and $\frac{N}{\lceil Nt \rceil}$.

Definition 1. *Define A_1, A_2, and A_3 as the set of integers $N \geq 5$ such that $\frac{1}{t'(N)}$ is $\frac{N}{\lceil Nt \rceil + 1}$, $\frac{N}{\lceil Nt \rceil}$, and $\frac{\lceil Nt \rceil}{N - \lceil Nt \rceil + 1}$, respectively. That is to say,*

$$A_1 = \left\{N \,\middle|\, \frac{\lceil Nt \rceil}{N - \lceil Nt \rceil + 1} \leq \frac{N}{\lceil Nt \rceil + 1}\right\} = \{N | 8, 16, 21, ...\},$$

$$A_2 = \left\{N \,\middle|\, \frac{N}{\lceil Nt \rceil} \leq \frac{\lceil Nt \rceil}{N - \lceil Nt \rceil + 1}\right\} = \{N | 5, 7, 9, ...\},$$

$$\text{and } A_3 = \left\{N \,\middle|\, \frac{N}{\lceil Nt \rceil + 1} < \frac{\lceil Nt \rceil}{N - \lceil Nt \rceil + 1} < \frac{N}{\lceil Nt \rceil}\right\} = \{N | 6, 11, 14...\}.$$

Now, we mention Lemmas 4 and 5, which will play an important role in the proof of competitiveness. Recall $t'(N)$ defined in (11). In the following lemmas, we write the value of $t'(N)$ simply as t'.

Lemma 4. *For all $N \geq 5$,*

$$2t'\lceil Nt' \rceil - 2t' - \lceil Nt' \rceil \geq 0 \tag{12}$$

holds true, where t' is the value of $t'(N)$.

Proof. We show the lemma according to the classification of N.
(i) Case $N \in A_1 = \{8, 16, 21, \ldots\}$:

$$\begin{aligned}
&2t'\lceil Nt' \rceil - 2t' - \lceil Nt' \rceil \\
&= 2 \cdot \frac{\lceil Nt \rceil + 1}{N} \left\lceil N \cdot \frac{\lceil Nt \rceil + 1}{N} \right\rceil - 2 \cdot \frac{\lceil Nt \rceil + 1}{N} - \left\lceil N \cdot \frac{\lceil Nt \rceil + 1}{N} \right\rceil \\
&= \frac{2}{N}(\lceil Nt \rceil + 1)^2 - \frac{2}{N}(\lceil Nt \rceil + 1) - (\lceil Nt \rceil + 1) \\
&= \frac{\lceil Nt \rceil + 1}{N}(2\lceil Nt \rceil + 2 - 2 - N) \\
&= \frac{\lceil Nt \rceil + 1}{N}(2\lceil Nt \rceil - N) \\
&> \frac{Nt + 1}{N}(2Nt - N) \quad (\because Nt < \lceil Nt \rceil) \\
&= \frac{Nt + 1}{N}\{N(2t - 1)\} > 0.
\end{aligned}$$

(ii) Case $N \in A_2 = \{5, 7, 9, \ldots\}$: For the cases of $N = 5$ and $N = 7$, we directly confirm the inequality: For $N = 5$, we have

$$2 \cdot \frac{4}{5}\left\lceil 5 \cdot \frac{4}{5} \right\rceil - 2 \cdot \frac{4}{5} - \left\lceil 5 \cdot \frac{4}{5} \right\rceil = \frac{4}{5} > 0.$$

For $N = 7$, we have

$$2 \cdot \frac{5}{7}\left\lceil 7 \cdot \frac{5}{7} \right\rceil - 2 \cdot \frac{5}{7} - \left\lceil 7 \cdot \frac{5}{7} \right\rceil = \frac{5}{7} > 0.$$

We then evaluate the inequality generally for the case where $N \geq 9$. Let $d = \frac{\lceil Nt \rceil}{N}$. We see that $d = \frac{\lceil Nt \rceil}{N} > \frac{Nt}{N} = t$ holds. We derive

$$2t'\lceil Nt'\rceil - 2t' - \lceil Nt'\rceil$$

$$= 2 \cdot \frac{\lceil Nt\rceil}{N}\left[N \cdot \frac{\lceil Nt\rceil}{N}\right] - 2 \cdot \frac{\lceil Nt\rceil}{N} - \left[N \cdot \frac{\lceil Nt\rceil}{N}\right]$$

$$= 2 \cdot \frac{\lceil Nt\rceil^2}{N} - 2 \cdot \frac{\lceil Nt\rceil}{N} - \lceil Nt\rceil$$

$$= 2Nd^2 - 2d - Nd \quad \left(\because d = \frac{\lceil Nt\rceil}{N}\right)$$

$$= d(2d-1)N - 2d$$

$$\geq d(2d-1)9 - 2d \quad \left(\because d(2d-1) > 0 \text{ and } N \geq 9 \text{ from } d > t > \frac{1}{2}\right)$$

$$= 18d\left(d - \frac{11}{18}\right) > 0 \quad \left(\because d > t > \frac{11}{18}\right).$$

(iii) Case $N \in A_3 = \{6, 11, 14, \ldots\}$: For the cases of $N = 6$ and $N = 11$, we directly confirm the inequality: For $N = 6$, we have

$$2 \cdot \frac{3}{4}\left[6 \cdot \frac{3}{4}\right] - 2 \cdot \frac{3}{4} - \left[6 \cdot \frac{3}{4}\right] = 1 > 0.$$

For $N = 11$, we have

$$2 \cdot \frac{5}{7}\left[11 \cdot \frac{5}{7}\right] - 2 \cdot \frac{5}{7} - \left[11 \cdot \frac{5}{7}\right] = 2 > 0.$$

It remains to prove the inequality for $N \geq 14$. We know that the equality

$$\lceil Nt'\rceil = \lceil Nt\rceil + 1$$

holds from the definition of A_3. Using this, we have

$$\begin{aligned}
2t'\lceil Nt'\rceil - 2t' - \lceil Nt'\rceil &= 2t'(\lceil Nt\rceil + 1) - 2t' - (\lceil Nt\rceil + 1) \\
&= 2t'\lceil Nt\rceil + 2t' - 2t' - \lceil Nt\rceil - 1 \\
&= 2t'\lceil Nt\rceil - \lceil Nt\rceil - 1 \\
&= 2 \cdot \frac{N - \lceil Nt\rceil + 1}{\lceil Nt\rceil} \cdot \lceil Nt\rceil - \lceil Nt\rceil - 1 \\
&= 2N - 3\lceil Nt\rceil + 1 \\
&> 2N - 3(Nt + 1) + 1 \quad (\because Nt < \lceil Nt\rceil) \\
&= N(2 - 3t) - 2 \\
&\geq 14(2 - 3t) - 2 \quad (\because N \geq 14) \\
&= 42 \cdot \left(\frac{13}{21} - t\right) > 0 \quad \left(\because t < \frac{13}{21}\right).
\end{aligned}$$

$\square$

Lemma 5. *For all $N \geq 5$,*

$$N - \lceil Nt' \rceil - \lceil Nt' \rceil t' + t' + 2 \geq 0 \tag{13}$$

holds true, where t' is the value of $t'(N)$.

Proof. We show the lemma according to the classification of N.
(i) Case $N \in A_1 = \{8, 16, 21, \ldots\}$: Let $d = \frac{\lceil Nt \rceil}{N}$, then

$$d < d + \frac{1}{N} < \frac{1}{d} - 1 + \frac{1}{dN}$$

$$-d^2 - d + 1 > \frac{d}{N} - \frac{1}{N}$$

holds from the definition of A_1. Then, we have

$$N - \lceil Nt' \rceil - \lceil Nt' \rceil t' + t' + 2$$
$$= N - \left[N \cdot \frac{\lceil Nt \rceil + 1}{N} \right] - \left[N \cdot \frac{\lceil Nt \rceil + 1}{N} \right] \cdot \frac{\lceil Nt \rceil + 1}{N} + \frac{\lceil Nt \rceil + 1}{N} + 2$$
$$= N - \lceil Nt \rceil - \frac{(\lceil Nt \rceil + 1)^2}{N} + \frac{\lceil Nt \rceil + 1}{N} + 1$$
$$= N - Nd^2 - Nd - d + 1$$
$$= N(-d^2 - d + 1) - d + 1$$
$$> N \left(\frac{d}{N} - \frac{1}{N} \right) - d + 1 \quad \left(\because -d^2 - d + 1 > \frac{d}{N} - \frac{1}{N} \right)$$
$$= d - 1 - d + 1 = 0.$$

(ii) Case $N \in A_2 = \{5, 7, 9, \ldots\}$:

$$N - \lceil Nt' \rceil - \lceil Nt' \rceil t' + t' + 2$$
$$= N - \left[N \frac{\lceil Nt \rceil}{N} \right] - \left[N \frac{\lceil Nt \rceil}{N} \right] \cdot \frac{\lceil Nt \rceil}{N} + \frac{\lceil Nt \rceil}{N} + 2$$
$$= N - \lceil Nt \rceil - \frac{\lceil Nt \rceil^2}{N} + \frac{\lceil Nt \rceil}{N} + 2$$
$$= N - \lceil Nt \rceil + 2 - \frac{\lceil Nt \rceil}{N} (\lceil Nt \rceil - 1)$$
$$> N - (Nt + 1) + 2 - \frac{Nt + 1}{N} (Nt + 1 - 1) \quad (\because \lceil Nt \rceil < Nt + 1)$$
$$= N - Nt - Nt^2 - t + 1$$
$$= N(-t^2 - t + 1) + 1 - t$$
$$= 1 - t > 0 \quad (\because t < 0.7).$$

(iii) Case $N \in A_3 = \{6, 11, 14, \ldots\}$: From the definition of A_3,

$$\lceil Nt' \rceil = \lceil Nt \rceil + 1$$

holds. Then, we have

$$N - \lceil Nt' \rceil - \lceil Nt' \rceil t' + t' + 2$$
$$= N - (\lceil Nt \rceil + 1) - (\lceil Nt \rceil + 1) \cdot \frac{N - \lceil Nt \rceil + 1}{\lceil Nt \rceil} + \frac{N - \lceil Nt \rceil + 1}{\lceil Nt \rceil} + 2$$
$$= 0.$$

□

Our choice of the item size classification is as follows:

Definition 2. *For each $N \geq 5$, define a family of classes $\overline{X}$, $\overline{L}$, $\overline{M}$, and $\overline{S}$ as the following. The definition of classes depends on N. We below denote the value of $t'(N)$ by simply t'. First of all, regardlessly of the value of N, set*

$$\overline{X}_{min} = \lceil Nt' \rceil,$$
$$\overline{L}_{max} = \lceil Nt' \rceil - 1,$$
$$\overline{L}_{min} = N - \lceil Nt' \rceil + 2,$$
$$and \ \overline{M}_{max} = N - \lceil Nt' \rceil + 1.$$

(a) If $2\lceil Nt' \rceil - N - 3 > 0$, set

$$\overline{M}_{min} = 2\lceil Nt' \rceil - N - 2,$$
$$\overline{S}_{max} = 2\lceil Nt' \rceil - N - 3,$$
$$and \ \overline{S}_{min} = 1.$$

(b) Otherwise, the class S is empty and set

$$\overline{M}_{min} = 1.$$

Lemma 6. *For each $N \geq 5$, $\overline{X}$, $\overline{L}$, $\overline{M}$, $\overline{S}$, and $c = \frac{1}{t'(N)}$ satisfy the condition of Lemma 2.*

Proof. We show that the conditions for the item size classification in Lemma 2 are satisfied by substituting the value which were defined in Definition 2. Firstly, we mention common part of (a) and (b) in Definition 2. The condition of (2) is trivial. For the conditions of (3) and (4),

$$\overline{X}_{min} = \lceil Nt' \rceil \geq \left\lceil \frac{N}{\frac{1}{t'}} \right\rceil = \left\lceil \frac{N}{c} \right\rceil$$
$$and \ \overline{L}_{max} = \lceil Nt' \rceil - 1 = \overline{X}_{min} - 1.$$

The condition (5) is confirmed as:

$$N - \lceil Nt' \rceil - \lceil Nt' \rceil t' + t' + 2 \geq 0 \quad (\because (13))$$
$$(\lceil Nt' \rceil - 1)t' \leq N - \lceil Nt' \rceil + 2$$
$$\overline{L}_{max}t' \leq \overline{L}_{min}$$
$$\frac{\overline{L}_{max}}{\overline{L}_{min}} \leq \frac{1}{t'}.$$

The following derivation holds true by Lemma 4.

$$2\overline{L}_{min} \geq 2\overline{L}_{max}t' \quad (\because (13))$$
$$= 2(\lceil Nt' \rceil - 1)t'$$
$$\geq \lceil Nt' \rceil \quad (\because (12))$$
$$= \overline{X}_{min}.$$

The condition (7) holds true, since

$$\overline{M}_{max} = N - \lceil Nt' \rceil + 1 = N - (\lceil Nt' \rceil - 1) = N - \overline{L}_{max}.$$

Moreover, we mention each of the cases (a) and (b).
(a) If $2\lceil Nt' \rceil - N - 3 > 0$, the condition (8) holds true, since

$$\overline{M}_{min} = 2\lceil Nt' \rceil - N - 2 = \lceil Nt' \rceil - (N - \lceil Nt' \rceil + 2) \geq \overline{X}_{min} - \overline{L}_{min}.$$

The condition (9) holds true, since

$$\overline{S}_{max} = 2\lceil Nt' \rceil - N - 3 = \overline{M}_{min} - 1.$$

The condition (10) is trivial.
(b) Otherwise, the condition (8) holds true, since

$$\overline{M}_{min} = 1 = \lceil Nt' \rceil - \lceil Nt' \rceil + 1 = \lceil Nt' \rceil - (\lceil Nt' \rceil - 1) \geq \overline{X}_{min} - \overline{L}_{min}.$$

$\square$

5   Conclusion

Our result reveals that the competitive ratio of our problem does not increase monotonically as N grows. This anomaly may be somehow related to the validity of the competitive ratio. A natural extension is to consider the case where each item is integer-sized, but the profit is not equal to the size.

References

1. Borodin, A., El-Yaniv, R.: Online Computation and Competitive Analysis. Cambridge University Press, Cambridge (1998)
2. Fiat, A., Woeginger, G.J. (eds.): Online Algorithms, The State of the Art (the Book Grown Out of a Dagstuhl Seminar, June 1996). LNCS, vol. 1442. Springer, Heidelberg (1998). https://doi.org/10.1007/BFb0029561
3. Han, X., Kawase, Y., Makino, K., Guo, H.: Online removable knapsack problem under convex function. Theor. Comput. Sci. **540–541**, 62–69 (2014). Combinatorial Optimization: Theory of algorithms and Complexity
4. Iwama, K., Taketomi, S.: Removable online knapsack problems. In: Widmayer, P., Eidenbenz, S., Triguero, F., Morales, R., Conejo, R., Hennessy, M. (eds.) ICALP 2002. LNCS, vol. 2380, pp. 293–305. Springer, Heidelberg (2002). https://doi.org/10.1007/3-540-45465-9_26
5. Kellerer, H., Pferschy, U., Pisinger, D.: Knapsack Problems. Springer, Heidelberg (2004). https://doi.org/10.1007/978-3-540-24777-7
6. Komm, D.: An Introduction to Online Computation: Determinism, Randomization, Advice. Springer, Cham (2016). https://doi.org/10.1007/978-3-319-42749-2

An Improved Approximation Algorithm for the Prize-Collecting Red-Blue Median Problem

Zhen Zhang$^{(\boxtimes)}$, Yutian Guo, and Junyu Huang

School of Computer Science and Engineering, Central South University,
Changsha 410083, People's Republic of China
`csuzz@foxmail.com`

Abstract. The *red-blue median* problem considers a set of *red* facilities, a set of *blue* facilities, and a set of clients located in some metric space. The goal is to open k_r red facilities and k_b blue facilities such that the sum of the distance from each client to its nearest opened facility is minimized, where k_r, $k_b \geq 0$ are two given integers. Designing approximation algorithms for this problem remains an active area of research due to its applications in various fields. However, in many applications, the existence of noisy data poses a big challenge for the problem. In this paper, we consider the *prize-collecting red-blue median* problem, where the noisy data can be removed by paying a penalty cost. The current best approximation for the problem is a ratio of 24, which was obtained by LP-rounding. We deal with this problem using a local search algorithm. We construct a layered structure of the swap pairs, which yields a $(9 + \epsilon)$-approximation for the prize-collecting red-blue median problem. Our techniques generalize to a more general *prize-collecting τ-color median* problem, where the facilities have τ different types, and give a $(4\tau + 1 + \epsilon)$-approximation for the problem for the case where τ is a constant.

Keywords: Clustering · Approximation · Local search

1 Introduction

k-median is a widely studied clustering problem and finds applications in many fields related to unsupervised learning. Given a set $\mathcal{D}$ of clients and a set $\mathcal{F}$ of facilities in a metric space, the k-median problem is to open k facilities such that the sum of the distance from each client to its nearest opened facility is minimized.

In many applications, the clustering problem has different types of facilities and upper bound on the number of the opened facilities of each type. One such

This work was supported by National Natural Science Foundation of China (61672536, 61872450, 61828205, and 61802441), Hunan Provincial Key Lab on Bioinformatics, and Hunan Provincial Science and Technology Program (2018WK4001).

© Springer Nature Switzerland AG 2020
J. Chen et al. (Eds.): TAMC 2020, LNCS 12337, pp. 94–106, 2020.
https://doi.org/10.1007/978-3-030-59267-7_9

example is in the design of Content Distribution Networks [2], where a set of clients need to be connected to a set of servers with a few different types, and there is a budget constraint on the number of the arranged severs of each type. Motivated by such applications, Hajiaghayi *et al.* [9] introduced the *red-blue median* problem which involves two facility-types. They showed that local search yields a constant factor approximation for the problem. The current best approximation for the red-blue median problem is a ratio of $5 + \epsilon$ due to Friggstad and Zhang [8]. Inspired by the work on the red-blue median problem, Krishnaswamy *et al.* [12] introduced a more general *matroid median* problem, where the set of facilities has a matroid structure, and the set of the opened facilities should be an independent set in the matroid. The matroid median problem does not only generalize the red-blue median problem, but also the τ-*color median* problem where more than two facility-types are considered. Krishnaswamy *et al.* [12] gave a 16-approximation for the matroid median problem by an LP-rounding technique. The approximation guarantee was later improved by a series of work [4,14] to the current best ratio of $7.081 + \epsilon$ [13].

Although red-blue median and its related clustering problems have been extensively studied, algorithms developed for these problems could significantly deteriorate their performance when applied to real-world data. One reason is that these problems implicitly assume that all clients can be clustered into several distinct groups. However, real-world data are often contaminated with various types of noises, which need to be excluded from the solution [3,6,13]. To deal with such noisy data, Charikar *et al.* [3] introduced the problem of *prize-collecting clustering*. The problem is the same as the standard clustering problem, except that we can remove a set of distant clients and pay their penalty costs instead. By discarding the distant clients, one could significantly reduce the clustering cost and thus improve the quality of solution.

Hajiaghayi *et al.* [9] gave an $O(1)$-approximation for the prize-collecting red-blue median problem by local search technique. The approximation ratio is implicit but can be easily shown not better than 30. Krishnaswamy *et al.* [12] later gave a 360-approximation algorithm for the prize-collecting red-blue median problem based on an LP-rounding technique. They showed that the guarantee of 360-approximation generalizes to the prize-collecting matroid median problem. The approximation ratio was recently improved to 24 by a novel rounding procedure [14]. This is also the current best approximation guarantee for both the problems of prize-collecting red-blue median and prize-collecting τ-color median. The noisy data appear frequently in the clustering problems, and the prize-collecting versions of other clustering problems have also been extensively studied [5,7,11,15].

We curtly remark on the commonly used approaches for clustering to show the obstacles in obtaining better ratios than 24 for the problems of prize-collecting red-blue median and prize-collecting τ-color median.

1). Hajiaghayi *et al.* [9] gave a constant factor approximation for the prize-collecting red-blue median problem by local search. Their analysis is based on a technique for dividing the facilities into *blocks* with certain properties.

This provides a clear way to get the approximation guarantee for the local search algorithm. However, getting such well structured blocks induces a large approximation ratio for the problem. It seems quite difficult to apply the technique given in [9] to beat the 24-approximation. Moreover, the method for constructing blocks relies heavily on the fact that the facilities have no more than two different types, which cannot be applied to the τ-color median problem. Indeed, it is still an open problem that whether local search works for the τ-color median problem for the case where τ is a constant, see discussions in [8, 10].

2). LP-rounding has been shown to be an effective technique for the problems of red-blue median and τ-color median [4, 12–14]. However, it was known that the existence of the penalized clients has a strong impact on the performance of the algorithms based on LP-rounding [12, 14]. For instance, the standard LP relaxation of the red-blue median problem has a variable x_{ij} associated with each facility i and client j, which indicates that whether j is connected to i. A constraint $\sum_i x_{ij} = 1$ is given for each client j to ensure that j is connected to a facility. Unfortunately, in the prize-collecting red-blue median problem, the sum $\sum_i x_{ij}$ is not guaranteed to be an integer since not all the clients should be connected. It is unclear whether the LP-rounding approach for clustering with outliers given in [13] can be adapted to clustering with penalties and beat the 24 approximation ratio.

3). The technique of primal-dual has been widely applied for the problem of prize-collecting clustering [3, 7, 11]. However, it is difficult to use this technique to deal with the red-blue median problem, as discussed in [9, 10]. This is further compounded in the prize-collecting red-blue median problem since there is an additional task of identifying the penalized clients.

1.1   Our Results

We use a local search algorithm to deal with the prize-collecting τ-color median problem. Starting with an arbitrary feasible solution, the algorithm tries to swap no more than $O(\tau)$ facilities of each type. It terminates if no such swap yields an improved solution. Otherwise, it iterates with the improved solution. The solution given by the algorithm is called *local optimum*.

Theorem 1. *The local optimum for the prize-collecting τ-color median problem is a $(4\tau + 1)$-approximation solution.*

On the basis of standard techniques [1] (which is curtly described in Sect. 2), the runtime of the local search algorithm can be polynomially bounded for the case where τ is a constant, which induces an arbitrarily small loss in the approximation ratio.

Theorem 2. *For any $\epsilon > 0$, there is a $(4\tau + 1 + \epsilon)$-approximation algorithm for the prize-collecting τ-color median problem that runs in polynomial time for the case where τ is a constant.*

Theorem 2 implies a $(9 + \epsilon)$-approximation for the prize-collecting red-blue median problem, which improves the previous best approximation ratio of 24 given by Swamy [14]. Note that for the prize-collecting τ-color median problem, we only obtain improved approximation guarantee for the case where $\tau \leq 5$, and the ratio of 24 given in [14] is still the best guarantee for the general prize-collecting τ-color median problem. Indeed, Krishnaswamy *et al.* [12] showed that local search cannot yield constant factor approximation for the τ-color median problem with polynomial time. However, this negative result does not rule out the possibility of obtaining a local search-based $O(1)$-approximation for the problem in polynomial time for the case where τ is a constant. Theorem 2 shows that local search actually yields an $O(1)$-approximation for this special case.

1.2 Our Techniques

The local search algorithms are commonly analyzed by considering a set of swap pairs where some facilities in the local optimum are swapped with some facilities from an optimal solution. The desired approximation guarantee is obtained by the fact that no such swap pair can improve the local optimum. However, in the prize-collecting τ-color median problem, the swap pairs may violate the constraint on the number of the opened facilities of each type. For instance, after closing a red facility and opening a blue facility, we are forced to swapping another pair of facilities to balance the number of the facilities of each type. This makes the analysis of the cost induced by the local optimum much more complex.

For each to-be-clustered client j, let O_j and S_j be the costs of j induced by the local optimum and optimal solution, respectively. Our analysis starts with carefully constructing a set of *feasible swap pairs* with some special properties (see Sect. 3.1). By estimating the increased cost induced by the constructed swap pairs, we obtain a set of inequalities that involve some "$+O_j$"terms, some "$-S_j$" terms, and some "$+S_j$" terms for each client j (see Sect. 3.2). We want to add these inequalities together to get $O(1)\sum_j O_j - O(1)\sum_j S_j \geq 0$, based on which the desired approximation ratio can be obtained. The challenge is how to eliminate each "$+S_j$" term. To overcome this challenge, we prove the existence of a layered structure of our constructed swap pairs (see Sect. 3.3). It is shown that for each swap pair, the "$+S_j$" terms induced by it can be counteracted by repeatedly using the swap pairs in the lower layers. These ideas lead to the proof of the $(4\tau + 1 + \epsilon)$-approximation ratio.

2 Preliminaries

The prize-collecting τ-color median problem can be defined as follows.

Definition 1 (prize-collecting τ-color median). *Given a set $\mathcal{C}$ of clients and τ disjoint sets $\mathcal{F}^1, \ldots, \mathcal{F}^\tau$ of facilities in a metric space, τ positive integers*

$k_1, \ldots, k_\tau$, and a penalty function p defined over the clients in $\mathcal{C}$, where $p(j) \geq 0$ for each $j \in \mathcal{C}$, the goal is to identify a subset $\mathcal{S}^t \subseteq \mathcal{F}^t$ of no more than k_t facilities for each $t \in \{1, \ldots, \tau\}$, such that the objective function

$$\sum_{j \in \mathcal{C}} \min\{d(j, \bigcup_{1 \leq t \leq \tau} \mathcal{S}^t), p(j)\}$$

is minimized, where $d(j, \bigcup_{1 \leq t \leq \tau} \mathcal{S}^t)$ denotes the distance from j to its nearest facility in $\bigcup_{1 \leq t \leq \tau} \mathcal{S}^t$.

The special case of $\tau = 2$ corresponds to the prize-collecting red-blue median problem. Given τ sets $\mathcal{S}^1, \ldots, \mathcal{S}^\tau$ of facilities, where $\mathcal{S}^t \subseteq \mathcal{F}^t$ for each $1 \leq t \leq \tau$, we call $\mathbb{S} = (\mathcal{S}^1, \ldots, \mathcal{S}^\tau)$ a feasible solution if $|\mathcal{S}^t| = k_t$ for each $1 \leq t \leq \tau$, and let $\Phi(\mathbb{S}) = \sum_{j \in \mathcal{C}} \min\{d(j, \bigcup_{1 \leq t \leq \tau} \mathcal{S}^t), p(j)\}$ denote its cost. Let OPT denote the cost of an optimal solution. The local search algorithm for the problem is described in Algorithm 1.

Algorithm 1: Local search for the prize-collecting τ-color median problem

---

Input: An instance $(\mathcal{C}, \mathcal{F}^1, \ldots, \mathcal{F}^\tau, k_1, \ldots, k_\tau, p)$ of the prize-collecting τ-color median problem;

Output: A local optimum $\mathbb{S} = (\mathcal{S}^1, \ldots, \mathcal{S}^\tau)$;

1 Let $\mathbb{S} = (\mathcal{S}^1, \ldots, \mathcal{S}^\tau)$ be an arbitrary feasible solution;

2 **while** there exists a feasible solution $\widetilde{\mathbb{S}} = (\widetilde{\mathcal{S}}^1, \ldots, \widetilde{\mathcal{S}}^\tau)$ such that $|\widetilde{\mathcal{S}}^t - \mathcal{S}^t| \leq 2\tau$ for each $1 \leq t \leq \tau$ and $\Phi(\widetilde{\mathbb{S}}) < \Phi(\mathbb{S})$ **do**

3 $\quad \lfloor \ \mathbb{S} \Leftarrow \widetilde{\mathbb{S}};$

4 **return** $\mathbb{S}$.

Each iteration of Algorithm 1 takes $O(|\mathcal{C}| \prod_{1 \leq t \leq \tau} (|\mathcal{F}^t| k_t)^{2\tau})$ time, which is polynomial in the input size for the case where τ is a constant. However, it may be the case that the number of the iterations exponentially depends on the input size. We can use a well-known trick to ensure that the algorithm terminates in a polynomial number of steps. The idea is to execute a swap only if $\Phi(\widetilde{\mathbb{S}}) \leq (1 - \frac{\epsilon}{\Delta})\Phi(\mathbb{S})$, where the value of Δ is polynomial in the input size. Our analysis is compatible with this trick: it can be verified that the total weight of all the inequalities we consider is polynomially bounded. See [1] for details of the trick.

3   Analysis

We introduce some notations to help analyze Algorithm 1. Let $\mathcal{F} = \bigcup_{1 \leq t \leq \tau} \mathcal{F}^t$. Let $d(i, j)$ denote the distance from i to j for each $i, j \in \mathcal{C} \cup \mathcal{F}$. Let $\mathbb{S} = (\mathcal{S}^1, \ldots, \mathcal{S}^\tau)$ denote the local optimum and $\mathbb{O} = (\mathcal{O}^1, \ldots, \mathcal{O}^\tau)$ be an optimal solution. Define $\mathcal{S} = \bigcup_{1 \leq t \leq \tau} \mathcal{S}^t$ and $\mathcal{O} = \bigcup_{1 \leq t \leq \tau} \mathcal{O}^t$. Let $\mathcal{P}$ and $\mathcal{P}^*$ denote the sets of the clients that are penalized when opening the facilities from $\mathcal{S}$ and $\mathcal{O}$, respectively. For each $j \in \mathcal{C} \backslash \mathcal{P}$, let s_j denote the nearest facility to j in $\mathcal{S}$, and define $S_j = d(j, s_j)$. Similarly, for each $j \in \mathcal{C} \backslash \mathcal{P}^*$, let o_j be the nearest facility to j in $\mathcal{O}$, and define $O_j = d(j, o_j)$. For each $i \in \mathcal{O}$ and $i' \in \mathcal{S}$, define $\mathcal{N}^*(i) = \{j \in \mathcal{C} \backslash \mathcal{P}^* : o_j = i\}$ and $\mathcal{N}(i') = \{j \in \mathcal{C} \backslash \mathcal{P} : s_j = i'\}$. For each $\mathcal{O}' \subseteq \mathcal{O}$ and $\mathcal{S}' \subseteq \mathcal{S}$, let $\mathcal{N}^*(\mathcal{O}') = \bigcup_{i \in \mathcal{O}'} \mathcal{N}^*(i)$ and $\mathcal{N}(\mathcal{S}') = \bigcup_{i \in \mathcal{S}'} \mathcal{N}(i)$. Given an integer $1 \leq t \leq \tau$ and a facility $i \in \mathcal{F}^t$, define $T(i) = t$ as its type. Given two integers t_1 and t_2, if $t_1 = t_2$, then let $\delta(t_1, t_2) = 1$. Otherwise, let $\delta(t_1, t_2) = 0$.

3.1   A Set of Swap Pairs

Algorithm 1 closes a set $\mathcal{A}_{out}$ of facilities and opens a set $\mathcal{A}_{in}$ of facilities in each iteration, let $\mathcal{A} = (\mathcal{A}_{out} \mid \mathcal{A}_{in})$ denote this swap pair. We call $\mathcal{A}$ a *feasible swap pair* if $|\mathcal{A}_{out}| = |\mathcal{A}_{in}| \neq 0$, and $|\mathcal{A}_{out} \cap \mathcal{F}^t| = |\mathcal{A}_{in} \cap \mathcal{F}^t|$ holds for each $1 \leq t \leq \tau$. It is easy to show that after performing a feasible swap pair, a feasible solution to the problem is still feasible. We also use a notation of *almost-feasible pairs*. Given a swap pair $\mathcal{B} = (\mathcal{B}_{out} \mid \mathcal{B}_{in})$ that is not feasible, if there exist a facility $i_1 \in \mathcal{B}_{out}$ and a facility $i_2 \in \mathcal{B}_{in}$, such that either $\mathcal{B}_{out} \backslash \{i_1\} = \mathcal{B}_{in} \backslash \{i_2\} = \emptyset$, or $(\mathcal{B}_{out} \backslash \{i_1\} \mid \mathcal{B}_{in} \backslash \{i_2\})$ is a feasible swap pair, then we call $\mathcal{B}$ an almost-feasible pair, and define $T(\mathcal{B}_{out}) = T(i_1)$ and $T(\mathcal{B}_{in}) = T(i_2)$. The following proposition follows directly from the definition of the almost-feasible pairs.

Proposition 1. *Given an almost-feasible pair $\mathcal{B}$ and two facilities $i_1 \in \mathcal{S}$, $i_2 \in \mathcal{O}$, swap pair $(\mathcal{B}_{out} \cup \{i_1\} \mid \mathcal{B}_{in} \cup \{i_2\})$ is a feasible swap pair iff $T(\mathcal{B}_{out}) = T(i_2)$ and $T(\mathcal{B}_{in}) = T(i_1)$.*

It can be seen that no feasible swap pair $\mathcal{A}$ with $|\mathcal{A}_{out} \cap \mathcal{F}^t| = |\mathcal{A}_{in} \cap \mathcal{F}^t| \leq 2\tau$ for each $t \in \{1, \ldots, \tau\}$ can be performed to reduce the cost of the local optimum. We consider a set of such swap pairs to show that the local optimum has small cost. These swap pairs close some facilities from $\mathcal{S}$ and open some facilities from $\mathcal{O}$. The swap pairs are selected based on the following mapping relationships.

Definition 2 $(\varphi(*), \eta(*))$. *For each $i \in \mathcal{O}$, let $\varphi(i)$ denote the nearest facility to i in $\mathcal{S}$. For each $i' \in \mathcal{S}$ with $\varphi^{-1}(i') \neq \emptyset$, let $\eta(i')$ denote the nearest facility to i' in $\varphi^{-1}(i')$. Given a set $\mathcal{S}' \subseteq \mathcal{S}$, define $\varphi^{-1}(\mathcal{S}') = \bigcup_{i \in \mathcal{S}'} \varphi^{-1}(i)$.*

Define $\mathcal{S}_1 = \{i \in \mathcal{S} : \varphi^{-1}(i) \neq \emptyset\}$, $\mathcal{O}_1 = \{\eta(i) : i \in \mathcal{S}_1\}$, $\mathcal{S}_2 = \mathcal{S} \backslash \mathcal{S}_1$, and $\mathcal{O}_2 = \mathcal{O} \backslash \mathcal{O}_1$. The procedure for selecting the swap pairs is given in Algorithm 2. Note that this procedure is only used in the analysis. The algorithm yields a set $\mathbb{A}$ of feasible swap pairs, which is empty initially. In the process of the algorithm, we say that a facility is unpaired if it is not yet involved in a swap pair from $\mathbb{A}$. Each facility is involved in at most one swap pair in $\mathbb{A}$.

In the first loop (steps 2 and 3), Algorithm 2 considers each subset $\mathcal{S}' \subseteq \mathcal{S}_1$ of size no more than τ, and adds $(\mathcal{S}' \mid \bigcup_{i \in \mathcal{S}'} \{\eta(i)\})$ to $\mathbb{A}$ if it is a feasible swap pair. The algorithm then constructs a set $\mathbb{B}$ of almost-feasible pairs. By the termination condition of the first loop, for each unpaired facility $i \in \mathcal{S}_1$, $(\{i\} \mid \{\eta(i)\})$ is not a feasible swap pair and can be viewed as an almost-feasible

---

Algorithm 2: Selecting a set of swap pairs

Input: The local optimum $(\mathcal{S}^1, \dots, \mathcal{S}^\tau)$ and an optimal solution $(\mathcal{O}^1, \dots, \mathcal{O}^\tau)$;

Output: A set $\mathbb{A}$ of swap pairs;

1 $\mathbb{A} \Leftarrow \emptyset$, $\mathbb{B} \Leftarrow \emptyset$, $\mathcal{S}_1' \Leftarrow \{i \in \bigcup_{1 \leq t \leq \tau} \mathcal{S}^t : \varphi^{-1}(i) \neq \emptyset\}$, $\mathcal{S}_2' \Leftarrow (\bigcup_{1 \leq t \leq \tau} \mathcal{S}^t) \backslash \mathcal{S}_1'$, $\mathcal{O}_1' \Leftarrow \{\eta(i) : i \in \mathcal{S}_1'\}$, $\mathcal{O}_2' \Leftarrow (\bigcup_{1 \leq t \leq \tau} \mathcal{O}^t) \backslash \mathcal{O}_1'$;

2 **while** $\exists \mathcal{S}' \subseteq \mathcal{S}_1'$ with $1 \leq |\mathcal{S}'| \leq \tau$, such that $\mathcal{A} = (\mathcal{S}' \mid \bigcup_{i \in \mathcal{S}'} \{\eta(i)\})$ is a feasible swap pair **do**

3 $\quad \lfloor \ \mathbb{A} \Leftarrow \mathbb{A} \cup \{\mathcal{A}\}, \mathcal{S}_1' \Leftarrow \mathcal{S}_1' \backslash \mathcal{S}', \mathcal{O}_1' \Leftarrow \mathcal{O}_1' \backslash \bigcup_{i \in \mathcal{S}'} \{\eta(i)\};$

4 **for** each $i \in \mathcal{S}_1'$ **do**

5 $\quad \lfloor \ \mathbb{B} \Leftarrow \mathbb{B} \cup \{(i \mid \eta(i))\};$

6 **while** $\exists \mathcal{B}^1, \mathcal{B}^2 \in \mathbb{B}$ such that $\mathcal{B} = (\mathcal{B}_{out}^1 \cup \mathcal{B}_{out}^2 \mid \mathcal{B}_{in}^1 \cup \mathcal{B}_{in}^2)$ is an almost-feasible pair **do**

7 $\quad \lfloor \ \mathbb{B} \Leftarrow \mathbb{B} \cup \{\mathcal{B}\} \backslash \{\mathcal{B}^1, \mathcal{B}^2\};$

8 **while** $\exists \ \mathbb{B}' \subseteq \mathbb{B}$, $\mathcal{H}_1 \subseteq \mathcal{S}_2'$, and $\mathcal{H}_2 \subseteq \varphi^{-1}(\bigcup_{\mathcal{B} \in \mathbb{B}'} \mathcal{B}_{out}) \cap \mathcal{O}_2'$, such that $1 \leq |\mathbb{B}'| = |\mathcal{H}_1| = |\mathcal{H}_2| \leq \tau$ and $\mathcal{A} = (\bigcup_{\mathcal{B} \in \mathbb{B}'} \mathcal{B}_{out} \cup \mathcal{H}_1 \mid \bigcup_{\mathcal{B} \in \mathbb{B}'} \mathcal{B}_{in} \cup \mathcal{H}_2)$ is a feasible swap pair **do**

9 $\quad \lfloor \ \mathbb{A} \Leftarrow \mathbb{A} \cup \{\mathcal{A}\}, \mathbb{B} \Leftarrow \mathbb{B} \backslash \mathbb{B}', \mathcal{S}_2' \Leftarrow \mathcal{S}_2' \backslash \mathcal{H}_1, \mathcal{O}_2' \Leftarrow \mathcal{O}_2' \backslash \mathcal{H}_2;$

10 **while** $\exists \mathcal{B} \in \mathbb{B}, i_1 \in \mathcal{S}_2'$, and $i_2 \in \mathcal{O}_2'$, such that $\mathcal{A} = (\mathcal{B}_{out} \cup \{i_1\} \mid \mathcal{B}_{in} \cup \{i_2\})$ is a feasible swap pair **do**

11 $\quad \lfloor \ \mathbb{A} \Leftarrow \mathbb{A} \cup \{\mathcal{A}\}, \mathbb{B} \Leftarrow \mathbb{B} \backslash \{\mathcal{B}\}, \mathcal{S}_2' \Leftarrow \mathcal{S}_2' \backslash \{i_1\}, \mathcal{O}_2' \Leftarrow \mathcal{O}_2' \backslash \{i_2\};$

12 **while** $\exists i_1 \in \mathcal{S}_2'$ and $i_2 \in \mathcal{O}_2'$ such that $\mathcal{A} = (\{i_1\} \mid \{i_2\})$ is a feasible swap pair **do**

13 $\quad \lfloor \ \mathbb{A} \Leftarrow \mathbb{A} \cup \{\mathcal{A}\}, \mathcal{S}_2' \Leftarrow \mathcal{S}_2' \backslash \{i_1\}, \mathcal{O}_2' \Leftarrow \mathcal{O}_2' \backslash \{i_2\};$

14 **return** $\mathbb{A}$.

---

pair. The algorithm adds all such almost-feasible pairs to $\mathbb{B}$ in the second loop (steps 4 and 5). In the third loop (steps 6 and 7), it combines two almost-feasible pairs from $\mathbb{B}$ if this yields a new almost-feasible pair. The combination is performed iteratively until no two pairs in $\mathbb{B}$ can form an almost-feasible pair. After that, the algorithm combines the almost-feasible pairs in $\mathbb{B}$ with some facilities from $\mathcal{S}_2$ and $\mathcal{O}_2$ to obtain a set of feasible swap pairs in the fourth and fifth loops (steps 8, 9, 10, and 11). In the fourth loop (steps 8 and 9), the algorithm iteratively determines whether there exist g almost-feasible pairs $\mathcal{B}^1, \dots, \mathcal{B}^g$ in $\mathbb{B}$, g unpaired facilities in $\varphi^{-1}(\bigcup_{1 \leq t \leq g} \mathcal{B}_{out}^t) \cap \mathcal{O}_2'$, and g unpaired facilities in $\mathcal{S}_2$ that can form a feasible swap pair, where $1 \leq g \leq \tau$. If such a feasible swap pair exists, then the algorithm adds the swap pair to $\mathbb{A}$ and deletes $\mathcal{B}^1, \dots, \mathcal{B}^g$ from $\mathbb{B}$. In the fifth loop (steps 10 and 11), for each remained almost-feasible pair $\mathcal{B}$ in $\mathbb{B}$, the algorithm finds two unpaired facilities $i_1 \in \mathcal{S}_2$ and $i_2 \in \mathcal{O}_2$, such that $(\mathcal{B}_{out} \cup \{i_1\} \mid \mathcal{B}_{in} \cup \{i_2\})$ is a feasible swap pair and can be added to $\mathbb{A}$. Finally, the remained unpaired facilities in $\mathcal{S}_2$ and $\mathcal{O}_2$ are assigned to single-swap pairs and added to $\mathbb{A}$ in the last loop (steps 12 and 13).

See Figs. 1 and 2 for an example. By the definitions of $\mathcal{S}_1$, $\mathcal{S}_2$, $\mathcal{O}_1$, and $\mathcal{O}_2$, we have $\{r_1, r_2, r_3, r_4, b_1\} = \mathcal{S}_1$, $\{b_2, b_3, b_4\} = \mathcal{S}_2$, $\{r_1^*, r_2^*, b_1^*, b_2^*, b_3^*\} = \mathcal{O}_1$, and

$\{r_3^*, r_4^*, b_4^*\} = \mathcal{O}_2$. It can be seen that $\mathcal{A}_1 = (\{r_1\} \mid \{\eta(r_1)\})$ and $\mathcal{A}_2 = (\{r_2, b_1\} \mid \{\eta(r_2), \eta(b_1)\})$ are two feasible swap pairs and should be added to $\mathbb{A}$ in the first loop of Algorithm 2 (steps 2 and 3). The algorithm then considers two almost-feasible pairs $\mathcal{B}_1 = (\{r_3\} \mid \{\eta(r_3)\})$ and $\mathcal{B}_2 = (\{r_4\} \mid \{\eta(r_4)\})$. In the fourth loop (steps 8 and 9), it combines $\mathcal{B}_1$ with two facilities $r_3^* \in \varphi^{-1}(r_3) \cap \mathcal{O}_2$ and $b_2 \in \mathcal{S}_2$ to obtain a feasible swap pair $\mathcal{A}_3 = (\{r_3, b_2\} \mid \{r_3^*, \eta(r_3)\})$. In the fifth loop (steps 10 and 11), a feasible swap pair $\mathcal{A}_4 = (\{r_4, b_3\} \mid \{r_4^*, \eta(r_4)\})$ is obtained by combining $\mathcal{B}_2$ with two facilities $r_4^* \in \mathcal{O}_2$ and $b_3 \in \mathcal{S}_2$. Finally, the remained unpaired facilities b_4 and b_4^* are combined into a feasible swap pair in the last loop (steps 12 and 13). The constructed swap pairs are shown in Fig. 2.

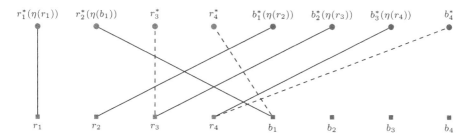

Fig. 1. $\mathcal{O} = \{r_1^*, r_2^*, r_3^*, r_4^*, b_1^*, b_2^*, b_3^*, b_4^*\}$ and $\mathcal{S} = \{r_1, r_2, r_3, r_4, b_1, b_2, b_3, b_4\}$ are the sets of the facilities opened in the optimal solution and local optimum respectively, where $\{r_1^*, r_2^*, r_3^*, r_4^*, r_1, r_2, r_3, r_4\} \subseteq \mathcal{F}^1$ and $\{b_1^*, b_2^*, b_3^*, b_4^*, b_1, b_2, b_3, b_4\} \subseteq \mathcal{F}^2$. For each $i \in \mathcal{S}$ and $i^* \in \varphi^{-1}(i) \backslash \{\eta(i)\}$, we joint i and i^* with a dashed line. For each $i \in \mathcal{S}$ with $\varphi^{-1}(i) \neq \emptyset$, we joint i and $\eta(i)$ with a solid line.

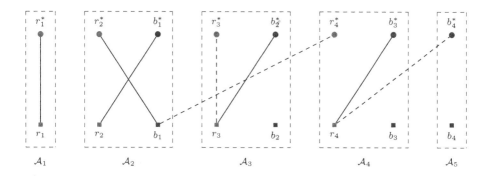

Fig. 2. The constructed swap pairs.

Let $\mathbb{A}$ denote the set of the swap pairs given by Algorithm 2. Let $\mathbb{B}$ be the set of the almost-feasible pairs obtained after the third loop of the algorithm (steps 6 and 7). We now give some useful properties of Algorithm 2.

Proposition 2. *We have $\sum_{\mathcal{B} \in \mathbb{B}} \delta(T(\mathcal{B}_{out}), t) \cdot \sum_{\mathcal{B} \in \mathbb{B}} \delta(T(\mathcal{B}_{in}), t) = 0$ for each $t \in \{1, \ldots, \tau\}$.*

Proposition 3. *For each $\mathcal{A} \in \mathbb{A}$ and $i \in \mathcal{A}_{out}$ with $\varphi^{-1}(i) \neq \emptyset$, $\eta(i) \in \mathcal{A}_{in}$.*

Proposition 4. *For each $\mathcal{A} \in \mathbb{A}$ and $t \in \{1, \ldots, \tau\}$, $\left| \mathcal{A}_{out} \cap \mathcal{S}^t \right| = \left| \mathcal{A}_{in} \cap \mathcal{O}^t \right| \leq 2\tau$.*

Proposition 5. *Each facility $i \in \mathcal{S} \cup \mathcal{O}$ appears exactly one time in the swap pairs from $\mathbb{A}$.*

3.2   An Upper Bound on the Cost Increase

In this section, we present a strategy for reconnecting clients after performing the swap pairs from $\mathbb{A}$ on the local optimum. This gives an upper bound on the increased cost induced by a swap pair. Consider a swap pair $\mathcal{A} \in \mathbb{A}$, we close the facilities in $\mathcal{A}_{out}$ and open the facilities in $\mathcal{A}_{in}$. We reconnect the clients from $\mathcal{N}(\mathcal{A}_{out}) \cup \mathcal{N}^*(\mathcal{A}_{in})$. Each $j \in \mathcal{N}^*(\mathcal{A}_{in})$ is reconnected to o_j (o_j is guaranteed to be opened by the definition of $\mathcal{N}^*(*)$). We pay the penalty costs of the clients from $\mathcal{N}(\mathcal{A}_{out}) \cap \mathcal{P}^*$. The clients from $\mathcal{N}(\mathcal{A}_{out}) \backslash \mathcal{P}^*$ should be reconnected to a nearby opened facility. By Proposition 3, $\eta(i)$ is opened for each $i \in \mathcal{A}_{out}$. This motivates the following strategy for reconnecting each $j \in \mathcal{N}(\mathcal{A}_{out}) \backslash \mathcal{P}^*$. See Fig. 3 for an example of the strategy.

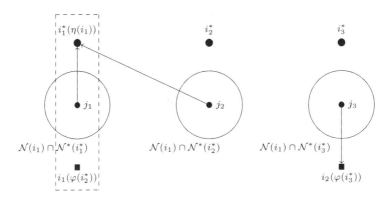

Fig. 3. Reconnection of the clients after performing the swap pair $\mathcal{A} = (\{i_1\} \mid \{i_1^*\})$. The solid lines indicate the connection of the clients after the swap. For each $j \in \mathcal{N}(i_1) \backslash \mathcal{P}^*$, j is reconnected to o_j, if $j \in \mathcal{N}(i_1) \cap \mathcal{N}^*(i_1^*)$; $\varphi(o_j)$, if $j \in \mathcal{N}(i_1) \cap \mathcal{N}^*(i_3^*)$; and $\eta(\varphi(o_j))$, if $j \in \mathcal{N}(i_1) \cap \mathcal{N}^*(i_2^*)$.

- If $o_j \in \mathcal{A}_{in}$, then j is reconnected to o_j.
- If $o_j \notin \mathcal{A}_{in}$ and $\varphi(o_j) \notin \mathcal{A}_{out}$, then j is reconnected to $\varphi(o_j)$.
- If $o_j \notin \mathcal{A}_{in}$ and $\varphi(o_j) \in \mathcal{A}_{out}$, then j is reconnected to $\eta(\varphi(o_j))$.

The following lemma shows that the reconnection cost of each client $j \in \mathcal{N}(\mathcal{A}_{out}) \backslash \mathcal{P}^*$ can be bounded by a combination of O_j and S_j.

Lemma 1. *For each* $j \in \mathcal{C} \backslash (\mathcal{P}^* \cup \mathcal{P})$, *we have* $d(j, \varphi(o_j)) \leq 2O_j + S_j$ *and* $d(j, \eta(\varphi(o_j))) \leq 3O_j + 2S_j$.

The following lemma follows from an upper bound on the increased cost induced by performing a swap pair from $\mathbb{A}$.

Lemma 2. *For each swap pair* $\mathcal{A} \in \mathbb{A}$, *we have*

$$
0 \leq \sum_{j \in \mathcal{N}^*(\mathcal{A}_{in}) \cap \mathcal{P}} (O_j - p(j)) + \sum_{j \in \mathcal{N}^*(\mathcal{A}_{in}) \backslash \mathcal{P}} (O_j - S_j)
$$
$$
+ \sum_{j \in \mathcal{N}(\mathcal{A}_{out}) \cap \mathcal{P}^*} (p(j) - S_j) + \sum_{j \in [\mathcal{N}(\mathcal{A}_{out}) \backslash \mathcal{P}^*] \backslash \mathcal{N}^*(\varphi^{-1}(\mathcal{A}_{out}) \cup \mathcal{A}_{in})} 2O_j
$$
$$
+ \sum_{j \in \mathcal{N}^*(\varphi^{-1}(\mathcal{A}_{out}) \backslash \mathcal{A}_{in}) \cap \mathcal{N}(\mathcal{A}_{out})} (3O_j + S_j). \tag{1}
$$

3.3   A Layered Structure of the Swap Pairs

Observe that inequality (1) contains "$+O_j$" terms for some clients from $\mathcal{C} \backslash \mathcal{P}^*$, "$-S_j$" terms for some clients from $\mathcal{C} \backslash \mathcal{P}$, "$+p(j)$" terms for some clients from $\mathcal{P}^*$, and "$-p(j)$" terms for some clients from $\mathcal{P}$. Our idea for obtaining the approximation guarantee for the local optimum is to add together some inequalities of this type such that we can get $O(1)(\sum_{j \in \mathcal{C} \backslash \mathcal{P}^*} O_j + \sum_{j \in \mathcal{P}^*} p(j)) - O(1)(\sum_{j \in \mathcal{C} \backslash \mathcal{P}} S_j + \sum_{j \in \mathcal{P}} p(j)) \geq 0$, which directly implies the desired approximation ratio. The challenge is that inequality (1) also involves a "$+S_j$" term for a client from $\mathcal{N}^*(\varphi^{-1}(\mathcal{A}_{out}) \backslash \mathcal{A}_{in}) \cap \mathcal{N}(\mathcal{A}_{out})$. In this case, we have to repeatedly use another inequality which contains a "$-S_j$" term to counteract the "$+S_j$" term. To obtain a "$-S_j$" term for each $j \in \mathcal{C} \backslash \mathcal{P}$, we prove the existence of a layered structure of the swap pairs from $\mathbb{A}$. Note that this structure is only used in the analysis.

Lemma 3. $\mathbb{A}$ *can be partitioned into* f *disjoint sets* $\mathbb{A}_1, \ldots, \mathbb{A}_f$ *satisfying the following properties.*

- $3 \leq f \leq \tau + 1$.
- $\bigcup_{\mathcal{A} \in \mathbb{A}_t} \varphi^{-1}(\mathcal{A}_{out}) \backslash \mathcal{A}_{in} \subseteq \bigcup_{\mathcal{A} \in \mathbb{A}_t^-} \mathcal{A}_{in}$ *for each* $t \in \{1, \ldots, f-1\}$, *where* $\mathbb{A}_t^- = \bigcup_{t < t' \leq f} \mathbb{A}_{t'}$.
- $\bigcup_{\mathcal{A} \in \mathbb{A}_f} \varphi^{-1}(\mathcal{A}_{out}) = \emptyset$.

Before proving Lemma 3, we first show its implication. Given a swap pair $\mathcal{A} \in \mathbb{A}$, inequality (1) involves a "$+S_j$" term for each $j \in \mathcal{N}^*(\varphi^{-1}(\mathcal{A}_{out}) \backslash \mathcal{A}_{in}) \cap \mathcal{N}(\mathcal{A}_{out})$ and a "$-S_j$" term for each $j \in \mathcal{N}^*(\mathcal{A}_{in})$. Using Lemma 3, we know that $\bigcup_{\mathcal{A} \in \mathbb{A}_1} \varphi^{-1}(\mathcal{A}_{out}) \backslash \mathcal{A}_{in} \subseteq \bigcup_{\mathcal{A} \in \mathbb{A}_1^-} \mathcal{A}_{in}$ and $\bigcup_{\mathcal{A} \in \mathbb{A}_2} \varphi^{-1}(\mathcal{A}_{out}) \backslash \mathcal{A}_{in} \subseteq \bigcup_{\mathcal{A} \in \mathbb{A}_2^-} \mathcal{A}_{in}$. Thus, we can multiply inequality (1) by factor 2 for each $\mathcal{A} \in \mathbb{A}_1^-$ to counteract the "$+S_j$" terms induced by the swap pairs from $\mathbb{A}_1$. Now each swap pair from $\mathbb{A}_2$ induces some "$+2S_j$" terms, which can be counteracted by multiplying inequality (1) by factor 3 instead of 2 for each $\mathcal{A} \in \mathbb{A}_2^-$. By a similar argument, we can counteract all the "$+S_j$" terms through using the swap pairs from $\mathbb{A}_t$ for t times, for each $t \in \{1, \ldots, f\}$. We will later show that this yields an $O(f)$-approximation guarantee for the prize-collecting τ-color median problem.

Proof [of **Lemma** 3]. Let $\mathbb{A}'$ denote the set of the swap pairs added to $\mathbb{A}$ in the fifth loop of Algorithm 2 (steps 10 and 11). It can be seen that each $\mathcal{A} \in \mathbb{A}'$ is a combination of an almost-feasible pair and two facilities from $\mathcal{S}_2 \cup \mathcal{O}_2$. For each $\mathcal{A} \in \mathbb{A}'$, define $T(\mathcal{A}) = T(\mathcal{B}'_{out})$, where $\mathcal{B}'$ denotes the almost-feasible pair involved in $\mathcal{A}$. For each $t \in \{1, \ldots, \tau\}$, define $\mathbb{A}'_t = \{\mathcal{A} \in \mathbb{A}' : T(\mathcal{A}) = t\}$. We construct a directed graph G as follows: A vertex v_t is constructed for each $t \in \{1, \ldots, \tau\}$ with $\mathbb{A}'_t \neq \emptyset$; For any two vertices v_{t^1}, v_{t^2} of G, there is a directed edge (simply called arc) from v_{t^1} to v_{t^2} if there exist a swap pair $\mathcal{A} \in \mathbb{A}'_{t^1}$ and a facility $i \in \varphi^{-1}(\mathcal{A}_{out}) \backslash \mathcal{A}_{in}$ such that $T(i) = t^2$. We have the following claim.

Claim 1. *G is a directed acyclic graph, whose vertices are no more than $\tau - 1$.*

Define $\mathcal{V}$ as the vertex set of G. Given two vertices $v, v' \in \mathcal{V}$, if there exists a path from v to v' in G, then let $L(v, v')$ denote the number of the vertices in a longest path from v to v'. Otherwise, let $L(v, v') = 2$. Let $\mathcal{V}_0$ denote the set of the vertices in G whose in-degrees are 0. For each $v \in \mathcal{V}_0$, define $L(v) = 2$. For each $v \in \mathcal{V} \backslash \mathcal{V}_0$, define $L(v) = \max_{v' \in \mathcal{V}_0} L(v', v) + 1$. We have $2 \leq L(v) \leq |\mathcal{V}| + 1$ for each $v \in \mathcal{V}$. Let $f = \max_{v \in \mathcal{V}} L(v) + 1$. We have $3 \leq f \leq |\mathcal{V}| + 2 \leq \tau + 1$ by Claim 1. We partition $\mathbb{A}$ into f disjoint sets $\mathbb{A}_1, \ldots, \mathbb{A}_f$ as follows.

- Let $\mathbb{A}_1$ be the set of the swap pairs added to $\mathbb{A}$ in the first and fourth loops of Algorithm 2 (steps 2, 3, 8, and 9).
- For each integer $1 < g < f$, let $\mathbb{A}_g = \bigcup_{L(v_t)=g} \mathbb{A}'_t$.
- Let $\mathbb{A}_f$ be the set of the swap pairs added to $\mathbb{A}$ in the last loop of Algorithm 2 (steps 12 and 13).

Recall that $\mathbb{A}_t^- = \bigcup_{t < t' \leq f} \mathbb{A}_{t'}$ for each $t \in \{1, \ldots, f - 1\}$. Based on Claim 1 and the properties of the swap pairs from $\mathbb{A}$, we have the following result.

Claim 2. *For each $t \in \{1, \ldots, f - 1\}$, $\bigcup_{\mathcal{A} \in \mathbb{A}_t} \varphi^{-1}(\mathcal{A}_{out}) \backslash \mathcal{A}_{in} \subseteq \bigcup_{\mathcal{A} \in \mathbb{A}_t^-} \mathcal{A}_{in}$.*

For each swap pair $\mathcal{A}$ added to $\mathbb{A}$ in the last loop of Algorithm 2, we have $\mathcal{A}_{out} \subseteq \mathcal{S}_2$. The definition of $\mathcal{S}_2$ implies that $\varphi^{-1}(\mathcal{S}_2) = \emptyset$, which in turn implies that $\varphi^{-1}(\mathcal{A}_{out}) = \emptyset$ for each $\mathcal{A} \in \mathbb{A}_f$. By the fact that $3 \leq f \leq \tau + 1$ and Claim 2, we complete the proof of Lemma 3. □

3.4   Bound the Cost of the Local Optimum

We are now ready to bound the cost of the local optimum. Adding together several inequalities (1), we obtain the following result.

Lemma 4. $\sum_{j \in \mathcal{C} \setminus \mathcal{P}} S_j + \sum_{j \in \mathcal{P} \setminus \mathcal{P}^*} p(j) \leq (4\tau + 1) \sum_{j \in \mathcal{C} \setminus \mathcal{P}^*} O_j + (\tau + 1) \sum_{j \in \mathcal{P}^* \setminus \mathcal{P}} p(j)$.

Adding $\sum_{j \in \mathcal{P}^* \cap \mathcal{P}} p(j)$ to both sides of the inequality in Lemma 4 and simplifying, we have

$$\sum_{j \in \mathcal{C} \setminus \mathcal{P}} S_j + \sum_{j \in \mathcal{P}} p(j) \leq (4\tau + 1) \sum_{j \in \mathcal{C} \setminus \mathcal{P}^*} O_j + (\tau + 1) \sum_{j \in \mathcal{P}^*} p(j) \leq (4\tau + 1)OPT,$$

which implies that the local optimum is a $(4\tau + 1)$-approximation solution to the prize-collecting τ-color median problem.

References

1. Arya, V., Garg, N., Khandekar, R., Meyerson, A., Munagala, K., Pandit, V.: Local search heuristics for k-median and facility location problems. SIAM J. Comput. **33**(3), 544–562 (2004)
2. Bateni, M., Hajiaghayi, M.: Assignment problem in content distribution networks: unsplittable hard-capacitated facility location. ACM Trans. Algorithms **8**(3), 20:1–20:19 (2012)
3. Charikar, M., Khuller, S., Mount, D.M., Narasimhan, G.: Algorithms for facility location problems with outliers. In: Proceedings of the 12th ACM-SIAM Symposium on Discrete Algorithms, pp. 642–651 (2001)
4. Charikar, M., Li, S.: A dependent LP-rounding approach for the k-median problem. In: Czumaj, A., Mehlhorn, K., Pitts, A., Wattenhofer, R. (eds.) ICALP 2012. LNCS, vol. 7391, pp. 194–205. Springer, Heidelberg (2012). https://doi.org/10.1007/978-3-642-31594-7_17
5. Cohen-Addad, V., Feldmann, A.E., Saulpic, D.: Near-linear time approximation schemes for clustering in doubling metrics. In: Proceedings of the 60th IEEE Symposium on Foundations of Computer Science, pp. 540–559 (2019)
6. Feng, Q., Zhang, Z., Huang, Z., Xu, J., Wang, J.: Improved algorithms for clustering with outliers. In: Proceedings of the 30th International Symposium on Algorithms and Computation, pp. 61:1–61:12 (2019)
7. Feng, Q., Zhang, Z., Shi, F., Wang, J.: An improved approximation algorithm for the k-means problem with penalties. In: Chen, Y., Deng, X., Lu, M. (eds.) FAW 2019. LNCS, vol. 11458, pp. 170–181. Springer, Cham (2019). https://doi.org/10.1007/978-3-030-18126-0_15
8. Friggstad, Z., Zhang, Y.: Tight analysis of a multiple-swap heurstic for budgeted red-blue median. In: Proceedings of the 43rd International Colloquium on Automata, Languages, and Programming, pp. 75:1–75:13 (2016)
9. Hajiaghayi, M.T., Khandekar, R., Kortsarz, G.: Budgeted red-blue median and its generalizations. In: de Berg, M., Meyer, U. (eds.) ESA 2010. LNCS, vol. 6346, pp. 314–325. Springer, Heidelberg (2010). https://doi.org/10.1007/978-3-642-15775-2_27
10. Hajiaghayi, M., Khandekar, R., Kortsarz, G.: Local search algorithms for the red-blue median problem. Algorithmica **63**(4), 795–814 (2012). https://doi.org/10.1007/s00453-011-9547-9
11. Jain, K., Mahdian, M., Markakis, E., Saberi, A., Vazirani, V.V.: Greedy facility location algorithms analyzed using dual fitting with factor-revealing LP. J. ACM **50**(6), 795–824 (2003)

12. Krishnaswamy, R., Kumar, A., Nagarajan, V., Sabharwal, Y., Saha, B.: The matroid median problem. In: Proceedings of 22nd Annual ACM-SIAM Symposium on Discrete Algorithms, pp. 1117–1130 (2011)
13. Krishnaswamy, R., Li, S., Sandeep, S.: Constant approximation for k-median and k-means with outliers via iterative rounding. In: Proceedings of the 50th ACM Symposium on Theory of Computing, pp. 646–659 (2018)
14. Swamy, C.: Improved approximation algorithms for matroid and knapsack median problems and applications. ACM Trans. Algorithms **12**(4), 49:1–49:22 (2016)
15. Zhang, D., Hao, C., Wu, C., Xu, D., Zhang, Z.: Local search approximation algorithms for the k-means problem with penalties. J. Comb. Optim. **37**(2), 439–453 (2019)

LP-Based Algorithms for Computing Maximum Vertex-Disjoint Paths with Different Colors

Yunyun Deng[1], Yi Chen[2], Kewen Liao[3], and Longkun Guo[4(✉)]

[1] Officers College of Chinese People's Armed Police Force, Chengdu 610213,
People's Republic of China
[2] College of Mathematics and Computer Science, Fuzhou University, Fuzhou 350116,
People's Republic of China
[3] Peter Faber Business School, Australian Catholic University, Sydney, Australia
kewen.liao@acu.edu.au
[4] School of Computer Science and Technology, Qilu University of Technology
(Shandong Academy of Sciences), Jinan 250353, People's Republic of China
longkun.guo@gmail.com

Abstract. Booming applications in wireless networks have imposed a great growth in data transmission together with stricter requirements of bandwidth and load balancing. In order to capture and meet the requirements, we consider a new problem of computing maximum disjoint paths with different colors (*MDPDC*) in networks. In *MDPDC*, transmission frequencies are modeled as different colors and the aim is to find a maximum number of constrained node-disjoint paths where nodes in any disjoint path share the same color, while colors are different among paths. Observing the $\mathcal{NP}$-completeness of *MDPDC*, the paper proposes two linear programming based algorithms as generic solutions.

Keywords: Disjoint paths with different frequencies · Wireless network · $\mathcal{NP}$-complete · Linear programming

1 Introduction

In the past decades, wireless communication technology has brought tremendous changes to the Internet, telecom and data networks while dramatically boosting the development of passive and ubiquitous computing. Along also comes with people's rich communication contents which, in addition to text data, contain photos, videos, large multimedia files *etc*. As a result, data traffic has explosively increased, bringing in unprecedented challenges in network congestion and packet loss and consequently unsatisfactory user experience provided limited network resources.

Multi-path routing, which leverages multiple paths in a network for data transmission, is generally considered as a preferred routing method. It usually comes with a much better data transmission quality than the widely deployed

© Springer Nature Switzerland AG 2020
J. Chen et al. (Eds.): TAMC 2020, LNCS 12337, pp. 107–118, 2020.
https://doi.org/10.1007/978-3-030-59267-7_10

single shortest path routing, e.g. *the 802.1D* protocols [1]. In particular, disjoint paths routing is designated to provide better bandwidth and load balancing. However, in the context of wireless networks, two nodes emitting data at the same frequency (say *5180* MHz in *5G* channels) can interfere with each other and result significant packet losses at an end node receiving signals from both nodes simultaneously. Therefore, in order to greatly reduce the interference, disjoint paths can be further restricted to transfer data end-to-end at different frequencies. For this purpose, we model each frequency as a color, and introduce the problem of computing maximum disjoint paths with different colors (*MDPDC*) in the following:

Definition 1. *(Maximum disjoint paths with different colors, MDPDC) Given a universal set of colors R (representing different frequency bands), a network graph $G = (V, E)$ and a pair of distinct vertices $s, t \in V$, where each $v_i \in V \setminus \{s, t\}$ is assigned with a set of colors $R_i \subseteq R$. The MDPDC problem is to maximize the number of vertex-disjoint st-paths $P_1, P_2, \ldots, P_k$, such that for each path P_j, there exists a color $col(P_j)$ which satisfies two conditions: (1) Each $v_i \in P_j$ can be colored with $col(P_j)$, i.e. $col(P_j) \in R_i$ holds for every vertex $v_i \in P_j$; (2) For any $j \neq j'$, P_j and $P_{j'}$ must have different colors, i.e. $col(P_j) \neq col(P_{j'})$ must hold.*

In the paper, we develop LP-based algorithms for finding near-optimal solutions for MDPDC, since optimally solving the problem requires exponential time unless $P = NP$ according to the theorem as below (The formal proof will be given in later sections):

Theorem 2. *The MDPDC problem is $\mathcal{NP}$-complete in either directed or undirected graphs.*

1.1   Related Work

To the best of our knowledge, the problem of finding k-disjoint paths with different colors (*k-DPDC*) was introduced by Zhang *et al.* [20], who addressed the restricted *k-DPDC* problem subjected to MinSum and MinMax objectives in a directed acyclic graph (*DAG*), and respectively presented two fully polynomial-time approximation scheme (*FPTAS*). Previous to Zhang *et al.*, Wu [18] has studied another color-disjoint variant of *MDPDC*, namely the *Maximum Colored Disjoint Path* (*Max CDP*) problem. Different with *MDPDC*, *Max CDP* allows two or more disjoint paths sharing one identical color, while it has the similar aim of maximizing the number of vertex-disjoint paths between two specified vertices, where each path consists of edges of the same colors. Wu [18] has developed both approximation algorithm with a factor c and exact algorithm for *Max CDP* in the paper, where c is the number of colors. In line with [2], Bonizzoni *et al.* shown *Max CDP* is not approximable within factor $c^{1-\epsilon}$ for any constant $\epsilon > 0$, where c is number of colors. In [7], Dondi *et al.* introduced *Max CDDP*, a new variant of *Max CDP* that aims to find a maximum number of uni-color paths that are both vertex-disjoint and color-disjoint in an edge-colored graph.

They show the problem can be solved within a factor $\frac{1}{2}$ by a parameterized approximation algorithm, where the parameter is the size of the vertex cover.

Besides the studies of k-$DPDC$, the classical disjoint paths (DP) problem has a rich research history. The DP problem is known as $\mathcal{NP}$-complete in directed graphs for both vertex-disjoint and edge-disjoint versions, even for the task of computing only two paths [8]. In contrast, DP was shown polynomial in undirected graphs when $k = 2$. Shiloach [16] proposed a polynomial-time algorithm for the two vertex-disjoint paths problem in undirected graphs with a runtime $O(mn)$. The result was generalized that the problem is shown can be solved in polynomial time for any k in arbitrary undirected graphs [14], while the runtime was later improved to $O(n + m\alpha(m,n))$ by Tholey [17], where α is the inverse of the Ackermann function. For general k, DP was shown polynomial solvable in some special graphs. Frank proposed an algorithm for finding k edge-disjoint paths in a planar undirected graph with a runtime $O(|n|^3 \log|n|)$ [9]. Later, Schrijver [15] showed the k disjoint paths problem in directed planar graphs is solvable in polynomial time $|n| + \sigma + k$ based on cohomology over free groups method. More recently, because of its advantage in reliability as well as congestion avoidance, some variants of DP has recently witnessed renewed interests from researchers in the networking community. In particular, Challal et $al.$ proposed a highly reliable intrusion fault-tolerant routing scheme, through a secure multi-path routing structure [3]. Moreover, disjoint paths were also used for minimizing energy consumption in networks due to Lin et $al.$, where they designed a fast heuristic solution called Two Disjoint Paths by Shortest Path (2DP-SP)[13]. Hou et $al.$ employed its vertex disjoint version, namely BS-2NDP, to formulate the generic Bandwidth Scheduling problem to support big data transfer [11]. Besides, a similar but more complicated problem, named the δ V-$kPESP$ $problem$, aiming to compute k edge-disjoint paths sharing at most δ common vertices, was first studied in [19] for $k = 2$ and shown solvable within a time complexity $O\left(mn^2 + n^3 \cdot \log n\right)$. Later, the runtime was improved to $O\left(\delta m + n \log n\right)$[10]. Moreover, the edge version of δ V-$kPESP$ of computing k partial edge-disjoint paths sharing at most δ common edges, was shown solvable within a runtime $O\left(mn \log_{(1+m/n)} n + \delta n^2\right)$ for $k = 2$ by Deng et $al.$ [6].

When the input graph contains exactly one color, $MDPDC$ can be considered as a network flow problem. Flow and disjoint paths problems, along with several variants and special cases, have been extensively studied. While allowing fractional paths within the flow and with each connection requirement to be completely satisfied or none, the all-or-nothing flow problem was introduced in [4] together with approximation algorithms. For the node-capacitated all-or-nothing flow problem, Chekuri et $al.$ obtain a ratio of $O(\log^4 k \log n)$ for general graph and $O(\log^2 k \log n)$ for planar graph [5]. For the edge capacitated all-or-nothing flow problem, Chekuri et $al.$ obtain an $O(\log^2 k)$ approximation in general graphs and an $O(\log k)$ approximation in planar graphs [5]. These respectively improve the earlier ratios of $O(\log^3 n \log \log n)$ and $O(\log^2 n \log \log n)$ obtained in [4]. Later, Kawarabayashi and Kobyashi gave an $O(1)$-approximation algorithm and showed that the integrality gap is $O(1)$ for the all-or-nothing multicommodity flow problem in planar graphs in [12].

1.2 Our Results

The main results of this paper can be summarized as follows:

- We introduce a novel *MDPDC* problem/model for the big data challenges in wireless networks such as increased network routing congestion and packet loss due to signal interference.
- We prove the $\mathcal{NP}$-Completeness of *MDPDC* in directed graphs via a simple reduction from the directed *Disjoint Path* (*DP*) problem. For the harder case of undirected graphs, as *DP* becomes polynomial-time solvable, we instead construct a more complicated but essential reduction from the fundamental 3*SAT* problem that is well known to be NP-complete.
- After settling the computational complexity, linear program (*LP*) models are proposed for *MDPDC*. Based on the *LP*s, two algorithms are invented to solve *MDPDC* via rounding fractional *LP* solutions.

In addition, numerical experiments manifested that the *LP*-based algorithm outputs integral solutions at a relatively high probability.

1.3 Organization

The reminder of the paper is organized as follows: Sect. 2 gives $\mathcal{NP}$-completeness proofs for MDPDC in both directed and undirected graphs; Sect. 3 presents *LP*-based algorithms for solving *MDPDC*; Sect. 4 concludes the paper.

2 Proof of Theorem 2

In this section, we shall show the correctness of Theorem 2 via proving the $\mathcal{NP}$-completeness of the two different color disjoint paths *(2DPDC)* problem, which is a simpler case of the maximum different color disjoint paths (*MDPDC*) problem and determines only whether there exist at least two different color disjoint paths. Apparently, the $\mathcal{NP}$-hardness of *MDPDC* can immediately be reduced from the $\mathcal{NP}$-completeness of *2DPDC*.

We first prove the $\mathcal{NP}$-completeness of *2DPDC* in *digraphs* by simply reducing from the disjoint paths (*DP*) problem that is known $\mathcal{NP}$-complete. Notably, the proof is only valid for *digraphs* and can not be extended to undirected graph, because *DP* is polynomially solvable in undirected graph [8]. So instead we propose another reduction from *3SAT* with more complicated details to show *MDPDC* remains $\mathcal{NP}$-complete in undirected graph.

2.1   The $\mathcal{NP}$-completeness Proof for $2DPDC$ in Digraphs

Apparently, $2DPDC$ is in $\mathcal{NP}$, since for a given disjoint paths pair (P_1,P_2), we can check whether $col(P_1) \neq col(P_2)$ holds or not in polynomial time. $2DPDC$ is apparently $\mathcal{NP}$-complete in a directed graph by constructing a reduction from the *Disjoint Path* (DP) Problem which is known $\mathcal{NP}$-complete in digraph [8]. Assume that we are given an instance of the DP problem, which is, a directed graph $G(V,E)$ and four distinct vertices s_1, t_1, s_2 and t_2. In the polynomial-time reduction, we create a directed graph $G_1(V_1,E_1)$, where $V_1 = V \cup \{s,t\}$ in that s and t respectively denote the source and destination vertices of the $2DPDC$ problem, while $E_1 = E \cup \{(s,s_1),(s,s_2),(t_1,t),(t_2,t)\}$. An example of constructing auxiliary graph $G_1(V_1,E_1)$ is as illustrated in Fig. 1a. It remains to show the following lemma to complete the $\mathcal{NP}$-hardness proof.

Lemma 3. *There exists a solution for 2DPDC iff there exists a solution for the corresponding DP instance.*

Proof. Suppose $2DPDC$ in G_1 is solvable, *i.e.*, there is a pair of disjoint paths $\{P_1,P_2\}$ from s to t with different colors in G_1. Note that following the construction vertices s_1 and t_1 are both in red, while vertices s_2 and t_2 are blue. Then $P_1 = s \rightarrow s_1 \rightarrow \ldots \rightarrow t_1 \rightarrow t$ and $P_2 = s \rightarrow s_2 \rightarrow \ldots \rightarrow t_2 \rightarrow t$ are two disjoint paths with different colors, in which $P_1' = s_1 \rightarrow \ldots \rightarrow t_1$ and $P_2' = s_2 \rightarrow \ldots \rightarrow t_2$ immediately compose a feasible solution to DP. Conversely, assume $\{P_1'(s_1,t_1), P_2'(s_2,t_2)\}$ is a solution to DP. Then $P_1 = s \rightarrow s_1 \rightarrow \ldots \rightarrow t_1 \rightarrow t$ and $P_2 = s \rightarrow s_2 \rightarrow \ldots \rightarrow t_2 \rightarrow t$ immediately compose a solution to $2DPDC$, as every vertex on P_1 can be colored as red while P_2 blue. This completes the proof. □

However, we cannot extend the above $\mathcal{NP}$-completeness proof to undirected graphs, because interestingly DP is polynomial-solvable in an undirected graph. In the following, $MDPDC$ is shown to remain $\mathcal{NP}$-complete in undirected graphs, which is different to the case of DP.

2.2   The $\mathcal{NP}$-completeness Proof for $2DPDC$ in Undirected Graphs

Lemma 4. *2DPDC is $\mathcal{NP}$-complete in an undirected graph.*

Similar to the directed case, emph2DPDC is obviously in $\mathcal{NP}$. Then, we will prove the remaining part of Lemma 4 by reducing from $3SAT$ that is known to be NP-complete. An instance of $3SAT$ has n variables $\{x_1, \ldots, x_n\}$, with each variable x_i giving rise to *literals* x_i, $\overline{x}_i$. It also has m clauses $\{C_1, \ldots, C_m\}$, each a 3-subset of literals. A true assignment is a function $\tau : \{x_i\} \rightarrow \{\text{true, false}\}$. We say that C_j is satisfied under τ if C_j contains a literal x_i with $\tau(x_i) =$true, or a literal $\overline{x}_i$ with $\tau(x_i) =$false. The $3SAT$ problem is to determine whether there is an assignment satisfying all the m clauses.

For any given instance of $3SAT$, the key idea of our reduction is to construct an auxiliary graph G, such that G contains two different disjoint paths P_1 and

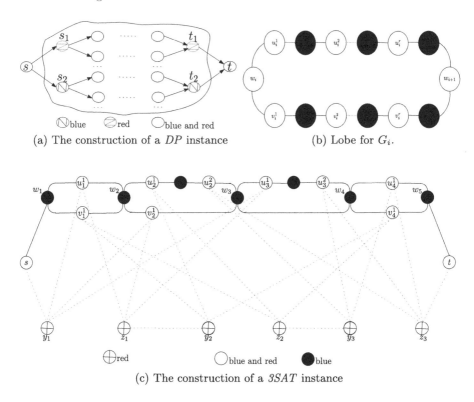

(a) The construction of a DP instance   (b) Lobe for G_i.

(c) The construction of a $3SAT$ instance

Fig. 1. The auxiliary graph

P_2 with $col(P_1) \neq col(P_2)$ iff the instance of $3SAT$ is satisfiable. The construction is composed with the following three parts. First, for each variable x_i with p_i occurrences of x_i and q_i occurrences of $\overline{x}_i$ in the clauses, we add a vertex w_i with color blue and construct a lobe G_i in Fig. 1b which contains two paths from w_i to w_{i+1}, say $P_1(w_i, w_{i+1})$ and $P_2(w_i, w_{i+1})$ respectively. Formally, we set:

$$P_1(w_i, w_{i+1}) = w_i \rightarrow u_i^1 \rightarrow b \rightarrow \cdots \rightarrow u_i^j \rightarrow b \rightarrow \cdots \rightarrow u_i^{p_i} \rightarrow w_{i+1},$$

$$P_2(w_i, w_{i+1}) = w_i \rightarrow v_i^1 \rightarrow b \rightarrow \cdots \rightarrow v_i^j \rightarrow b \rightarrow \cdots \rightarrow v_i^{q_i} \rightarrow w_{i+1},$$

where b is a vertex of color blue and p_i and q_i are respectively the numbers of occurrences of x_i and $\overline{x}_i$.

Then, for each clause C_j, add two vertices y_i and z_i, as well as edge (z_i, y_{i+1}), $1 \leq i \leq m-1$, with the color of vertices are red. In addition, we add s and t together with four edges (s, y_1), (s, w_1), (z_m, t), (w_{n+1}, t).

Last but not the least, for the relationship between the variables and the clauses, say variable x_i occurs in clause C_j as its pth occurrence, if the occurrence is $\overline{x}_i$, we add two edges (y_j, u_j^p), (u_j^p, z_j); Otherwise, C_j contains $\overline{x}_i$, then we add two edges (y_j, v_j^p), (v_j^p, z_j).

For example, graph G constructed for the instance $x_1 \vee \overline{x}_2 \vee x_3$, $\overline{x}_1 \vee x_2 \vee \overline{x}_4$, $x_2 \vee x_3 \vee x_4$ is as illustrated in Fig. 1c.

Then since $2DPDC$ is clearly in $\mathcal{NP}$, the correctness of Lemma 4 can be immediately obtained from the following lemma:

Lemma 5. *An instance of 3SAT is satisfiable iff in its corresponding auxiliary graph G there exist two different disjoint paths P_1 and P_2 but $col(P_1) \neq col(P_2)$.*

Proof. The proof is omitted due to the length constraint. □

Following Lemma 3 and 4, $2DPDC$ is $\mathcal{NP}$-complete in both directed and undirected graphs. Therefore, we have the $\mathcal{NP}$-hardness of $MDPDC$ and then the correctness of Theorem 2.

3 *LP*-Based Algorithms for *MDPDC*

In this section, we will first transform the vertex colored $MDPDC$ problem to its equivalent edge colored version, then give a linear program (LP) relaxation for computing maximum disjoint paths with different colors, and at last propose two algorithms by employing flow decomposition.

3.1 The Transformation

For an vertex-colored graph $G(V, E)$, we convert it into an edge-colored graph by assigned the edges in the graph with colors. For an edge $e_i = (u, v) \in E$, we set $R_{e_i} = R(u) \cap R(v)$ as the set of colors for the edge e_i. An example of the transformation is as shown in Fig. 2, where Fig. 2a is an original vertex-colored graph and Fig. 2b is the corresponding edge-colored graph, where r, g, b are the colors of vertices and edges while $\emptyset$ represents that no color is assigned to the edge.

3.2 The *LP* Formula

We denote by y_r the total value of flow accommodated by the paths of color $r \in R$, and denoted by x_e^r the flow of color r accommodated by the edge e. Note that, $x_e^r = 1$ means edge e is used completely for a flow of color r, while $x_e^r = 0$ indicates otherwise. Moreover, Inequality (1) ensures that each edge is at most in one color and Inequality (2) guarantees each *color-edge* can only have one occurrence. Besides, Inequality (4) guarantees that each vertex (except s and t) can be used once at most. Then the integer linear programming (ILP) formula for $MDPDC$ is as below ($ILP(1)$):

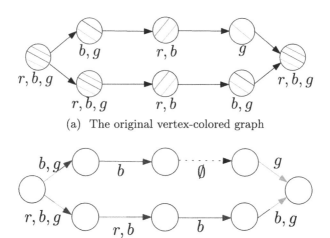

(a) The original vertex-colored graph

(b) The edge-colored graph

Fig. 2. An example for the transformation

$$\max \quad \sum_{r \in R} y_r$$

$$s.t. \sum_{e \in \delta^+(v)} x_e^r - \sum_{e \in \delta^-(v)} x_e^r = 0 \; \forall r \in R, \; v \in V \backslash \{s, t\}$$

$$\sum_{e \in \delta^+(s)} x_e^r \leq 1 \qquad \forall r \in R \qquad (1)$$

$$\sum_{r \in R} x_e^r \leq 1 \qquad \forall e \in E \qquad (2)$$

$$\sum_{e \in \delta^+(s)} x_e^r - y_r \geq 0 \qquad \forall r \in R \qquad (3)$$

$$\sum_{e \in \delta^+(v)} \sum_{r \in R} x_e^r \leq 1 \qquad \forall v \in V \setminus \{s, t\} \qquad (4)$$

$$x_e^r, y_r \in \{0,1\} \qquad \forall r \in R, \forall e \in E$$

where $\delta^+(v)$ and $\delta^-(v)$ denote the two sets of edges leaving and entering v in G, respectively.

Lemma 6. *ILP(1) correctly models the MDPDC problem.*

Proof. Given a solution for *ILP*, we can construct a corresponding solution to *MDPDC* and *vice versa*. Suppose there exists a solution for *ILP(1)*, say $Y = \{y_1, \ldots, y_r, \ldots, y_{|R|}\}$ with $\sum_{i=1}^{|R|} y_i = g$. First of all, Y apparently indicates an *st*-flow of value g, say f, since s and t are respectively with a degree g and $-g$ while each other vertex of $E(f) \backslash \{s, t\}$ is with degree 0 or 2. Then because Inequality (1) guarantees that a feasible solution of *ILP(1)* contains at most one edge leaving s for each color. That is, each edge leaving s has different color with any other edges. Moreover, each edge in f can only be assigned with one color (*i.e.*, Inequality 2). That is, there is no shared edge in the solution. Besides, Inequality (4) indicates that each vertex except s and t in f can be selected for only once. In other word, Inequality (4) ensures vertex-disjoint. Thus, f is an *st*-flow of value

---

Algorithm 1. A heuristic algorithm for *MDPDC*

Input: A graph G, specified vertices s and t, a color set R;
Output: A solution to *MDPDC*.
0: Set $\mathcal{P} := \emptyset$;
1: Solve *LP(1)* and obtain $\boldsymbol{x}$ be the solution to *LP(1)*;
2: **If** $\boldsymbol{x}$ is integral (i.e. each dimension of $\boldsymbol{x}$ is an integer) **then**
    return $\mathcal{P} \cup \boldsymbol{x}$;
3: **If** $\boldsymbol{x}$ is not integral **then**
4:    Decompose $\boldsymbol{x}$ to a set of flow-paths Q by Algorithm 2;
5:    Select a path P_j with maximal flow value from Q;
6:    Set $\mathcal{P} := \mathcal{P} \cup \{P_j\}$;
7:    $G := G/P_j$ and $R := R \setminus col(P_j)$;
8:    **If** there exists st-path in G with colors of R **then**
        Go to Step 1 against *LP(1)* with new G and R;
9: Return $\mathcal{P}$.

---

g, and is a set of disjoint paths with different colors in G because each edge in f is integral. Therefore, Y is indeed a solution to *MDPDC*. Conversely and obviously, from a solution to *MDPDC*, we can immediately construct a feasible solution to its corresponding formula of *ILP(1)*. Combining the two directions, we immediately complete the proof. □

From the above lemma, an optimum solution to *ILP(1)* is exactly an optimum solution to the corresponding *MDPDC* problem and *vice versa*. We denote the *LP* relaxation for *MDPDC* as *LP(1)*, whose difference with *ILP(1)* is that x_e^r and y_r are relaxed to be real numbers between 0 and 1.

3.3 A Greedy-Based Iterative *LP* Rounding Algorithm

Our algorithm is composed of iterations, where each iteration is to select and round a maximum path-flow from a computed optimum solution to *LP(1)*, and return the remained graph of removing the rounded path for the next iteration. Apparently, the aim of the entire rounding is to maximize the number of the possible remained paths.

The main steps of each iteration proceed as below: First, solve *LP(1)* against the remained graph (which is initially the original graph) to obtain an st-flow; Second, decompose the flow to a set of flow-paths Q by employing Algorithm 2, which repeats peeling a flow-path with maximum value from the st-flow until the decomposition is done; Third, select and round a flow-path $P_j \in Q$ with the maximum flow value. At last, remove the edges on the rounded path from the graph, and use the remained graph for the next iteration. The algorithm terminates when the st-flow computed via *LP(1)* is integral. Formally, the whole heuristic algorithm is as in Algorithm 1.

It remains to give the edge-peeling algorithm for flow decomposition. The key idea is to repeatedly peel off each path-flow of a minimum flow value of an edge thereon, from the flow produced by solving the *LP* formula, until the value of the

---

Algorithm 2. The edge-peeling algorithm

Input: Specified vertices s and t, a set of colors R, a solution of $LP(1)$ Sol;
Output: *A set of flow-paths* $\mathcal{P}$.
0: Set $\mathcal{P} := \emptyset$;
1: Decompose Sol to a collection of sets of flow-paths, say $\{f_1, \ldots, f_h\}$ where f_i is the
set of flow with color i;
2: **For** f_i in $\{f_1, \ldots, f_h\}$ **do**
3:    **While** the value of f_j is not zero **do**
4:       Find an st-path P_j in the graph for f_j and add to $\mathcal{P}$;
5:       Decrease each edge in P_j with minimum flow value;
6:    **EndWhile**
7: Return $\mathcal{P}$.

---

flow is zero. The algorithm is mainly composed by three phases: Firstly, compute an optimal solution to *LP(1)*, and obtain an st flow; Secondly, decompose the solution of *LP(1)* to a collection of subflow, say$\{f_1, \ldots, f_h\}$, where f_i is the set of flow with color i; Thirdly, for each subflow, repeatedly peel a portion of the subflow until the value of the remaining subflow is 0. In the phase, we find an st-path P_j in the graph exactly accommodating the subflow, and then reduce the flow value of each edge in P_j by Δ, where Δ is the minimum value of the edges on P_j. The full layout of the algorithm is as described in Algorithm 2.

3.4   An *LP*-rounding Algorithm Based on a Second *ILP*

The disadvantage of Algorithm 1 is its high runtime, because it runs iterations. So we propose another algorithm which round a fractional solution to an integral one in only one iteration. The key idea is to use another *ILP* to round the solution of *LP(1)*, i.e. to select proper flow-paths to compose the approximate solution. The *ILP* is formally as below *(ILP2)*:

$$
\begin{aligned}
\max \quad & \sum_{r \in R} y_r \\
\text{s.t.} \quad & \sum_{j:\, e \in F_j} x_j \leq 1 && \forall e \in E \\
& \sum_{j:col(F_j)=r} x_j - y_r \geq 0 && \forall r \in R \\
& \sum_{j:col(F_j)=r} x_j \leq 1 && \forall r \in R \\
& \sum_{j:\, v \in F_j} x_j \leq 1 && \forall v \in V \setminus \{s, t\} \\
& x_j, y_r \in \{0,1\} && 1 \leq j \leq h, \forall r \in R
\end{aligned}
$$

Similarly, we denote by y_r the total value of flow accommodated by the paths of color $r \in R$, and denote by x_j the j-th flow in $\mathcal{F} := (F_1, \ldots, F_h)$, where $x_j = 1$ means the j-th flow is used completely in the solution, while $x_j = 0$ indicates otherwise. Furthermore, the first Inequality in *ILP(2)* ensures that each edge can be selected for at most once. The second guarantees that the value of each

---

Algorithm 3. A rounding algorithm based on another *LP*

Input: A general graph G, specified vertices s and t, a set of colors R;
Output: A solution to *MDPDC*.
0: Solve *LP(1)* and obtain x an optimal solution;
1: **If** x is integral (i.e. each dimension of x is an integer) **then**
   return x;
2: Decompose x to flow-paths $\{F_1, \ldots, F_h\}$ by Algorithm 2;
3: Solve *ILP(2)* against $(F_1, \ldots, F_h)$ to obtain $\{y_1, \ldots, y_g\}$.

---

flow for each color can not exceed value one; the third Inequality guarantees at most one path is selected for each color. At last, the fourth Inequality ensures each vertex can only be assigned with one color.

The main steps of the rounding proceed as below: Firstly, solve *LP(1)* to obtain an st-flow that is then decomposed to a set of path-flow $\mathcal{F}$ by employing *the edge peeling* Algorithm 2, when the solution is not integer; Secondly, run *ILP(2)* against the path-flow $\mathcal{F} := (F_1, \ldots, F_h)$ to obtain a rounded integral solution $\{y_1, \ldots, y_g\}$. The detailed algorithm is as depicted in Algorithm 3.

4   Conclusion

In this paper, we first proved the NP-completeness of *MDPDC* in both directed and undirected graphs, by giving reductions from the directed disjoint paths problem and the *3SAT* problem, respectively. Then, we gave an integer linear program (*ILP*) for modeling *MDPDC*, and developed two algorithms based on rounding fractional solution to the accordingly relaxed linear program (*LP*). In future, we shall evaluate performance gains of the algorithms by comparing them with exact solutions from *ILP* and *LP* solvers through numerical experiments.

Acknowledgements. This work is supported by Natural Science Foundation of China (No. 61772005) and Natural Science Foundation of Fujian Province (No. 2017J01753). Part of the research was done when the first and forth authors were with College of Mathematics and Computer Science, Fuzhou University, P.R. China.

References

1. https://standards.ieee.org/standard/802_1D-2004.html
2. Bonizzoni, P., Dondi, R., Pirola, Y.: Maximum disjoint paths on edge-colored graphs: approximability and tractability. Algorithms **6**(1), 1–11 (2013)
3. Challal, Y., Ouadjaout, A., Lasla, N., Bagaa, M., Hadjidj, A.: Secure and efficient disjoint multipath construction for fault tolerant routing in wireless sensor networks. J. Netw. Comput. Appl. **34**(4), 1380–1397 (2011)
4. Chekuri, C., Khanna, S., Shepherd, F.B.: The all-or-nothing multicommodity flow problem. In: Proceedings of the Thirty-sixth Annual ACM Symposium on Theory of Computing, pp. 156–165. ACM (2004)

5. Chekuri, C., Khanna, S., Shepherd, F.B.: Multicommodity flow, well-linked terminals, and routing problems. In: Proceedings of the Thirty-seventh Annual ACM Symposium on Theory of Computing, pp. 183–192. ACM (2005)
6. Deng, Y., Guo, L., Huang, P.: Exact algorithms for finding partial edge-disjoint paths. In: Wang, L., Zhu, D. (eds.) COCOON 2018. LNCS, vol. 10976, pp. 14–25. Springer, Cham (2018). https://doi.org/10.1007/978-3-319-94776-1_2
7. Dondi, R., Sikora, F.: Finding disjoint paths on edge-colored graphs: more tractability results. J. Comb. Optim. **36**(4), 1315–1332 (2017). https://doi.org/10.1007/s10878-017-0238-6
8. Fortune, S., Hopcroft, J., Wyllie, J.: The directed subgraph homeomorphism problem. Theoret. Comput. Sci. **10**(2), 111–121 (1980)
9. Frank, A.: Edge-disjoint paths in planar graphs. J. Comb. Theory Ser. B **39**(2), 164–178 (1985)
10. Guo, L., Deng, Y., Liao, K., He, Q., Sellis, T., Hu, Z.: A fast algorithm for optimally finding partially disjoint shortest paths. In: IJCAI, pp. 1456–1462 (2018)
11. Hou, A., Wu, C.Q., Fang, D., Wang, Y., Wang, M.: Bandwidth scheduling for big data transfer using multiple fixed node-disjoint paths. J. Network Comput. Appl. **85**, 47–55 (2017)
12. Kawarabayashi, K.I., Kobayashi, Y.: All-or-nothing multicommodity flow problem with bounded fractionality in planar graphs. In: IEEE Symposium on Foundations of Computer Science (2013)
13. Lin, G., Soh, S., Chin, K.-W., Lazarescu, M.: Energy aware two disjoint paths routing. J. Netw. Comput. Appl. **43**, 27–41 (2014)
14. Robertson, N., Seymour, P.D.: Graph minors. XIII. The disjoint paths problem. J. Combin. Theory Ser. B **63**, 65–110 (1995)
15. Schrijver, A.: Finding k disjoint paths in a directed planar graph. SIAM J. Comput. **23**(4), 780–788 (1994)
16. Shiloach, Y.: A polynomial solution to the undirected two paths problem. J. the ACM (JACM) **27**(3), 445–456 (1980)
17. Tholey, T.: Solving the 2-disjoint paths problem in nearly linear time. Theory Comput. Syst. **39**(1), 51–78 (2006)
18. Wu, B.Y.: On the maximum disjoint paths problem on edge-colored graphs. Discrete Optim. **9**(1), 50–57 (2012)
19. Yallouz, J., Rottenstreich, O., Babarczi, P., Mendelson, A., Orda., A.: Optimal link-disjoint node-"somewhat disjoint" paths. In: 2016 IEEE 24th International Conference on Network Protocols (ICNP), pp. 1–10. IEEE (2016)
20. Zhang, S., Chen, L., Yang, W.: On fault-tolerant path optimization under QoS constraint in multi-channel wireless networks. Theoret. Comput. Sci. **695**, 74–82 (2017)

A Constant Factor Approximation for Lower-Bounded k-Median

Yutian Guo, Junyu Huang, and Zhen Zhang[✉]

School of Computer Science and Engineering, Central South University,
Changsha 410083, People's Republic of China
csuzz@foxmail.com

Abstract. The lower-bounded k-median problem considers a set C of clients, a set F of facilities, and a parameter B, the goal is to open k facilities and connect each client to an opened facility, such that each opened facility is connected with at least B clients and the total connection cost is minimized. The problem is known to admit an $O(1)$-approximation algorithm, while the constant is implicit and seems to be a very large constant. In this paper, we give an approach that converts the lower-bounded k-median problem to the capacitated facility location problem, which yields a $(516 + \epsilon)$-approximation for the lower-bounded k-median problem.

Keywords: Approximation algorithm · k-median

1 Introduction

k-median is a widely studied clustering problem and has applications in many fields related to computer science. Given a set C of clients and a set F of facilities located in a metric space, the k-median problem aims to open a set $S \subseteq F$ of at most k facilities, such that the objective function $\sum_{j \in C} d(j, S)$ is minimized, where $d(j, S)$ denotes the distance from $j \in C$ to its nearest facility in S. This problem is known to be NP-hard, which leads to a lot of efforts devoted to obtaining its approximation algorithms [5, 7, 11, 13, 16]. The current best approximation guarantee for the problem is $2.675 + \epsilon$ [5], which is obtained Li and Svensson [16].

The clustering problem has an inherent assumption that each client can be optionally connected to any opened facility. However, many real world scenarios associate a notion of *lower bound* with each facility, and the number of the clients connected to each facility should not be less than the lower bound associated with it. For example, in the design of buy-at-bulk network, a set of demands needs to be connected to a set of servers, and each server is required to have

This work was supported by National Natural Science Foundation of China (61672536, 61872450, 61828205, and 61802441), Hunan Provincial Key Lab on Bioinformatics, and Hunan Provincial Science and Technology Program (2018WK4001).

© Springer Nature Switzerland AG 2020
J. Chen et al. (Eds.): TAMC 2020, LNCS 12337, pp. 119–131, 2020.
https://doi.org/10.1007/978-3-030-59267-7_11

a minimum amount of demand assigned to it. Karger and Minkoff [12] and Guha *et al.* [10] introduced the lower-bounded facility location problem to deal with such constraints. They presented constant-factor bi-criteria approximation algorithms for the problem, which violate the lower bound of the facilities. On the basis of the techniques given in [12] and [10], Svitkina [17] gave a $(448 + \epsilon)$-approximation without violating the lower bound constraint. The ratio was later improved to $82.6 + \epsilon$ by Ahmadian and Swamy [1]. Most recently, Li [15] gave a $(3926 + \epsilon)$-approximation for the facility location problem with non-uniform lower bounds.

In this paper, we consider the lower-bounded k-median problem.

Definition 1 (lower-bounded k-median). *Given a set C of clients and a set F of facilities located in a metric space, an integer k, and a parameter B, the lower-bounded k-median problem is to open a set S of at most k facilities, and identify a connection function σ, such that the number of the clients connected to each facility is no less than B, and the objective $cost(S, \sigma) = \sum_{j \in C} d(j, \sigma(j))$ is minimized, where $\sigma(j) = i$ denotes client j is connected to facility $i \in S$ for each $j \in C$, and $d(j, \sigma(j))$ denotes the distance from j to $\sigma(j)$.*

In Euclidean space, Ding and Xu [9] gave a $(1 + \epsilon)$-approximation for the lower-bounded k-median problem with running time $O(n^2 d \cdot (\log n)^{k+2} \cdot 2^{poly(k/\epsilon)})$. Bhattacharya *et al.* [4] later improved the running time of the algorithm in [9] to $O(n^2 d \cdot (\log n)^2 \cdot (\frac{k}{\epsilon})^{O(k/\epsilon)})$. Ahmadian and Swamy [2] gave that the problem admits an $O(1)$-approximation algorithm that runs in polynomial time. The approximation ratio is implicit but seems to be a very large number.

1.1   Our Results

In this paper, we obtain the following result for the lower-bounded k-median problem.

Theorem 1. *There exists a $(516 + \epsilon)$-approximation algorithm for the lower-bounded k-median problem that runs in polynomial time.*

We now give the high level idea of our approach. Given an instance of the lower-bounded k-median problem, it can be seen that the instance has a feasible solution if $|C| \geq B$. We present a bi-criteria approximation algorithm for the problem. The algorithm yields a constant factor approximation solution which violates the lower bound of the facilities. To convert such a bi-criteria approximation solution to a feasible solution, we reconnect some clients and close some facilities so that the lower bound constraint of each facility can be satisfied. We consider an instance of the capacitated facility location problem to minimize the loss in the cost induced by the converting. This instance is constructed by interchanging the roles of the clients and the facilities. A set of clients connected to a same location is now viewed as a facility whose capacity is the same as the size of the client set, and a facility whose lower bound is violated now becomes a set T of clients with $|T| = (1 - \lambda)B$, where λB is the number of the clients connected to this facility in the original bi-criteria approximation solution. Based on a known $O(1)$-approximation algorithm for the capacitated facility location problem, we convert the bi-criteria solution to a solution satisfying the lower bound, which induces a constant factor loss in the approximation ratio.

1.2 Other Related Work

A commonly studied extension of the clustering problem is the capacitated clustering problem, which can be viewed as the opposite of the lower-bounded clustering problem in some sense. In this problem, each facility is associated with a capacity, and the number of the clients connected to a facility should be less than its capacity. The capacitated facility location problem can be formally defined as follows.

Definition 2 (Capacitated facility location). *Given a set C of clients and a set F of facilities located in a metric space, an opening cost f_i and a capacity u_i associated with each $i \in F$, the capacitated facility location problem is to open a set S of facilities, and identify a connection function σ, such that the number of the clients connected to each facility $i \in S$ is no more than u_i, and the objective $cost(S, \sigma) = \sum_{j \in C} d(j, \sigma(j)) + \sum_{i \in S} f_i$ is minimized, where $\sigma(j) = i$ denotes client j is connected to facility $i \in S$ for each $j \in C$, and $d(j, \sigma(j))$ denotes the distance from j to $\sigma(j)$.*

The capacitated clustering problem is significantly harder than the ordinary clustering problem. There are several known $O(1)$-approximation algorithms for the capacitated facility location problem. The current best approximation guarantee for the problem is $5 + \epsilon$ [3], which was obtained based on a local search algorithm. However, constant factor approximation algorithms for the capacitated k-median problem only exist for the case where the capacity constraint or the number of clusters can be violated [8,14].

2 A Bi-criteria Approximation

In this section, we give a constant factor bi-criteria approximation for the lower-bounded k-median problem by constructing an instance of the k-facility location problem. This problem can be formally defined as follows.

Definition 3 (k-facility location). *Given a set C of clients and a set F of facilities located in a metric space, an integer k, and an opening cost f_i associated with each $i \in F$, the k-facility location problem aims to open a set S of at most k facilities, such that the objective $\sum_{j \in C} d(j, S) + \sum_{i \in S} f_i$ is minimized, where $d(j, S)$ denotes the distance from j to its nearest facility in S.*

Let $I = (C, F, k, B)$ be an instance of the lower-bounded k-median problem. Given a facility $i \in F$, let J_i denote the set of the B clients in C closest to i. Given $\beta \in (0, 1)$ and a solution (S, σ) of I, we call (S, σ) a β-covered solution if it connects no less than βB clients to each $i \in F$, and let $cost_I(S,\sigma)$ denote its cost of the problem. We construct an instance $I' = (C, F, f)$ for the k-facility location problem, where $f_i = \frac{2\beta}{1-\beta} \sum_{j \in J_i} d(i,j)$ for each $i \in F$. We solve the constructed instance by the algorithm given in [6], which gives a 3.25-approximation solution for the k-facility location problem. Let S' be the resulted set of the open facilities. Let $cost_{I'}(S')$ denote the cost of S' for the k-facility location problem.

We now show that any λ-approximation solution of I' can be converted to an $O(\lambda)$-approximation solution of I, which induces a constant factor violation in the lower bound of the facilities.

Lemma 1. *For an arbitrary solution (S,σ) of I, the solution is also a feasible for I', and we have*

$$\sum_{i \in S} f_i \leq \frac{2\beta}{1-\beta} cost(S,\sigma).$$

Proof. Since (S,σ) is a feasible solution of I, for each $i \in S$, i is connected with at least B clients. This implies that $\sum_{j \in J_i} d(i,j) \leq \sum_{j \in \sigma^{-1}(i)} d(i,j)$. Thus, we have

$$\sum_{i \in S} f_i = \sum_{i \in S} \left(\frac{2\beta}{1-\beta} \sum_{j \in J_i} d(i,j) \right)$$

$$\leq \frac{2\beta}{1-\beta} \sum_{i \in S, j \in \sigma^{-1}(i)} d(i,j)$$

$$= \frac{2\beta}{1-\beta} \sum_{j \in C} d(j, \sigma(j))$$

$$= \frac{2\beta}{1-\beta} cost_I(S,\sigma),$$

where the first step follows from the the definition of f_i.   □

Lemma 1 implies that for any solution (S,σ) of instance I, we have $cost_{I'}(S) \leq \frac{1+\beta}{1-\beta} cost_I(S)$. We proceed by showing that a solution S' to I' can be converted to a β-covered solution (S,σ) of I.

Lemma 2. *Given a solution S' of I', we can find a β-covered solution (S,σ) of I such that $cost_I(S,\sigma) \leq cost_{I'}(S')$.*

Proof. We prove the lemma by giving an algorithm that yields the desired β-covered solution. Based on instance I', we construct a new instance I^k for the k-median problem by removing all facility open costs. Let $S = S'$ initially. While there exists some $i \in S$ such that $cost_{I'}(S\setminus\{i\}) \leq cost_{I'}(S)$, let $S = S\setminus\{i\}$. The final solution (S,σ) is called a minimal feasible solution of instance I^k, where each $j \in C$ is assigned to its nearest facility in S by function σ. It is easy to show that $cost_{I^k}(S,\sigma) \leq cost_{I'}(S)$, which implies that $cost_{I^k}(S,\sigma) \leq cost_{I'}(S) \leq cost_{I'}(S')$.

Now we want to show that in the solution (S, σ), each facility $i \in S$ is connected by at least βB clients. For the sake of contradiction, assume that there exists a facility $i \in S$ such that $|\sigma^{-1}(i)| \leq \beta B$. This implies that $|J_i \backslash \sigma^{-1}(i)| \geq (1 - \beta)B$. We know that there exists a client $j' \in J_i \backslash \sigma^{-1}(i)$, such that

$$d(i, j') \leq \frac{1}{(1-\beta)B} \sum_{j \in J_i \backslash \sigma^{-1}(i)} d(i, j) \leq \frac{1}{(1-\beta)B} \sum_{j \in J_i} d(i, j).$$

Since j' is not connected to i in the solution (S, σ), it is connected to some other facility $i' \in S$ with $d(j', i') \leq d(j', i)$. Thus, we have

$$
\begin{aligned}
\sum_{j \in \sigma^{-1}(i)} d(j, i') &\leq \sum_{j \in \sigma^{-1}(i)} (d(j, i) + d(i, j') + d(j', i')) \\
&\leq \sum_{j \in \sigma^{-1}(i)} d(j, i) + |\sigma^{-1}(i)| \times 2d(i, j') \\
&\leq \sum_{j \in \sigma^{-1}(i)} d(j, i) + \beta B \times \frac{2}{(1-\beta)B} \sum_{j \in J_i} d(i, j) \\
&= \sum_{j \in \sigma^{-1}(i)} d(j, i) + \frac{2\beta}{(1-\beta)} \sum_{j \in J_i} d(i, j).
\end{aligned}
$$

If we close i and reconnect each client from i to i', then the increment in the connection cost is no more than $\frac{2\beta}{(1-\beta)} \sum_{j \in J_i} d(i, j)$, which is bounded by f'_i. We have $cost_{I'}(S \backslash \{i\}) \leq cost_{I'}(S)$, contradicting that (S, σ) is a minimal feasible solution of instance I^k. Thus (S, σ) is a β-covered solution of I. □

Based on Lemma 1 and Lemma 2, we get the following approximation guarantee.

Theorem 2. *There exists a $3.25\frac{1+\beta}{1-\beta}$-approximation algorithm for the lower-bounded k-median problem which violates the lower bound by a factor β.*

Proof. We first get a solution S' of I' using the 3.25-approximation algorithm for the k-facility location problem. We denote the set of the opened facilities in an optimal solution of I by S^*. We have $cost_{I'}(S') \leq 3.25cost_{I'}(S^*)$. Thus, we get

$$
\begin{aligned}
cost_{I'}(S') &\leq 3.25 \Big(\sum_{i \in S^*} f_i + \sum_{j \in C} d(j, S^*) \Big) \\
&\leq 3.25 \Big(\frac{2\beta}{1-\beta} cost_I(S^*, \sigma^*) + \sum_{j \in C} d(j, S^*) \Big) \\
&\leq 3.25 \Big(\frac{2\beta}{1-\beta} cost_I(S^*, \sigma^*) + cost_I(S^*, \sigma^*) \Big) \\
&= 3.25 \frac{1+\beta}{1-\beta} cost_I(S^*, \sigma^*),
\end{aligned}
$$

where the second step follows from Lemma 1. By Lemma 2, we can obtain a β-covered solution (S^o, σ^o) of I such that

$$cost_I(S^o, \sigma^o) \leq cost_{I'}(S') \leq 3.25(\frac{1+\beta}{1-\beta})cost_I(\sigma^*).$$

$\square$

By Theorem 2, we can get a pseudo-solution for the lower-bounded k-median problem, which violates the lower bound restriction by the a constant $\beta \in (0, 1)$. This implies that we find a solution where each opened facility is connected with at least βB clients instead of B clients. In the following we will show how to make such a solution feasible for the lower-bounded k-median problem.

3   The Approximation Algorithm

3.1   Aggregating Clients

Given an instance $I = (C, F, B, k)$ of the lower-bounded k-median problem, by Lemma 1 and Lemma 2, we can obtain a bi-criteria approximation solution (S^o, σ^o) which violates the constraint of lower bound by a factor β. We construct a new instance I^1 for the lower-bounded k-median problem, where C, F, and B are the same as that of I, but the metric is different from I. In instance I^1, each client $j \in C$ is moved to $\sigma^o(j)$. Then, for each $i_1, i_2 \in F$ and $j_1, j_2 \in C$, we have $d^1(i_1, i_2) = d(i_1, i_2)$, $d^1(i_1, j_1) = d(i_1, \sigma^o(j_1))$, and $d^1(j_1, j_2) = d(\sigma^o(j_1), \sigma^o(j_2))$. For arbitrary $i \in F$ and $j \in C$, using triangle inequality, we get

$$d^1(i, j) = d(i, \sigma^o(j)) \leq d(i, j) + d(j, \sigma^o(j)). \tag{1}$$

For an optimal solution (S^*, σ^*) of instance I, we have

$$\begin{aligned} cost_{I^1}(S^*, \sigma^*) &\leq cost_I(S^*, \sigma^*) + cost_I(S^o, \sigma^o) \\ &\leq cost_I(S^*, \sigma^*) + 3.25\frac{1+\beta}{1-\beta}cost_I(S^*, \sigma^*) \\ &= \left(1 + 3.25\frac{1+\beta}{1-\beta}\right)cost_I(S^*, \sigma^*), \end{aligned} \tag{2}$$

where the first step follows from inequality (1), the second step follows from Theorem 2. We have the following result based on the methods in [17].

Theorem 3. *If there is an α_1-approximation solution of I^1, we can efficiently find an α-approximation solution of I, where $\alpha = \alpha_1(1 + 3.25\frac{1+\beta}{1-\beta}) + 3.25\frac{1+\beta}{1-\beta}$.*

3.2 Contracting Facility Set

We now focus on instance $I^1 = (C, F, B, k)$. For each $i \in S^o$, define $\gamma_i = \{j | \sigma^{o-1}(i)\}$ as the set of clients connected to i. We have $|\gamma_i| \geq \beta B$ for each $i \in S^o$. An instance $I^2 = (C, S^o, B, k)$ is constructed by removing each facility in $F \backslash S^o$ from I^1.

Lemma 3. *If there is a solution (S^1, σ^1) of I^1, then we can efficiently find a solution (S^2, σ^2) of I^2 such that $cost_{I^2}(S^2, \sigma^2) \leq 2cost_{I^1}(S^1, \sigma^1)$.*

Proof. For each $i \in F \backslash S^o$, let i' denote the facility in S^o nearest to i. We construct a solution (S^2, σ^2) of I^2 by opening each $i \in S^1 \cap S^o$ and facility i' for each $i \in F \cap (S^1 \backslash S^o)$. For a facility $i \in F \cap (S^1 \backslash S^o)$, the clients connected to i in solution (S^1, σ^1) are reconnected to i'. By triangle inequality and the definition of $d^1(*)$, the increased cost induced by a client j is bounded by $d^1(i, i') = d(i, i') \leq d(i, \sigma^{o-1}(j)) = d^1(i, j)$. Summing the inequality over each $j \in C$, we get that the total increased cost is no more than $cost_{I^1}(S^1, \sigma^1)$, which implies that $cost_{I^2}(S^2, \sigma^2) \leq 2cost_{I^1}(S^1, \sigma^1)$. □

Lemma 3 implies that a solution of instance I^1 can be converted to a feasible solution of instance I^2. It is easy to see that a solution (S, σ) of instance I^2 is also feasible of instance I^1 and satisfies $cost_{I^1}(S, \sigma) = cost_{I^2}(S, \sigma)$. Thus, we get the following result for I^1 and I^2.

Theorem 4. *Given an α_2-approximation solution of I^2, we can find an α_1-approximation solution of I^1, where $\alpha_1 = 2\alpha_2$.*

Proof. Let (S^{*1}, σ^{*1}) be an optimal solution of instance I^1. Using Lemma 3, we can get a solution (S^2, σ^2) of I^2 that satisfies $cost_{I^2}(S^2, \sigma^2) \leq 2cost_{I^1}(S^{*1}, \sigma^{*1})$. Let (S, σ) denote an α_2-approximation solution of I^2, we have $cost_{I^2}(S, \sigma) \leq 2\alpha_2 cost_{I^1}(S^{*1}, \sigma^{*1})$. Recall that (S, σ) is also feasible for I^1 and $cost_{I^1}(S, \sigma) = cost_{I^2}(S, \sigma) \leq 2\alpha_2 cost_{I^1}(S^{*1}, \sigma^{*1})$. □

Now we focus on instance I^2. We only consider one facility for each position in I^2.

3.3 Adding Penalties to Instance I^2

Based on instance I^2, we construct a new instance I^3 by considering penalties for closing the facilities from S^o. For each $i \in S^o$, if i is closed in the solution, then a penalty cost $Pc_{I^3}(i) = \frac{2\beta-1}{\beta}|\gamma_i|\ell_i$ should be paid, where ℓ_i denotes the distance from i to its nearest facility in $S^o \backslash \{i\}$. For a solution (S, σ) of I^3, define $Pc_{I^3}(S, \sigma) = \sum_{i \in S^o \backslash S} Pc_{I^3}(i)$ as the total penalty cost of (S, σ).

Lemma 4. *For any solution (S, σ) of I^2 and I^3, we have*

$$cost_{I^2}(S, \sigma) \leq cost_{I^3}(S, \sigma) \leq \frac{3\beta - 1}{\beta} cost_{I^2}(S, \sigma).$$

Proof. The cost of solution (S, σ) of I^3 consists of the connection cost and the penalty cost of the closed facilities, where the connection cost is equal to $cost_{I^2}(S, \sigma)$. Thus, $cost_{I^2}(S, \sigma) \leq cost_{I^3}(S, \sigma)$. We have

$$\sum_{i \in S^o \setminus S} Pc_{I^3}(i) = \sum_{i \in S^o \setminus S} \frac{2\beta - 1}{\beta} |\gamma_i| \ell_i$$

$$\leq \frac{2\beta - 1}{\beta} \sum_{i \in S^o \setminus S} \sum_{j \in \gamma_i} d^1(j, \sigma(j))$$

$$\leq \frac{2\beta - 1}{\beta} \sum_{j \in C} d^1(j, \sigma(j))$$

$$= \frac{2\beta - 1}{\beta} cost_{I^2}(S, \sigma).$$

This implies that

$$cost_{I^3}(S, \sigma) = cost_{I^2}(S, \sigma) + Pc_{I^3}(S, \sigma) \leq \frac{3\beta - 1}{\beta} cost_{I^2}(S, \sigma).$$

$\square$

Lemma 4 implies that I^2 can be converted to I^3 with a constant factor loss in the approximation ratio.

Theorem 5. *Given an α_3-approximation solution of I^3, we can find an α_2-approximation solution of I^2, where $\alpha_2 = \frac{3\beta - 1}{\beta} \alpha_3$.*

Proof. Let (S^{*2}, σ^{*2}) be an optimal solution of instance I^2. By Lemma 4, there exists a solution (S^3, σ^3) of I^3 such that $cost_{I^3}(S^3, \sigma^3) \leq \frac{3\beta - 1}{\beta} cost_{I^2}(S^{*2}, \sigma^{*2})$. Let (S, σ) denote an α_3-approximation solution of I^3, we have $cost_{I^2}(S, \sigma) \leq cost_{I^3}(S, \sigma) \leq \frac{3\beta - 1}{\beta} \alpha_3 cost_{I^2}(S^{*2}, \sigma^{*2})$. $\square$

3.4   Constructing an Instance of Capacitated Facility Location

In this section, we show how to convert I^3 to an instance of the capacitated facility location problem (CFL). Recall that we only consider one facility for each position in the lower bounded k-median problem. For each $i \in S^o$, let $\Delta_i^1 = |\gamma_i|$ and $\Delta_i^2 = |\gamma_i| - B$, we define a variable $\Delta_i \in \{\Delta_i^1, \Delta_i^2\}$. In addition, we define P_i as the position of i for any $i \in S^o$. If $\Delta_i = \Delta_i^1$, then we close facility i. In such case, Δ_i^1 clients should be reconnected. For the case where $\Delta_i = \Delta_i^2$, we open facility i, if $\Delta_i^2 > 0$ then $|\Delta_i^2|$ clients from γ_i can be reconnected without violating the lower bound of i. Otherwise, $|\Delta_i^2|$ clients should be reconnected to i. It can be seen that instance I^3 is to identify the value of Δ_i for each P_i where $i \in S^o$ such that $\sum_{i \in S^o} \Delta_i \geq 0$.

We now show how to construct an instance I^4 of CFL based on I^3. To avoid confusion, each facility in the CFL instance is called a C-facility, and each client in CFL instance is called a C-client. The total cost of an instance of CFL is the sum of the open cost of C-facilities and connection cost of C-clients. For each P_i where $i \in S^o$, we construct a C-facility with open cost $\frac{2\beta-1}{\beta}|\gamma_i|\ell_i$ and capacity $\Delta_i^1 - \Delta_i^2$. Moreover, a set of C-clients or a C-facility in P_i are constructed depending on the value of Δ_i^2. If $\Delta_2^i < 0$, then we construct a set of $|\Delta_i^2|$ C-clients. If $\Delta_i^2 > 0$, then we construct a C-facility with open cost 0 and capacity Δ_i^2. Note that some locations may have more than one C-facilities in I^4. Given a solution (S, σ) of I^4, let $f_{I^4}(S, \sigma)$ denote the open cost of C-facilities and $\theta_{I^4}(S, \sigma)$ denote the connection cost of C-clients.

Lemma 5. *Given any solution (S, σ) of instance I^3, we can find a solution (S^c, σ^c) of instance I^4 of CFL such that $cost_{I^4}(S^c, \sigma^c) \leq cost_{I^3}(S, \sigma)$.*

Proof. We identify the value of Δ_i for each P_i where $i \in S^o$ based on solution (S, σ). We have $\sum_{i \in S^o} \Delta_i \geq 0$ due to (S, σ) is a feasible solution of I^3. As mentioned above, in instance I^4, for each P_i where $i \in S^o$, there are $|\Delta_i^2|$ C-clients in P_i which need to be connected if $\Delta_i^2 < 0$. Otherwise we open the C-facility with open cost 0 and capacity Δ_i^2 at this position. Moreover, for each P_i where $i \in S^o \setminus S$, we open the C-facility at this position, whose open cost and capacity are $\frac{2\beta-1}{\beta}|\gamma_i|\ell_i$ and $\Delta_i^1 - \Delta_i^2$, respectively.

Now, we get a set S^c of opened C-facility for instance I^4. Note that if a position P_i where $i \in S$ is located two opened C-facilities, then in this position the total capacity is Δ_i^1. If there are $|\Delta_i^2|$ C-clients in P_i where $i \in S^o \setminus S$, then the C-facility in the same position has the priority of connecting these clients. Recall that the capacity of such a C-facility is $\Delta_i^1 - \Delta_i^2$, which implies that the $|\Delta_i^2|$ C-clients can be connected to it without violating the capacity. Thus, in instance I^4, we have $\Delta_i = \Delta_i^2$ for each P_i where $i \in S$ and $\Delta_i = \Delta_i^1$ for each P_i where $i \in S^o \setminus S$. Recall that $\sum_{i \in S^o} \Delta_i \geq 0$. Thus we can find a feasible solution for I^4 based on S^c.

Let (S^c, σ^c) denote the constructed solution of I^4, where σ^c is obtained by "switching" the direction of σ. Assume that η clients located in position P_1 are connected to a facility located in P_2 by σ for some $\eta > 0$. For the instance I^4, if there exists C-clients in P_2, then η C-clients in P_2 are connected to the C-facilities in P_1 by σ^c. Otherwise we will do nothing and this is feasible for I^4. It can be seen that the connection cost of solution (S^c, σ^c) on instance I^4 is no more than the connection cost of (S, σ) on I^3.

In solution (S, σ) to instance I^3, if a facility $i \in S^o$ is not opened, then a penalty cost $\frac{2\beta-1}{\beta}|\gamma_i|\ell_i$ should be paid. The penalty cost is equal to the open cost of a C-facility with capacity $\Delta_i^1 - \Delta_i^2$ in instance I^4. Thus, we get

$$cost_{I^4}(S^c, \sigma^c) = \theta_{I^4}(S^c, \sigma^c) + f_{I^4}(S^c, \sigma^c)$$
$$\leq \theta_{I^3}(S, \sigma) + Pc_{I^3}(S, \sigma)$$
$$= cost_{I^3}(S, \sigma).$$

$\square$

Lemma 5 implies that a solution (S, σ) of I^3 can be converted to a solution (S^c, σ^c) of I^4. We now show how a solution of I^4 can be converted to a solution of I^3.

Lemma 6. *Given a solution (S^c, σ^c) of instance CFL, we can find a solution (S, σ) of instance I^3 such that $cost_{I^3}(S, \sigma) \leq \frac{2\beta}{2\beta-1} cost_{I^4}(S^c, \sigma^c)$.*

Proof. We construct a solution (S, σ) of I^3 based on solution (S^c, σ^c). Given a position, if a C-facility with open cost $\frac{2\beta-1}{\beta}|\gamma_i|\ell_i$ and capacity $\Delta_i^1 - \Delta_i^2$ in the position is opened in I^4, then no facility in the position is opened in I^3. Otherwise, the facility in the position is opened in I^3. For a position where a C-facility i with open cost $\frac{2\beta-1}{\beta}|\gamma_i|\ell_i$ and capacity $\Delta_i^1 - \Delta_i^2$ is opened in instance I^4, we have the following two cases: (1) $\Delta_i^2 > 0$, and (2) $\Delta_i^2 < 0$. For case (1), a C-facility with open cost 0 and capacity Δ_i^2 is located in the position. For case (2), there are $|\Delta_i^2|$ C-clients in the position that can be connected with the C-facilities located in the same position. In both cases, the C-facilities in the position can still be connected with Δ_i^1 C-clients. For each position where no facility is opened in instance I^3, then Δ_i^1 clients in the position should be reconnected, and the total capacity in instance I^4 is Δ_i^1. For other positions, we have $\Delta_i = \Delta_i^2$ in both instances I^3 and I^4. Since (S^c, σ^c) is feasible for I^4, we have $\sum_{i \in S^o} \Delta_i \geq 0$, which implies that we can find a feasible solution for I^3 based on S.

We now find the connection function σ for instance I^3. Such a function cannot be simply identified by "switching" the direction of σ^c. Indeed, the number of the C-clients connected with each C-facility is not guaranteed to be equal to the capacity of the C-facility. However, all the clients need to be connected in instance I^3. This implies that connecting the clients in I^3 by "switching" the direction of σ^c may cause some clients unconnected. It can be seen that such unconnected clients are located in the positions where no facility is opened in solution (S, σ) and the number of the C-clients which are connected to this position is not equal to the sum of the capacities of the C-facilities located in the same position.

For each P_i where $i \in S^o$, let δ_{P_i} be the set of the unconnected clients located in P_i. For the case where $\delta_{P_i} > 0$, we first attempt to connect each unconnected client to the nearest facility i' to i. If i' is opened, then we connect each client in δ_{P_i} to i', and the connection cost is at most $B\ell_i \leq \frac{|\gamma_i|}{\beta}\ell_i$. If i' is not opened, we further consider the following two cases: (1) $|\delta_{P_i}| + |\delta_{P_{i'}}| \geq B$, and (2) $|\delta_{P_i}| + |\delta_{P_{i'}}| < B$. For case (1), we open i' and connect each client in δ_{P_i} to i'. The condition of case (1) implies that i' can be opened without violating its lower bound. For case (2), we move each client in δ_i to i' and let $\delta_{P_{i'}} = \delta_{P_i} \cup \delta_{P_{i'}}$. We now perform the same operation described above on $P_{i'}$.

The challenge is that the procedure may be caught in several facilities, and we cannot open a facility to satisfy the lower bound. For instance, it may be the case that the clients are moved to a facility i' for more than one time, and the walk forms a cycle. Let i denote the facility in the previous position of i' in the walk. Our approach to deal with this issue is to connect these clients to the opened facility i^o that minimizes the connection cost. We have $|\delta_{P_i}| < B$ and $|\gamma_{i'}| + |\gamma_i| \geq 2\beta B$. So there are at least $|\gamma_{i'}| + |\gamma_{i^*}| - |\delta_{P_{i^*}}| \geq (2\beta - 1)B$ connected clients and we can bound the connection cost as

$$\sum_{j \in \gamma_{i'} \cup \gamma_i \setminus \delta_{P_i}} d^1(j, \sigma(j)) \geq (2\beta - 1)Bd^1(i^o, \{i', i\}). \tag{3}$$

Using triangle inequality and inequality (3), we have

$$\sum_{j \in \delta_{P_i}} d^1(j, i^o) \leq Bd^1(i, i^o) \leq B(d^1(i^o, \{i', i\}) + \ell_i)$$

$$\leq \frac{1}{2\beta - 1} \sum_{j \in \gamma_{i'} \cup \gamma_i \setminus \delta_{P_i}} d^1(j, \sigma(j)) + \frac{\ell_i |\gamma_i|}{\beta},$$

which implies that the total increased cost induced by the unconnected clients is no more than

$$\frac{1}{2\beta - 1} Pc_{I^3}(S, \sigma) + \frac{1}{2\beta - 1} \theta_{I^3}(S, \sigma).$$

Thus, we have

$$cost_{I^3}(S, \sigma) \leq \theta_{I^3}(S, \sigma) + Pc_{I^3}(S, \sigma) + \frac{1}{2\beta - 1} Pc_{I^3}(S, \sigma) + \frac{1}{2\beta - 1} \theta_{I^3}(S, \sigma)$$

$$= \frac{2\beta}{2\beta - 1} Pc_{I^3}(S, \sigma) + \frac{2\beta}{2\beta - 1} \theta_{I^3}(S, \sigma)$$

$$= \frac{2\beta}{2\beta - 1} cost_{CFL}(S^c, \sigma^c).$$

$\square$

Theorem 6. *Given an α_4-approximation solution of CFL, we can find an α_3-approximation solution of I^3, where $\alpha_3 = \frac{2\beta}{2\beta-1} \alpha_4$.*

Proof. Let (S^{*3}, σ^{*3}) be an optimal solution of instance I^3. Using Lemma 5, there exists a solution (S^c, σ^c) of I^4 such that $cost_{I^4}(S^c, \sigma^c) \leq cost_{I^3}(S^{*3}, \sigma^{*3})$. Given an α_4-approximation solution (S', σ') of I^4, we have $cost_{I^4}(S', \sigma') \leq \alpha_{I^4} cost_{I^3}(S^{*3}, \sigma^{*3})$. By Lemma 6, we can get a solution (S, σ) of I^3 that satisfies $cost_{I^3}(S, \sigma) \leq \frac{2\beta}{2\beta-1} cost_{I^4}(S', \sigma') \leq \frac{2\beta}{2\beta-1} \alpha_4 cost_{I^3}(S^{*3}, \sigma^{*3})$.

3.5 Combining Everything

Using the algorithm for the capacitated facility location problem given in [1], we get a $(1 + \sqrt{2})$-approximation solution of I^4. Let $\beta = \frac{2}{3}$. By Theorem 6, we get $\alpha_3 = \frac{2\beta}{2\beta-1}(1 + \sqrt{2}) = 4(1 + \sqrt{2})$. By Theorems 5, 4, and 3, we get $\alpha_2 = \frac{3\beta-1}{\beta} \alpha_3 = \frac{3\beta-1}{\beta} \times 4(1+\sqrt{2}) = 6(1+\sqrt{2})$, $\alpha_1 = 2\alpha_2 = 2 \times 6(1+\sqrt{2}) = 12(1+\sqrt{2})$, $\alpha = \alpha_1(1 + 3.25\frac{1+\beta}{1-\beta}) + 3.25\frac{1+\beta}{1-\beta} = 12(1 + \sqrt{2}) \times (1 + 3.25\frac{1+\beta}{1-\beta}) + 3.25\frac{1+\beta}{1-\beta} \approx 516$.

References

1. Ahmadian, S., Swamy, C.: Improved approximation guarantees for lower-bounded facility location. In: Erlebach, T., Persiano, G. (eds.) WAOA 2012. LNCS, vol. 7846, pp. 257–271. Springer, Heidelberg (2013). https://doi.org/10.1007/978-3-642-38016-7_21

2. Ahmadian, S., Swamy, C.: Approximation algorithms for clustering problems with lower bounds and outliers. In: Proceedings of the 43rd International Colloquium on Automata, Languages, and Programming, pp. 69:1–69:15 (2016)

3. Bansal, M., Garg, N., Gupta, N.: A 5-approximation for capacitated facility location. In: Epstein, L., Ferragina, P. (eds.) ESA 2012. LNCS, vol. 7501, pp. 133–144. Springer, Heidelberg (2012). https://doi.org/10.1007/978-3-642-33090-2_13

4. Bhattacharya, A., Jaiswal, R., Kumar, A.: Faster Algorithms for the Constrained k-means Problem. Theory of Comput. Syst. $62(1)$, 93–115 (2017). https://doi.org/10.1007/s00224-017-9820-7

5. Byrka, J., Pensyl, T., Rybicki, B., Srinivasan, A., Trinh, K.: An improved approximation for k-median and positive correlation in budgeted optimization. ACM Trans. Algorithms $13(2)$, 23:1–23:31 (2017)

6. Charikar, M., Li, S.: A dependent LP-rounding approach for the k-median problem. In: Czumaj, A., Mehlhorn, K., Pitts, A., Wattenhofer, R. (eds.) ICALP 2012. LNCS, vol. 7391, pp. 194–205. Springer, Heidelberg (2012). https://doi.org/10.1007/978-3-642-31594-7_17

7. Cohen-Addad, V., Klein, P.N., Mathieu, C.: Local search yields approximation schemes for k-means and k-median in Euclidean and minor-free metrics. In: Proceedings of the 57th IEEE Symposium on Foundations of Computer Science, pp. 353–364 (2016)

8. Demirci, H.G., Li, S.: Constant approximation for capacitated k-median with $(1+\epsilon)$-capacity violation. In: Proceedings of the 43rd International Colloquium on Automata, Languages, and Programming, pp. 73:1–73:14 (2016)

9. Ding, H., Xu, J.: A unified framework for clustering constrained data without locality property. In: Proc. 26th Annual ACM-SIAM Symposium on Discrete Algorithms, pp. 1471–1490 (2015)

10. Guha, S., Meyerson, A., Munagala, K.: Hierarchical placement and network design problems. In: Proceedings of the 41st Annual Symposium on Foundations of Computer Science, pp. 603–612 (2000)

11. Jain, K., Vazirani, V.V.: Approximation algorithms for metric facility location and k-median problems using the primal-dual schema and Lagrangian relaxation. J. ACM $48(2)$, 274–296 (2001)

12. Karger, D.R., Minkoff, M.: Building Steiner trees with incomplete global knowledge. In: Proceedings of the 41st Annual Symposium on Foundations of Computer Science, pp. 613–623 (2000)

13. Kumar, A., Sabharwal, Y., Sen, S.: Linear-time approximation schemes for clustering problems in any dimensions. J. ACM $57(2)$, 1–32 (2010)

14. Li, S.: Approximating capacitated k-median with $(1 + \epsilon)k$ open facilities. In: Proceedings of the 27th Annual ACM-SIAM Symposium on Discrete Algorithms, pp. 786–796 (2016)

15. Li, S.: On facility location with general lower bounds. In: Proceedings of the 13th Annual ACM-SIAM Symposium on Discrete Algorithms, pp. 2279–2290 (2019)
16. Li, S., Svensson, O.: Approximating k-median via pseudo-approximation. SIAM J. Comput. **45**(2), 530–547 (2016)
17. Svitkina, Z.: Lower-bounded facility location. ACM Trans. Algorithms **6**(4), 69:1–69:16 (2010)

Reverse Mathematics, Projective Modules and Invertible Modules

Huishan Wu$^{(\boxtimes)}$

School of Information Science, Beijing Language and Culture University,
15 Xueyuan Road, Haidian District, Beijing 100083, China
`huishanwu@blcu.edu.cn`

Abstract. We study projective modules and invertible modules by techniques of reverse mathematics. Dual Basis Lemma provides an equivalent characterization of projective R-modules M via their dual $Hom_R(M, R)$. It is a useful tool to prove various theorems about projective modules. We first formalize and prove the Dual Basis Lemma in RCA$_0$. Then we study Kaplansky's Theorem, which says that every submodule of a free module over a hereditary ring is projective. We show that RCA$_0$ proves that a submodule of a free module over a Σ_1^0-hereditary ring is a direct sum of projective modules, and thus projective. By defining invertible R-submodules of an extension ring of R via Σ_2^0 formulas, we show that RCA$_0$ proves the statement that invertible R-modules are finitely generated projective R-modules. Modified Projectivity Test and Modified Injectivity Test are basic tests for determining projective modules and injective modules, respectively. Lastly, we show that the Modified Projectivity Test and the Modified Injectivity Test are provable in ACA$_0$ and RCA$_0$, respectively.

Keywords: Reverse mathematics · Projective module · Invertible module · Hereditary ring

1 Introduction

The logical strength of properties and theorems related to abelian groups and vector spaces are well-learned in reverse algebra (see [1,2]). Since modules over rings are mathematical structures which are more general than abelian groups as well as vector spaces, people began to study modules by means of reverse mathematics. For example, Yamazaki have investigated the proof-theoretic strength of results of modules over general rings by techniques of reverse mathematics (refer to [3] and [4]). Particularly, he investigated proofs of theorems about semisimple modules, projective modules as well as injective modules. The author have examined the existence of radicals and socles of modules in reverse mathematics

This work is supported by the National Natural Science Foundation of China (No. 61972052) and the Discipline Team Support Program of Beijing Language and Culture University (No. GF201905).

© Springer Nature Switzerland AG 2020
J. Chen et al. (Eds.): TAMC 2020, LNCS 12337, pp. 132–143, 2020.
https://doi.org/10.1007/978-3-030-59267-7_12

in a recent paper [5]. Here, we study projective modules and invertible modules over commutative rings form the viewpoint of reverse mathematics (for the background on projective modules and invertible modules, we refer to [6]).

We briefly review two subsystems of second order arithmetic. The weakest subsystem we work in is RCA_0, which contains basic axioms, Σ^0_1-induction and Δ^0_1-comprehension. The next subsystem is ACA_0, that is, RCA_0 plus Σ^0_k-comprehension for all $k \geq 1$. In the remaining, modules are all over commutative rings.

1.1 Projective Modules

Definition 1. (RCA_0) *An R-module F is free if it has a basis, that is, a maximal linearly independent set which generates F.*

Lemma 1. *(Yamazaki [3]) Let M be a module over a ring R. The following assertions are equivalent over RCA_0.*

(1) M is a direct summand of a free R-module.
(2) For any surjective R-homomorphism $g : N_1 \to N_2$ and R-homomorphism $h : M \to N_2$, there is an R-homomorphism $\psi : M \to N_1$ such that $h = g\psi$. ψ is said to be a lifting of h along g.

Projective modules are often defined by (2) of Lemma 1 in standard algebra text books, such as [6,7]. The author used this classical definition, and studied basic properties of projective modules in her PhD thesis [8]. For instance, she showed that ACA_0 proves that submodules of projective modules over Σ^0_1-PIDs (principal ideal domains) are projective (Corollary 7.1, [8]). Here, we follow Yamazaki and adopt (1) of Lemma 1 as the definition for projective module; then we can show that RCA_0 proves that submodules of projective modules over Σ^0_1-PIDs are projective (see Corollary 3 in Sect. 3).

Definition 2. *(Yamazaki [4]) (RCA_0) An R-module M is projective if it is a direct summand of a free R-module F. That is, there is a submodule K of F such that $F = M \bigoplus K$.*

Proposition 1. *(Yamazaki [4]) The following assertions are equivalent over RCA_0.*

(1) For each $i \in \mathbb{N}$, P_i is projective.
(2) $P = \bigoplus_{i \in \mathbb{N}} P_i$ is projective.

For projective modules, we will study the proof-theoretic strength of results like Dual Basis Lemma and Kaplansky's Theorem. Classically, the Dual Basis Lemma characterizes projective modules via their dual modules, where the dual of an R-module M is the set of R-homomorphisms from M to R, i.e., $Hom_R(M, R)$.

- **Dual Basis Lemma:** An R-module M is projective if and only if there exists a family of elements $\langle x_i : i \in I \rangle$ of M and linear functionals $\langle h_i : i \in I \rangle$ of $Hom_R(M, R)$ such that for any $x \in M$, $h_i(x) = 0$ for almost all $i \in I$, and $x = \sum_{i \in I} h_i(x) x_i$.

In Sect. 2, we formalize the Dual Basis Lemma in RCA$_0$, and show that an R-module M is projective if and only if there exists a Σ_1^0-sequence $\langle x_i : i \in \mathbb{N} \rangle$ of elements of M and a uniform sequence $\langle f_i : i \in \mathbb{N} \rangle$ of R-homomorphisms from M to R such that for any $x \in M$, $f_i(x) = 0$ for almost all i, and $x = \sum_{i \in \mathbb{N}} f_i(x) x_i$.

Free modules over commutative rings are projective, the converse fails in general. For hereditary rings R, as stated in the Kaplansky's Theorem below, projective R-modules have a close relationship with free modules.

- **Kaplansky's Theorem:** Any submodule of a free module over a hereditary ring is a direct sum of projective modules; in particular, the submodule is projective.

In Sect. 3, we will study the Kaplansky's Theorem. We first define Σ_1^0-hereditary rings and then show that RCA$_0$ proves that submodules of projective modules over Σ_1^0-hereditary rings are projective. Since Σ_1^0-PIDs are automatically hereditary, RCA$_0$ proves that submodules of projective modules over Σ_1^0-PIDs are projective.

1.2   Invertible Modules

Let S be a commutative extension ring of a commutative ring R. Then an R-submodule M of S is invertible if and only if there exists an R-submodule N such that the multiplication of M and N is R if and only if the multiplication of M and the quotient of R modulo M (denoted by $R : M$) is R. Classically, invertible modules are finitely generated projective modules. In Sect. 4, we will define invertible modules via Σ_2^0-formulas and show that RCA$_0$ proves that invertible R-submodules of an extension ring of R are finitely generated projective R-modules.

We are also interested in invertible modules inside a special kind of ring extensions of commutative rings, namely, the localization of a ring at its regular elements.

Definition 3. (RCA$_0$) *Let R be a commutative ring. An element x of R is called regular if $\forall y \in R[xy = 0_R \rightarrow y = 0_R]$. That is, x is a non zero-divisor of R.*

Let R be a commutative ring and $Reg(R)$ be the set of all regular elements of R. $Reg(R)$ is closed under multiplication.

Definition 4. (RCA$_0$) *The localization of R with respect to the multiplicative set $Reg(R)$ is called the total ring of quotients of R, denoted by $Q(R)$.*

When R is a domain, $Q(R)$ is just the quotient field of R. We mention that Sato in his PhD thesis (Theorem 6.8, [9]) proved that "for a commutative ring R, $Q(R)$ exists" is equivalent to ACA_0 over RCA_0. Classically, an R-submodule M of $Q(R)$ is invertible if and only if M is projective and $M \cap Reg(R) \neq \emptyset$. In Sect. 4, we will prove this equivalence within RCA_0 under the assumption that $Q(R)$ exists.

1.3 Modified Projectivity and Injectivity Test

Definition 5. (RCA_0) *Let M be a module over a ring R. M is called injective if for any one-to-one R-homomorphism $\iota : N_1 \to N_2$ and R-homomorphism $g : N_1 \to M$, there is an R-homomorphism $h : N_2 \to M$ such that $g = h\iota$.*

The notion of injective modules is dual to that of projective modules. It is interesting that injective modules can be used to determine the projectivity of modules. We now introduce a test for projectivity. Let $\mathcal{P}$ be a property of R-modules such that any R-module M can be embedded into an R-module M' possessing property $\mathcal{P}$.

- **Modified Projectivity Test:** An R-module M is projective if and only if for any surjective R-homomorphism $g : N_1 \to N_2$ with N_1 possessing property $\mathcal{P}$ and R-homomorphism $h : M \to N_2$, there is an R-homomorphism $\psi : M \to N_1$ such that $h = g\psi$.

In classical algebra, an R-module M can be embedded into an injective $\mathbb{Z}$-module, A say. Then M can also be embedded into an injective R-module, namely, $Hom_{\mathbb{Z}}(R, A)$. Observe that $Hom_{\mathbb{Z}}(R, A)$ is a third order object, thus can not be studied in subsystems of second-order arithmetic. However, for $\mathbb{Z}$-modules, by Proposition 2, the property $\mathcal{P}$ in the Modified Projectivity Test can be taken to be the injectivity.

Proposition 2. *(Yamazika [4]) RCA_0 proves that every $\mathbb{Z}$-module can be embedded into an injective $\mathbb{Z}$-module.*

In general, as long as the property $\mathcal{P}$ in the Modified Projectivity Test is provided, we can eventually prove it in ACA_0. This will be done in Sect. 5.

Note that we have the well-known Baer's Test for injectivity. Yamazika have already obtained the reverse mathematical result of it.

Theorem 1. *(Yamazika [3]) The following are equivalent over RCA_0.*

(1) ACA_0.
(2) Let R be a commutative ring. If for any ideal I of R, and R-homomorphism $g : I \to M$, there is an R-homomorphism $h : R \to M$ such that $g(x) = h(x)$ for all $x \in I$, then M is injective.

For the detail proof of Theorem 1 above, please refer to the author's Ph.D thesis (Theorem 7.4, [8]).

Dual to the Modified Projectivity Test, we also have the Modified Injectivity Test for testing injective modules. We now introduce this test for injective modules and study the proof-theoretic strength of it. Let $\mathcal{Q}$ be a property of R-modules such that for any R-module M, there is a surjective R-homomorphism from an R-module N possessing property $\mathcal{Q}$ to M.

- **Modified Injectivity Test:** An R-module M is injective if and only if for any one-to-one R-homomorphism $\iota : N_1 \to N_2$ with N_2 possessing property $\mathcal{Q}$ and R-homomorphism $g : N_1 \to M$, there is an R-homomorphism $h : N_2 \to M$ such that $g = h\iota$.

Note that RCA_0 proves that an R-module is a surjective image of a free R-module, the property $\mathcal{Q}$ can be taken as the freeness of modules. We will study the Modified Injectivity Test and prove it in RCA_0. Although Baer's Test is more effective for testing injectivity in classical algebra, the proof of it requires ACA_0, which is more harder than the Modified Injectivity Test.

The rest of the paper is organized as follows. We first formalize and prove the Dual Basis Lemma and the Kaplansky's Theorem in RCA_0 in Sects. 2 and 3, respectively. We then define invertible modules via Σ_2^0 formulas and study properties of invertible R-submodules of the total ring of quotients of R in Sect. 4. Finally, in Sect. 5, we formalize and prove the Modified Projectivity Test and the Modified Injectivity Test within ACA_0 and RCA_0, respectively.

2   Dual Basis Lemma

Dual Basis Lemma provides a nice characterization for projective modules, it is useful to prove properties of invertible modules in Sect. 4. We first study this lemma.

Theorem 2. *(Dual Basis Lemma) The following are equivalent over* RCA_0.

(1) An R-module M is projective.
(2) There exists a Σ_1^0-sequence $\langle x_i : i \in \mathbb{N} \rangle$ of elements of M and a uniform sequence $\langle f_i : i \in \mathbb{N} \rangle$ of R-homomorphisms from M to R such that for any $x \in M$, $f_i(x) = 0$ for almost all i, and $x = \sum_{i \in \mathbb{N}} f_i(x) x_i$.

Proof. For $(1) \Rightarrow (2)$. Suppose that M is projective. RCA_0 proves that there is a free module F with a basis $\{b_i : i \in \mathbb{N}\}$ and a onto R-homomorphism $g : F \to M$. For the identity R-homomorphism $id_M : M \to M$, by Lemma 1, there is a homomorphism $\psi : M \to F$ such that $id_M = g\psi$. That is, the diagram commutes:

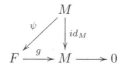

For any $x \in M$, $x = g(\psi(x))$ and $\psi(x)$ is a unique finite R-linear sum of base elements. Let $\psi(x) = \sum_{i \in I} r_i b_i$ be the unique finite sum. For any $x \in M$, define $f(i, x) = r_i$ if $i \in I$ and 0 otherwise. As we can define the unique expression $\sum_{i \in I} r_i b_i$ of $\psi(x)$ effectively from x, $f : \mathbb{N} \times M \to R$ exists in RCA_0.

For each $i \in \mathbb{N}$, let $f_i(x) = f(i, x), x \in M$. By definition, $f_i(x) = 0$ except finitely many $i \in \mathbb{N}$. It can be checked directly that f_i is an R-homomorphism from M to R. Now $\langle f_i : i \in \mathbb{N} \rangle$ is a uniform sequence of R-homomorphisms. For each $i \in \mathbb{N}$, let $x_i = g(b_i)$, then $\langle x_i : i \in \mathbb{N} \rangle$ is a Σ_1^0-sequence of M. Moreover,

$$x = g(\psi(x)) = g(\sum_{i \in I} r_i b_i) = \sum_{i \in I} r_i g(b_i) = \sum_{i \in I} f_i(x) x_i = \sum_{i \in \mathbb{N}} f_i(x) x_i.$$

For (2) $\Rightarrow$ (1). Suppose that the desired Σ_1^0-sequence $\langle x_i : i \in \mathbb{N} \rangle$ and the desired uniform sequence $\langle f_i : i \in \mathbb{N} \rangle$ exist. Let $F = \bigoplus_{i \in \mathbb{N}} R b_i$ be the free R-module with basis $\{b_i : i \in \mathbb{N}\}$. On the one hand, we can define an R-homomorphism $g : F \to M$ by setting $g(b_i) = x_i, i \in \mathbb{N}$.

On the other hand, as $x = \sum_{i \in \mathbb{N}} f_i(x) x_i$ for each $x \in M$, we can also define an R-homomorphism $\psi : M \to F$ by mapping x_i to b_i, that is,

$$\psi(x) = \psi(\sum_{i \in \mathbb{N}} f_i(x) x_i) = \sum_{i \in \mathbb{N}} f_i(x) \psi(x_i) = \sum_{i \in \mathbb{N}} f_i(x) b_i.$$

For each $x \in M$,

$$g(\psi(x)) = g(\sum_{i \in \mathbb{N}} f_i(x) b_i) = \sum_{i \in \mathbb{N}} f_i(x) g(b_i) = \sum_{i \in \mathbb{N}} f_i(x) x_i = x.$$

So $g\psi = id_M$, and ψ is one-to-one. Then $F \cong M \bigoplus ker(g)$. Indeed,

$$\Psi : M \bigoplus ker(g) \to F; \langle x, y \rangle \mapsto \psi(x) + y$$

is an R-module isomorphism, we omit the direct checking here. Now $M \bigoplus ker(g)$ is a free R-module. That is, M is a direct summand of a free module, and thus projective. $\qquad \square$

3 Kaplansky's Theorem

Classically, a commutative ring R is hereditary if every ideal of it is projective as R-modules. Note that Downey, Lempp and Mileti proved in [10] that WKL_0 is equivalent to the statement that a commutative ring which is not a field has a nontrivial proper ideal. So a commutative ring may have no ideals in RCA_0. Like Σ_1^0-principal ideal domains (such domains are already studied in literature, such as [8,9]), to study properties of hereditary rings, we will focus on Σ_1^0-hereditary rings. Recall that a sequence $\langle r_i : i \in \mathbb{N} \rangle$ of a commutative ring R is called a Σ_1^0-ideal of R if $\forall i, j \in \mathbb{N} \exists k \in \mathbb{N}[r_k = r_i + r_j]$ and $\forall i \in \mathbb{N} \forall r \in R \exists k \in \mathbb{N}[r_k = r r_i]$.

Definition 6. (RCA$_0$) *A commutative ring R is called Σ_1^0-hereditary if every Σ_1^0-ideal of R is projective as R-modules.*

Σ_1^0-principal ideal domains are natural examples of Σ_1^0-hereditary domains.

Theorem 3. (RCA$_0$) *(Kaplansky's Theorem) Let R be a Σ_1^0-hereditary ring. Then any submodule M of a free R-module is a direct sum of projective R-modules; in particular, M is projective.*

Proof. Let F be a countable free R-module. Then F is isomorphic to $\bigoplus\limits_{0 \le i \le n} R$ for some $n \in \mathbb{N}$ or $\bigoplus\limits_{i \in \mathbb{N}} R$. Without loss of generality, assume that $F = \bigoplus\limits_{i \in \mathbb{N}} R$. Elements of F are finite tuples of the form $\langle r_0, \cdots, r_n \rangle$ with $r_0, \cdots, r_n \in R$. For each $n \in \mathbb{N}$, set $F_n = \bigoplus\limits_{0 \le i \le n} R$, and let $F_{-1} = 0$, the zero R-module.

For a submodule M of F, we can define an R-homomorphism

$$f_n : M \cap F_n \to R; \; x = \langle r_0, \cdots, r_n \rangle \mapsto r_n.$$

Then $ker(f_n) = M \cap F_{n-1}$ and $im(f_n) = M \cap F_n / M \cap F_{n-1}$. We view elements of $im(f_n)$ as least $\le$-representatives under the equivalence relation $\sim$:

- for any $x, y \in M \cap F_n$, $x \sim y \Leftrightarrow x - y \in M \cap F_{n-1}$,

where $\le$ is a fixed linear order on F. So $im(f_n)$ is indeed Δ_1^0 with parameter M. Now $im(f_n)$ is a Δ_1^0 ideal of R. As R is Σ_1^0-hereditary, $im(f_n)$ is projective as R-modules.

For simplicity, we adopt symbol A_n to denote $im(f_n)$. As A_n is projective, for the identity homomorphism $id_{A_n} : A_n \to A_n$ and the surjective homomorphism $f_n : M \cap F_n \to A_n$, there is an R-homomorphism $g_n : A_n \to M \cap F_n$ such that $id_{A_n} = f_n g_n$. Then we have the commutative diagram:

Let $B_n = \{y \in M \cap F_n : \exists x \in A_n [g_n(x) = y]\}$ be the image of g_n. Then $g_n : A_n \to B_n$ is an isomorphism. As A_n is projective, so is B_n. At first glance, B_n is Σ_1^0. By the following claim, B_n is indeed Δ_1^0. So it exists in RCA$_0$.

Claim. $M = \bigoplus\limits_{n \in \mathbb{N}} B_n$.

Proof of the Claim. First, we show that for each $n \in \mathbb{N}$,

$$M \cap F_n = (M \cap F_{n-1}) \bigoplus B_n.$$

Let $x = \langle r_0, \cdots, r_n \rangle \in M \cap F_n$. Then $f_n(x) = r_n = f_n(g_n(r_n))$, and then

$$y = x - g_n(r_n) \in ker(f_n) = M \cap F_{n-1}.$$

So $x = y + g_n(r_n) \in M \cap F_{n-1} + B_n$. This shows that $M \cap F_n = (M \cap F_{n-1}) + B_n$. If $y \in (M \cap F_{n-1}) \cap B_n$, let $y = g_n(r)$ with $r \in A_n$, then

$$r = f_n(g_n(r)) = f_n(y) = 0,$$

and then $y = g_n(r) = 0$. So $(M \cap F_{n-1}) \cap B_n = \{0\}$.

Second, we prove that $M = \sum_{n \in \mathbb{N}} B_n$. Suppose otherwise, then $M \supsetneq \sum_{n \in \mathbb{N}} B_n$. Let N be the least number such that there is a $x \in (M \cap F_N) \setminus \sum_{n \in \mathbb{N}} B_n$. By $M \cap F_N = (M \cap F_{N-1}) \oplus B_N$, there are $y \in M \cap F_{N-1}$ and $z \in B_N$ such that $x = y + z$. If $y \in \sum_{n \in \mathbb{N}} B_n$, then $x \in \sum_{n \in \mathbb{N}} B_n$, which is a contradiction. We have $y \notin \sum_{n \in \mathbb{N}} B_n$. But then $y \in (M \cap F_{N-1}) \setminus \sum_{n \in \mathbb{N}} B_n$, this also contradicts the choice of N. So $M = \sum_{n \in \mathbb{N}} B_n$.

Third, $M = \bigoplus_{n \in \mathbb{N}} B_n$. Let $x_0 + \cdots + x_n = 0$ with $x_i \in B_i$ for $i = 0, \cdots, n$. Now

$$x_n = -(x_0 + \cdots + x_{n-1}) \in (M \cap F_{n-1}) \cap B_n,$$

so $x_n = 0$, and $x_0 + \cdots + x_{n-1} = 0$. By induction, we see that $x_{n-1} = \cdots = x_0 = 0$. So the expression of zero as sums of elements in B_n with $n \in \mathbb{N}$ is unique, and thus $M = \bigoplus_{n \in \mathbb{N}} B_n$.

This completes the proof of the claim.

Since B_n is projective for all $n \in \mathbb{N}$, $M = \bigoplus_{n \in \mathbb{N}} B_n$ is a direct sum of projective R-modules. By Proposition 1, M is projective.   □

Corollary 1. (RCA$_0$) *Let R be a Σ_1^0-hereditary ring and M an R-module. Then M is projective iff it can be embedded in to a free R-module.*

Corollary 2. (RCA$_0$) *A ring R is Σ_1^0-hereditary iff Σ_1^0-submodules of projective R-modules are projective.*

Proof. ($\Rightarrow$) Let M be a projective R-module and N a Σ_1^0-submodule of M. By definition, M is a direct summand of a free R-module. Then N is a Σ_1^0-submodule of a free R-module. By almost the same proof in the Kaplansky's Theorem above, the Σ_1^0-submodule N is projective.

($\Leftarrow$) Let I be a Σ_1^0-ideal of R. Then I is a Σ_1^0-submodule of the regular module R. As the regular module R is projective, I is projective as R-modules.   □

Corollary 3. (RCA$_0$) *Let R be a Σ_1^0-principal ideal domain. Then every Σ_1^0-submodule of a projective R-module is projective.*

4   Invertible Modules

Definition 7. (RCA$_0$) *Let $R \subseteq S$ be commutative rings. S is a natural R-module. For two R-submodules M, N of S, the multiplication of M, N, denoted by MN, is the R-submodule of S generated by $\{xy : x \in M, y \in N\}$.*

Classically, an R-submodule M of an extension ring S of R is invertible if and only if the multiplication of M and M^{-1} is R (i.e., $MM^{-1} = R$), where $M^{-1} = \{y \in S : My \subseteq R\}$ and $My = \{xy : x \in M\}$. Observe that $MM^{-1} = R$ is equivalent to $1_R \in MM^{-1}$, invertible submodules can be defined by Σ_2^0-formulas.

Definition 8. (RCA$_0$) *An R-submodule M of a commutative ring S is invertible if there are $x_1, \cdots, x_n \in M$ and $y_1, \cdots, y_n \in S$ such that for all $x \in M$, we have $xy_1, \cdots, xy_n \in R$ and $1_R = x_1 y_1 + \cdots x_n y_n$.*

Invertible modules are closely related to projective modules. We now study properties of invertible modules.

Proposition 3. (RCA$_0$) *An invertible R-submodule of a commutative ring S is finitely generated and projective.*

Proof. Let M be an invertible R-submodule of S. By definition, $1_R = x_1 y_1 + \cdots + x_n y_n$ for some $x_1, \cdots, x_n \in M$ and $y_1, \cdots, y_n \in S$ with $xy_1, \cdots xy_n \in R$ for all $x \in M$. To show that M is projective, we will apply the Dual Basis Lemma (i.e., Theorem 2 in Sect. 2). As S is commutative, for any $z \in M$, we have

$$z = zx_1 y_1 + \cdots + zx_n y_n$$
$$= zy_1 x_1 + \cdots + zy_n x_n$$

Now $\{x_1 \cdots, x_n\}$ is a finite sequence of elements of M. Note that for $1 \leq i \leq n$, $zy_i \in R$, we can define R-homomorphisms $f_i : M \to R$ by setting $f_i(z) = zy_i$. Then $z = f_1(z)x_1 + \cdots + f_n(z)x_n$, and then M is projective by Dual Basis Lemma. Moreover, M is generated by $\{x_1, \cdots, x_n\}$. $\qquad \square$

We now study invertible submodules inside the total ring of quotients of commutative rings. That is, the localization of a ring at its regular elements, where an element x of R is regular if $\forall y \in R[xy = 0_R \to y = 0_R]$.

Let R be a commutative ring and

$$Reg(R) = \{x \in R : \forall y \in R[xy = 0_R \to y = 0_R]\}.$$

The total ring of quotients of R is just $Q(R) = R \times Reg(R)/ \sim$, where for any $(r, s), (r', s') \in R \times Reg(R)$,

$$(r, s) \sim (r', s') \Leftrightarrow \exists u \in Reg(R) [urs' = ur's].$$

As usual, the equivalence class of (r, s) under the equivalence relation $\sim$ are denoted by rs^{-1}, and elements of $Q(R)$ are of the form rs^{-1} with $r \in R, s \in Reg(R)$.

Quotient fields are standard examples of ring extension of domains. The process of constructing quotient fields from a given domain is effective, so RCA_0 proves that every domain have a quotient field. Unlike quotient fields, Sato proved in his PhD thesis (Theorem 6.8, [9]) that the statement that every commutative ring has the total ring of quotients is equivalent to ACA_0 over RCA_0.

Proposition 4. *Assume that R is a commutative ring such that $Q(R)$ exists. Let M be an R-submodule of $Q(R)$, then the following are equivalent over RCA_0.*

(1) M is invertible.
(2) M is projective and $M \cap Reg(R) \neq \emptyset$.

Proof. For $(1) \Rightarrow (2)$. By Proposition 3, M is projective and $1_R = x_1 y_1 + \cdots x_n y_n$ for some $x_1, \cdots, x_n \in M$ and $y_1, \cdots, y_n \in Q(R)$. For $i = 1, \cdots, n$, by taking a common denominator z, we can write $x_i = a_i z^{-1}$ and $y_i = b_i z^{-1}$ where $a_i, b_i \in R, z \in Reg(R)$. Then $a_i = zx_i \in M$, $b_i = zy_i \in R$ for $1 \leq i \leq n$, and then

$$z^2 = zx_1 zy_1 + \cdots + zx_n zy_n = a_1 b_1 + \cdots + a_n b_n \in M.$$

As z is regular, so is z^2. Then $z^2 \in M \cap Reg(R)$, and $M \cap Reg(R) \neq \emptyset$.

For $(2) \Rightarrow (1)$. Assume that M is projective. By the Dual Basis Lemma, there is a Σ_1^0-sequence $\langle x_i : i \in \mathbb{N} \rangle$ of elements of M and a uniform sequence $\langle f_i : i \in \mathbb{N} \rangle$ of R-homomorphisms from M to R such that for any $x \in M$, $f_i(x) = 0$ for almost all $i \in \mathbb{N}$ and $x = \sum_{i \in \mathbb{N}} f_i(x)x_i$. Fix an element $z \in M \cap Reg(R)$. Then $f_i(z) = 0$ for almost all i. Let $I = \{i \in \mathbb{N} : f_i(z) \neq 0\}$. I is a finite set and $z = \sum_{i \in I} f_i(z)x_i$.

For each $i \in I$, let $y_i = f_i(z)z^{-1}$. Then $y_i \in Q(R)$ and $f_i(z) = y_i z$. Moreover, for all $x \in M$, let $x = rs^{-1}$ where $r \in R, s \in Reg(R)$, we have

$$szf_i(x) = f_i(szx) = f_i(szrs^{-1}) = f_i(rz) = rf_i(z),$$

then

$$f_i(x) = z^{-1}s^{-1}rf_i(z) = z^{-1}f_i(z)x = y_i x.$$

So $y_i x \in R$ for all $x \in M$. By $z = \sum_{i \in I} f_i(z)x_i = \sum_{i \in I} zy_i x_i = z(\sum_{i \in I} x_i y_i)$, We have $1_R = \sum_{i \in I} x_i y_i$.

Hence, there exist finite sequences $\langle x_i \in M : i \in I \rangle$ and $\langle y_i \in Q(R) : i \in I \rangle$ such that for each $i \in I$, $xy_i \in R$ for all $x \in M$ and $1_R = \sum_{i \in I} x_i y_i$. By definition, M is invertible. $\square$

5 Modified Projectivity and Injectivity Test

Proposition 5. *(ACA_0) Modified Projectivity Test: Let $\mathcal{P}$ be a property such that an R-module can be embedded into an R-module with property $\mathcal{P}$. Then M is a projective R-module if and only if for any surjective R-homomorphism $g : N_1 \to N_2$ with N_1 possessing property $\mathcal{P}$ and R-homomorphism $h : M \to N_2$, there is an R-homomorphism $\psi : M \to N_1$ such that $h = g\psi$.*

Proof. ($\Rightarrow$) If M is projective, that is, M is a direct summand of a free R-module, then by Lemma 1, the conclusion is clear.

($\Leftarrow$) We reason in ACA_0. Let $g : N_1 \to N_2$ be a surjective R-homomorphism and $h : M \to N_2$ an R-homomorphism, we need to show that there is an R-homomorphism $\psi : M \to N_1$ such that $h = g\psi$.

First, N_1 can be embedded into an R-module satisfying $\mathcal{P}$, M_1 say, and let $\lambda : N_1 \hookrightarrow M_1$ be the embedding. Form the direct sum $M_1 \oplus N_2$ of M_1, N_2. For convenience, assume that M_1, N_2 are subsets of $\mathbb{N}$ (generally, elements of modules are encoded by natural numbers). The elements of $M_1 \oplus N_2$ are pairs $\langle x, y \rangle$ with $x \in M_1, y \in N_2$, where $\langle \cdot, \cdot \rangle : \mathbb{N} \times \mathbb{N} \to \mathbb{N}$ is a fixed bijection. $M_1 \oplus N_2$ is the R-module with usual point wise operations, so it exists in RCA_0. By a direct checking, $A = \{\langle \lambda(x), -g(x) \rangle : x \in N_1\}$ is a submodule of $M_1 \oplus N_2$. Since

$$\langle y, z \rangle \in A \Leftrightarrow \exists x \in N_1[y = \lambda(x) \wedge z = -g(x)],$$

A is Σ_1^0 and thus exists by Σ_1^0-comprehension. Let B be the quotient module of $M_1 \oplus N_2$ modulo A. Elements of B are of the form $\langle x, y \rangle$ with additional equality relation $=_B$: $\langle x, y \rangle =_B \langle x', y' \rangle \Leftrightarrow \langle x - x', y - y' \rangle \in A$. Then B exists in ACA_0. Define R-homomorphisms $\alpha : M_1 \to B; x \mapsto \langle x, 0 \rangle$ and $\beta : N_2 \to B; y \mapsto \langle 0, y \rangle$.

(i) α is onto. Let $\langle x, y \rangle \in B$. As $y \in N_2$, and g is onto, there is a $z \in N_1$ such that $y = g(z)$. Then $\langle 0, y \rangle =_B \langle 0, g(z) \rangle =_B \langle \lambda(z), 0 \rangle$ and then

$$\langle x, y \rangle =_B \langle x, 0 \rangle + \langle 0, y \rangle =_B \langle x, 0 \rangle + \langle \lambda(z), 0 \rangle =_B \langle x + \lambda(z), 0 \rangle = \alpha(x + \lambda(z)).$$

(ii) β is one-to-one. Let $y \in N_2$ with $\beta(y) = \langle 0, 0 \rangle$. Then $\langle 0, y \rangle \in A$, and then $y = -g(0) = 0$.

For the surjective $\alpha : M_1 \to B$ and the homomorphism $\beta h : M \to B$, as M_1 has the property $\mathcal{P}$, by the assumption, there is a homomorphism $\psi : M \to M_1$ such that $\alpha\psi = \beta h$. Now we have the following commutative diagram:

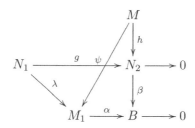

Note that $im(\lambda) = \{y \in M_1 : \exists x \in N_1[\lambda(x) = y]\}$. We show that for any $x \in M$, $\psi(x) \in im(\lambda)$. By

$$\alpha(\psi(x)) = \langle \psi(x), 0 \rangle = \beta(h(x)) = \langle 0, h(x) \rangle,$$

$\langle \psi(x), -h(x) \rangle \in A$. That is, there is a $y \in N_1$ such that $\psi(x) = \lambda(y)$, and $-h(x) = -g(y)$. In particular, $\psi(x) \in im(\lambda)$. So ψ is also an R-homomorphism from M to $im(\lambda)$. As λ is one-to-one, the inverse of λ is an R-homomorphism from $im(\lambda)$ to N_1.

Let $\varphi = \lambda^{-1}\psi : M \to N_1$ be an R-homomorphism. For any $x \in M$, we have

$$\beta h(x) = \alpha(\psi(x)) = \alpha\lambda(\lambda^{-1}\psi(x)) = \alpha\lambda(\varphi(x)) = \beta(g(\varphi(x))).$$

Since β is one-to-one, for any $x \in M$, $h(x) = g(\varphi(x))$. That is, $h = g\varphi$. We have shown that for any surjective R-homomorphism $g : N_1 \to N_2$ and any R-homomorphism $h : M \to N_2$, there is an R-homomorphism $\varphi : M \to N_1$ such that $h = g\varphi$. By Lemma 1, M is projective. □

Unlike the Modified Projectivity Test, the Modified Injectivity Test for injective modules is indeed provable in RCA_0, we omit the detail proof here.

Proposition 6. (RCA_0) *Modified Injectivity Test: Let $\mathcal{Q}$ be a property of R-modules such that for any R-module M, there is an R-module N possessing property $\mathcal{Q}$ and also a surjective R-homomorphism from N to M. Then M is an injective R-module if and only if for any one-to-one R-homomorphism $\iota : N_1 \to N_2$ with N_2 possessing property $\mathcal{Q}$ and any R-homomorphism $g : N_1 \to M$, there is an R-homomorphism $h : N_2 \to M$ such that $g = h\iota$.*

References

1. Friedman, H.M., Simpson, S.G., Smith, R.L.: Countable algebra and set existence axioms. Ann. Pure Appl. Logic **25**, 141–181 (1983)
2. Simpson, S.G.: Subsystems of Second Order Arithmetic. Springer, Heidelberg (1999)
3. Yamazaki, T.: Reverse mathematics and commutative ring theory. Computability Theory and Foundations of Mathematics, Tokyo Institute Of Technology, 18–20 February 2013
4. Yamazaki, T.: Homological algebra and reverse mathematics (a middle report). In: Second Workshop on Mathematical Logic and its Applications in Kanazawa (2018)
5. Wu, H.: The complexity of radicals and Socles of modules. Notre Dame J. Formal Logic **61**, 141–153 (2020)
6. Lam, T.Y.: Lectures on Modules and Rings. Graduate Texts in Mathematics. Springer, New York (1998). https://doi.org/10.1007/978-1-4612-0525-8
7. Hilton, P.J., Stammbach, U.: A Course in Homological Algebra. Graduate Texts in Mathematics, 2nd edn. Springer, New York (1997). https://doi.org/10.1007/978-1-4684-9936-0
8. Wu, H.: Computability theory and algebra. Ph.D thesis, Nanyang Technological University, Singapore (2017)
9. Sato, T.: Reverse mathematics and countable algebraic systems. Ph.D thesis, Tohoku University, Sendai, Japan (2016)
10. Downey, R.G., Lempp, S., Mileti, J.R.: Ideals in commutative rings. J. Algebra **314**, 872–887 (2007)

Two-Stage Submodular Maximization Problem Beyond Non-negative and Monotone

Zhicheng Liu[1], Hong Chang[1], Ran Ma[2], Donglei Du[3], and Xiaoyan Zhang[1(✉)]

[1] School of Mathematical Science and Institute of Mathematics, Nanjing Normal University, Nanjing 210023, People's Republic of China
zhangxiaoyan@njnu.edu.cn
[2] School of Management Engineering, Qingdao University of Technology, Qingdao 266520, People's Republic of China
[3] Faculty of Management, University of New Brunswick, Fredericton, New Brunswick E3B 5A3, Canada

Abstract. Two-stage submodular maximization problems have been recently applied in machine learning, economics and engineering. In this paper, we consider a two-stage submodular problem subject to cardinality constraint and matroid constraint. Previous work for this problem usually assume that the objective functions are non-negative and monotone. Our focus in this work relaxes these assumptions by considering an objective function which is the expected difference of a non-negative monotone submodular function and a non-negative monotone modular function, and hence neither non-negative nor monotone. We present strong approximation guarantees by offering two bi-factor approximation algorithms for this problem. The first is a deterministic $\left(\frac{1}{2}\left(1 - e^{-2}\right), 1\right)$-approximation algorithm, and the second is a randomized $\left(\frac{1}{2}\left(1 - e^{-2}\right) - \epsilon, 1\right)$-approximation algorithm with improved time efficiency. Moreover, we generalize the matroid constraint to k-matroid constraint and also give the corresponding approximation algorithms.

Keywords: Submodular maximization · Greedy algorithm · Matroid

1 Introduction

We consider a two-stage submodular maximization problem. Given a ground set $V = \{1, \ldots, n\}$, let $F = (f_1, f_2, \ldots, f_m)$ be a set of functions such that each $f_j : 2^V \to \mathbb{R}_+$ $(j = 1, \ldots, m)$ is a nonnegative monotone submodular function, and let $\ell = (\ell_1, \ldots, \ell_m)$ be a set of functions such that each $\ell_j : 2^V \to \mathbb{R}_+$ $(j = 1, \ldots, m)$ is a nonnegative monotone modular function. For a given nonnegative integer p, the two-stage problem is as follows

$$\max_{S \subseteq V : |S| \leq p} F(S) := \sum_{j=1}^{m} \max_{T \in \mathcal{I}(S)} (f_j(T) - \ell_j(T)), \tag{1.1}$$

© Springer Nature Switzerland AG 2020
J. Chen et al. (Eds.): TAMC 2020, LNCS 12337, pp. 144–155, 2020.
https://doi.org/10.1007/978-3-030-59267-7_13

where $\mathcal{I}(\mathcal{S})$ is the family of the common independent sets of matroid $\mathcal{M} = (S, \mathcal{I}(S))$ over the same ground set $S \subseteq V$. Any matriod satisfies three properties: (i) $\emptyset \in \mathcal{I}$; (ii) If $J' \subseteq J \in \mathcal{I}(S)$, then $J' \in \mathcal{I}(S)$; and (iii) $\forall A, B \in \mathcal{I}(S)$, if $|A| < |B|$, then there exists an element $u \in B \setminus A$ such that $A + u \in I$.

A set function $f : 2^V \to \mathbb{R}_+$ is non-decreasing if $f(S) \leq f(T), \forall S \subseteq T \subseteq V$. It is submodular if $f(S) + f(T) \geq f(S \cap T) + f(S \cup T), \forall S, T \subseteq V$. It is supermodular if its negative is submodular. It is modular if it is both submodular and supermodular.

Problem (1.1) is two-stage in nature because of the two optimization phases involved therein: (1) the first stage finds the optimal $S \subseteq V$ of size at most p to maximize the average of the first stage objective values over $j = 1, \ldots, m$; and (2) the second stage (inner problem) maximizes the difference between a monotone submodular function and a monotone modular function subject to an arbitrary matriod constraint on the ground $S \subseteq V$ and $j = 1, \ldots, m$, resulting in the optimal objective value being a function of S.

Problem (1.1) has various interpretations which lead to many applications [2, 18, 22]. For instance, the modular function ℓ may be considered as a penalty when choosing elements in existing submodular maximization problems. In facility location games, we can assume the function f represents the revenue and the function ℓ represents the cost of opening a facility. Our contribution is to offer two bi-factor approximation algorithms to solve Problem (1.1).

First, we present a deterministic algorithm that returns a set $S \subseteq V$ such that

$$F(S) \geq \frac{1}{2}\left(1 - \frac{1}{e^2}\right) \sum_{j=1}^{m} f_j(S_j^*) - \sum_{j=1}^{m} \ell_j(S_j^*),$$

where S_j^* $(j = 1, \ldots, m)$ is the optimal solution of the following problem

$$S_j^* = \arg \max_{T \in I(S^*)} (f_j(T) - \ell_j(T)),$$

and S^* is the optimal solution of $F(S)$.

This algorithm involves $O(rmpn)$ function evaluations, where $r = r_{\mathcal{M}}(V)$ is the rank of the ground set (The rank function of a matriod $\mathcal{M} = (V, \mathcal{I})$ is defined as $r_{\mathcal{M}}(S) = \max\{|T| : T \subseteq S, T \in I\}, \forall S \subseteq V$.)

The second algorithm is randomized with improved function calls $O(rm \log \frac{1}{\epsilon} n)$, while sacrificing only an $\epsilon > 0$ in the approximation ratio.

Finally, we generalize the matroid constraint to k-matroid constraint and give two algorithms. The first is a deterministic algorithm that returns a set $S \subseteq V$ such that

$$F(S) \geq \frac{1}{k+1}\left(1 - \frac{1}{e^{k+1}}\right) \sum_{j=1}^{m} f_j(S_j^*) - \sum_{j=1}^{m} \ell_j(S_j^*),$$

and the second algorithm is randomized with improved time efficiency.

The main challenges in this problem are as follows:

(1) for $i \in \{1, \ldots, p\}$, we need to construct functions

$$\Phi_i(S) = \left(1 - \frac{2}{p}\right)^{p-i} \sum_{j=1}^{m} f_j(T_j) - \sum_{j=1}^{m} \ell_j(T_j);$$

(2) the objective functions are not monotone.

The rest of this paper is organized as follows. In Sect. 2, we review relevant literature. In Sect. 3 we introduce some definitions and properties of submodular function and matroid. In Sects. 4.1 and 4.3, respectively, we consider the case when constraints are cardinality constraint and matroid constraint, and present a deterministic algorithm and a randomized algorithm, along with their approximation ratios analysis. In Sect. 4, we consider the k-matroid constraint. Finally, we offer concluding remarks in Sect. 6.

2   Related Work

The literature on (single-stage) submodular maximization is extensive due to its wide applications in machine learning, economics, and engineering, among many others. For example, feature/variable selection [12], crowd teaching [21], recommender systems [3], and influence maximization [11]. Especially, submodularity has been used in a lot of summarization settings. The authors in [17] selected representative images using an exemplarbased clustering approach. The authors in [7,9] worked on submodular image summarization directly, while the authors in [19] investigate document summarization. Since the modern data sets is very large, much work pays attention to solving submodular maximization at large scale. This work ranges from distributed submodular optimization [4,13,16,17], to streaming approaches [1,10], as well as algorithms based on filtering the ground set in multiple stages [6].

The most relevant work to our problem is the two-stage submodular maximization problem investigated in [2,18,22], where the objective function is a monotone submodular function and satisfies some distribution $\mathbb{D}$. In [2], they give two algorithms based on techniques of continuous optimization and local search, although their runtimes are very expensive. The authors in [18] develop the first streaming and distributed algorithms to this problem. The authors in [22] use a simple greedy algorithm to improve the approximation ratio from $\frac{1}{2}(1 - \frac{1}{e})$ to $\frac{1}{2}(1 - \frac{1}{e^2})$.

Our problem is also related to the problem of the maximization of the difference between γ-weakly submodular and modular function [8], where their objective function is $f - \ell$ with f being a monotone γ-weakly submodular function and ℓ being a non-negative modular function. They present an algorithm with approximation ratio $1 - \frac{1}{e}$.

Another relevant problem considers an objective function that is the sum of a submodular function and a modular function [5,23]. In [23], they consider the problem of maximizing $f + \ell$ where f is a monotone submodular function and l is an arbitrary modular function subject to a solvable polytope constraint. They use the multilinear extension technique to present an $(1 - 1/e)$-approximation algorithm, assuming the optimal value is known. In [5], a new greedy algorithm is designed to remove assumption of knowing the optimal value.

3 Preliminaries

Given a ground set $V = \{1, \ldots, n\}$ and a family of subsets $\mathcal{I}$ of V, a matroid $\mathcal{M} = (V, \mathcal{I})$ satisfies the following properties

(1) If $A \subseteq B \in \mathcal{I}$, then $A \in \mathcal{I}$;
(2) If $A, B \in \mathcal{I}$ and $|A| \leq |B|$, then there exists an element $u \in B \backslash A$ for which $A + u \in \mathcal{I}$.

Next, we introduce the equivalent definition of submodular function and properties of submodular function.

Definition 1. *A function $f \colon 2^V \to \mathbb{R}_+$ is submodular if for every $X \subseteq Y \subseteq V, a \in V \setminus Y$,*

$$f(X \cup \{a\}) - f(X) \geq f(Y \cup \{a\}) - f(Y).$$

Property 1. *For any submodular function $f \colon 2^V \to \mathbb{R}_+$ and $X, Y \subseteq V$, we have*

$$\sum_{u \in X} (f(Y \cup \{u\}) - f(Y)) \geq f(X \cup Y) - f(Y).$$

4 Two-Stage Submodular Maximization Subject to Cardinality and Matroid Constraints

4.1 The Deterministic Algorithm

The algorithm works in p rounds. In every round, we use a distorted greedy method to get the maximum increment and our algorithm only adds the element with positive marginal value. It starts with an empty set $S^0 = \emptyset$ and uses exchange property of matroid to avoid violating the matroid constraints.

For convenience, we let $\Delta_j^f(x, A) = f_j(\{x\} \cup A) - f_j(A) \geq 0$, and $\Delta_j^l(x, A) = \ell_j(\{x\} \cup A) - \ell_j(A) \geq 0$ denote the marginal contributions of an element x to the set A of the functions f_j, and ℓ_j, respectively, where these marginal contributions are nonnegative due to the monotonicity of f_j and ℓ_j. We use $\nabla_j^f(x, y, A) = f_j(\{x\} \cup A \setminus y) - f_j(A)$, and $\nabla_j^l(x, y, A) = \ell_j(\{x\} \cup A \setminus y) - \ell_j(A)$ to define the corresponding gains of removing a set $y \in A$ and replacing it with x, respectively. These two quantities' signs are not restricted. In addition, we also denote

$$\Delta_j(x, A) = \left(1 - \frac{2}{p}\right)^{p-(i+1)} \Delta_j^f(x, A) - \Delta_j^l(x, A),$$

$$\nabla_j(x, y, A) = \left(1 - \frac{2}{p}\right)^{p-(i+1)} \nabla_j^f(x, y, A) - \nabla_j^l(x, y, A).$$

Algorithm 1: Distorted Replacement Greedy

1: $S^0 \leftarrow \emptyset, T_j^0 \leftarrow \emptyset \ (\forall 1 \leq j \leq m)$
2: **for** $0 \leq i \leq p-1$ **do**
3:     $e_i \leftarrow \arg\max_{e \in V} \sum_{j=1}^{m} \left[\left(1 - \frac{2}{p}\right)^{p-(i+1)} \nabla_j^f(e, T_j^i) - \nabla_j^l(e, T_j^i) \right]$
4:     **if** $\sum_{j=1}^{m} \left[\left(1 - \frac{2}{p}\right)^{p-(i+1)} \nabla_j^f(e_i, T_j^i) - \nabla_j^l(e_i, T_j^i) \right] > 0$ **then**
5:         $S^{i+1} \leftarrow S^i \cup \{e_i\}$
6:         **for all** $1 \leq j \leq m$ **do**
7:             **if** $\left(1 - \frac{2}{p}\right)^{p-(i+1)} \nabla_j^f(e_i, T_j^i) - \nabla_j^l(e_i, T_j^i) > 0$ **then**
8:                 $T_j^{i+1} \leftarrow T_j^i \cup \{e_i\} \setminus \mathsf{Rep}_j(e_i, T_j^i)$
9:             **else**
10:                 $T_j^{i+1} \leftarrow T_j^i$
11:             **end if**
12:         **end for**
13:     **else**
14:         $S^{i+1} \leftarrow S^i, T_1^{i+1} \leftarrow T_1^i, T_2^{i+1} \leftarrow T_2^i, \cdots, T_m^{i+1} \leftarrow T_m^i$
15:     **end if**
16: **end for**
17: Return sets S^p and $T_1^p, T_2^p, \cdots, T_m^p$

Consider the set A. According to the exchange property of matroid, our algorithm does not violate the matroid constraint, i.e. $\mathcal{I}(x, A) = \{y \in A : A \cup \{x\} \setminus \{y\} \in \mathcal{I}\}$. So, we define the replacement gain of x w.r.t. a set A as follows:

$$\left(1 - \frac{2}{p}\right)^{p-(i+1)} \nabla_j^f(x, A) - \nabla_j^l(x, A) = \begin{cases} \max\{0, \Delta_j(x, A)\}, & \text{if } A \cup \{x\} \in \mathcal{I}, \\ \max\{0, \max_{y \in \mathcal{I}(x,A)} \nabla_j(x, y, A)\}, & \text{otherwise.} \end{cases}$$

Finally, we use $\mathsf{Rep}_j(x, A)$ to denote the element that should be replaced by x:

$$\mathsf{Rep}_j(x, A) = \begin{cases} \emptyset, & \text{if } A \cup \{x\} \in \mathcal{I}, \\ \arg\max_{y \in \mathcal{I}(x,A)} \nabla_j(x, y, A), & \text{otherwise.} \end{cases}$$

4.2   The Analysis

Define S^* as the optimal solution

$$S^* = \arg\max_{|S| \leq p} \sum_{j=1}^{m} \max_{T \in I(S)} (f_j(T) - \ell_j(T)).$$

Denote S_j^* as the optimal solution of $f_j - \ell_j$

$$S_j^* = \arg\max_{T \in I(S^*)} (f_j(T) - \ell_j(T)).$$

According to the algorithm, S_i is the solution of function $F(S)$ obtained by the greedy heuristic in the ith round, and T_j^i is the set which was chosen by $f_j(T)$.

Our analysis relies on the following potential function. Let p denote the cardinality constraint. For any S^i in the algorithm $(i = 0, 1, \cdots, p)$, define

$$\Phi_i(S^i) = \left(1 - \frac{2}{p}\right)^{p-i} \sum_{j=1}^m f_j(T_j^i) - \sum_{j=1}^m \ell_j(T_j^i).$$

Define further

$$\tilde{\nabla}(e_i, T^i) = \max\left\{\sum_{j=1}^m \left[\left(1 - \frac{2}{p}\right)^{p-(i+1)} \nabla_j^f(e_i, T_j^i) - \nabla_j^l(e_i, T_j^i)\right], 0\right\}.$$

Our analysis includes three lemmas. First, we use Lemma 1 below to bound the distorted objective function Φ.

Lemma 1. *In each iteration of the Distorted Replacement Greedy algorithm,*

$$\Phi_{i+1}(S^{i+1}) - \Phi_i(S^i) = \tilde{\nabla}(e_i, T^i) + \frac{2}{p}\left(1 - \frac{2}{p}\right)^{p-(i+1)} \sum_{j=1}^m f_j(T_j^i).$$

We present the complete proof in the appendix.

The second lemma gives a lower bound of $\left(1 - \frac{2}{p}\right)^{p-(i+1)} \nabla_j^f(e_i, T_j^i) - \nabla_j^l(e_i, T_j^i)$.

Lemma 2

$$\left(1 - \frac{2}{p}\right)^{p-(i+1)} \nabla_j^f(e_i, T_j^i) - \nabla_j^l(e_i, T_j^i)$$

$$\geq \left(1 - \frac{2}{p}\right)^{p-(i+1)} [f_j(\{e_i\} \cup T_j^i \setminus y) - f_j(T_j^i)] - [\ell_j(\{e_i\} \cup T_j^i) - \ell_j(T_j^i)],$$

where $y \in T_j^i$.

Proof. If $T_j^i \cup \{e_i\} \in \mathcal{I}$, and $y \in T_j^i$, then

$$\left(1 - \frac{2}{p}\right)^{p-(i+1)} \nabla_j^f(e_i, T_j^i) - \nabla_j^l(e_i, T_j^i) \geq \Delta_j(e_i, T_j^i)$$

$$= \left(1 - \frac{2}{p}\right)^{p-(i+1)} \Delta_j^f(e_i, T_j^i) - \Delta_j^l(e_i, T_j^i)$$

$$= \left(1 - \frac{2}{p}\right)^{p-(i+1)} [f_j(e_i \cup T_j^i) - f_j(T_j^i)] - [\ell_j(e_i \cup T_j^i) - \ell_j(T_j^i)]$$

$$\geq \left(1 - \frac{2}{p}\right)^{p-(i+1)} [f_j(e_i \cup T_j^i \setminus y) - f_j(T_j^i)] - [\ell_j(e_i \cup T_j^i) - \ell_j(T_j^i)].$$

If $T_j^i \cup \{e_i\} \notin \mathcal{I}$, and $y \in T_j^i$, then

$$\left(1 - \frac{2}{p}\right)^{p-(i+1)} \nabla_j^f(e_i, T_j^i) - \nabla_j^l(e_i, T_j^i) \geq \nabla_j(e_i, y, T_j^i)$$

$$= \left(1 - \frac{2}{p}\right)^{p-(i+1)} \nabla_j^f(e_i, y, T_j^i) - \nabla_j^l(e_i, y, T_j^i)$$

$$= \left(1 - \frac{2}{p}\right)^{p-(i+1)} [f_j(e_i \cup T_j^i \setminus y) - f_j(T_j^i)] - [\ell_j(e_i \cup T_j^i \setminus y) - \ell_j(T_j^i)]$$

$$\geq \left(1 - \frac{2}{p}\right)^{p-(i+1)} [f_j(e_i \cup T_j^i \setminus y) - f_j(T_j^i)] - [\ell_j(e_i \cup T_j^i) - \ell_j(T_j^i)].$$

$\square$

In the following lemma, we show that the increment in each iteration is large enough to ensure the desired approximation ratio.

Lemma 3. *In each iteration of the Distorted Replacement Greedy algorithm,*

$$\tilde{\nabla}(e_i, T^i)$$

$$\geq \frac{1}{p}\left(1 - \frac{2}{p}\right)^{p-(i+1)} \sum_{j=1}^m f_j(S_j^*) - \frac{2}{p}\left(1 - \frac{2}{p}\right)^{p-(i+1)} \sum_{j=1}^m f_j(T_j^i) - \frac{1}{p}\sum_{j=1}^m \ell_j(S_j^*).$$

Proof. From the exchange property of matroid, there exists a mapping $\pi_j : S_j^* \setminus T_j^i \to T_j^i \setminus S_j^* \cup \{\emptyset\}$ such that $(T_j^i \setminus \pi_j(e)) \cup \{e\} \in \mathcal{I}$, where $e \in S_j^* \setminus T_j^i$.

$$\tilde{\nabla}(e_i, T^i) = \max\left\{\sum_{j=1}^m \left[\left(1 - \frac{2}{p}\right)^{p-(i+1)} \nabla_j^f(e_i, T_j^i) - \nabla_j^l(e_i, T_j^i)\right], 0\right\}$$

$$\geq \max_{e \in S^*}\left\{\sum_{j=1}^m \left[\left(1 - \frac{2}{p}\right)^{p-(i+1)} \nabla_j^f(e, T_j^i) - \nabla_j^l(e, T_j^i)\right], 0\right\}$$

$$\geq \max_{e \in S^*}\left\{\sum_{j=1}^m \left[\left(1 - \frac{2}{p}\right)^{p-(i+1)} \nabla_j^f(e, T_j^i) - \nabla_j^l(e, T_j^i)\right]\right\}$$

$$\geq \frac{1}{|S^*|} \sum_{e \in S^*}\sum_{j=1}^m \left[\left(1 - \frac{2}{p}\right)^{p-(i+1)} \nabla_j^f(e, T_j^i) - \nabla_j^l(e, T_j^i)\right]$$

$$\geq \frac{1}{p} \sum_{e \in S^*}\sum_{j=1}^m \left[\left(1 - \frac{2}{p}\right)^{p-(i+1)} \nabla_j^f(e, T_j^i) - \nabla_j^l(e, T_j^i)\right]$$

$$= \frac{1}{p} \sum_{j=1}^m \sum_{e \in S^*} \left[\left(1 - \frac{2}{p}\right)^{p-(i+1)} \nabla_j^f(e, T_j^i) - \nabla_j^l(e, T_j^i)\right]$$

$$\geq \frac{1}{p} \sum_{j=1}^m \sum_{e \in S_j^* \setminus T_j^i} \left[\left(1 - \frac{2}{p}\right)^{p-(i+1)} \nabla_j^f(e, T_j^i) - \nabla_j^l(e, T_j^i)\right]$$

$$\geq \frac{1}{p} \sum_{j=1}^m \sum_{e \in S_j^* \setminus T_j^i} \left\{\left(1 - \frac{2}{p}\right)^{p-(i+1)} [f_j(\{e\} \cup T_j^i \setminus \pi_t(e)) - f_j(T_j^i)] - [\ell_j(\{e\} \cup T_j^i) - \ell_j(T_j^i)]\right\}$$

$$= \frac{1}{p} \sum_{j=1}^m \sum_{e \in S_j^* \setminus T_j^i} \left(1 - \frac{2}{p}\right)^{p-(i+1)} [f_j(\{e\} \cup T_j^i \setminus \pi_t(e)) - f_j(T_j^i)] - \frac{1}{p} \sum_{j=1}^m \sum_{e \in S_j^* \setminus T_j^i} [\ell_j(\{e\} \cup T_j^i)$$

$$- \ell_j(T_j^i)]$$

$$= \frac{1}{p} \sum_{j=1}^{m} \sum_{e \in S_j^* \setminus T_j^i} \left(1 - \frac{2}{p}\right)^{p-(i+1)} \left[f_j(\{e\} \cup T_j^i \setminus \pi_t(e)) - f_j(\{e\} \cup T_j^i) + f_j(\{e\} \cup T_j^i) - f_j(T_j^i)\right]$$

$$- \frac{1}{p} \sum_{j=1}^{m} \sum_{e \in S_j^* \setminus T_j^i} \Delta_j^l(e, T_j^i)$$

$$= \frac{1}{p} \sum_{j=1}^{m} \sum_{e \in S_j^* \setminus T_j^i} \left(1 - \frac{2}{p}\right)^{p-(i+1)} \left[\Delta_j^f(e, T_j^i) - \Delta_j^f(\pi_t(e), \{e\} \cup T_j^i \setminus \pi_t(e))\right] - \frac{1}{p} \sum_{j=1}^{m} \sum_{e \in S_j^* \setminus T_j^i} \Delta_j^l(e, T_j^i)$$

$$\geq \frac{1}{p} \sum_{j=1}^{m} \sum_{e \in S_j^* \setminus T_j^i} \left(1 - \frac{2}{p}\right)^{p-(i+1)} \left[\Delta_j^f(e, T_j^i) - \Delta_j^f(\pi_t(e), T_j^i \setminus \pi_t(e))\right] - \frac{1}{p} \sum_{j=1}^{m} \sum_{e \in S_j^* \setminus T_j^i} \Delta_j^l(e, T_j^i)$$

$$\geq \frac{1}{p} \left(1 - \frac{2}{p}\right)^{p-(i+1)} \sum_{j=1}^{m} [f_j(S_j^* \cup T_j^i) - 2f_j(T_j^i)] - \frac{1}{p} \sum_{j=1}^{m} \ell_j(S_j^* \setminus T_j^i)$$

$$\geq \frac{1}{p} \left(1 - \frac{2}{p}\right)^{p-(i+1)} \sum_{j=1}^{m} [f_j(S_j^*) - 2f_j(T_j^i)] - \frac{1}{p} \sum_{j=1}^{m} \ell_j(S_j^*)$$

$$= \frac{1}{p} \left(1 - \frac{2}{p}\right)^{p-(i+1)} \sum_{j=1}^{m} f_j(S_j^*) - \frac{2}{p} \left(1 - \frac{2}{p}\right)^{p-(i+1)} \sum_{j=1}^{m} f_j(T_j^i) - \frac{1}{p} \sum_{j=1}^{m} \ell_j(S_j^*).$$

The fifth inequality follows because $\left[\left(1 - \frac{2}{p}\right)^{p-(i+1)} \nabla_j^f(e, T_j^i) - \nabla_j^l(e, T_j^i)\right] \geq 0$ and $S_j^* \setminus T_j^i \subseteq S^*$. The sixth inequality is due to Lemma 2. The seventh and eighth inequalities are from Property 1. $\qquad \square$

These three lemmas together imply the desired approximation ratio of this algorithm.

Theorem 1. *Algorithm 1 returns a set S^p with*

$$F(S^p) \geq \frac{1}{2} \left(1 - \frac{1}{e^2}\right) \sum_{j=1}^{m} f_j(S_j^*) - \sum_{j=1}^{m} \ell_j(S_j^*).$$

Proof. According to the definition of Φ, we obtain

$$\Phi_0(S_0) = \left(1 - \frac{2}{p}\right)^p \sum_{j=1}^{m} f_j(\emptyset) - \sum_{j=1}^{m} \ell_j(\emptyset) = 0,$$

$$\Phi_p(S_p) = \left(1 - \frac{2}{p}\right)^0 \sum_{j=1}^{m} f_j(T_j^p) - \sum_{j=1}^{m} \ell_j(T_j^p) \leq F(S_p).$$

Applying Lemmas 1 and 3, respectively, we have

$$\Phi_{i+1}(S^{i+1}) - \Phi_i(S^i) = \tilde{\nabla}(e_i, T^i) + \frac{2}{p}\left(1 - \frac{2}{p}\right)^{p-(i+1)} \sum_{j=1}^{m} f_j(T^i_j)$$

$$\geq \frac{1}{p}\left(1 - \frac{2}{p}\right)^{p-(i+1)} \sum_{j=1}^{m} f_j(S^*_j) - \frac{2}{p}\left(1 - \frac{2}{p}\right)^{p-(i+1)} \sum_{j=1}^{m} f_j(T^i_j)$$

$$- \frac{1}{p}\sum_{j=1}^{m} \ell_j(S^*_j) + \frac{2}{p}\left(1 - \frac{2}{p}\right)^{p-(i+1)} \sum_{j=1}^{m} f_j(T^i_j)$$

$$= \frac{1}{p}\left(1 - \frac{2}{p}\right)^{p-(i+1)} \sum_{j=1}^{m} f_j(S^*_j) - \frac{1}{p}\sum_{j=1}^{m} \ell_j(S^*_j).$$

Finally,

$$F(S_p) \geq \Phi_p(S_p) - \Phi_0(S_0) = \sum_{i=0}^{p-1}(\Phi_{i+1}(S^{i+1}) - \Phi_i(S^i))$$

$$\geq \sum_{i=0}^{p-1}\left[\frac{1}{p}\left(1 - \frac{2}{p}\right)^{p-(i+1)} \sum_{j=1}^{m} f_j(S^*_j) - \frac{1}{p}\sum_{j=1}^{m} \ell_j(S^*_j)\right]$$

$$\geq \frac{1}{2}\left(1 - \frac{1}{e^2}\right)\sum_{j=1}^{m} f_j(S^*_j) - \sum_{j=1}^{m} \ell_j(S^*_j).$$

$\square$

4.3 The Randomized Algorithm

In this section, we modify Algorithm 1 to obtain a randomized distorted greedy algorithm. It uses the same greedy objective as that in the previous algorithm, but enjoys an asymptotically faster running time by utilizing the sampling techniques in [15]. Instead of choosing the greedy element from the entire ground set in every round, Randomized Distorted Greedy uses a random sample $A_i \subseteq V$ of size $\lceil \frac{n}{p} \log \frac{1}{\epsilon} \rceil$ to replace the entire ground set. If the sample size is sufficiently large, A_i contains at least one element of S with high probability.

Theorem 2. *Algorithm 2 returns a set S^p with*

$$F(S^p) \geq \left(\frac{1}{2}(1 - \frac{1}{e^2}) - \epsilon\right)\sum_{j=1}^{m} f_j(S^*_j) - \sum_{j=1}^{m} \ell_j(S^*_j).$$

The proof is in the appendix.

Algorithm 2: Randomized Distorted Replacement Greedy

1: $S^0 \leftarrow \emptyset, T_j^0 \leftarrow \emptyset, s \leftarrow \lceil \frac{n}{p} \log \frac{1}{\epsilon} \rceil \ (\forall 1 \leq j \leq m)$

2: **for** $0 \leq i \leq p-1$ **do**

3:    $A_i \leftarrow$ sample s elements uniformly and independently from V

4:    $e_i \leftarrow \arg\max_{e \in A_i} \sum\limits_{j=1}^{m} \left[\left(1 - \frac{2}{p}\right)^{p-(i+1)} \nabla_j^f(e, T_j^i) - \nabla_j^l(e, T_j^i) \right]$

5:    **if** $\sum\limits_{j=1}^{m} \left[\left(1 - \frac{2}{p}\right)^{p-(i+1)} \nabla_j^f(e_i, T_j^i) - \nabla_j^l(e_i, T_j^i) \right] > 0$ **then**

6:       $S^{i+1} \leftarrow S^i \cup \{e_i\}$

7:       **for all** $1 \leq j \leq m$ **do**

8:          **if** $\left(1 - \frac{2}{p}\right)^{p-(i+1)} \nabla_j^f(e_i, T_j^i) - \nabla_j^l(e_i, T_j^i) > 0$ **then**

9:             $T_j^{i+1} \leftarrow T_j^i \cup \{e_i\} \setminus \mathsf{Rep}_j(e_i, T_j^i)$

10:          **else**

11:             $T_j^{i+1} \leftarrow T_j^i$

12:          **end if**

13:       **end for**

14:    **else**

15:       $S^{i+1} \leftarrow S^i, T_1^{i+1} \leftarrow T_1^i, T_2^{i+1} \leftarrow T_2^i, \cdots, T_m^{i+1} \leftarrow T_m^i$

16:    **end if**

17: **end for**

18: Return sets S^p and $T_1^p, T_2^p, \cdots, T_m^p$

---

5 Two-Stage Submodular Maximization Subject to Cardinality and k-matroid Constraints

In this section, we generalize the matroid constraint to k-matroid constraint. For this problem, we can modify Algorithm 1 to obtain Algorithm 3. The full details of Algorithm 3 and the corresponding proofs are deferred to the full version of the paper.

Theorem 3. *Algorithm 3 returns a set S^p with*

$$F(S^p) \geq \frac{1}{k+1}\left(1 - \frac{1}{e^{k+1}}\right) \sum_{j=1}^{m} f_j(S_j^*) - \sum_{j=1}^{m} \ell_j(S_j^*).$$

Algorithm 3: Distorted Replacement Greedy k-matroid

1: $S^0 \leftarrow \emptyset, T_j^0 \leftarrow \emptyset \ (\forall 1 \leq j \leq m)$

2: **for** $0 \leq i \leq p-1$ **do**

3:     $e_i \leftarrow \arg\max_{e \in V} \sum_{j=1}^{m} \left[\left(1 - \frac{k+1}{p}\right)^{p-(i+1)} \nabla_j^f(e, T_j^i) - \nabla_j^l(e, T_j^i) \right]$

4:     **if** $\sum_{j=1}^{m} \left[\left(1 - \frac{k+1}{p}\right)^{p-(i+1)} \nabla_j^f(e_i, T_j^i) - \nabla_j^l(e_i, T_j^i) \right] > 0$ **then**

5:         $S^{i+1} \leftarrow S^i \cup \{e_i\}$

6:         **for all** $1 \leq j \leq m$ **do**

7:             **if** $\left(1 - \frac{k+1}{p}\right)^{p-(i+1)} \nabla_j^f(e_i, T_j^i) - \nabla_j^l(e_i, T_j^i) > 0$ **then**

8:                 $T_j^{i+1} \leftarrow T_j^i \cup \{e_i\} \setminus \mathsf{Rep}_j(e_i, T_j^i)$

9:             **else**

10:                $T_j^{i+1} \leftarrow T_j^i$

11:            **end if**

12:        **end for**

13:    **else**

14:        $S^{i+1} \leftarrow S^i, T_1^{i+1} \leftarrow T_1^i, T_2^{i+1} \leftarrow T_2^i, \cdots, T_m^{i+1} \leftarrow T_m^i$

15:    **end if**

16: **end for**

17: Return sets S^p and $T_1^p, T_2^p, \cdots, T_m^p$

---

6 Conclusion

In this paper, we design two distorted greedy algorithms for solving the random two-stage submodular problem. Unlike the previous work [2,18,22], we do not assume the objective function being monotone. We leave two open problems for future research. One is to consider modular functions of not necessarily monotonic and another one is that if each f_j has curvature c, is it possible to use the results in [23] to improve the approximation ratio presented in this paper.

Acknowledgements. This research is supported or partially supported by the National Natural Science Foundation of China (Grant Nos. 11871280, 11501171, 11771251, 11971349, 11771386 and 11728104), the Natural Sciences and Engineering Research Council of Canada (NSERC) Grant 06446 and Qinglan Project.

References

1. Badanidiyuru, A., Mirzasoleiman, B., Karbasi, A., Krause, A.: Streaming submodular maximization: massive data summarization on the fly. In: KDD, pp. 671–680 (2014)
2. Balkanski, E., Mirzasoleiman, B., Krause, A., Singer, Y.: Learning sparse combinatorial representations via two-stage submodular maximization. In: ICML, pp. 2207–2216 (2016)
3. El-Arini, K., Veda, G., Shahaf, D., Guestrin, C.: Turning down the noise in the blogosphere. In KDD, pp. 289–298 (2009)

4. Epasto, A., Mirrokni, V.S., Zadimoghaddam, M.: Bicriteria distributed submodular maximization in a few rounds. In: SPAA, pp. 25–33 (2017)
5. Feldman, M.: Guess free maximization of submodular and linear sums. In: Friggstad, Z., Sack, J.-R., Salavatipour, M.R. (eds.) WADS 2019. LNCS, vol. 11646, pp. 380–394. Springer, Cham (2019). https://doi.org/10.1007/978-3-030-24766-9_28
6. Feldman, M., Harshaw, C., Karbasi, A.: Greed is good: near-optimal submodular maximization via greedy optimization. In: COLT, pp. 758–784 (2017)
7. Feldman, M., Karbasi, A., Kazemi, E.: Do less, get more: streaming submodular maximization with subsampling. CoRR, abs/1802.07098 (2018). http://arxiv.org/abs/1802.07098 maximization. In: FOCS, pp. 570–579 (2011)
8. Harshaw, C., Feldman, M., Ward, J., Karbasi, A.: Submodular maximization beyond non-negativity: guarantees, fast algorithms, and applications. In: ICML, pp. 2634–2643 (2019)
9. Kirchhoff, K., Bilmes, J.: Submodularity for data selection in statistical machine translation. In: EMNLP, pp. 131–141 (2014)
10. Krause, A., Gomes, R.G.: Budgeted nonparametric learning from data streams. In: ICML, pp. 391–398 (2010)
11. Kempe, D., Kleinberg, J.M., Tardos, á: Maximizing the spread of influence through a social network. In: KDD, pp. 137–146 (2003)
12. Krause, A., Guestrin, C.: Near-optimal nonmyopic value of information in graphical models. In: UAI, pp. 324–331 (2005)
13. Kumar, R., Moseley, B., Vassilvitskii, S., Vattani, A.: Fast greedy algorithms in MapReduce and streaming. TOPC $\mathbf{2}$(3), 141–1422 (2015)
14. Lee, J., Mirrokni, V.S., Nagarajan, V., Sviridenko, M.: Maximizing nonmonotone submodular functions under matroid or knapsack constraints. SIAM J. Discrete Math. $\mathbf{23}$(4), 2053–2078 (2010)
15. Mirzasoleiman, B., Badanidiyuru, A., Karbasi, A., Vondraák, J., Krause, A.: Lazier than lazy greedy. In: AAAI, pp. 1812–1818 (2015)
16. Mirzasoleiman, B., Karbasi, A., Badanidiyuru, A., Krause, A.: Distributed submodular cover: succinctly summarizing massive data. In: NIPS, pp. 2881–2889 (2015)
17. Mirzasoleiman, B., Karbasi, A., Sarkar, R., Krause, A.: Distributed submodular maximization: identifying representative elements in massive data. In: NIPS, pp. 2049–2057 (2013)
18. Mitrovic, M., Kazemi, E., Zadimoghaddam, M., Karbasi, A.: Data summarization at scale: a two-stage submodular approach. In: ICML, pp. 3593–3602 (2018)
19. Lin, H., Bilmes, J.A.: Multi-document summarization via budgeted maximization of submodular functions. In: HLT-NAACL, pp. 912–920 (2010)
20. Schrijver, A.: Combinatorial Optimization-Polyhedra and Efficiency. Springer, Berlin (2003)
21. Singla, A., Bogunovic, I., Bartok, G., Karbasi, A., Krause, A.: Near-optimally teaching the crowd to classify. In: ICML, pp. 154–162 (2014)
22. Stan, S., Zadimoghaddam, M., Krause, A., Karbasi, A.: Probabilistic submodular maximization in sub-linear time. In: ICML, pp. 3241–3250 (2017)
23. Sviridenko, M., Vondrák, J., Ward, J.: Optimal approximation for submodular and supermodular optimization with bounded curvature. Math. Oper. Res. $\mathbf{42}$(4), 1197–1218 (2017)

Optimal Matroid Bases with Intersection Constraints: Valuated Matroids, M-convex Functions, and Their Applications

Yuni Iwamasa[1]([⊠])[iD] and Kenjiro Takazawa[2][iD]

[1] National Institute of Informatics, Tokyo 101-8430, Japan
yuni_iwamasa@nii.ac.jp
[2] Hosei University, Tokyo 184-8584, Japan
takazawa@hosei.ac.jp

Abstract. For two matroids M_1 and M_2 with the same ground set V and two cost functions w_1 and w_2 on 2^V, we consider the problem of finding bases X_1 of M_1 and X_2 of M_2 minimizing $w_1(X_1) + w_2(X_2)$ subject to a certain cardinality constraint on their intersection $X_1 \cap X_2$. Lendl, Peis, and Timmermans (2019) discussed modular cost functions: They reduced the problem to weighted matroid intersection for the case where the cardinality constraint is $|X_1 \cap X_2| \leq k$ or $|X_1 \cap X_2| \geq k$; and designed a new primal-dual algorithm for the case where $|X_1 \cap X_2| = k$. The aim of this paper is to generalize the problems to have nonlinear convex cost functions, and to comprehend them from the viewpoint of discrete convex analysis. We prove that each generalized problem can be solved via valuated independent assignment, valuated matroid intersection, or M-convex submodular flow, to offer a comprehensive understanding of weighted matroid intersection with intersection constraints. We also show the NP-hardness of some variants of these problems, which clarifies the coverage of discrete convex analysis for those problems. Finally, we present applications of our generalized problems in matroid congestion games and combinatorial optimization problems with interaction costs.

Keywords: Valuated independent assignment · Valuated matroid intersection · M-convex submodular flow · Matroid congestion game · Combinatorial optimization problem with interaction costs

1 Introduction

Weighted matroid intersection is one of the most fundamental combinatorial optimization problems solvable in polynomial time. This problem generalizes

The first author was supported by JSPS KAKENHI Grant Number JP19J01302, Japan.
The second author was supported by JSPS KAKENHI Grant Numbers JP16K16012, JP26280004, Japan.

© Springer Nature Switzerland AG 2020
J. Chen et al. (Eds.): TAMC 2020, LNCS 12337, pp. 156–167, 2020.
https://doi.org/10.1007/978-3-030-59267-7_14

a number of tractable problems including maximum-weight bipartite matching and minimum-weight arborescence. The comprehension of mathematical structures of weighted matroid intersection, e.g., Edmonds' intersection theorem [4] and Frank's weight splitting theorem [5], contributes to the development of polyhedral combinatorial optimization as well as matroid theory.

Recently, Lendl, Peis, and Timmermans [9] have introduced the following variants of weighted matroid intersection, in which a cardinality constraint is imposed on the intersection. Let V be a finite set, n a positive integer, and $[n] := \{1, 2, \ldots, n\}$. For each $i \in [n]$, let $M_i = (V, \mathcal{B}_i)$ be a matroid with ground set V and base family of $\mathcal{B}_i$, and w_i a modular function on 2^V. Let k be a nonnegative integer. The problems are formulated as follows.

$$\text{Minimize } w_1(X_1) + w_2(X_2)$$
$$\text{subject to } X_i \in \mathcal{B}_i \quad (i = 1, 2), \tag{1.1}$$
$$|X_1 \cap X_2| = k.$$

$$\text{Minimize } \sum_{i=1}^{n} w_i(X_i)$$
$$\text{subject to } X_i \in \mathcal{B}_i \quad (i \in [n]), \tag{1.2}$$
$$\left| \bigcap_{i=1}^{n} X_i \right| \le k.$$

They further discussed the following problem for polymatroids. Let $B_1, B_2 \subseteq \mathbf{Z}^V$ be the base polytopes of some polymatroids on the ground set V. Here, w_1 and w_2 are linear functions on $\mathbf{Z}^V$.

$$\text{Minimize } w_1(x_1) + w_2(x_2)$$
$$\text{subject to } x_i \in B_i \quad (i = 1, 2), \tag{1.3}$$
$$\sum_{v \in V} \min\{x_v, y_v\} \ge k.$$

Lendl et al. [9] showed that the problems (1.1)–(1.3) are polynomial-time solvable. They developed a new primal-dual algorithm for the problem (1.1), and reduced the problems (1.2) and (1.3) to existing tractable problems of weighted matroid intersection and *polymatroidal flow*, respectively. By this result, they affirmatively settled an open question on the polynomial-time solvability of the *recoverable robust matroid basis problem* [7].

The aim of this paper is to provide a comprehensive understanding of the result of Lendl et al. [9] in view of *discrete convex analysis* (*DCA*) [14,17], particularly focusing on *M-convexity* [11]. DCA offers a theory of convex functions on the integer lattice $\mathbf{Z}^V$, and M-convexity, a quantitative generalization of matroids, plays the central roles in DCA. M-convex functions naturally appear in combinatorial optimization, economics, and game theory [18,19].

The formal definition of M-convex functions is given as follows. A function $f : \mathbf{Z}^V \to \mathbf{R} \cup \{+\infty\}$ is said to be *M-convex* if it satisfies the following generalization of the matroid exchange axiom: for all $x = (x_v)_{v \in V}$ and $y = (y_v)_{v \in V}$ with $x, y \in \mathrm{dom} f$, and all $v \in V$ with $x_v > y_v$, there exists $u \in V$ with $x_u < y_u$ such that $f(x) + f(y) \geq f(x - \chi_v + \chi_u) + f(y + \chi_v - \chi_u)$, where $\mathrm{dom} f$ denotes the effective domain $\{x \in \mathbf{Z}^V \mid f(x) < +\infty\}$ of f and χ_v the v-th unit vector for $v \in V$. In particular, if $\mathrm{dom} f$ is included in the hypercube $\{0, 1\}^V$, then f is called a *valuated matroid*[1] [2,3].

We address M-convex (and hence nonlinear) generalizations of the problems (1.1)–(1.3). Let $\omega_1, \omega_2, \ldots, \omega_n$ be valuated matroids on 2^V, where we identify 2^V with $\{0, 1\}^V$ by the natural correspondence between $X \subseteq V$ and $x \in \{0, 1\}^V$; $x_v = 1$ if and only if $v \in X$.

- For the problem (1.1), by generalizing the modular cost functions w_1 and w_2 to valuated matroids, we obtain:

$$\text{Minimize } \omega_1(X_1) + \omega_2(X_2)$$
$$\text{subject to } |X_1 \cap X_2| = k. \tag{1.4}$$

- For the problem (1.2), as well as generalizing $w_1, w_2, \ldots, w_n$ to valuated matroids, we generalize the cardinality constraint $|\bigcap_{i=1}^n X_i| \leq k$ to a matroid constraint. Namely, let $M = (V, \mathcal{I})$ be a new matroid, where $\mathcal{I}$ denotes its independent set family, and generalize (1.2) as follows.

$$\text{Minimize } \sum_{i=1}^n \omega_i(X_i)$$
$$\text{subject to } \bigcap_{i=1}^n X_i \in \mathcal{I}. \tag{1.5}$$

- It is also reasonable to take the cardinality constraint into the objective function. Let $w \colon V \to \mathbf{R}$ be a weight function. The next problem is a variant of the above problem.

$$\text{Minimize } \sum_{i=1}^n \omega_i(X_i) + w\left(\bigcap_{i=1}^n X_i\right). \tag{1.6}$$

- Let f_1 and f_2 be M-convex functions on $\mathbf{Z}^V$ such that $\mathrm{dom} f_1$ and $\mathrm{dom} f_2$ are included in $\mathbf{Z}_+^V$, where $\mathbf{Z}_+$ is the set of nonnegative integers. Also let $w \colon \mathbf{Z}^V \to \mathbf{R}$ be a linear function. The problem (1.3) is generalized as follows.

$$\text{Minimize } f_1(x) + f_2(y) + w(\min\{x, y\})$$
$$\text{subject to } \sum_{v \in V} \min\{x_v, y_v\} \geq k, \tag{1.7}$$

where $\min\{x, y\} := (\min\{x_v, y_v\})_{v \in V}$.

---

[1] The original definition of a valuated matroid is an *M-concave function*, i.e., the negative of an M-convex function, such that its effective domain is included in the hypercube.

Our main contribution is to show the tractability of the generalized problems (1.4)–(1.7):

Theorem 1. *There exist polynomial-time algorithms to solve the problems (1.4), (1.5), (1.6) for $w \geq 0$, and (1.7) for $w \leq 0$.*

The algorithm for the problem (1.4) is based on *valuated independent assignment* [12,13], that for (1.5) and (1.6) on *valuated matroid intersection* [12,13], and that for (1.7) on $M^\natural$-*convex submodular flow* [15]. It would be noteworthy that we essentially require the concept of valuated matroid intersection to solve the problem (1.6) even if ω_i is a modular function for each $i \in [n]$. That is, the problem (1.6) with modular functions ω_i ($i \in [n]$) is an interesting example which only requires matroids to define, but requires valuated matroids to solve. It might also be interesting that the problem (1.5) can be solved in polynomial time when $n \geq 3$, in spite of the fact that matroid intersection for more than two matroids is NP-hard.

We also demonstrate that the tractability of the problems (1.6) and (1.7) relies on the assumptions on w ($w \geq 0$ and $w \leq 0$, respectively), by showing the NP-hardness of the problems.

Theorem 2. *The problems (1.6) and (1.7) are NP-hard in general even if $w \leq 0$ and $m \geq 3$ for (1.6), and $w \geq 0$ and $k = 0$ for (1.7).*

We then present applications of our generalized problems to *matroid congestion games* [1] and *combinatorial optimization problems with interaction costs* (*COPIC*) [8]. We show that computing the socially optimal state in a certain generalized model of matroid congestion games can be reduced to (a generalized version of) the problem (1.6), and thus can be done in polynomial time. We also reduce a certain generalized case of the COPIC with *diagonal costs* to (1.6) and (1.7), to provide a generalized class of COPIC which can be solved in polynomial time.

The proofs are omitted due to space constraint; see the full version for the proofs.

2   Algorithms

In this section, we provide polynomial-time algorithms for the problems (1.4)–(1.7) to prove Theorem 1. Theorem 2 is also shown in this section.

We first prepare several facts and terminologies on M-convex functions. For an M-convex function f, all members in dom$f$ have the same "cardinality," that is, there exists some integer r such that $\sum_{v \in V} x_v = r$ for all $x \in$ dom$f$. We call r the *rank* of f. An $M^\natural$-*convex function* [20] is a function $f : \mathbf{Z}^V \to \mathbf{R} \cup \{+\infty\}$ defined by the following weaker exchange axiom: for all $x = (x_v)_{v \in V}$ and $y = (y_v)_{v \in V}$ with $x, y \in$ dom f, and all $v \in V$ with $x_v > y_v$, it holds that $f(x) + f(y) \geq f(x - \chi_v) + f(y + \chi_v)$, or there exists $u \in V$ with $x_u < y_u$ such that $f(x) + f(y) \geq f(x - \chi_v + \chi_u) + f(y + \chi_v - \chi_u)$. From the definition, it is clear that $M^\natural$-convex functions are a slight generalization of M-convex functions. Meanwhile, $M^\natural$-convexity and M-convexity are essentially equivalent concepts (see e.g., [17]).

2.1 Reduction of (1.4) to Valued Independent Assignment

This subsection provides a polynomial-time algorithm for solving the problem (1.4). In [9], the authors developed a new algorithm specific to (1.1). In this article, we give a reduction of the generalized problem (1.4) to a known tractable problem of *valued independent assignment* [12,13], building upon the DCA perspective.

Let $G = (V, V'; E)$ be a bipartite graph, ω and ω' valuated matroids on 2^V and on $2^{V'}$, respectively, and w a weight function on E. Then the *valuated independent assignment problem* parameterized by an integer k, referred to as VIAP(k), is described as follows.

$$\text{Minimize } \omega(X) + \omega'(X') + w(F)$$
$$\text{VIAP}(k) \quad \text{subject to } F \subseteq E \text{ is a matching of } G \text{ with } \partial F \subseteq X \cup X',$$
$$|F| = k,$$

where ∂F denote the set of endpoints of F.

We first consider the following variant of the problem (1.4):

$$\text{Minimize } \omega_1(X_1) + \omega_2(X_2)$$
$$\text{subject to } |X_1 \cap X_2| \geq k, \tag{2.1}$$

in which the constraint $|X_1 \cap X_2| = k$ is replaced by $|X_1 \cap X_2| \geq k$. The problem (2.1) can be naturally reduced to VIAP(k) as follows. Set the input bipartite graph G of VIAP(k) by $(V_1, V_2; \{\{v_1, v_2\} \mid v \in V\})$, where V_i is a copy of V and $v_i \in V_i$ is a copy of $v \in V$ for $i = 1, 2$. We regard ω_i as a valuated matroid on 2^{V_i}. Set $w := 0$ for all edges. Then consider VIAP(k) for such G, ω_1, ω_2, and w. One can see that, if (X_1, X_2) is feasible for the problem (2.1), i.e., $|X_1 \cap X_2| \geq k$, then there is a matching F of G with $\partial F \subseteq X_1 \cup X_2$ and $|F| = k$, i.e., (X_1, X_2, F) is feasible for VIAP(k). On the other hand, if (X_1, X_2, F) is feasible for VIAP(k), then (X_1, X_2) is feasible for the problem (2.1). Moreover the objective value of a feasible solution (X_1, X_2) for the problem (2.1) is equal to that of any corresponding feasible solution (X_1, X_2, F) for VIAP(k) since w is identically zero.

Thus the problem (2.1) can be solved in polynomial time in the following way based on the augmenting path algorithm for VIAP(k) [12,13]; see also [16, Theorem 5.2.62]. Let X_1 and X_2 be the minimizers of ω_1 and ω_2, respectively, which can be found in a greedy manner.

Step 1: If $|X_1 \cap X_2| \geq k$, then output them and stop. Otherwise, let $X_1^j := X_1$ and $X_2^j := X_2$, where $j := |X_1 \cap X_2| < k$.

Step 2: Execute the augmenting path algorithm for VIAP(k). Then we obtain a sequence $\left((X_1^j, X_2^j), (X_1^{j+1}, X_2^{j+1}), \ldots, (X_1^\ell, X_2^\ell)\right)$ of solutions, where $\left|X_1^{j'} \cap X_2^{j'}\right| = j'$ for $j' = j, j+1, \ldots, \ell$. If $\ell < k$, then output "the problem (2.1) is infeasible." If $\ell \geq k$, then output (X_1^k, X_2^k).

The above approach directly leads to the following algorithm for the problem (1.4). Again let X_1 and X_2 be the minimizers of w_1 and w_2, respectively.

Case 1: If $|X_1 \cap X_2| \leq k$, then execute the augmenting path algorithm for VIAP(k), and let $\left((X_1^j, X_2^j), (X_1^{j+1}, X_2^{j+1}), \ldots, (X_1^\ell, X_2^\ell)\right)$ be the sequence of solutions obtained in the algorithm. If $\ell < k$, then output "the problem (1.4) is infeasible." If $\ell \geq k$, then output (X_1^k, X_2^k).

Case 2: If $|X_1 \cap X_2| > k$, then let r be the rank of w_2 and $\overline{w_2}(X) := w_2(V \setminus X)$ for $X \subseteq V$, which is the *dual* valuated matroid of w_2. Then apply Case 1, where VIAP(k) is replaced by VIAP($r - k$) for $(G, w, w_1, \overline{w_2})$.

Remark 1. If we are given at least three valuated matroids, then the problems (2.1) (and hence (1.4)) will be NP-hard, since it can formulate the matroid intersection problem for three matroids. ∎

2.2   Reduction of (1.5) and (1.6) to Valuated Matroid Intersection

In this subsection, we give polynomial-time algorithms for solving the problems (1.5) and (1.6) by reducing them to *valuated matroid intersection*. This is the following generalization of weighted matroid intersection problem: Given two valuated matroids ω and ω' on 2^V, minimize the sum $\omega(X) + \omega'(X)$ for $X \subseteq V$. It is known [12,13] that valuated matroid intersection is polynomially solvable.

In order to reduce our problems to valuated matroid intersection, we need to prepare two valuated matroids for each problem. One valuated matroid is common in the reductions of the problems (1.5) and (1.6), which is defined as follows. Let $\coprod_{i \in [n]} V$ be the discriminated union of n copies of V. We denote by $(X_1, X_2, \ldots, X_n)$ a subset $\coprod_{i \in [n]} X_i$ of $\coprod_{i \in [n]} V$, where $X_i \subseteq V$ for each $i \in [n]$. Let us define $\tilde{w}$ by the disjoint sum of $w_1, w_2, \ldots, w_n$. That is, $\tilde{w}$ is a function on $2^{\coprod_{i \in [n]} V}$ defined by $w(X_1, X_2, \ldots, X_n) := w_1(X_1) + w_2(X_2) + \cdots + w_n(X_n)$ for $(X_1, X_2, \ldots, X_n) \subseteq \coprod_{i \in [n]} V$. It is a valuated matroid with rank $r := \sum_{i=1}^n r_i$, where r_i is the rank of w_i.

We then provide the other valuated matroid used in the reduction of the problem (1.5). Define a set system $\tilde{M} = (\coprod_{i \in [n]} V, \tilde{\mathcal{B}})$ by

$$\tilde{\mathcal{B}} = \left\{ (X_1, X_2, \ldots, X_n) \,\middle|\, X_i \subseteq V (i \in [n]), \bigcap_{i=1}^n X_i \in \mathcal{I}, \sum_{i=1}^n |X_i| = r \right\}.$$

It is clear that (1.5) is equivalent to the problem of minimizing the sum of $\tilde{w}$ and $\delta_{\tilde{\mathcal{B}}}$, where $\delta_{\tilde{\mathcal{B}}}$ denotes the indicator function of $\tilde{\mathcal{B}}$, namely, $\delta_{\tilde{\mathcal{B}}}(X_1, X_2, \ldots, X_n) := 0$ if $(X_1, X_2, \ldots, X_n) \in \tilde{\mathcal{B}}$ and $\delta_{\tilde{\mathcal{B}}}(X_1, X_2, \ldots, X_n) := +\infty$ otherwise. We now have the following lemma.

Lemma 1. $\tilde{M}$ *is a matroid with the base family* $\tilde{\mathcal{B}}$.

It follows from Lemma 1 that the function $\delta_{\tilde{\mathcal{B}}}$ is a valuated matroid, and we thus conclude that the problem (1.5) can be reduced to valuated matroid intersection.

Remark 2. If we replace the constraint $\bigcap_{i=1}^{n} X_i \in \mathcal{I}$ in (1.5) by $\bigcap_{i=1}^{n} X_i \in \mathcal{B}$, where $\mathcal{B}$ is the base family of some matroid, then the problem will be NP-hard even if $m = 2$, since it can formulate the matroid intersection problem for three matroids.   ■

We next provide another valuated matroid that is used in the reduction of the problem (1.6). A *laminar convex function* [17, Section 6.3], which is a typical example of an M♮-convex function, plays a key role here. A function $f : \mathbf{Z}^V \to \mathbf{R} \cup \{+\infty\}$ is said to be *laminar convex* if f is representable as

$$f(x) = \sum_{X \in \mathcal{L}} g_X \left(\sum_{v \in X} x_v \right) \quad (x \in \mathbf{Z}^V),$$

where $\mathcal{L} \subseteq 2^V$ is a laminar family on V, and for each $X \in \mathcal{L}, g_X : \mathbf{Z} \to \mathbf{R} \cup \{+\infty\}$ is a univariate discrete convex function, i.e., $g_X(k+1) + g_X(k-1) \geq 2g_X(k)$ for every $k \in \mathbf{Z}$. As mentioned above, a laminar convex function is M♮-convex.

Now define a function $\tilde{w}$ on $2^{\coprod_{i \in [n]} V}$ by $\tilde{w}(X_1, X_2, \ldots, X_n) := w(\bigcap_{i=1}^{n} X_i)$ for $(X_1, X_2, \ldots, X_n) \subseteq \coprod_{i \in [n]} V$. It is clear that (1.6) is equivalent to the problem of minimizing the sum of $\tilde{\omega}$ and the restriction of $\tilde{w}$ to $\{(X_1, X_2, \ldots, X_n) \subseteq \coprod_{i \in [n]} V \mid \sum_{i=1}^{n} |X_i| = r\}$. For $\tilde{w}$, the following holds.

Lemma 2. *The function $\tilde{w}$ is a laminar convex function on $2^{\coprod_{i \in [n]} V}$ if $w \geq 0$.*

By Lemma 2, the restriction of $\tilde{w}$ to $\{(X_1, X_2, \ldots, X_n) \subseteq \coprod_{i \in [n]} V \mid \sum_{i=1}^{n} |X_i| = r\}$ is a valuated matroid on $2^{\coprod_{i \in [n]} V}$ if $w \geq 0$. Indeed, it is known [20] that, for an M♮-convex function f and an integer r, the restriction of f to a hyperplane $\{x \in \mathbf{Z}^V \mid \sum_{v \in V} x_v = r\}$ is an M-convex function with rank r, if its effective domain is nonempty. Thus the problem (1.6) can be formulated as the valuated matroid intersection problem for $\tilde{\omega}$ and $\tilde{w}$, establishing the tractability of the problem (1.6) in case of $w \geq 0$.

On the other hand, if $w \leq 0$ and $m \geq 3$, then the problem (1.6) is NP-hard, since it can formulate the matroid intersection problem for three matroids.

Remark 3. As mentioned in Sect. 1, the problem (1.6) with $w \geq 0$ does not fall into the weighted matroid intersection framework even if all functions are modular, while it can be reduced to valuated matroid intersection. That is, the concept of M-convexity is crucial for capturing the tractability of (1.6) even when all functions are modular.   ■

2.3 Reduction of (1.7) to M♮-Convex Submodular Flow

In this subsection, we prove that the problem (1.7) with $w \leq 0$ can be solved in polynomial time by reducing it to *M♮-convex submodular flow*, which is defined as follows. Let f be an M♮-convex function on $\mathbf{Z}^V$ and $G = (V, A)$ a directed graph endowed with an upper capacity $\bar{c} : A \to \mathbf{R} \cup \{+\infty\}$, a lower capacity $\underline{c} : A \to \mathbf{R} \cup \{-\infty\}$, and a weight function $w : A \to \mathbf{R}$. For a flow $\xi \in \mathbf{R}^A$, define its boundary $\partial\xi \in \mathbf{R}^V$ by $\partial\xi(v) := \sum\{\xi(a) \mid a \in A, \ a \text{ enters } v \text{ in } G\} - \sum\{\xi(a) \mid a \in A, \ a \text{ leaves } v \text{ in } G\}$ for $v \in V$. The $M^{\natural}$-convex submodular flow problem for (f, G) is the following problem with variable $\xi \in \mathbf{R}^A$:

$$\text{Minimize} \quad f(\partial\xi) + \sum_{a \in A} w(a)\xi(a)$$

$$\text{subject to} \quad \underline{c}(a) \le \xi(a) \le \overline{c}(a),$$
$$\partial\xi \in \mathrm{dom}\, f.$$

It is known [15] that the $M^{\natural}$-convex submodular flow problem can be solved in polynomial time.

The problem (1.7) with $w \le 0$ can be reduced to $M^{\natural}$-convex submodular flow in the following way. Let r_1 and r_2 be the rank of f_1 and f_2, respectively. We define univariate functions g_1 and g_2 on $\mathbf{Z}$ by

$$g_1(p) := \begin{cases} 0 & \text{if } p \le r_2 - k, \\ +\infty & \text{otherwise,} \end{cases} \qquad g_2(q) := \begin{cases} 0 & \text{if } q \le r_1 - k, \\ +\infty & \text{otherwise.} \end{cases}$$

Then define the function h on $\mathbf{Z}^{V \sqcup \{p\} \sqcup V \sqcup \{q\}}$ by the disjoint sum of f_1, g_1 with the simultaneous coordinate inversion and f_2, g_2, i.e.,

$$h(x, p, y, q) := (f_1(-x) + g_1(-p)) + (f_2(y) + g_2(q)) \quad x, y \in \mathbf{Z}^V \ (\text{and } p, q \in \mathbf{Z}).$$

It is not difficult to see that h is $M^{\natural}$-convex. We then construct a directed bipartite graph $G = (\{x_v\}_{v \in V} \cup \{p\}, \{y_v\}_{v \in V} \cup \{q\}; A)$ endowed with a weight function $\hat{w}: A \to \mathbf{R}$ defined by

$$A := \{(x_v, y_v) \mid v \in V\} \cup \{(p, y_v) \mid v \in V\} \cup \{(x_v, q) \mid v \in V\},$$

$$\hat{w}(a) := \begin{cases} w(v) & \text{if } a = (x_v, y_v), \\ 0 & \text{otherwise,} \end{cases} \qquad (a \in A).$$

Here we identify the vertices of G with the variables of f. Consider the following instance of the $M^{\natural}$-convex submodular flow problem:

$$\text{Minimize} \quad h(\partial\xi) + \sum_{a \in A} \hat{w}(a)\xi(a)$$

$$\text{subject to} \quad \xi(a) \ge 0 \quad (a \in A), \qquad (2.2)$$
$$\partial\xi \in \mathrm{dom}\, h.$$

The following lemma shows that the problem (1.7) with $w \le 0$ is reduced to the problem (2.2), and thus establishes its tractability.

Lemma 3. *The problem (1.7) with $w \le 0$ is equivalent to the problem (2.2).*

The NP-hardness of the problem (1.7) with $w \ge 0$ and $k = 0$ follows from the fact that it can formulate the problem of minimizing $f_1(x_1) + f_2(x_2)$ subject to $\sum_{v \in V} \min\{x_v, y_v\} = 0$, whose NP-hardness has been shown in [9].

3 Applications

In this section, we present two applications of our generalized problems (1.6) and (1.7). One is for *matroid congestion games*, and the other for *combinatorial optimization problems with interaction costs*.

3.1 Socially Optimal States in Valuated Matroid Congestion Games

A *congestion game* [21] is represented by a tuple $(N, V, (\mathcal{B}_i)_{i \in N}, (c_v)_{v \in V})$, where $N = \{1, 2, \ldots, n\}$ is a set of players, V is a set of resources, $\mathcal{B}_i \subseteq 2^V$ is the set of strategies of a player $i \in N$, and $c_v \colon \mathbf{Z}_+ \to \mathbf{R}_+$ is a nondecreasing cost function associated with a resource $v \in V$. Here $\mathbf{R}_+$ is the set of nonnegative real numbers. A *state* $\mathcal{X} = (X_1, X_2, \ldots, X_n)$ is a collection of strategies of all players, i.e., $X_i \in \mathcal{B}_i$ for each $i \in N$. For a state $\mathcal{X} = (X_1, X_2, \ldots, X_n)$, let $x_v(\mathcal{X})$ denote the number of players using v, i.e., $x_v(\mathcal{X}) = |\{i \in N \mid v \in X_i\}|$. If $\mathcal{X}$ is clear from the context, $x_v(\mathcal{X})$ is abbreviated as x_v. In a state $\mathcal{X}$, every player using a resource $v \in V$ should pay $c_v(x_v)$ to use v, and thus the total cost paid by a player $i \in N$ is $\sum_{v \in X_i} c_v(x_v)$.

The importance of congestion games is appreciated through the fact that the class of congestion games coincides with that of *potential games*. Rosenthal [21] proved that every congestion game is a potential game, and conversely, Monderer and Shapley [10] proved that every potential game is represented by a congestion game with the same potential function.

We show that, in a certain generalized model of *matroid congestion games with player-specific costs*, computing *socially optimal states* reduces to (a generalized version of) the problem (1.6). A state $\mathcal{X}^* = (X_1^*, X_2^*, \ldots, X_n^*)$ is called *socially optimal* if the sum of the costs paid by all the players is minimum, i.e.,

$$\sum_{i \in N} \sum_{v \in X_i^*} c_v(x_v(\mathcal{X}^*)) \leq \sum_{i \in N} \sum_{v \in X_i} c_v(x_v(\mathcal{X}))$$

for any state $\mathcal{X} = (X_1, X_2, \ldots, X_n)$. In a *matroid congestion game*, the set $\mathcal{B}_i \subseteq 2^V$ of the strategies of each player $i \in N$ is the base family of a matroid on V. For matroid congestion games, a socially optimal state can be computed in polynomial time if the cost functions are *weakly convex* [1,23], while it is NP-hard for general nondecreasing cost functions [1]. A function $c \colon \mathbf{Z}_+ \to \mathbf{R}$ is called *weakly convex* if $(x + 1) \cdot c(x + 1) - x \cdot c(x)$ is nondecreasing for each $x \in \mathbf{Z}_+$. In a *player specific-cost* model, the cost paid by a player $i \in N$ for using $v \in V$ is represented by a function $c_{i,e} \colon \mathbf{Z}_+ \to \mathbf{R}_+$, which may vary with each player.

We consider the following generalized model of congestion games with player-specific costs. In a state $\mathcal{X} = (X_1, X_2, \ldots, X_n)$, the cost paid by a player $i \in N$ is

$$\omega_i(X_i) + \sum_{v \in X_i} d_v(x_v),$$

where $w_i\colon 2^V \to \mathbf{R}_+$ is a monotone set function and $d_v\colon \mathbf{Z}_+ \to \mathbf{R}_+$ is a non-decreasing function for each $v \in V$. This model represents a situation where a player $i \in N$ should pay $w_i(X_i)$ regardless of the strategies of the other players, as well as $d_v(x_v)$ for every resource $v \in X_i$, which is an additional cost resulting from the congestion on v. It is clear that the standard model of congestion games is a special case where $w_i(X_i) = \sum_{v \in X_i} c_v(1)$ for every $i \in N$ and every $X_i \in \mathcal{B}_i$, and

$$
d_v(x) = \begin{cases} 0 & (x = 0), \\ c_v(x) - c_v(1) & (x \geq 1). \end{cases}
$$

In this model, the sum of the costs paid by all the players is equal to

$$
\sum_{i \in N} w_i(X_i) + \sum_{v \in V} x_v \cdot d_v(x_v). \tag{3.1}
$$

The following lemma is also straightforward to see.

Lemma 4. *The following are equivalent.*

- c_v *is weakly convex.*
- d_v *is weakly convex.*
- $x \cdot d_v$ *is discrete convex.*

By Lamma 4, if c_v (or d_v) is weakly convex, then the function $\sum_{v \in V} x_v \cdot d_v(x_v)$ is laminar convex.

The solution for the problem (1.6), or the DCA perspective for (1.6), provides a new insight on this model of cost functions in matroid congestion games. In addition to the weak convexity of d_v ($v \in V$), this model allows us to introduce some convexity of the cost function w_i. Namely, we assume that w_i is a valuated matroid for every $i \in N$. Then, computing the optimal state, i.e., minimizing (3.1), is naturally viewed as the valuated matroid intersection problem for the valuated matroid $\sum_{i \in N} w_i(X_i)$ and the laminar convex function $\sum_{v \in V} x_v \cdot d_v(x_v)$ as in the problem (1.6). Thus it can be done in polynomial time.

3.2 Combinatorial Optimization Problem with Interaction Costs

Lendl, Ćustić, and Punnen [8] introduced a framework of *combinatorial optimization with interaction costs (COPIC)*, described as follows. For two sets V_1 and V_2, we are given cost functions $w_1\colon V_1 \to \mathbf{R}$ and $w_2\colon V_2 \to \mathbf{R}$, as well as *interaction costs* $q\colon V_1 \times V_2 \to \mathbf{R}$. The objective is to find a pair of feasible sets $X_1 \subseteq V_1$ and $X_2 \subseteq V_2$ minimizing

$$
w_1(X_1) + w_2(X_2) + \sum_{u \in X_1} \sum_{v \in X_2} q_{uv}.
$$

We focus on the *diagonal COPIC*, where V_1 and V_2 are identical and $q_{uv} = 0$ if $u \neq v$. We further assume that the feasible sets are the base families of matroids. That is, we are given two matroids $(V, \mathcal{B}_1)$ and $(V, \mathcal{B}_2)$, and a pair (X_1, X_2) of subsets of V is feasible if and only if $X_1 \in \mathcal{B}_1$ and $X_2 \in \mathcal{B}_2$. In summary, the problem is formulated as

$$\text{Minimize } w_1(X_1) + w_2(X_2) + q(X_1 \cap X_2)$$
$$\text{subject to } X_i \in \mathcal{B}_i \quad (i = 1, 2). \tag{3.2}$$

If w_1 and w_2 are identically zero and $q \geq 0$, then the problem (3.2) amounts to finding a socially optimal state in a two-player matroid congestion game, and thus can be solved in polynomial time [1]. Lendl et al. [8] showed the solvability on the case where the interaction cost q may be arbitrary.

Now we can discuss another direction of generalization; the costs $w_1(X_1)$ and $w_2(X_2)$ are valuated matroids. This is a special case of the problem (1.6) or the problem (1.7), and thus can be solved in polynomial time when $q \geq 0$ or $q \leq 0$.

4   Discussions

In this paper, we have analyzed the complexity of several types of minimization of the sum of valuated matroids (or M-convex functions) under intersection constraints. For the following standard problem of this type, its complexity is still open even when the cardinality constraint $|X_1 \cap X_2| = k$ is removed and w_1, w_2 are modular functions on the base families of some matroids:

$$\text{Minimize } \omega_1(X_1) + \omega_2(X_2) + w(X_1 \cap X_2)$$
$$\text{subject to } |X_1 \cap X_2| = k,$$

where ω_1 and ω_2 are valuated matroids on 2^V, w is a modular function on 2^V, and k is a nonnegative integer.

The above problem seems similar to VIAP(k), but is essentially different; the problem of this type formulated by VIAP(k) is

$$\text{Minimize } \omega_1(X_1) + \omega_2(X_2) + w(F)$$
$$\text{subject to } F \subseteq X_1 \cap X_2,$$
$$|F| = k.$$

Only the following cases are known to be tractable:

- w is identically zero. This case is equivalent to the problem (1.4).
- $w \geq 0$ and the cardinality constraint $|X_1 \cap X_2| = k$ is removed. This case is a subclass of the problem (1.6).
- $w \leq 0$ and $|X_1 \cap X_2| = k$ is replaced by $|X_1 \cap X_2| \geq k$. This case is a subclass of the problem (1.7).
- $|X_1 \cap X_2| = k$ is removed and ω_1, ω_2 are the indicator functions of the base families of some matroids. This case has been dealt with Lendl et al. [8]; see Sect. 3.2.

Another possible direction of research would be to generalize our framework so that it includes computing the socially optimal state of *polymatroid congestion games* [6,22], as we have done for matroid congestion games in Sect. 3.1.

References

1. Ackermann, H., Röglin, H., Vöcking, B.: On the impact of combinatorial structure on congestion games. J. ACM **55**(6), 25:1–25:22 (2008)
2. Dress, A.W.M., Wenzel, W.: Valuated matroids: a new look at the greedy algorithm. Appl. Math. Lett. **3**(2), 33–35 (1990)
3. Dress, A.W.M., Wenzel, W.: Valuated matroids. Adv. Math. **93**, 214–250 (1992)
4. Edmonds, J.: Submodular functions, matroids, and certain polyhedra. In: Jünger, M., Reinelt, G., Rinaldi, G. (eds.) Combinatorial Optimization — Eureka, You Shrink!. LNCS, vol. 2570, pp. 11–26. Springer, Heidelberg (2003). https://doi.org/10.1007/3-540-36478-1_2
5. Frank, A.: A weighted matroid intersection algorithm. J. Algorithms **2**, 328–336 (1981)
6. Harks, T., Klimm, M., Peis, B.: Sensitivity analysis for convex separable optimization over integral polymatroids. SIAM J. Optim. **28**, 2222–2245 (2018)
7. Hradovich, M., Kasperski, A., Zieliński, P.: The recoverable robust spanning tree problem with interval costs is polynomialy solvable. Optim. Lett. **11**(1), 17–30 (2016). https://doi.org/10.1007/s11590-016-1057-x
8. Lendl, S., Ćustić, A., Punnen, A.P.: Combinatorial optimization with interaction costs: complexity and solvable cases. Discrete Optim. **33**, 101–117 (2019)
9. Lendl, S., Peis, B., Timmermans, V.: Matroid bases with cardinality constraints on the intersection (2019). arXiv:1907.04741v1
10. Monderer, D., Shapley, L.S.: Potential games. Games Econ. Behav. **14**, 124–143 (1996)
11. Murota, K.: Convexity and Steinitz's exchange property. Adv. Math. **124**, 272–311 (1996)
12. Murota, K.: Valuated matroid intersection, I: optimality criteria. SIAM J. Discrete Math. **9**, 545–561 (1996)
13. Murota, K.: Valuated matroid intersection, II: algorithms. SIAM J. Discrete Math. **9**, 562–576 (1996)
14. Vygen, J.: Discrete convex analysis. Math. Intell. **26**(3), 74–76 (2004). https://doi.org/10.1007/BF02986756
15. Murota, K.: Submodular flow problem with a nonseparable cost function. Combinatorica **19**, 87–109 (1999). https://doi.org/10.1007/s004930050047
16. Murota, K.: Matrices and Matroids for Systems Analysis. Springer, Heidelberg (2000)
17. Murota, K.: Discrete Convex Analysis. SIAM, Philadelphia (2003)
18. Murota, K.: Recent developments in discrete convex analysis. In: Cook, W., Lovász, L., Vygen, J. (eds.) Research Trends in Combinatorial Optimization, pp. 219–260. Springer, Heidelberg (2009). https://doi.org/10.1007/978-3-540-76796-1_11
19. Murota, K.: Discrete convex analysis: a tool for economics and game theory. J. Mech. Inst. Des. **1**(1), 151–273 (2016)
20. Murota, K., Shioura, A.: M-convex function on generalized polymatroid. Math. Oper. Res. **24**(1), 95–105 (1999)
21. Rosenthal, R.W.: A class of games possessing pure-strategy Nash equilibria. Int. J. Game Theory **2**, 65–67 (1973). https://doi.org/10.1007/BF01737559
22. Takazawa, K.: Generalizations of weighted matroid congestion games: pure Nash equilibrium, sensitivity analysis, and discrete convex function. J. Comb. Optim. **38**, 1043–1065 (2019). https://doi.org/10.1007/s10878-019-00435-9
23. Werneck, R.F.F., Setubal, J.C.: Finding minimum congestion spanning trees. ACM J. Exp. Algorithmics **5**, 11:1–11:22 (2000)

On the Complexity of Acyclic Modules in Automata Networks

Kévin Perrot[1], Pacôme Perrotin[2(✉)], and Sylvain Sené[1]

[1] Université publique, Marseille, France
[2] Aix-Marseille Univ., Univ. de Toulon, CNRS, LIS, UMR 7020, Marseille, France
pacome.perrotin@lis-lab.fr

Abstract. Modules were introduced as an extension of Boolean automata networks. They have inputs which are used in the computation said modules perform, and can be used to wire modules with each other. In the present paper we extend this new formalism and study the specific case of acyclic modules. These modules prove to be well described in their limit behavior by functions called output functions. We provide other results that offer an upper bound on the number of attractors in an acyclic module when wired recursively into an automata network, alongside a diversity of complexity results around the difficulty of deciding the existence of cycles depending on the number of inputs and the size of said cycle.

1 Introduction

Automata networks (ANs) are a generalisation of Cellular automata (CAs). While classical CAs require a n-dimensional lattice with uniform local functions, ANs can be built on any graph structure, and with any function at each vertex of the graph. They have been applied to the study of genetic regulation networks [7,9,15,16,23] where the influence of different genes (inhibition, activation) are represented by automata whose functions mirror together the global dynamics of the network. This application in particular motivates the development of tools to understand, predict and describe the dynamics of ANs in an efficient way. In the worst case, studying the dynamics of an AN (*i.e.* analysing the behavior of all possible configurations of the system) will always take an exponential amount of time in the size of the network. Attempts using mainly combinatorics have been made to predict and count specific limit behavior of the system without enumerating the entire network's dynamics [2,4,10]. Other studies focused on understanding the dynamics of such complex systems by considering them as compositions of bricks simpler to analyse [5,8,22] and propose to study manners of controlling these bricks and/or systems [6,19]. In line with such approaches and [12] the authors developed in [20] the formalism of modules. They are ANs with inputs, and operators called wirings that allow modules to be composed into larger modules, and eventually into ANs. In this paper we propose an exploration of a specific type of modules, namely acyclic modules,

© Springer Nature Switzerland AG 2020
J. Chen et al. (Eds.): TAMC 2020, LNCS 12337, pp. 168–180, 2020.
https://doi.org/10.1007/978-3-030-59267-7_15

which do not include cycles in their interaction graph. The present paper also introduces output functions, which characterise the behavior of an acyclic module as a function of the inputs of the network over time. Output functions allow us to characterise the dynamics of a network while forgetting its inner structure, illustrated by Theorem 1, which shows that if two acyclic modules have equivalent output functions, they also have isomorphic attractors.

In Sect. 2 we propose definitions of ANs, modules and wirings. Section 3 presents definitions of acyclicity in modules and related concepts and results. Finally in Sect. 4 we explore complexity results around acyclic modules and their inputs. The demonstrations of all results are available in the arXiv version of this paper.

General Notations. We denote $\mathbb{B}$ the set of Booleans $\mathbb{B} = \{0, 1\}$. For Λ an alphabet, we denote Λ^n the set of vectors of size n with values in Λ. For $x \in \Lambda^n$, we might denote x by $x_1 x_2 \dots x_n$. For example, a vector $x \in \mathbb{B}^3$ defined such that $x_1 = 1$, $x_2 = 0$, $x_3 = 1$ can alternatively be denoted by $x = 101$. For S an ordered set of labels, $x \in \Lambda^S$, s in S, and f a function which takes x as an input, we might denote $f(x) = s$ as a simplification of $f(x) = x_s$. For G a digraph, we denote by $V(G)$ the set of its vertices and by $A(G)$ the set of its arcs. Let G, G' be two digraphs, we denote $G \subseteq G'$ if and only if G is an induced subdigraph of G', that is $V(G) \subseteq V(G')$ and $u, v \in V(G)$ implies $(u, v) \in A(G) \Leftrightarrow (u, v) \in A(G')$. For $f : A \to B$, and $C \subseteq A$, we denote $f|_C$ the function defined over $f|_C : C \to B$ such that $f|_C(x) = f(x)$ for all $x \in C$. For $x \in \Lambda^S$, for any function $f : R \to S$ (for some set R), we define $x \circ f$ as $(x \circ f)_r = x_{f(r)}$, for all $r \in R$. For $X = (x_1, x_2, \dots, x_k)$ a sequence of $x_i \in \Lambda^S$, we define $X \circ f$ as the sequence $(x_1 \circ f, x_2 \circ f, \dots, x_k \circ f)$. In most of our examples, the alphabet Λ will be $\mathbb{B}$ and the set S finite, hence $x \in \Lambda^S$ will be considered as a Boolean vector (according to some order on S).

2 Definitions

2.1 Automata Networks

ANs are composed by a set S of automata. Each automaton in S, or node, is at any time in a state in Λ. Gathering those isolated states into a vector of dimension $|S|$ provides us with a configuration of the network. More formally, a *configuration* of S over Λ is a vector in Λ^S. The state of every automaton is bound to evolve as a function of the configuration of the entire network. Each node has a unique function, called a local function that is predefined and does not change over time. A *local function* is thus a function f defined over $f : \Lambda^S \to \Lambda$. An AN is described as a set which provides a local function to every node in the network. Formally, an *automata network* F is a set of local functions f_s over S and Λ for every $s \in S$.

Example 1. For $\Lambda = \mathbb{B}$, and $S = \{a, b, c\}$, let F be a Boolean AN with local functions $f_a(x) = \neg a$, $f_b(x) = a \vee \neg c$, and $f_c(x) = \neg c \wedge \neg a$.

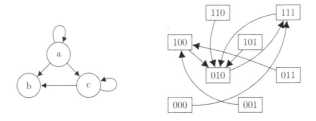

Fig. 1. (Left) Interaction digraph and of (right) dynamics of the network of Example 1

The configuration of an AN is updated using the local functions. The protocol by which the local functions are applied is called its update schedule. Many different update schedules exist (actually, there are an infinite number of these), and it is well known that changing the update schedule of ANs can change the obtained dynamics [3,14,17,21]. The update schedule used in this paper is the parallel update schedule, in which every node udpates its value according to its local function at each time step. Thus, considering a configuration x of an AN F, the *update* $F(x)$ of F over x is the configuration such that for all $s \in S$, $F(x)_s = f_s(x)$, where f_s is the local function assigned to s in F.

Example 2. Following the previous example, we can see that $F(000) = 111$, $F(010) = 111$ and that $F(111) = 010$.

ANs are usually represented by the influence that automata hold on each other. As such the visual representation of an AN is a directed graph, called an interaction digraph, whose nodes are the automata of the network, and arcs are the influences that link the different automata. Formally, s *influences* s' if and only if there exist two configurations x, x' such that $\forall r \in S, r \neq s \iff x_r = x'_r$, and $F(x)_{s'} \neq F(x')_{s'}$. From this, we define the *interaction digraph* of F as the directed graph with nodes S such that (s, s') is an arc of the digraph if and only if s influences s'. For instance the interaction digraph of the network developed in Example 1 is depicted in Fig. 1.

To encapsulate the entire behavior of the network, one needs to enumerate all the possible configurations the network, namely the elements of Λ^S, and describe the global update function upon this set. This is often done via another graphical representation, which is another digraph, called the dynamics of the network. Intuitively, this graph defines an arc from x to x' if and only if the update of the network over the configuration x results in the configuration x'. Formally, the *dynamics* of F can be represented as the digraph G with vertex set Λ^S, such that (x, x') is an arc in G if and only if $F(x) = x'$. The dynamics of the network developed in Example 1 is presented in Fig. 1.

The dynamics of a network is a large object. A commonly studied part of this object is called the attractors of the networks. An attractor is a sequence of configurations which constitutes a cycle in the dynamics of the network. Alternatively, the attractors of a network can be defined as the set of non trivial strongly connected components of its dynamics. Formally, an *attractor* of F is a connected component of the subdigraph $G_L \subseteq G$, such that x is a node in G_L if and only if there exists $k \in \mathbb{N} \setminus \{0\}$ such that $F^k(x) = x$. Notice that, classically in the domain of ANs, An attractor of size one is called a *fixed point*, whereas an attractor of size greater than one is called a *limit cycle*.

Example 3. In our example, the attractors of F are the configurations 010 and 111 since they verify $F^2(010) = 010$ and $F^2(111) = 111$. For any other configuration, updating the network more than two times changes the state of the network to 010 or 111. Alternatively, the configuration 010 and 111 form the only non trivial strongly connected component of the dynamics of this network.

2.2   Modules

Informally, modules can be described as ANs with inputs. More formally, for a given module, we introduce a new set of labels, usually denoted I, which contains the inputs of the module. By convention, inputs will be denoted with Greek letters. A local function of a module does not only depend on the states of the automata of the network, but also on the evaluations of the inputs. Inputs are not automata, and do not have a state; but it is interesting to suggest that inputs are added nodes of the network that do not admit local functions. Formally, by considering S and I as sets of labels, and Λ as an alphabet, a *module* is a set which, for every $s \in S$, defines a local function $f_s : \Lambda^{S \cup I} \to \Lambda$.

Example 4. For $\Lambda = \mathbb{B}$, $S = \{a, b, c\}$ and $I = \{\alpha, \beta, \gamma\}$ let M be a module with local functions $f_a(x, i) = \neg b \vee \alpha$, $f_b(x, i) = a \vee \neg c \vee \beta \vee \neg \alpha$, and $f_c(x, i) = \neg c \wedge \neg \gamma$.

The digraph representation of a module is similar to that of an AN; the inputs are added for clarity as incident arrows to the nodes they influence. For instance, the module of Example 4 is illustrated in Fig. 3. As well, updating a module over the parallel update schedule is similar to updating an AN. The inputs are introduced with specific notations which are detailed below. Let x and i be configurations over S and I respectively. The update of a module M over x and i, denoted $M(x, i)$, is defined as a configuration over S such that, for all $s \in S$, $M(x, i)_s = f_s(x, i)$, where f_s is the local function assigned to s in M.

Example 5. Let us update the module M over the node configuration $x = 011$ and the input configuration $i = 000$. We compute $f_a(x, i) = \neg 1 \vee 0 = 0$, $f_b(x, i) = 0 \vee \neg 1 \vee 0 \vee \neg 0 = 1$ and $f_c(x, i) = \neg 1 \wedge \neg 0 = 0$, thus giving $M(011, 000) = 010$.

Since it will be convenient to update a module over multiple iterations at once, we will generally consider a sequence of input configurations of the form $(i_1, i_2, \ldots, i_m)$. For $\alpha, \beta, \ldots$ the inputs of the considered module, we will denote for convenience $\alpha_1, \beta_1, \ldots$ the evaluation of those inputs in the configuration i_1, and so on, denoting $\alpha_k, \beta_k, \ldots$ the evaluation of the respective inputs in the configuration i_k. We will denote by $M(x, (i_1, i_2, \ldots, i_m))$ the execution of m updates of the module M starting with configuration x, taking the input configuration i_k at update number k. Formally, it is defined recursively as:

$$M(x, (i_1, i_2, \ldots i_m)) = M(M(x, i_1), (i_2, \ldots, i_m)), \text{ with } M(x, \varnothing) = x.$$

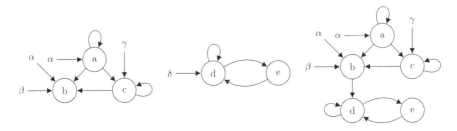

Fig. 2. Illustration of the wiring of Example 7. Interaction digraphs of the modules (left) M, (center) M' and (right) M''.

2.3   Wirings

Modules are a formalism of composition and decomposition of ANs. As such, we define the process of composing modules together as wiring. Wirings exist in two forms. One is recursive, and proposes the rearrangement of a single module by connecting inputs of the module to itself. The second type of wiring is non-recursive, and defines the combination of two modules into one, connecting inputs of one module to the nodes of the other. When an input is connected, any function depending on the value of that input relies on the state of the connected node instead. Those two sorts of wirings were proven to be universal to compose any network from elementary parts [20]. Wiring operations are defined upon an object that specifies the operated connections, usually denoted ω which is a partial function defined from a subset of inputs of the second module to nodes of the first.

Recursive Wiring. *Let M be a module with label sets S and I which, for every $s \in S$, defines the local function f_s. For $\omega : I \nrightarrow S$ a partial function, we define $\circlearrowright_\omega M$ the module which, for every $s \in S$, defines the local function f'_s such that:*

$$\forall x \in \Lambda^{S \cup I \setminus \mathrm{dom}(\omega)}, \ f'_s(x) = f_s(x \circ \hat{\omega}), \ \text{with } \hat{\omega}(k) = \begin{cases} \omega(k) & \text{if } k \in \mathrm{dom}(\omega) \\ k & \text{if } k \in S \cup I \setminus \mathrm{dom}(\omega) \end{cases}.$$

Example 6. For $\Lambda = \mathbb{B}$, $S = \{a, b, c\}$ and $I = \{\alpha, \beta, \gamma\}$ let M be a module with local functions $f_a(x, i) = \neg b \lor \alpha$, $f_b(x, i) = a \lor \neg c \lor \beta \lor \neg \alpha$, and $f_c(x, i) = \neg c \land \neg \gamma$. Let us define a partial function $\omega : I \to S$ such that $\mathrm{dom}(\omega) = \{\alpha, \gamma\}$, and $\omega(\alpha) = c$ and $\omega(\gamma) = a$. The result of the recursive wiring $\circlearrowright_\omega M$ is a module with label sets $S' = S$ and $I' = \{\beta\}$ with local functions $f'_a(x, i) = \neg b \lor c$, $f'_b(x, i) = a \lor \neg c \lor \beta \lor \neg c$, and $f'_c(x, i) = \neg c \land \neg a$.

Non-recursive Wiring. *Let M and M' be two modules with respective label sets S, I, and S', I'. We denote f_s and $f'_{s'}$ the local functions defined respectively in M and M' for every $s \in S$ aand $s' \in S'$. For $\omega : I' \nrightarrow S$ a partial function, we define $M \rightarrowtail_\omega M'$ the module with label sets $S \cup S'$ and $I \cup I' \setminus \mathrm{dom}(\omega)$ which, for every $s \in S \cup S'$, defines the local function f''_s such that:*

$$\forall x \in \Lambda^{\mathscr{S}}, \; f_s''(x) = \begin{cases} f_s(x|_{S \cup I}) & \text{if } s \in S \\ f_s'(x \circ \hat{\omega}) & \text{if } s \in S' \end{cases} \quad \text{with } \hat{\omega}(k) = \begin{cases} \omega(k) & \text{if } k \in \mathrm{dom}(\omega) \\ k & \text{if } k \in \mathscr{S} \end{cases},$$

$$\text{for } \mathscr{S} = S \cup S' \cup I \cup I' \setminus \mathrm{dom}(\omega).$$

Example 7. For $\Lambda = \mathbb{B}$, $S = \{a, b, c\}$ and $I = \{\alpha, \beta, \gamma\}$, let M be a module with local functions $f_a(x, i) = \neg b \vee \alpha$, $f_b(x, i) = a \vee \neg c \vee \beta \vee \neg \alpha$, and $f_c(x, i) = \neg c \wedge \neg \gamma$. Let also be $S' = \{d, e\}$, $I' = \{\delta\}$ and M' another module with local functions $f_d'(x, i) = \neg d \vee e \vee \delta$ and $f_e'(x, i) = \neg e \vee d$. Let $\omega : I' \to S$ be the function such that $\omega(\delta) = b$. The result of the non-recursive wiring $M \rightarrowtail_\omega M'$ is the module with sets $S'' = \{a, b, c, d, e\}$ and $I'' = \{\alpha, \beta, \gamma\}$ with local functions $f_a''(x, i) = \neg b \vee \alpha$, $f_b''(x, i) = a \vee \neg c \vee \beta \vee \neg \alpha$, $f_c''(x, i) = \neg c \wedge \neg \gamma$, $f_d''(x, i) = \neg d \vee e \vee b$ and $f_e''(x, i) = \neg e \vee d$. (See an illustration in Fig. 2.)

3 Acyclicity

3.1 Acyclic Automata Networks

Acyclicity is a property of the interaction digraph of the considered AN; it means that no node of the network influences itself, neither by a direct loop nor through the action of any cycle that would include this node. Acyclic ANs have been one of the first families of ANs to be studied and characterised [21]. Their dynamical behavior is trivial: there is only one fixed point, which attracts every other configuration. This is true under the parallel update schedule as well as any other schedule which would eventually update every node a minimum amount of time for the stabilisation to propagate. This early result led to the simple conclusion that cycles are essential to the complexity of their dynamics.

3.2 Acyclic Modules

Acyclicity. *A module M is acyclic if its interaction digraph is acyclic.*

Example 8. For $\Lambda = \mathbb{B}$, $S = \{a, b, c\}$ and $I = \{\alpha, \beta, \gamma\}$ let M be a module with local functions $f_a(x, i) = \alpha$, $f_b(x, i) = a \vee \beta \vee \neg \alpha$, and $f_c(x, i) = \neg b \wedge a \wedge \neg \gamma$. M is acyclic. (See an illustration in Fig. 3.)

The dynamics of this family of objects is simple enough to be studied, and complex enough to provide insights into the general dynamics of ANs. It is indeed clear that every AN can be decomposed into a recursively wired acyclic module. This can be done by taking a feedback arc set of the interaction digraph of the network, and producing a module that replaces every arc in the set by an input.

As an acyclic module has no loop or cycle in its influences, it can support no long lasting memory used for computation. As such the behavior of any node in the network can be understood as a function of only the evaluation of the inputs in its last iterations. This function is called an output function and how much it must look in the past to make its prediction is called the delay of the function.

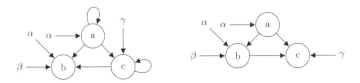

Fig. 3. Interaction digraph of (left) the module of Example 4, (right) the acyclic module of Example 8.

For M a module with k inputs, an output function O with delay m is a function defined over a sequence of inputs $(i_1, i_2, \ldots, i_m)$. Each node of a network defines its own output function, similarly to how it defines a local function. The output function of a node always has minimal delay and will depend on the output functions defined by the nodes which influence it. In other terms, if node a influences node b, then whatever output function which predicts the value of a based only on inputs will be useful to predict the evaluation of b one iteration later. As such b does not directly depend on the output function of a, but on the output function of a with incremented delay.

Output functions are sufficient to describe the behavior of the entire module from the inputs after a given amount of time. This fact is illustrated by the Property 1 below.

Node Output. *Let M be an acyclic module. For every $s \in S$, we define the output function of s, denoted O_s, as the output function with minimal delay m such that for any sequence of inputs $J = (i_1, i_2, \ldots, i_m)$ and any configuration x, $M(x, J)_s = O_s(J)$.*

Example 9. For $\Lambda = \mathbb{B}$, $S = \{a, b, c\}$ and $I = \{\alpha, \beta, \gamma\}$ let M be a module with local functions $f_a(x, i) = \alpha$, $f_b(x, i) = a \vee \beta \vee \neg\alpha$, and $f_c(x, i) = \neg b \wedge a \wedge \neg\gamma$. The module M verifies the following output functions : $O_a = \alpha_1$, which has delay 1, $O_b = \alpha_2 \vee \beta_1 \vee \neg\alpha_1$, which has delay 2, and $O_c = \neg\alpha_3 \wedge \neg\beta_2 \wedge \alpha_2 \wedge \alpha_2 \wedge \neg\gamma_1$, which has delay 3.

Property 1. Let M be an acyclic module. For every $s \in S$, s has one and only one output function O_s.

Property 1 can be further refined to propose the following result, which states that two networks have the same attractors if and only if the modules they can be decomposed into have the same number of inputs and the same output functions on the nodes on which those inputs are wired. As such, modules can be considered as black boxes which are to be considered equivalent in their limit behavior, as long as they share the same output functions, according to some bijection between their inputs.

Theorem 1. *Let M and M' be two acyclic modules, with T and T' subsets of their nodes such that $|T| = |T'|$. If there exists g a bijection from I to I' and h a* *bijection from T to T' such that for every $s \in T$, O_s and $O'_{h(s)}$ have same delay, and for every input sequence J with length the delay of O_s,*

$$O_s(J) = O'_{h(s)}(J \circ g^{-1})$$

then for any function $\omega : I \to T$, the networks $\circlearrowleft_\omega M$ and $\circlearrowleft_{h \circ \omega \circ g^{-1}} M'$ have isomorphic attractors (up to the renaming of automata given by h).

Output functions are a characterisation of the behavior of acyclic modules which is enough to understand their limit dynamics under parallel schedule. This characterisation behaves in expected ways under non-recursive wirings. Taking two acyclic modules and wiring them non-recursively makes a module whose output functions are deducible from the output functions of the initial acyclic module. We now state a result which provides an upper bound on the number of attractors of each size of an AN, which is wired from a module with k inputs.

Theorem 2. *Taking an acyclic module with k inputs and wiring all inputs recursively gives an AN. Let us denote $a(k, c)$ the number of attractors of size c of its dynamics. We state $a(k, c) \leq A(k, c)$, with:*

$$A(k, 1) = |\Lambda|^k \qquad and \qquad A(k, c) = |\Lambda|^{kc} - \sum_{c' < c, \ c' | c} A(k, c').$$

The smallest k which can be provided for any AN is equal to the minimum feedback arc set of the network. As such this result is very similar to a previous result of [2,4], which states an upper bound on the total number of attractors depending on the size of a positive feedback arc set. Though the global bound with a positive feedback arc set would be stronger, the present result is different as it operates on parallel update schedule and provides different bounds on different sizes of attractors, where the previous result offered a bound on the total count of attractors under asynchronous update schedule.

3.3   One-to-One Modules

A module with only one input has the particularity of being recursively wired in a linear amount of possible ways. That is, the only degree of freedom in the wiring is the choice of the node which will serve as output. Let us consider a module with only one input, and let us consider $e \in S$ as the designated output node of the module. In this context we will denote $\circlearrowleft M$ as the AN obtained by wiring the input of the module to its designated output. Furthermore, the output function O_e will sometimes be denoted O, as the designated output function of the module. Such an acyclic module with only one input and a designated output is called a one-to-one module.

Theorem 3. *Let M be a one-to-one module. The one-to-one module M_{min} with a minimum number of nodes and which defines the same output function as M is of size d, for d the delay of the output function of M.*

An example of the application of Theorem 3 is illustrated in Fig. 4. This construction is polynomial in time, and bears strong resemblances with the objects known as Feedback Shift Registers [11].

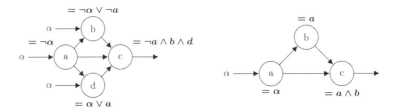

Fig. 4. Illustration of Theorem 3. Both modules consider c as their output node, and display the same output function $O = \alpha_2 \wedge \alpha_3$ The module on the right is optimal, as 3 is the delay of its output function.

4   Complexity Results

This section presents complexity results that have been obtained around output functions, and the difficulty of the analysis of the dynamics of acyclic modules after being recursively wired into a complete network. Remark that these questions have been widely addressed in the context of threshold Boolean ANs [1,13,18]. Such a wiring will sometimes be denoted as a complete recursive wiring of the module. A module is encoded into the input of a decision problem as the list of its local functions written in propositional logic. As such the computation of the output functions of an acyclic module is comparable to the computation of a circuit.

Let us provide a few decision problems on the dynamics of a network obtained from a recursively wired acyclic module.

▸ ACYCLIC MODULE ATTRACTOR PROBLEM
   Input: An acyclic module M with k inputs and n nodes, a function ω which defines a complete recursive wiring over M, and a number c encoded in unary.
   Question: Does there exist an attractor of size c in the dynamics of $\circlearrowleft_\omega M$?

▸ ONE-TO-ONE MODULE ATTRACTOR PROBLEM
   Input: A one-to-one module M with n nodes, a function ω which defines a complete recursive wiring over M, and a number c encoded in unary.
   Question: Does there exist an attractor of size c in the dynamics of $\circlearrowleft_\omega M$?

▸ ACYCLIC MODULE FIXED POINT PROBLEM
   Input: An acyclic module M with k inputs and n nodes, and a function ω which defines a complete recursive wiring over M.
   Question: Does there exist a configuration x such that $\circlearrowleft_\omega M(x) = x$?

▸ ONE-TO-ONE MODULE FIXED POINT PROBLEM
   Input: A one-to-one module M with n nodes, and a function ω which defines a complete recursive wiring over M.
   Question: Does there exist a configuration x such that $\circlearrowleft_\omega M(x) = x$?

Those four problems are variations of the same question under different sets of constraints. The first problem, the Acyclic Module Attractor Problem, generalises the other three decision problems, while the One-to-one Module Fixed Point Problem is a specific case of the other three decision problems. We provide our complexity analysis of those problems in a way that mirrors this diamond-like structure.

Theorem 4. *The Acyclic Module Attractor Problem can be solved in time $\mathcal{O}(f(k \times c)q(n))$ for some function f and q a polynomial, i.e. it is fixed parameter tractable.*

Sketch of Proof. We obtain this by iterating all of the possible input sequences of size c. We execute the network on each sequence and check if the outputs correspond to the given input. This process scales polynomially with the size of the network, but exponentially with the size of the attractor and the number of inputs.   □

Theorem 5. *The One-to-one Module Attractor Problem is NP-complete.*

Sketch of Proof. We provide a reduction from the SAT problem which for any formula with m variables, constructs a module of size $3m + 1$. The first $3m$ nodes encode the input and the last node checks the evaluation. If at any point the formula is evaluated at false or if the encoding is wrong, the whole network stabilises to a fixed point. If the encoding is correct and the evaluation positive, the configuration will shift in the network, providing an attractor of size $3m + 1$. The existence of this attractor is proven equivalent to the satisfiability of the formula.   □

Theorem 6. *The Acyclic Module Fixed Point Problem is NP-complete.*

Sketch of Proof. We provide a reduction from the SAT problem. In this reduction, the obtained module will stabilise only if a given node, which computes a SAT formula, has constant value 1.   □

Corollary 1. *The One-to-one Module Fixed Point Problem is in P.*

Sketch of Proof. This is an application of Theorem 4.   □

The above stated results imply that the size of the network is not a meaningful parameter in the difficulty of the task of finding attractors. Thereom 4 shows that the two parameters which apply this effect are the size of the desired attractor and the number of inputs the network bears when seen as an acyclic module. In other terms this second parameter is the level of interconnectivity of the network. Theorems 5 and 6 prove that this caracterisation is tight. Together, these four theorems provide a new perspective on a known fact; that cycles in ANs are crucial for complexity to arise.

► ACYCLIC MODULE OUTPUT CONSTRUCTION PROBLEM

Input: A set $\{M_1, M_2, \ldots, M_\ell\}$ of acyclic modules, and O an output function encoded in a lookup table.

Question: Does there exist a set of non-recursive wirings ω which can construct an acyclic module from $M_1, M_2, \ldots, M_\ell$ such that O is an output function of the obtained module?

Theorem 7. *The Acyclic Module Output Construction Problem is NP-complete.*

Sketch of Proof. We provide a reduction from the SAT problem. We ask for the construction of an output function via the wiring of two modules with a unique constant function '0' and '1' respectively, and a bigger module which executes a computation from its inputs based on the formula, such that the target output function is obtained by non-recursive wirings if and only if the formula is satisfiable.   □

5   Conclusion

Automata Networks are complex systems, the exhaustive study of which requires an amount of resources exponential in the size of the network. By defining and studying acyclic modules we propose an innovative way of approaching this question. Theorem 1 proposes the reduction of the limit dynamic of a network to the output functions of an acyclic module which composes it. We think that this result, alongside with Theorem 3 which is a direct application of it, provides an interesting way of categorising networks depending on their output functions. Also presented are Theorem 2 which proposes a bound on the total number of attractors depending on the number of inputs in an acyclic module, and the results listed in Sect. 4, which state a range of complexity results on acyclic modules. The set of results proposed in this paper describe, in our opinion, a good picture of the limits and possibilities that come from studying acyclic modules.

In future works, we plan to expand this formalism to more general update schedules, and to propose a version of Theorem 3 which would generalise to modules with more than one input and one output. We also plan to apply those tools to optimise large automata networks, such as those designed and studied in biology applications.

Acknowledgments. The works of Kévin Perrot and Sylvain Sené were funded mainly by their salaries as French State agents, affiliated to Aix-Marseille Univ., Univ. de Toulon, CNRS, LIS, UMR 7020, Marseille, France (both) and to Univ. Côte d'Azur, CNRS, I3S, UMR 7271, Sophia Antipolis, France (KP), and secondarily by ANR-18-CE40-0002 FANs project, ECOS-Sud C16E01 project, STIC AmSud CoDANet 19-STIC-03 (Campus France 43478PD) project.

References

1. Alon, N.: Asynchronous threshold networks. Graphs Combin. **1**, 305–310 (1985). https://doi.org/10.1007/BF02582959

2. Aracena, J.: Maximum number of fixed points in regulatory Boolean networks. Bull. Math. Biol. **70**, 1398–1409 (2008). https://doi.org/10.1007/s11538-008-9304-7
3. Aracena, J., Gómez, L., Salinas, L.: Limit cycles and update digraphs in Boolean networks. Discrete Appl. Math. **161**, 1–12 (2013)
4. Aracena, J., Richard, A., Salinas, L.: Number of fixed points and disjoint cycles in monotone Boolean networks. SIAM J. Discrete Math. **31**, 1702–1725 (2017)
5. Bernot, G., Tahi, F.: Behaviour preservation of a biological regulatory network when embedded into a larger network. Fund. Inform. **91**, 463–485 (2009)
6. Biane, C., Delaplace, F.: Causal reasoning on Boolean control networks based on abduction: theory and application to cancer drug discovery. IEEE/ACM Trans. Comput. Biol. Bioinform. **16**, 1574–1585 (2019)
7. Davidich, M.I., Bornholdt, S.: Boolean network model predicts cell cycle sequence of fission yeast. PLoS One **3**, e1672 (2008)
8. Delaplace, F., Klaudel, H., Melliti, T., Sené, S.: Analysis of modular organisation of interaction networks based on asymptotic dynamics. In: Gilbert, D., Heiner, M. (eds.) CMSB 2012. LNCS, pp. 148–165. Springer, Heidelberg (2012). https://doi.org/10.1007/978-3-642-33636-2_10
9. Demongeot, J., Goles, E., Morvan, M., Noual, M., Sené, S.: Attraction basins as gauges of robustness against boundary conditions in biological complex systems. PLoS One **5**, e11793 (2010)
10. Demongeot, J., Noual, M., Sené, S.: Combinatorics of Boolean automata circuits dynamics. Discr. Appl. Math. **160**, 398–415 (2012)
11. Elspas, B.: The theory of autonomous linear sequential networks. IRE Trans. Circuit Theory **6**(1), 45–60 (1959)
12. Feder, T.: Stable networks and product graphs. Ph.D thesis, Stanford Univ. (1990)
13. Floreen, P., Orponen, P.: Counting stable states and sizes of attraction domains in Hopfield nets is hard. In: Proceedings of the of IJCNN 1989, pp. 395–399 (1989)
14. Goles, E., Salinas, L.: Comparison between parallel and serial dynamics of Boolean networks. Theor. Comput. Sci. **396**, 247–253 (2008)
15. Kauffman, S.A.: Metabolic stability and epigenesis in randomly constructed genetic nets. J. Theor. Biol. **22**, 437–467 (1969)
16. Mendoza, L., Alvarez-Buylla, E.R.: Dynamics of the genetic regulatory network for Arabidopsis thaliana flower morphogenesis. J. Theor. Biol. **193**, 307–319 (1998)
17. Noual, M., Sené, S.: Synchronism versus asynchronism in monotonic Boolean automata networks. Nat. Comput. **17**(2), 393–402 (2017). https://doi.org/10.1007/s11047-016-9608-8
18. Orponen, P.: Neural networks and complexity theory. In: Havel, I.M., Koubek, V. (eds.) MFCS 1992. LNCS, vol. 629, pp. 50–61. Springer, Heidelberg (1992). https://doi.org/10.1007/3-540-55808-X_5
19. Pardo, J., Ivanov, S., Delaplace, F.: Sequential reprogramming of biological network fate. In: Bortolussi, L., Sanguinetti, G. (eds.) CMSB 2019. LNCS, vol. 11773, pp. 20–41. Springer, Cham (2019). https://doi.org/10.1007/978-3-030-31304-3_2
20. Perrot, K., Perrotin, P., Sené, S.: A framework for (de)composing with boolean automata networks. In: Durand-Lose, J., Verlan, S. (eds.) MCU 2018. LNCS, vol. 10881, pp. 121–136. Springer, Cham (2018). https://doi.org/10.1007/978-3-319-92402-1_7

21. Robert, F.: Discrete Iterations: A Metric Study. Springer, Heidelberg (1986). https://doi.org/10.1007/978-3-642-61607-5
22. Siebert, H.: Dynamical and structural modularity of discrete regulatory networks. In: Proceedings of COMPMOD 2009, volume 6 of EPTCS, pp. 109–124 (2009)
23. Thomas, R.: Boolean formalization of genetic control circuits. J. Theor. Biol. **42**, 563–585 (1973)

Eternal Connected Vertex Cover Problem

Toshihiro Fujito$^{(\boxtimes)}$ and Tomoya Nakamura

Toyohashi University of Technology, Toyohashi 441-8580, Japan
`fujito@cs.tut.ac.jp, w-nakamura@algo.cs.tut.ac.jp`

Abstract. A new variant of the classic problem of VERTEX COVER (VC) is introduced. Either CONNECTED VERTEX COVER (CVC) or ETERNAL VERTEX COVER (EVC) is already a variant of VC, and ETERNAL CONNECTED VERTEX COVER (ECVC) is an eternal version of CVC, or a connected version of EVC. A *connected vertex cover* of a connected graph G is a vertex cover of G inducing a connected subgraph in G. CVC is the problem of computing a minimum connected vertex cover. For EVC a multi-round game on G is considered, and in response to every attack on some edge e of G, a guard positioned on a vertex of G must cross e to repel it. EVC asks the minimum number of guards to be placed on the vertices of a given G, that is sufficient to repel any sequence of edge attacks of an arbitrary length.

ECVC is EVC in which any configuration of guards, that is the set of vertices occupied by them, needs to be kept connected in each round. This paper presents, besides some basic structural properties of ECVC, 1) a polynomial time algorithm for ECVC on chordal graphs, 2) NP-completeness of ECVC on locally connected graphs, 3) a complete characterization of ECVC on cactus graphs, block graphs, or any graphs in which every block is either a simple cycle or a clique, and 4) a 2-approximation algorithm for ECVC on general graphs.

Keywords: Eternal vertex cover · Connected vertex cover · Chordal graphs

1 Introduction

In this paper we introduce a new computational problem called ETERNAL CONNECTED VERTEX COVER (ECVC), and as the name suggests, it is a variant of both ETERNAL VERTEX COVER (EVC) and CONNECTED VERTEX COVER (CVC), where each of them is in turn a variant of the classic problem of VERTEX COVER (VC). A vertex subset $S \subseteq V$ of a graph $G = (V, E)$ is called a *vertex cover* of G when every edge in G is incident to some vertex of S. VC is the problem of computing the size of a minimum vertex cover of G, called the *vertex cover number* $\tau(G)$ of G.

EVC is a variant of VC with game-theoretic flavor. The game begins with some number of guards placed on vertices of an input graph $G = (V, E)$ (possibly

This work is supported in part by JSPS KAKENHI under Grant Number 17K00013.

© Springer Nature Switzerland AG 2020
J. Chen et al. (Eds.): TAMC 2020, LNCS 12337, pp. 181–192, 2020.
https://doi.org/10.1007/978-3-030-59267-7_16

more than one guard on one vertex), thereafter proceeding in rounds without changing the number of guards being used. In each round, any (but only one) edge $e \in E$ is subject to an attack by the enemy, and each guard positioned at any $u \in V$ may remain there or move to a vertex v adjacent to u passing an edge $\{u, v\}$ in response to such an attack. Any attack on an edge can be defended if at least one guard moves to a neighboring vertex by passing over the attacked edge within the corresponding round, but when no guard can do so, the attack on G succeeds and the game is over. The minimum number of guards sufficient to keep on protecting G by indefinitely repelling any arbitrarily long sequence of edge attacks is called the *eternal vertex cover number* $\tau^\infty(G)$ of G. Following the presentation of [3], let $\mathcal{C}$ be a family of vertex covers of G of the same cardinality. It is intended that each $C \in \mathcal{C}$ specifies a *configuration* of guards, that is the set of vertices occupied by guards. If a team of guards can successfully keep on defending G from an indefinitely long sequence of arbitrary edge attacks by changing its configurations chosen only from $\mathcal{C}$, starting with an arbitrary one, $\mathcal{C}$ is called an *eternal vertex cover class* and a vertex cover in $\mathcal{C}$ an *eternal vertex cover*, of G. The eternal vertex cover number $\tau^\infty(G)$ of G is then the minimum cardinality of an eternal vertex cover of G. EVC was introduced by Klostermeyer and Mynhardt [16] and was further studied by Fomin et al. [11] and others. Closely related, earlier introduced, and more extensively studied is the *eternal domination* (or *eternal security*) problem, where an arbitrary *vertex* is attacked, instead of an edge, and it is required that some guard moves onto the vertex to repel it in each round; see for instance [1,5,6,13,14,16,18].

CONNECTED VERTEX COVER (CVC) is another variant of VC, where a vertex cover inducing a connected subgraph of G is sought in a given *connected* graph G. Such a vertex cover is called a *connected vertex cover (cvc)*, and the *connected vertex cover number* $\tau_c(G)$ is the minimum cardinality of a cvc of G. CVC was introduced by Garey and Johnson [12] along with its NP-hardness. The problem is known to have some applications in the domain of wireless network design, and it was also indicated that CVC can be used to solve the Top Right Access point minimum length corridor problem (see [9]).

The new problem which we call ETERNAL CONNECTED VERTEX COVER (ECVC) is a variation of EVC where every configuration in an eternal vertex cover class must be a cvc of G for a *connected* graph G, and such a cvc is called an *eternal connected vertex cover (ecvc)* of G. The *eternal connected vertex cover number* $\tau_c^\infty(G)$ of G is then the minimum cardinality of an ecvc of G. ECVC is supposed to arise as a natural problem in the context of graph protection where tight communication is needed among all the participating mobile guards for their coordinated moves. In fact a connected version of the eternal dominating set is already introduced, and some upper bounds on the minimum size of an *eternal connected dominating set* were obtained in [19].

1.1   Previous Related Work

Klostermeyer and Mynhardt showed in [17] that

- $\tau^\infty(C_n) = \tau(C_n) = \lceil n/2 \rceil$ for a cycle C_n on n vertices with $n \geq 3$,
- When G is a tree on n vertices with $n \geq 2$, $\tau^\infty(G)$ is one more than the number of its internal vertices,
- $\tau(G) \leq \tau^\infty(G) \leq 2\tau(G)$ for any graph G, and both of these bounds are tight,
- $\tau^\infty(G) = \tau(G)$ when G is a complete graph, Peterson graph, $K_m \square K_n$, and $C_m \square C_n$ (where $\square$ represents the box product),
- $\tau^\infty(G) = \tau(G)$ or $\tau^\infty(G) = \tau(G) + 1$ for $n \times m$ grid graphs depending on n and m.

while giving a characterization for graphs G for which $\tau^\infty(G) = 2\tau(G)$ holds (see [15] for more bounds on $\tau^\infty(G)$ and related graph parameters). The computational complexity of EVC (for general graphs) was investigated by Fomin et al. [11], and it was shown that

- the decision version of EVC is NP-hard but fixed parameter tractable when parameterized by the number of guards,
- EVC is approximable within a factor of 2 in polynomial time.

Following the work of Klostermeyer and Mynhardt [17], it was further pursued and achieved by Babu et al. [3] to characterize a class of graphs G having $\tau^\infty(G) = \tau(G)$, and it was shown that

- one can decide whether $\tau^\infty(G) = \tau(G)$ or not in polynomial time when G is a chordal graph,
- one can compute $\tau^\infty(G)$ in polynomial time when G is a biconnected chordal graph, and
- when graphs are locally connected, the decision version of EVC is NP-complete, and it is NP-hard to approximate $\tau^\infty(G)$ within any factor less than $10\sqrt{5} - 21 \approx 1.36$.

As for CVC,

- it is NP-hard even if graphs are planar bipartite of maximum degree 4 [10], planar biconnected of maximum degree 4 [21], 3-connected [24], or k-regular with $k \geq 4$ [20], and
- nontrivial cases in which $\tau_c(G)$ is polynomially computable are known only for graphs of maximum degree 3 [23] and for chordal graphs [9].

1.2   Our Contribution

This paper introduces a new variant of VERTEX COVER, CONNECTED VERTEX COVER, and ETERNAL VERTEX COVER, which we call ETERNAL CONNECTED VERTEX COVER (ECVC). We present, besides some basic structural properties of ECVC, 1) a polynomial time algorithm for ECVC on chordal graphs, 2) NP-completeness of the decision version of ECVC on locally connected graphs, 3) a complete characterization of ECVC on cactus graphs, block graphs, or any graphs in which every block is either a simple cycle or a clique, and 4) a 2-approximation algorithm for ECVC on general graphs.

Notation and Definitions. Extending the notation $\tau(G), \tau_c(G), \tau^\infty(G)$, and $\tau_c^\infty(G)$ introduced earlier for a graph $G = (V, E)$, let $\tau(G; U)$ ($\tau_c(G; U)$, resp.) for $U \subseteq V$ denote the minimum cardinality of a vc (cvc, resp.) C of G such that $U \subseteq C$; that is, $\tau(G; U) = \min\{|C| \mid C$ is a vc of G with $U \subseteq C\}$ and $\tau_c(G; U) = \min\{|C| \mid C$ is a cvc of G with $U \subseteq C\}$. Notice that $\tau_c(G; U) \geq \tau_c(G)$ but $\tau_c(G; U) \neq \tau_c(G)$ in general when $U \neq \emptyset$. Similarly, $\tau_c^\infty(G; U)$ is the minimum cardinality of an eternal connected vertex cover of G in a class $\mathcal{C}$ such that $U \subseteq C$ for any $C \in \mathcal{C}$. Thus, $\tau_c^\infty(G; U)$ is the minimum number of guards needed to protect all the edges of G indefinitely while maintaining their configuration always connected and all the vertices of U always occupied by them.

2   Some Basic Properties of Eternal Connected Vertex Cover

We start with a fundamental relation between CVC and ECVC. It was shown in [17] that, for any cvc C of G, all the edges of G can be eternally protected by $(|C| + 1)$ guards.

Proposition 1. *For any connected vertex cover C of any connected graph G, all the edges of G can be eternally protected by $(|C| + 1)$ guards in such a way that all the vertices in C are always occupied by guards. In particular,*

$$\tau_c^\infty(G) \leq \tau_c(G) + 1.$$

So, since $\tau_c^\infty(G) \geq \tau_c(G)$ obviously, the question to answer is whether $\tau_c^\infty(G) = \tau_c(G)$ or $\tau_c^\infty(G) = \tau_c(G) + 1$. It is also rather obvious that, if $\tau_c^\infty(G) = \tau_c(G)$, there must exist a minimum cvc containing v for each vertex v (i.e., $\tau_c(G; \{v\}) = \tau_c(G)$). More generally,

Lemma 1. *Let $G = (V, E)$ be a connected graph and $U \subseteq V$. If $\tau_c^\infty(G; U) = \tau_c(G; U)$, then $\tau_c(G; U) = \tau_c(G; U \cup \{v\})$ for every vertex $v \in V - U$.*

Proof. Let $\mathcal{C}$ be an ECVC class such that every configuration of $\mathcal{C}$ is an ecvc of cardinality $\tau_c(G; U)$ containing U. There must exist a configuration $C_v \in \mathcal{C}$ containing v for any $v \in V - U$ since, otherwise, an attack on an edge e incident to v cannot be defended by $\mathcal{C}$; some guard must pass e to protect e, implying a guard must be positioned on v after (or before) the attack. Therefore, there must exist a cvc of size $\tau_c(G; U)$ containing v.   □

Let us call a vertex of a graph G *forbidden* if it does not belong to any minimum cvc of G. In other words, v is a forbidden vertex iff $\tau_c(G; \{v\}) \neq \tau_c(G)$, and by Lemma 1 with $U = \emptyset$, the existence of a forbidden vertex gives an immediate evidence that $\tau_c^\infty(G) = \tau_c(G) + 1$.

Let us now introduce a basic structural property of connected vertex covers in light of block decomposition of graphs. For this purpose, let $\text{Cut}(G)$ denote the set of cut vertices in G and $V(B)$ the set of vertices in a block B of G.

Lemma 2 ([9]). *Let G be a connected graph. Then, $C \subseteq V$ is a cvc of G if and only if for each block B of G, $C \cap V(B)$ is a cvc of B containing $Cut(G) \cap V(B)$.*

Corollary 1. $\tau_c(G) = \tau_c(G; Cut(G))$ *and* $\tau_c^\infty(G) = \tau_c^\infty(G; Cut(G))$.

Lemma 2 also implies that any cut vertex belongs to every cvc. It follows that computation of a minimum cvc of G can be reduced to that of a smallest cvc of each block B containing all the cut vertices existent in B. That is,

Corollary 2 ([9]). *Let G be a connected graph and let $Cut(B) = Cut(G) \cap V(B)$ for a block B of G. Then, $C \subseteq V$ is a minimum cvc of G if and only if for each block B of G, $C \cap V(B)$ is a cvc of B of minimum size containing $Cut(B)$.*

From Lemma 1 it is clear that for a graph G to have $\tau_c(G) = \tau_c^\infty(G)$, it is necessary that every vertex of G belongs to some minimum cvc of G, whereas all of cut vertices must be included in any cvc according to Lemma 2. Therefore, Lemmas 1 and 2 imply the following necessary condition for $\tau_c^\infty(G) = \tau_c(G)$:

Lemma 3. *Let $G = (V, E)$ be a connected graph. If $\tau_c^\infty(G) = \tau_c(G)$, then $\tau_c(G) = \tau_c(G; \{v\})$ for every vertex $v \in V - Cut(G)$.*

Proof. If $\tau_c^\infty(G; Cut(G)) = \tau_c(G; Cut(G))$, then $\tau_c(G; Cut(G)) = \tau_c(G; Cut(G) \cup \{v\})$ for every vertex $v \in V - Cut(G)$, by Lemma 1. Since $\tau_c(G; Cut(G)) = \tau_c(G)$ and $\tau_c(G; Cut(G) \cup \{v\}) = \tau_c(G; \{v\})$ for any $v \in V - Cut(G)$ according to Lemma 2, $\tau_c^\infty(G) = \tau_c(G)$ implies that $\tau_c(G) = \tau_c(G; \{v\})$, $\forall v \in V - Cut(G)$.

It is interesting to see whether this necessary condition is also sufficient or not for $\tau_c^\infty(G) = \tau_c(G)$. In fact, in case between $\tau(G)$ and $\tau^\infty(G)$, they are not necessarily equal to each other even if every vertex belongs to some minimum vertex cover of G; it was instead shown to become sufficient when it is additionally assumed that every minimum vertex cover of G containing all cut vertices is connected [3].

Theorem 1 ([3]). *Let $G = (V, E)$ be a connected graph with $|V| \geq 2$. Suppose every vertex cover of of size $\tau(G; Cut(G))$ containing $Cut(G)$ is connected. Then $\tau^\infty(G) = \tau(G)$ if and only if $\tau(G; Cut(G) \cup \{v\}) = \tau(G)$ for every vertex $v \in V - Cut(G)$.*

Under the same assumption, it is possible to show that the necessary condition given in Lemma 3 is sufficient for the case between $\tau_c(G)$ and $\tau_c^\infty(G)$:

Theorem 2. *Let $G = (V, E)$ be a connected graph with $|V| \geq 2$. Suppose every vertex cover of of size $\tau(G; Cut(G))$ containing $Cut(G)$ is connected. Then, $\tau_c^\infty(G) = \tau_c(G)$ if and only if $\tau_c(G; \{v\}) = \tau_c(G)$ for every vertex $v \in V - Cut(G)$.*

Proof. Necessity is granted by Lemma 3 without any supposition on G.

So, assume that every vc containing $Cut(G)$ of size $\tau(G; Cut(G))$ is a cvc of G. Notice that the size of such a vc also equals to $\tau_c(G)$ since $\tau_c(G) = \tau_c(G; Cut(G))$. To prove sufficiency, take a family $\mathcal{C}$ of all minimum vc's (cvc's) containing

$\mathrm{Cut}(G)$. We will prove in what follows that $\mathcal{C}$ is indeed an ECVC class for G, which implies that $\tau_c^\infty(G) = \tau_c(G)$, if $\tau_c(G) = \tau_c(G; \{v\})$ for every vertex $v \in V - \mathrm{Cut}(G)$. Let C be any configuration in $\mathcal{C}$, that is, C is a cvc with $\mathrm{Cut}(G) \subseteq C$ of cardinality $\tau_c(G)$. Suppose an edge $e = \{u, v\}$ is attacked. If $e \subseteq C$ then it is easy to repel the attack by exchanging the positions of guards located on u and v within C without changing the current configuration. Of course $e \cap C \neq \emptyset$ as C is a cvc of G, and it suffices to show that G can be protected from an attack on e when $u \in C$ and $v \notin C$. Let $\mathcal{C}_v$ be a family of all cvc's containing $\mathrm{Cut}(G) \cup \{v\}$ of cardinality $\tau_c(G)$. Then, $\mathcal{C}_v$ is a nonempty subfamily of $\mathcal{C}$ since $\tau_c(G) = \tau_c(G; \{v\}), \forall v \in V - \mathrm{Cut}(G)$. Choose C_v within $\mathcal{C}_v$ such that C_v maximizes $|C \cap C_v|$ among those in $\mathcal{C}_v$. Let X, Y, and Z denote $C - C_v, C_v - C$, and $C \cap C_v$, respectively. Then $\mathrm{Cut}(G) \subseteq Z$, and either of X and Y is an independent set in G since each of C and C_v is a vertex cover of G. Also notice that $|X| = |Y| \geq 1$ since $|C| = |C_v|$ and $C \neq C_v$. Therefore, the subgraph $H = G[X \cup Y]$ of G induced by $X \cup Y$ is a bipartite graph of color classes X and Y of equal size.

Recall that $v \in Y$ and consider the bipartite subgraph H_x of H for $x \in X$, obtained by removing x and v from H:

Claim. H_x has a perfect matching for any $x \in X$.

Proof (of Claim). For any $Y' \subseteq Y$ consider $N_G(Y')$, the set of vertices adjacent in G with a vertex in Y'. Then, $N_G(Y') \subseteq X \cup Z$ since $C = X \cup Z$ is a vc of G. Consider next $N_H(Y') = N_G(Y') \cap X$. If $|N_H(Y')| < |Y'|$, then $N_H(Y') \cup (Y - Y') \cup Z$ is a vc of G of size smaller than $|C_v| = \tau_c(G)$, which contradicts our assumption on G. Therefore, we have $|N_H(Y')| \geq |Y'|, \forall Y' \subseteq Y$, and hence, H has a perfect matching by Hall's theorem [7].

Let us consider H_x now. The claim holds trivially if H_x is empty. Take nonempty $Y' \subseteq Y - \{v\}$, and then $|N_H(Y')| \geq |Y'|$ as already observed for H. Suppose $|N_H(Y')| = |Y'|$ and take $C_v' = N_H(Y') \cup (Y - Y') \cup Z$. Then, C_v' is a vc containing $\mathrm{Cut}(G)$ of size $|C_v| = \tau_c(G)$, and hence, it must be a cvc by the assumption on G. The existence of C_v' however contradicts our choice of C_v since C_v' has a larger overlap with C than C_v does. Therefore, we may conclude that $|N_H(Y')| \geq |Y'| + 1$ for any $Y' \subseteq Y - \{v\}$, and hence, $|N_{H_x}(Y')| \geq |Y'|$ and H_x has a perfect matching for any $x \in X$.   □

We are now ready to describe how to defend the attack on edge $\{u, v\}$ within G, where $u \in C$ and $v \in Y = C_v - C$, meanwhile changing configurations from C to C_v:

Case 1. $u \in X$: Let M denote a perfect matching existent within H_u. The attack on edge $\{u, v\}$ can be repelled by moving the guard on u to v and all the other guards within X to $Y - v$ along the edges of M.

Case 2. $u \in Z$: Since C is a cvc of G there exists a path between u and any vertex in X in G. Let P be a shortest path within $G[C] = G[X \cup Z]$ connecting u with some vertex in X, and let $x \in X$ be the one nearest to u among the vertices in X. Then, all the vertices of P other than x lie within Z, and suppose P starts at x and ends at u. Let M now denote a perfect matching existent within H_x. In response to the attack on $\{u, v\}$, move the guard on u to v, shift the positions of all the guards on $P - u$ including x by one vertex toward the end vertex u along P. All the guards in X other than x are at the same time moved to $Y - \{v\}$ along the edges of M.   □

Remark 1. The sufficient condition for $\tau_c^\infty(G) = \tau_c(G)$ proven above can be also seen to hold by following the proof of Theorem 1 given in [3] and observing that the EVC class used in it is an ECVC class for G.

The next corollary follows immediately from Theorem 2 (and Proposition 1) because $\text{Cut}(G) = \emptyset$ when G is biconnected:

Corollary 3. *Let $G = (V, E)$ be a biconnected graph such that every minimum vertex cover is connected. Then,*

$$\tau_c^\infty(G) = \begin{cases} \tau_c(G) & \text{if every vertex of } G \text{ belongs to some minimum cvc,} \\ \tau_c(G) + 1 & \text{otherwise.} \end{cases}$$

To illustrate the usefulness of Theorem 1 some graph classes for which every vertex cover containing all cut vertices is a cvc were focused in [3], and one of them is a class of *locally connected* graphs. A graph G is locally connected if for every vertex v of G, its open neighborhood $N_G(v) = \{u \in V \mid \{u, v\} \in E\}$ induces a connected subgraph in G. Some well-known graph classes are examples of locally connected graphs such as biconnected chordal graphs and biconnected internally triangulated planar graphs. It was observed in [3] that for a connected graph G every vertex cover S of G with $\text{Cut}(G) \subseteq S$ is connected if every block of G is locally connected. Therefore, the following is immediate from Theorem 2 and Corollary 3:

Corollary 4. *Let $G = (V, E)$ be a connected graph with $|V| \geq 2$ in which every block is locally connected. Then,*

$$\tau_c^\infty(G) = \begin{cases} \tau_c(G) & \text{if } \tau_c(G; \{v\}) = \tau_c(G) \text{for every vertex} v \in V - \text{Cut}(G), \\ \tau_c(G) + 1 & \text{otherwise.} \end{cases}$$

3   Polynomially Solvable Cases

For graphs of some simple structure it is easy to know the value of $\tau_c^\infty(G)$. For instance, it is an easy exercise to check that

- $\tau_c(C_n) = \tau_c^\infty(C_n) = n - 1$, and $\tau_c(K_n) = \tau_c^\infty(K_n) = n - 1$, where C_n is a cycle and K_n is a complete graph, both on n vertices, and
- for a tree T on n vertices with $n \geq 3$, $\tau_c(T) = \#$ of internal nodes in T, and $\tau_c^\infty(T) = \tau_c(T) + 1 = (\#$ of internal nodes in $T) + 1$,

and we will consider graphs of more nontrivial structures in what follows.

We have the following corollary of Lemma 1 for cases when forbidden vertices can be easily identified.

Corollary 5. $\tau_c^\infty(G) = \tau_c(G) + 1$ *if*

- *G has a leaf (i.e., a degree 1 vertex), or*
- *G has a block having exactly one non-cut vertex, or more generally,*
- *G has a block B s.t. $V(B) \not\subseteq Cut(G)$ and $V(B) \cap Cut(G)$ is a cvc of B.*

Proof. Will explain only the last case. Suppose G has a block B s.t. $V(B) \not\subseteq Cut(G)$ and $V(B) \cap Cut(G)$ is a cvc of B. Then there exist a vertex $v \in V(B) - Cut(G)$ and it must be forbidden. To see it suppose there exists a minimum cvc C of G containing v. Every minimum cvc of G must contain $Cut(G)$ entirely, and $V(B) \cap Cut(G)$ is a cvc of B by the assumption. Thus, $(C - \{v\}) \cap V(B)$ is still a cvc of B, and $C - \{v\}$ must be a cvc of G, contradicting the minimality of C. $\qquad\square$

A *cactus* is such a graph in which every edge belongs to at most one simple cycle. It can be also defined to be a graph in which every block is either a single edge or a simple cycle. When every block is a clique such a graph is called a *block graph*. The ETERNAL DOMINATION problem on cactus graphs was recently studied in [4], and an upper bound on the size of a minimum eternal domination was obtained together with a linear time algorithm for a special cases of cactus graphs where each vertex is in at most two biconnected components. Whether each block B is a cycle or a clique, ECVC is easier than ETERNAL DOMINATION because a vertex v can be forbidden in such a block B iff v is the only non-cut vertex in B. Therefore, we have

Theorem 3. *Let G be a cactus graph, a block graph, or any graph in which every block is either a simple cycle or a clique. Then,*

$$\tau_c^\infty(G) = \begin{cases} \tau_c(G) + 1 & \text{if G has a block containing exactly one non-cut vertex in it,} \\ \tau_c(G) & \text{otherwise.} \end{cases}$$

Proof. Notice that every vertex subset of $V(B)$ of size $|V(B)| - 1$ is a minimum cvc of B when B is a cycle or a clique. Hence, for any $X \subsetneq V(B)$ and $w \in V(B) - X$, there must exist a minimum cvc C of B such that $X \cup \{w\} \subseteq C$ as long as $|X| < |V(B)| - 1$ while such C does not exist in case $|X| = |V(B)| - 1$. Therefore, a vertex v can be forbidden iff it is the only non-cut vertex within some block of G. It follows from Lemma 3 that $\tau_c^\infty(G) = \tau_c(G) + 1$ in case when G has a block containing exactly one non-cut vertex.

Conversely, suppose every block contains at least two non-cut vertices or none. Any edge attack occurs within a block B of G, and if $V(B) \subseteq Cut(G)$, $V(B) \subseteq C$ for any minimum cvc C of G, and any edge attack within B can be certainly defended by the guards located on C *without* changing its configuration. For the case when $V(B) \not\subseteq Cut(G)$, B contains at least two non-cut vertices and consider the set $\mathcal{D}$ of all the subsets of $V(B)$ of size $|V(B)| - 1$ containing

$V(B) \cap \text{Cut}(G)$. It can be observed that, for any $D, D' \in \mathcal{D}$ and any minimum cvc C of G such that $C \cap V(B) = D$, $C - D \cup D'$ is another minimum cvc of G and that any edge attack within B can be defended by changing a guard configuration of some minimum cvc C only within B from $C \cap V(B) = D \in \mathcal{D}$ to another $D' \in \mathcal{D}$ (possibly $D = D'$). Therefore, we may conclude that $\tau_c^\infty(G) = \tau_c(G)$ in this case. □

It should also be observed that, for any graph $G = (V, E)$ considered in Theorem 3, $C \subseteq V$ is a minimum cvc of G iff 1) $\text{Cut}(G) \subseteq C$ and 2) for any block B of G, $|C \cap V(B)| = |V(B)| - 1$ whenever $V(B) \nsubseteq \text{Cut}(G)$. Therefore, we have explicit formulae for $\tau_c(G)$ and $\tau_c^\infty(G)$ for such a graph G:

Corollary 6. *Let G be a graph such that every block of G is either a cycle or a clique, and let b_{nc} denote the number of blocks of G containing a non-cut vertex in G. Then,*

$$\tau_c(G) = |V| - b_{\text{nc}}$$

and

$$\tau_c^\infty(G) = \begin{cases} |V| - b_{\text{nc}} + 1 & \text{if } G \text{ has a block containing exactly one non-cut vertex in it,} \\ |V| - b_{\text{nc}} & \text{otherwise.} \end{cases}$$

Although Corollary 4 does not give a polynomial algorithm for computing $\tau_c^\infty(G)$ for free, even if every block of G is locally connected, it is worth considering applications of Theorem 2 to derive more polynomially solvable cases of $\tau_c^\infty(G)$. For a graph $G = (V, E)$ and its vertex $v \in V$, consider the operation of attaching a new vertex w to G via new edge $\{v, w\}$. Let us call this a *pendant operation* on v of G, and denote the resulting graph by $G \circ v$.

Theorem 4. *Let $\mathcal{H}$ be a graph class closed under the pendant operation such that*

1. *every minimum vertex cover S of G with $\text{Cut}(G) \subseteq S$ is connected for any $G \in \mathcal{H}$, and*
2. *the minimum connected vertex cover number can be computed in polynomial time.*

Then $\tau_c^\infty(G)$ can be computed in polynomial time for any $G \in \mathcal{H}$.

Proof. Because of condition 1, $\tau_c^\infty(G) = \tau_c(G)$ if $\tau_c(G; \{v\}) = \tau_c(G)$ for every $v \in V - \text{Cut}(G)$, and otherwise, $\tau_c^\infty(G) = \tau_c(G) + 1$ by Theorem 2. The set $\text{Cut}(G)$ of cut vertices can be computed in polynomial time, and so is $\tau_c(G)$ according to condition 2. Thus, what remains is only to compute $\tau_c(G; \{v\})$ for each $v \in V - \text{Cut}(G)$. But this can be also done in polynomial time because clearly $\tau_c(G; \{v\}) = \tau_c(G \circ v)$ and $\mathcal{H}$ is closed under the pendant operation. □

A graph is *chordal* if it contains no induced cycle of length 4 or more. A class of chordal graphs is an example satisfying all the preconditions given in Theorem 4; namely, it is closed under the pendant operation, the minimum connected vertex cover number is polynomially computable [9], and condition 1 is satisfied because, as easily verified, the biconnected chordal graphs are locally connected. Therefore,

Corollary 7. *For any chordal graph G, $\tau_c^\infty(G)$ can be computed in polynomial time.*

4   NP-Completeness and Approximation

EVC was shown to be NP-hard in general [11], and it is NP-complete when graphs are locally connected [3]. ECVC can be shown equally hard but belonging to NP, due to Corollary 4:

Theorem 5. *For locally connected graphs (the decision version of) ECVC is NP-complete and it is NP-hard to approximate $\tau_c^\infty(G)$ to within a factor smaller than $10\sqrt{5} - 21$.*

Proof. According to Corollary 4, $\tau_c^\infty(G) = \tau_c(G)$ if $\tau_c(G; \{v\}) = \tau_c(G)$ for every $v \in V$, and $\tau_c^\infty(G) = \tau_c(G) + 1$ otherwise, when G is a (biconnected) locally connected graph. Thus, to decide whether $\tau_c^\infty(G) \leq k$ or not given a locally connected G and k, all we need to do is to guess a cvc of smallest size containing each vertex $v \in V$ and decide that $\tau_c^\infty(G) \leq k$ iff all the cvc's computed are of size k or less. Since ECVC on locally connected graphs is this way decidable nondeterministically in polynomial time, it belongs to NP.

To show NP-hardness, let us first observe that VC can be reduced to CVC on locally connected graphs in the same way as the standard reduction of VC to CVC. Given a connected graph $G = (V, E)$ as a VC instance, add a new vertex z and connect it to every vertex of G, and let G' denote the resulting graph. Then, G' is locally connected and $\tau_c(G') = \tau(G) + 1$. Thus, CVC on locally connected graphs is NP-hard. Moreover, since it is known to be NP-hard to approximate $\tau(G)$ on connected graphs to within a factor smaller than $10\sqrt{5} - 21$ [8], so is it to approximate $\tau_c(G)$ on locally connected graphs within the same factor. Because $\tau_c(G) \leq \tau_c^\infty(G) \leq \tau_c(G) + 1$ in general (by Proposition 1), it is also NP-hard to approximate $\tau_c^\infty(G)$ on locally connected graphs to within a factor smaller than $10\sqrt{5} - 21$.   □

It was shown in [11] that EVC is approximable within 2, using a maximal matching based approximation algorithm. As such this algorithm does not produce a connected solution in general, and in case of ECVC it is more natural to consider utilizing an approximation algorithm for CVC. Or, it is more convenient in fact to divert approximate solutions for TREE COVER into eternal cvc's. Suppose a subgraph $T = (V_T, E_T)$ of a connected graph $G = (V, E)$ is a tree. An edge set $E_T \subseteq E$ is called a *tree cover* of G if either e is an edge of T or e is adjacent to some edge of T for all edges $e \in E$. TREE COVER (TC) is the problem of computing a minimum cardinality tree cover of G. Clearly, the edge set $F \subseteq E$ of a tree is a tree cover if the set of vertices induced by F (i.e., $\{v \in V \mid v$ is an end vertex of some $e \in F\}$) is a cvc, and conversely, $C \subseteq V$ is a cvc if any spanning tree of $G[C]$, the subgraph of G induced by C, is a tree cover. It was shown in [2,22] that either CVC or TC is approximable within a factor of 2, and we may use any 2-approximation algorithm for TC for our purpose:

Theorem 6. *ECVC is approximable within a factor of 2. More specifically, an ecvc for G of size bounded by $2\tau_c(G)$ can be computed in polynomial time.*

Proof. The algorithm first computes a cvc C of a connected graph $G = (V, E)$. An ecvc is constructed by adding one more vertex to C, and we already know that all the edges of G can be eternally defended by a connected team of $|C| + 1$ mobile guards as described in Proposition 1.

To compute a cvc C, one can use any 2-approximation algorithm for TC. Let $\mathrm{OPT_{TC}}(G)$ denote the minimum size of a TC solution for G, and $F \subseteq E$ an approximate tree cover solution thus computed. Set $C = V[F]$, the set of vertices induced by F. Then, $|F| \leq 2 \cdot \mathrm{OPT_{TC}}(G)$, and because of the equivalence between CVC and TC solutions, $\tau_c(G) = \mathrm{OPT_{TC}}(G) + 1$. The size of an ecvc to be computed is $|C| + 1$ and it is bounded by

$$|V[F]| + 1 = |F| + 2 \leq 2 \cdot \mathrm{OPT_{TC}}(G) + 2 = 2(\mathrm{OPT_{TC}}(G) + 1) = 2\tau_c(G). \qquad \square$$

5   Concluding Remarks

A new variant of VC, CVC, and EVC, called ETERNAL CONNECTED VERTEX COVER (ECVC), has been introduced, and we have studied various aspects of ECVC mainly building on top of existing results concerning CVC and EVC.

Still some questions concerning core properties of ECVC remain unanswered. One of them is the question of whether the necessary condition for $\tau_c(G) = \tau_c^\infty(G)$ given in Lemma 3 is sufficient or not. As noted earlier, such sufficiency was denied by Babu et al. by some of concrete instances in case between $\tau(G)$ and $\tau^\infty(G)$ [3]. It is also possible to construct an instance denying such sufficiency even in the context of ETERNAL CONNECTED DOMINATING SET, but it remains open whether $\tau_c(G; \{v\}) = \tau_c(G)$, $\forall v \in \mathrm{Cut}(G)$, implies $\tau_c(G) = \tau_c^\infty(G)$ or not.

Another question of interest is whether ECVC can be solved in polynomial time or not for graphs of maximum degree 3. The class of graphs of maximum degree 3 is one of a few for which CVC is known solvable in polynomial time. Since this graph class is not closed under the pendant operation, however, it is not possible to apply the same approach as we used for chordal graphs.

References

1. Anderson, M., Barrientos, C., Brigham, R.C., Carrington, J.R., Vitray, R.P., Yellen, J.: Maximum-demand graphs for eternal security. J. Combin. Math. Combin. Comput. **61**, 111–127 (2007)
2. Arkin, E.M., Halldórsson, M.M., Hassin, R.: Approximating the tree and tour covers of a graph. Inform. Process. Lett. **47**(6), 275–282 (1993)
3. Babu, J., Chandran, L.S., Francis, M., Prabhakaran, V., Rajendraprasad, D., Warrier, J.N.: On graphs with minimal eternal vertex cover number. In: Pal, S.P., Vijayakumar, A. (eds.) CALDAM 2019. LNCS, vol. 11394, pp. 263–273. Springer, Cham (2019). https://doi.org/10.1007/978-3-030-11509-8_22

4. Blažej, V., Křišt'an, J.M., Valla, T.: On the m-eternal domination number of cactus graphs. In: Filiot, E., Jungers, R., Potapov, I. (eds.) RP 2019. LNCS, vol. 11674, pp. 33–47. Springer, Cham (2019). https://doi.org/10.1007/978-3-030-30806-3_4
5. Burger, A.P., Cockayne, E.J., Gründlingh, W.R., Mynhardt, C.M., van Vuuren, J.H., Winterbach, W.: Infinite order domination in graphs. J. Combin. Math. Combin. Comput. **50**, 179–194 (2004)
6. Cockayne, E.J., Dreyer Jr., P.A., Hedetniemi, S.M., Hedetniemi, S.T.: Roman domination in graphs. Discrete Math. **278**(1–3), 11–22 (2004)
7. Diestel, R.: Graph Theory. GTM, vol. 173. Springer, Heidelberg (2017). https://doi.org/10.1007/978-3-662-53622-3
8. Dinur, I., Safra, S.: On the hardness of approximating minimum vertex cover. Ann Math (2) **162**(1), 439–485 (2005)
9. Escoffier, B., Gourvès, L., Monnot, J.: Complexity and approximation results for the connected vertex cover problem in graphs and hypergraphs. J. Discrete Algorithms **8**(1), 36–49 (2010)
10. Fernau, H., Manlove, D.F.: Vertex and edge covers with clustering properties: complexity and algorithms. J. Discrete Algorithms **7**(2), 149–167 (2009)
11. Fomin, F.V., Gaspers, S., Golovach, P.A., Kratsch, D., Saurabh, S.: Parameterized algorithm for eternal vertex cover. Inf. Process. Lett. **110**(16), 702–706 (2010)
12. Garey, M.R., Johnson, D.S.: The rectilinear Steiner tree problem is NP-complete. SIAM J. Appl. Math. **32**(4), 826–834 (1977)
13. Goddard, W., Hedetniemi, S.M., Hedetniemi, S.T.: Eternal security in graphs. J. Combin. Math. Combin. Comput. **52**, 169–180 (2005)
14. Goldwasser, J.L., Klostermeyer, W.F.: Tight bounds for eternal dominating sets in graphs. Discrete Math. **308**(12), 2589–2593 (2008)
15. Klostermeyer, W.F.: An eternal vertex cover problem. J. Combin. Math. Combin. Comput. **85**, 79–95 (2013)
16. Klostermeyer, W.F., MacGillivray, G.: Eternal dominating sets in graphs. J. Combin. Math. Combin. Comput. **68**, 97–111 (2009)
17. Klostermeyer, W.F., Mynhardt, C.M.: Edge protection in graphs. Australas. J. Combin. **45**, 235–250 (2009)
18. Klostermeyer, W.F., Mynhardt, C.M.: Graphs with equal eternal vertex cover and eternal domination numbers. Discrete Math. **311**(14), 1371–1379 (2011)
19. Klostermeyer, W.F., Mynhardt, C.M.: Eternal total domination in graphs. Ars Combin. **107**, 473–492 (2012)
20. Li, Y., Wang, W., Yang, Z.: The connected vertex cover problem in k-regular graphs. J. Comb. Optim. **38**(2), 635–645 (2019). https://doi.org/10.1007/s10878-019-00403-3
21. Priyadarsini, P.L.K., Hemalatha, T.: Connected vertex cover in 2-connected planar graph with maximum degree 4 is NP-complete. Int. J. Math. Phys. Eng. Sci. **2**(1), 51–54 (2008)
22. Savage, C.: Depth-first search and the vertex cover problem. Inform. Process. Lett. **14**(5), 233–235 (1982)
23. Ueno, S., Kajitani, Y., Gotoh, S.: On the nonseparating independent set problem and feedback set problem for graphs with no vertex degree exceeding three. Discrete Math. **72**, 355–360 (1988)
24. Watanabe, T., Kajita, S., Onaga, K.: Vertex covers and connected vertex covers in 3-connected graphs. In: IEEE International Symposium on Circuits and Systems, pp. 1017–1020 (1991)

Parametric Streaming Two-Stage Submodular Maximization

Ruiqi Yang[1], Dachuan Xu[1], Longkun Guo[2], and Dongmei Zhang[3(✉)]

[1] Department of Operations Research and Scientific Computing, Beijing University of Technology, Beijing 100124, People's Republic of China
yangruiqi@emails.bjut.edu.cn, xudc@bjut.edu.cn
[2] School of Computer Science and Technology, Qilu University of Technology (Shandong Academy of Sciences), Jinan 250353, People's Republic of China
longkun.guo@gmail.com
[3] School of Computer Science and Technology, Shandong Jianzhu University, Jinan 250101, People's Republic of China
zhangdongmei@sdjzu.edu.cn

Abstract. We study the submodular maximization problem in generalized streaming setting using a two-stage policy. In the streaming context, elements are released in a fashion that an element is revealed at one time. Subject to a limited memory capacity, the problem aims to sieve a subset of elements with a sublinear size ℓ, such that the expecting objective value of all utility functions over the summarized subsets has a performance guarantee. We present a generalized one pass, $\left(\gamma_{\min}^5 / (5 + 2\gamma_{\min}^2) - O(\epsilon) \right)$-approximation, which consumes $O(\epsilon^{-1}\ell \log(\ell\gamma_{\min}^{-1}))$ memory and runs in $O(\epsilon^{-1}kmn \log(\ell\gamma_{\min}^{-1}))$ time, where k, n, m and $\gamma_{\min}$ denote the cardinality constraint, the element stream size, the amount of the learned functions, and the minimum generic submodular ratio of the learned functions, respectively.

Keywords: Submodular maximization · Streaming algorithm · Submodular ratio · Approximation ratio

1 Introduction

Submodular maximization has many applications, such as influence maximization [5], document summarization [7] and network monitoring [6]. The model can be formally described as

$$\max_{S \subseteq \mathcal{V}: S \in \mathcal{I}} f(S),$$

where $\mathcal{V}$ denotes the element ground set and $\mathcal{I}$ represents some specific-defined constraints, such as cardinality, knapsack and matroid constraints. The above problem can be treated as a single stage optimization and solved by the greedy method devised in [9], which starts with an empty set and picks an element with the maximum marginal gain in each iteration. Next we study a generalized set

© Springer Nature Switzerland AG 2020
J. Chen et al. (Eds.): TAMC 2020, LNCS 12337, pp. 193–204, 2020.
https://doi.org/10.1007/978-3-030-59267-7_17

function maximization, in which the true utility function follows some unknown distribution $\mathcal{D}$, but a family $\mathcal{F} = \{f_1, ..., f_m\}$ of these functions can be learned based on historical data. These learned functions are defined on the common ground set $\mathcal{V}$, such that each of them is equipped with a generic submodularity ratio γ_i, characterizing how close a normalized increasing set function is to be submodular. In addition, we assume the elements are revealed in a streaming fashion, in which an element is released at one time. The goal is to summarize a representative subset S with a size at most ℓ from the stream, such that each function f_i restricted on S has a high value. The problem is motivated by a ride-share application, in which there are hundreds of thousands of pick-up locations and the aim is to select a subset of the locations to maximize the utility in terms of total location value provided that the values vary over time. To assess the true utility in long-term, we can learn some functions that approximately calculate its value in expectation.

Streaming-based model is popular for handling optimization applications with a large scale. In the model, the elements are either stored in distribution, or generated in a place that has a storage capability. Moreover, we only have access to a limited memory at any time during the procession, while the goal is to assemble a small fraction of elements so as to maximize the utility. To evaluate the performance of algorithms for streaming submodular problems, Badanidiyuru et al. [1] introduced four parameters: passes, memory, running time, and approximation ratio. Threshold-based method plays an important role in the developing of streaming algorithms for submodular optimization. In such method, the main idea is to choose a proper threshold, according to which we determine whether the arrived element should be discarded or not. For streaming submodular maximization with a cardinality constraint, Badanidiyuru et al. [1] provided a threshold-based, one pass, $(0.5 - \epsilon)$-approximation, which consumes $O(k \log k / \epsilon)$ memory and runs in $O(n \log k / \epsilon)$ time, where n denotes the stream size. A survey of threshold-based methods for streaming submodular can be found in [12]. In addition, Buchbinder et al. [3] studied submodular maximization under online setting, in which the elements arrive online and the arriving element must be decided whether should be maintained or not before the next time slot. They presented a preemption-based algorithm, which can be extended with respect to streaming setting. In their algorithm, the arriving element is selected if it occurs a large utility increment via exchanging it with the elements of the current solution. Combining the two threshold-based and preemption-based methods, Mitrovic et al. [8] considered the two-stage submodular maximization under streaming, and provided a one pass, $1/(6 + \epsilon)$-approximation that uses $O(\ell \log \ell / \epsilon)$ memory and has $O(kmn \log \ell / \epsilon)$ running time, where ℓ, k, m, and n denote the subset size, the set cardinality, the learned functions and the element set size, respectively.

In this paper, we consider an extended version of the two-stage set function maximization problem, in which the utility functions are equipped with generic submodularity ratios. Inspired by the work of [8], we present a generalized streaming algorithm, which deserves an approximation ratio $(\gamma_{\min}^5/(5 +$

$2\gamma_{\min}^2) - O(\epsilon))$, uses $O(\ell \log(\ell\gamma_{\min}^{-1})/\epsilon)$ memory and runs in $O(kmn \log(\ell\gamma_{\min}^{-1})/\epsilon)$ time, where $\gamma_{\min} = \min_{i \in [m]} \gamma_i$ represents the minimum generic submodular ratio over all learned functions.

1.1 Related Work

To interpret the task involving multiple objectives, Balkanski et al. [2] introduced the two-stage submodular maximization problem, which is more expressive than the single stage submodular maximization. In the two-stage submodular maximization model, we are given an element ground set $\mathcal{V}$ of size n, a family $\mathcal{F} = \{f_1, ..., f_m\}$ of set functions in which each function $f : 2^{\mathcal{V}} \rightarrow R$ is submodular. The aim is to summarize a subset with size at most ℓ, which can serve as a new ground set of a high value for all functions. In particular, assuming $G(S) = 1/m \cdot \sum_{i=1}^{m} \max_{T \subseteq S:|T| \leq k} f_i(T)$, the two-stage submodular maximization problem can be casted as

$$\max_{S \subseteq \mathcal{V}:|S| \leq \ell} G(S).$$

Based on local search, Balkanski et al. [2] presented a $(1-1/e)/2$-approximation, which needs $O(km\ell n^2 \log(n))$ function evaluations (running time). Combining greedy and local search, Stan et al. [10] studied a more generalized two-stage submodular maximization with matroid constraint and improved the approximation ratio to $(1 - 1/e^2)/2$. In addition, they also decreased the amount of function evaluations to $O(m\ell nr)$, where r denotes the matroid rank. Recently, Yang et al. [11] considered the two-stage submodular maximization with more P-matroid constraint and devised a generalized $(1 - 1/e^2)/(P + 1)$-approximation with $O(m\ell nr^P)$ function evaluations. For the two-stage submodular maximization under distribution, Mitrovic et al. [8] proposed two approximations with ratios $(1 - 1/e^2)/4$ and 0.107, respectively.

1.2 Organizations

The rest of the paper is organized as follows. Section 2 provides some necessary preliminaries. Section 3 gives the main streaming algorithm whose theoretical analysis are presented in Sect. 4. At last, Sect. 5 concludes our work.

2 Preliminaries

Given an element ground set $\mathcal{V}$ of size n, a learned collection $\mathcal{F} = \{f_1, ..., f_m\}$ of set functions, where m represents an integer and all functions are defined on the same ground $\mathcal{V}$. Let $[m] = \{1, ..., m\}$. We have $f_i : 2^{\mathcal{V}} \rightarrow \mathbb{R}_+$ for any $i \in [m]$. A set function $f : 2^{\mathcal{V}} \rightarrow \mathbb{R}_+$ is *submodular*, if and only if for any $A, B \subseteq \mathcal{V}$, we have

$$f(A) + f(B) \geq f(A \cup B) + f(A \cap B).$$

For easy presentation, let $A + e = A \cup \{e\}$ and $A - e = A \setminus \{e\}$ for any $e \in \mathcal{V}$ and $A \subseteq \mathcal{V}$. We denote $\Delta(e, A) = f(A + e) - f(A)$ as the marginal gain of adding

e to A. A set function f is *normalized*, if $f(\emptyset) = 0$. Further, a normalized set function is *monotone* if $\Delta(e, A) \geq 0$. An equivalent definition of submodular can be represented as following: A normalized set function f is submodular if and only if for any $A \subseteq B \subseteq \mathcal{V}$ and $e \notin B$,

$$\Delta(e, A) \geq \Delta(e, B).$$

Following the work of [4], the definition of generic submodular ratio can be restated as follows.

Definition 1 *([4]). Given any normalized set function f, the generic submodular ratio is defined as the largest scalar $\gamma \in [0, 1]$ subject to*

$$\Delta(e, A) \geq \gamma \cdot \Delta(e, B), \forall A \subseteq B \subseteq \mathcal{V}, e \notin B.$$

Following by the above definition, we restate an observation as

$$\sum_{e \in S} f(e) \geq \gamma \cdot f(S), \forall S \subseteq \mathcal{V},$$

which has been proved by [4]. We consider a generalized two-stage submodular maximization problem under streaming setting. That is, elements are revealed in a streaming fashion. An element is released in each time slot. With respect to the limited memory capacity, the goal is to sieve a subset S of size at most $\ell \ll n$, such that the functions f_i, $i \in [m]$ over S have a good performance. Formally, the problem is defined as

$$\max_{S \subseteq \mathcal{V}: |S| \leq \ell} \sum_{i=1}^{m} \max_{T \subseteq S: |T| \leq k} f_i(T),$$

where k denotes an integer, each function f_i, $i \in [m]$ is equipped with a generic submodular ratio γ_i. Denote $G(S) = \sum_{i=1}^{m} \max_{T \subseteq S: |T| \leq k} f_i(T)$ for any $S \subseteq \mathcal{V}$, then the problem is to find $S \subseteq \mathcal{V}$ of size at most ℓ, such that $G(S)$ is maximized. Note that the objective function $G(S)$ loses submodularity even when all generic submodular ratios reduce to 1, as the counter-example given by [2].

In addition, we assume a value oracle is given, such that for any $A \subseteq \mathcal{V}$, we can have access to the value of $f_i(A)$, $i \in [m]$ in $O(1)$ time. We also assume the generic submodular ratios vary within the range of $(0, 1]$.

3   Algorithm Description

The main algorithm is based on two operations defined by local search: add and exchange. For any subset $A \subseteq \mathcal{V}$ and an element $e \in \mathcal{V}$, we reset

$$\Delta_i^A(e|A) = \Delta_i(e, A) = f_i(A + e) - f_i(A),$$

Algorithm 1. Framework of Replacement Streaming

Input: The distribution of functions, $\mathbb{D}$; The value of parameters, α, β.
1: Learn a collection $\mathcal{F} = \{f_1, ..., f_m\}$ of functions, where each function f_i is characterized by the generic submodular ratio γ_i;
2: Determine the current arriving element e_t if it should be maintained with the help of each threshold value $\tau \in O_t$;
3: Update the representative subset S_τ and surrogate sets $\{T_{\tau,i}\}_{i=1}^m$, with the help of add and exchange operations;
4: Stop at the time of meeting cardinality capacity or the stream is finished;
5: Return the summarization S_τ and surrogate subsets $\{T_{\tau,i}\}_{i=1}^m$ with the maximum objective value.

---

for any $i \in [m]$, as the gain of adding element e to A with respect to function f_i. In addition, we denote

$$\text{Exch}_i(e, A) = \arg \max_{e' \in A} f_i(A - e' + e) - f_i(A)$$

as the element of A with the maximum gain by exchanging it with elements in A. Denote the largest gain as

$$\Delta_i^E(e|A) = f_i(A - \text{Exch}_i(e, A) + e) - f_i(A).$$

We introduce symbol ∇ to show the true gain of an element. For any $A \subseteq \mathcal{V}$ and $e \in \mathcal{V}$, we have

$$\nabla_i(e, A) = \begin{cases} \mathbf{1}_{\{\Delta_i^A(e|A) \geq \frac{\alpha}{k\gamma_i} \cdot f_i(A)\}} \cdot \Delta_i^A(e|A), & \text{if } |S| < k; \\ \mathbf{1}_{\{\Delta_i^E(e|A) \geq \frac{\alpha}{k\gamma_i} \cdot f_i(A)\}} \cdot \Delta_i^E(e|A), & \text{otherwise,} \end{cases}$$

where $\mathbf{1}$ denotes the indicator function and α is a tunable parameter.

Then it remains to describe the main algorithm. Inspired by the work of [8], we apply their algorithm to more general set maximization scenario. The framework of our algorithm can be summarized as in Algorithm 1.

In the algorithm, we firstly learn a collection $\mathcal{F} = \{f_1, ..., f_m\}$ of functions, where each function f_i, $i \in [m]$, is measured by the generic submodular ratio γ_i. After the learning step, assume the element e_t arrives and we need to determine if the element e_t should be added to the solution. Note that, by monotonicity and the front observation, we can conclude

$$m_0 \leq f(S^*) \leq \frac{1}{m} \cdot \sum_{i=1}^m f_i(S_i^*) \leq \frac{1}{m} \cdot \sum_{i=1}^m \sum_{e \in S_i^*} \frac{f_i(e)}{\gamma_i} \leq \frac{\ell m_0}{\gamma_{\min}},$$

where $m_0 = 1/m \cdot \max_{e \in \mathcal{V}} \sum_{i=1}^m f_i(e)$ denotes the maximum single value of the objective function, $S^* = \arg \max_{S \subseteq \mathcal{V}: |S| \leq \ell} G(S)$ represents an optimal solution of the parametric two-stage problem and $S_i^* = \arg \max_{T \subseteq S^*: |T| \leq k} f_i(T)$ denotes an optimal solution of f_i restricted to S^*. Let $\tau^* = f(S^*)/(\beta \ell)$, where $\beta \geq 1$

represent a parameter which will be determined in the following section. We can construct

$$O = \left\{ (1+\epsilon)^l : \frac{m_0}{\beta \ell} \leq (1+\epsilon)^l \leq \frac{m_0}{\beta \gamma_{\min}} \right\}$$

and guess an approximation $\tau \in O$ of τ^* in $O(\log(\ell/\gamma_{\min}))$ time subject to

$$(1-\epsilon)\tau^* \leq \tau \leq \tau^*.$$

Although the true value of m_0 can be learned after the arrival of the whole stream, we can steadily approximate τ^* by constructing a relaxed candidate threshold set

$$O_t = \left\{ (1+\epsilon)^l : \frac{m_t}{\beta \ell} \leq (1+\epsilon)^l \leq \frac{m_t}{\gamma_{\min}} \right\},$$

where $m_t = 1/m \cdot \max_{t':t' \leq t} \sum_{i=1}^m f_i(e_{t'})$ denotes the maximum single value till time t. For each $\tau \in O_t$, provided τ as a new installed value, we start with $S_\tau = \emptyset$ and $T_{\tau,i}^0 = \emptyset$. Consider the arriving element e_t with $|S_\tau| < \ell$ for each $\tau \in O_t$, if $1/m \cdot \sum_{i=1}^m \nabla_i(e_t, T_{\tau,i}^{t-1}) \geq \tau$ for $T_{\tau,i}^{t-1}$ being the state of $T_{\tau,i}$ at the beginning of encountering e_t, then we update $S_\tau = S_\tau + e_t$. For each $i \in [m]$, if $\nabla_i(e_t, T_{\tau,i}^{t-1}) \geq 0$, we execute the following add and exchange operations, respectively.

- If $|T_{\tau,i}^{t-1}| < k$, update $T_{\tau,i}^t = T_{\tau,i}^{t-1} + e_t$;
- If $|T_{\tau,i}^{t-1}| = k$, update $T_{\tau,i}^t = T_{\tau,i}^{t-1} - \text{Exch}_i(e_t, T_{\tau,i}^{t-1}) + e_t$.

The procedure stops once the cardinality capacity of representative set is met or the stream finishes.

4   Theoretical Analysis

For briefness, we omit the index τ in S_τ and $\{T_{\tau,i}\}_{i=1}^m$. Let S^t be the state of S after the arriving of element e_t. For each $i \in [m]$, we denote T_i^t as the maintained set after the revealing of e_t. In addition, denote $A_i^t = \cup_{1 \leq j \leq t} T_i^j$ as the set of elements that ever appeared in T_i till time t. We yield a generalized lower bound of $\Delta_i^E(e_t|T_i^{t-1})$ as stated in the following lemma.

Lemma 1. *Consider the arriving element e_t is exchanged with $\text{Exch}_i(e_t, T_i^{t-1})$ for any $i \in [m]$, we have*

$$\Delta_i^E(e_t|T_i^{t-1}) \geq \gamma_i^2 \cdot \Delta_i^A(e_t|A_i^{t-1}) - \frac{1}{k\gamma_i} \cdot f_i(T_i^{t-1}).$$

Proof. Mainly following by the work of [3,8], we derive an inequality

$$\Delta_i^E(e_t|T_i^{t-1}) = f_i(T_i^{t-1} - \text{Exch}_i(e_t, T_i^{t-1}) + e_t) - f_i(T_i^{t-1})$$

$$\geq \frac{1}{k} \cdot \left[\sum_{e \in T_i^{t-1}} f_i(T_i^{t-1} - e + e_t) - f_i(T_i^{t-1}) \right]$$

$$= \frac{1}{k} \cdot \sum_{e \in T_i^{t-1}} f_i(T_i^{t-1} - e + e_t) - f_i(T_i^{t-1} - e)$$

$$- \frac{1}{k} \cdot \sum_{e \in T_i^{t-1}} f_i(T_i^{t-1}) - f_i(T_i^{t-1} - e).$$

The inequality is obtained by the fact of definition of $\text{Exch}_i(e_t, T_i^{t-1})$. Next we respectively bound the two terms of above inequality. By the definition of generic submodular ratio, for any $e \in T_i^{t-1}$, we have

$$\frac{1}{k} \cdot \sum_{e \in T_i^{t-1}} f_i(T_i^{t-1} - e + e_t) - f_i(T_i^{t-1} - e)$$

$$\geq \frac{1}{k} \cdot \sum_{e \in T_i^{t-1}} \gamma_i \cdot \left(f_i(T_i^{t-1} + e_t) - f_i(T_i^{t-1}) \right)$$

$$= \gamma_i \cdot \Delta_i^A(e_t|T_i^{t-1}). \tag{1}$$

To bound the second term, without loss of generality, we set $T_i^{t-1} = \{e_1, ..., e_k\}$. Reusing the definition of generic submodular ratio, we yield

$$\frac{1}{k} \cdot \sum_{e \in T_i^{t-1}} f_i(T_i^{t-1}) - f_i(T_i^{t-1} - e)$$

$$\leq \frac{1}{k\gamma_i} \cdot \sum_{z=1}^{k} f_i(\{e_1, ..., e_z\}) - f_i(\{e_1, ..., e_{z-1}\})$$

$$= \frac{1}{k\gamma_i} \cdot (f_i(T_i^{t-1}) - f_i(\emptyset))$$

$$\leq \frac{1}{k\gamma_i} \cdot f_i(T_i^{t-1}). \tag{2}$$

Combining inequalities (1) and (2), we conclude

$$\Delta_i^E(e_t|T_i^{t-1}) \geq \gamma_i \cdot \Delta_i^A(e_t|T_i^{t-1}) - \frac{1}{k\gamma_i} \cdot f_i(T_i^{t-1})$$

$$\geq \gamma_i^2 \cdot \Delta_i^A(e_t|A_i^{t-1}) - \frac{1}{k\gamma_i} \cdot f_i(T_i^{t-1}).$$

This completes the proof of this lemma.

Corollary 1. *If $\Delta_i^E(e_t|T_i^{t-1}) < \alpha/(k\gamma_i) \cdot f_i(T_i^{t-1})$, then we have*

$$\Delta_i^A(e_t|A_i^n) \leq \frac{\alpha+1}{k\gamma_i^4} \cdot f_i(T_i^n).$$

Proof. Following from the above lemma and the definition of generic submodular ratio, we yield

$$\gamma_i^3 \cdot \Delta_i^A(e_t|A_i^n) \leq \gamma_i^2 \cdot \Delta_i^A(e_t|A_i^{t-1})$$
$$\leq \Delta_i^E(e_t|T_i^{t-1}) + \frac{1}{k\gamma_i} \cdot f_i(T_i^{t-1})$$
$$< \frac{\alpha+1}{k\gamma_i} \cdot f_i(T_i^{t-1}) \leq \frac{\alpha+1}{k\gamma_i} \cdot f_i(T_i^n).$$

Since the utility f_i is nondecreasing over t, the last inequality is derived.

By the above lemma and the two type operations in Algorithm 1, we can estimate $f(T_i^t)$, $i \in [m]$ as in the next lemma.

Lemma 2. *Consider any arriving element e_t and any $i \in [m]$, we have*

$$f_i(T_i^t) \geq \frac{\alpha\gamma_i^2}{\alpha+1} \cdot f_i(A_i^t).$$

Proof. Note that $|T_i^t| < k$ means Algorithm 1 only executes add operation and we have $A_i^t = T_i^t$. Then the claim directly holds. Next we consider the case of $|T_i^t| = k$. The process can be completed by induction of iterations. We assume $f_i(T_i^{t-1}) \geq (\alpha\gamma_i^2)/(\alpha+1) \cdot f_i(A_i^{t-1})$ and the element e_t is exchanged to T_i^{t-1}. We already have $\Delta_i^E(e_t|T_i^{t-1}) \geq \alpha/(k\gamma_i) \cdot f_i(T_i^{t-1})$ in this case. Further, we acquire

$$f_i(T_i^t) = f_i(T_i^{t-1} - \text{Exch}_i(e_t, T_i^{t-1}) + e_t) = f_i(T_i^{t-1}) + \Delta_i(e_t, T_i^{t-1})$$
$$\geq f_i(T_i^{t-1}) + \max\left\{\gamma_i^2 \cdot \Delta_i(e_t|A_i^{t-1}) - \frac{1}{k\gamma_i} \cdot f_i(T_i^{t-1}), \frac{\alpha}{k\gamma_i} \cdot f_i(T_i^{t-1})\right\}$$
$$\geq f_i(T_i^{t-1}) + \max_{\lambda \in [0,1]}\left\{\lambda \cdot \left(\gamma_i^2 \cdot \Delta_i(e_t|A_i^{t-1}) - \frac{1}{k\gamma_i} \cdot f_i(T_i^{t-1})\right)\right.$$
$$\left. + (1-\lambda) \cdot \left(\frac{\alpha}{k\gamma_i} \cdot f_i(T_i^{t-1})\right)\right\}$$
$$\geq f_i(T_i^{t-1}) + \left\{\frac{\alpha}{\alpha+1} \cdot \left(\gamma_i^2 \cdot \Delta_i(e_t|A_i^{t-1}) - \frac{1}{k\gamma_i} \cdot f_i(T_i^{t-1})\right)\right.$$
$$\left. + \left(1 - \frac{\alpha}{\alpha+1}\right) \cdot \left(\frac{\alpha}{k\gamma_i} \cdot f_i(T_i^{t-1})\right)\right\}$$
$$= f_i(T_i^{t-1}) + \frac{\alpha\gamma_i^2}{\alpha+1} \cdot \Delta_i^A(e_t|A_i^{t-1})$$
$$\geq \frac{\alpha\gamma_i^2}{\alpha+1} \cdot f_i(A_i^t).$$

The first inequality comes from Lemma 1 and the factor of $\Delta_i^E(e_t|T_i^{t-1}) \geq \alpha/(k\gamma_i) \cdot f_i(T_i^{t-1})$. The second inequality is obtained by the definition of λ. The third inequality is derived by setting $\lambda := \alpha/(\alpha+1)$. The last inequality follows by the assumption of $f_i(T_i^{t-1})$.

Equipped with the above lemmas, we state the main result in the following theorem.

Theorem 1. *Given $\epsilon > 0$, setting $\alpha := 1$ and $\beta := (\alpha^2 + \alpha(3 + 2\gamma_{\min}^2) + 1)/(\alpha\gamma_{\min}^5)$, with $O(\ell\log(\ell/\gamma_{\min})/\epsilon)$ memory and $O(mnk\log(\ell/\gamma_{\min})/\epsilon)$ running time, Algorithm 1 makes one pass over the stream, and outputs a summarization $\widetilde{S}$ such that*

$$f(\widetilde{S}) \geq \left(\frac{\gamma_{\min}^5}{5 + 2\gamma_{\min}^2} - O(\epsilon)\right) \cdot f(S^*).$$

Proof. For briefness, we denote S and $\{T_i\}_{i=1}^m$ as the returned sets of Algorithm 1 according to threshold $\tau \in \cup_t O_t$ subject to $(1-\epsilon)\tau^* \leq \tau \leq \tau^*$. We distinguish two cases $|S| = \ell$ and $|S| < \ell$. For the case of $|S| = \ell$, we have

$$\frac{1}{m}\sum_{i=1}^m f_i(T_i^n)$$

$$= \frac{1}{m}\sum_{t=1}^n\sum_{i=1}^m \left[f_i(T_i^t) - f_i(T_i^{t-1})\right]$$

$$= \frac{1}{m}\sum_{i=1}^m \mathbb{1}_{e_t \in S} \cdot \nabla_i(e_t, T_i^{t-1})$$

$$\geq \ell\tau \geq \frac{1-\epsilon}{\beta} \cdot f(S^*). \tag{3}$$

Now we consider the case of $|S| < \ell$. Denote $S_i^* \setminus A_i^n = \{e_1, ..., e_s\}$, then we have

$$f_i(S_i^*) \leq f_i(A_i^n) + \Delta_i^A(S_i^*|A_i^n)$$

$$\leq f_i(A_i^n) + \frac{1}{\gamma_i} \cdot \sum_{t=1}^s [f_i(A_i^n \cup \{e_t\}) - f_i(A_i^n)]$$

$$= f_i(A_i^n) + \frac{1}{\gamma_i} \cdot \sum_{e_t \in S_i^*} \Delta_i^A(e_t|A_i^n),$$

where the first inequality follows by monotonicity and the second is achieved by the definition of generic submodular ratio. Summing up all inequalities of $i \in [m]$, we conclude

$$f(S^*) \leq \frac{1}{m}\sum_{i=1}^m f_i(A_i^n) + \frac{1}{\gamma_{\min}} \cdot \frac{1}{m}\sum_{i=1}^m\sum_{e_t \in S_i^*} \Delta_i^A(e_t|A_i^n)$$

$$\leq \frac{1}{m}\sum_{i=1}^m f_i(A_i^n) + \frac{1}{\gamma_{\min}} \cdot \frac{1}{m}\sum_{i=1}^m\sum_{e_t \in S^*} \mathbb{1}_{e_t \in S_i^*}\left[\mathbb{1}_{e_t \in D_i}\Delta_i^A(e_t|A_i^n)\right.$$

$$\left. + \mathbb{1}_{e_t \notin D_i}[\mathbb{1}_{e_t \in P_i}\Delta_i^A(e|A_i^n) + \mathbb{1}_{e_t \in Q_i}\Delta_i^A(e|A_i^n)]\right],$$

where D_i represents the set of elements of $S_i^* \setminus A_i^n$ encountered before setting threshold T_i^{t-1}, the other elements of $S_i^* \setminus A_i^n$ are partitioned into P_i and Q_i as follows:

$$P_i = \left\{ e_t : \Delta_i^E(e_t|T_i^{t-1}) < \frac{\alpha}{k\gamma_i} \cdot f_i(T_i^{t_z-1}) \right\}$$

and

$$Q_i = \left\{ e_t : \Delta_i^E(e_t|T_i^{t_z-1}) \geq \frac{\alpha}{k\gamma_i} \cdot f_i(T_i^{t_z-1}) \text{ and } \frac{1}{m} \cdot \sum_{i=1}^{m} \nabla_i(e_t|T_i^{t-1}) < \tau \right\}.$$

By Lemma 2, we bound $1/m \cdot \sum_{i=1}^m f_i(A_i^n)$ by

$$\frac{1}{m} \sum_{i=1}^m f_i(A_i^n) \leq \frac{1}{m} \sum_{i=1}^m \frac{\alpha+1}{\alpha\gamma_i^2} f_i(T_i^n) \leq \frac{\alpha+1}{\alpha\gamma_{\min}^2} \cdot \frac{1}{m} \sum_{i=1}^m f_i(T_i^n). \qquad (4)$$

Considering elements in D_i, we yield

$$\frac{1}{m} \sum_{i=1}^m \nabla_i(e_t, T_i^{t-1}) \leq \frac{1}{m} \sum_{i=1}^m \Delta_i^A(e_t|T_i^{t-1}) \leq \frac{1}{m} \sum_{i=1}^m \frac{f_i(e_t)}{\gamma_i} \leq \frac{m_t}{\gamma_{\min}} < \tau.$$

Then we can derive

$$\frac{1}{m} \sum_{i=1}^m \sum_{e_t \in S^*} \mathbf{1}_{e_t \in S_i^*} \mathbf{1}_{e_t \in D_i} \Delta_i^A(e_t|A_i^n) \leq \ell\tau \leq \frac{f(S^*)}{\beta}. \qquad (5)$$

Next we consider elements in P_i, then

$$\frac{1}{m} \sum_{i=1}^m \sum_{e_t \in S^*} \mathbf{1}_{e_t \in S_i^*} \mathbf{1}_{e_t \notin D_i} \mathbf{1}_{e_t \in P_i} \Delta_i^A(e_t|A_i^n) \leq \frac{\alpha+1}{\gamma_{\min}^4} \cdot \frac{1}{m} \sum_{i=1}^m f_i(T_i^n) \qquad (6)$$

follows by combining Corollary 1 and the fact of $|P_i| \leq k$. Considering elements in Q_i and following Lemma 1, we have

$$\frac{1}{m} \sum_{i=1}^m \mathbf{1}_{e_t \in S_i^*} \mathbf{1}_{e_t \notin D_i} \mathbf{1}_{e_t \in O_i} \left[\gamma_i^2 \Delta_i^A(e_t|A_i^{t-1}) - \frac{f_i(T_i^{t-1})}{k\gamma_i} \right]$$

$$\leq \frac{1}{m} \sum_{i=1}^m \mathbf{1}_{e_t \in S_i^*} \nabla_i(e_t, T_i^{t-1}) \leq \tau.$$

Then we yield

$$\frac{1}{m} \sum_{i=1}^m \mathbf{1}_{e_t \in S_i^*} \mathbf{1}_{e_t \notin D_i} \mathbf{1}_{e_t \in O_i} \Delta_i^A(e_t|A_i^n) \leq \frac{\tau}{\gamma_{\min}^2} + \frac{1}{k\gamma_{\min}^3} \cdot \frac{1}{m} \sum_{i=1}^m \mathbf{1}_{e_t \in S_i^*} f_i(T_i^n).$$

Based on the above inequality, we can derive

$$\frac{1}{m}\sum_{i=1}^{m}\sum_{e_t\in S^*\backslash A_i^n}\mathbf{1}_{e_t\in S_i^*}\mathbf{1}_{e_t\notin D_i}\mathbf{1}_{e_t\in O_i}\Delta_i^A(e|A_i^n)$$

$$\leq \frac{\ell\tau}{\gamma_{\min}^2}+\frac{1}{\gamma_{\min}^3}\cdot\frac{1}{m}\sum_{i=1}^{m}f_i(T_i^n)$$

$$\leq \frac{f(S^*)}{\beta\gamma_{\min}^2}+\frac{1}{\gamma_{\min}^3}\cdot\frac{1}{m}\sum_{i=1}^{m}f_i(T_i^n). \tag{7}$$

Equipped with the inequalities (4)–(7), we yield

$$f(S^*)\leq\left(\frac{\alpha+1}{\alpha\gamma_{\min}^2}+\frac{\alpha+1}{\gamma_{\min}^5}+\frac{1}{\gamma_{\min}^4}\right)\cdot\frac{1}{m}\sum_{i=1}^{m}f_i(T_i^n)+\frac{1}{\beta}\left(\frac{1}{\gamma_{\min}}+\frac{1}{\gamma_{\min}^3}\right)\cdot f(S^*)$$

$$\leq\frac{\alpha^2+3\alpha+1}{\alpha\gamma_{\min}^5}\cdot\frac{1}{m}\sum_{i=1}^{m}f_i(T_i^n)+\frac{2}{\beta\gamma_{\min}^3}\cdot f(S^*).$$

Rearranging the above inequality, we obtain

$$\frac{1}{m}\sum_{i=1}^{m}f_i(T_i^n)\geq\frac{\alpha\gamma_{\min}^5-2\alpha\gamma_{\min}^2/\beta}{\alpha^2+3\alpha+1}\cdot f(S^*). \tag{8}$$

To have Inequality (8), we need to guarantee

$$1-\frac{1}{\beta}\left(\frac{1}{\gamma_{\min}}+\frac{1}{\gamma_{\min}^3}\right)>0.$$

Setting $\beta>2/(\gamma_{\min}^3)$, we acquire

$$1-\frac{1}{\beta}\left(\frac{1}{\gamma_{\min}}+\frac{1}{\gamma_{\min}^3}\right)\geq1-\frac{2}{\beta\gamma_{\min}^3}>0.$$

Combining inequalities (3) and (8), we have

$$f(\tilde{S})\geq\frac{1}{m}\sum_{i=1}^{m}f_i(T_i^n)\geq\min\left\{\frac{1-\epsilon}{\beta},\frac{\alpha\gamma_{\min}^5-2\alpha\gamma_{\min}^2/\beta}{\alpha^2+3\alpha+1}\right\}\cdot f(S^*)$$

$$\geq\frac{\alpha\gamma_{\min}^5(1-\epsilon)}{\alpha^2+\alpha(3+2\gamma_{\min}^2)+1}\cdot f(S^*)$$

$$=\left(\frac{\gamma_{\min}^5}{5+2\gamma_{\min}^2}-O(\epsilon)\right)\cdot f(S^*).$$

By setting $\alpha:=1$ and $\beta:=(\alpha^2+\alpha(3+2\gamma_{\min}^2)+1)/(\alpha\gamma_{\min}^5)>2/\gamma_{\min}^3$, we can eventually derive the above inequality.

Corollary 2. *If all generic submodular ratios reduce to 1, by setting $\alpha=1$ and $\beta=7$, Algorithm 1 gets a $(1/7-O(\epsilon))$-approximation for the streaming two-stage submodular maximization problem.*

5 Conclusion

In this paper, we consider the problem of maximizing set functions with generic submodular ratios under a two-stage policy. Comparing to the work of [8], we extend the two-stage submodular model to a more general non-submodular setting and present a generalized streaming algorithm, whose performance guarantees are measured by the minimum generic submodular ratio. In future, we shall further study generalized two-stage submodular maximization with more practical constraints imposed by emerging applications.

Acknowledgements. The first two and fourth authors are supported by Natural Science Foundation of China (No. 11871081). The third author is supported by Natural Science Foundation of China (No. 61772005) and Natural Science Foundation of Fujian Province (No. 2017J01753).

References

1. Badanidiyuru, A., Mirzasoleiman, B., Karbasi, A., Krause, A.: Streaming submodular maximization: massive data summarization on the fly. In: Proceedings of KDD, pp. 671–680 (2014)
2. Balkanski, E., Mirzasoleiman, B., Krause, A., Singer, Y.: Learning sparse combinatorial representations via two-stage submodular maximization. In: Proceedings of ICML, pp. 2207–2216 (2016)
3. Buchbinder, N., Feldman, M., Schwartz, R.: Online submodular maximization with preemption. In: Proceedings of SODA, pp. 1202–1216 (2015)
4. Gong, S., Nong, Q., Liu, W., Fang, Q.: Parametric monotone function maximization with matroid constraints. J. Global Optim. **75**(3), 833–849 (2019). https://doi.org/10.1007/s10898-019-00800-2
5. Kempe, D., Kleinberg, J., Tardos, É.: Maximizing the spread of influence through a social network. In: Proceedings of KDD, pp. 137–146 (2003)
6. Krause, A., McMahan, H.B., Guestrin, C., Gupta, A.: Robust submodular observation selection. J. Mach. Learn. Res. **9**, 2761–2801 (2008)
7. Lin, H., Bilmes, J.: A class of submodular functions for document summarization. In: Proceedings of ACL, pp. 510–520 (2011)
8. Mitrovic, M., Kazemi, E., Zadimoghaddam, M., Karbasi, A.: Data summarization at scale: a two-stage submodular approach. In: Proceedings of ICML, pp. 3593–3602 (2018)
9. Nemhauser, G.L., Wolsey, L.A., Fisher, M.L.: An analysis of approximations for maximizing submodular set functions-I. Math. Program. **14**, 265–294 (1978)
10. Stan, S., Zadimoghaddam, M., Krause, A., Karbasi, A.: Probabilistic submodular maximization in sub-linear time. In: Proceedings of ICML, pp. 3241–3250 (2017)
11. Yang, R., Gu, S., Gao, C., Wu, W., Wang, H., Xu, D.: A two-stage constrained submodular maximization. In: Proceedings of AAIM, pp. 329–340 (2019)
12. Yang, R., Xu, D., Li, M., Xu, Y.: Thresholding methods for streaming submodular maximization with a cardinality constraint and its variants. In: Du, D.-Z., Pardalos, P.M., Zhang, Z. (eds.) Nonlinear Combinatorial Optimization. SOIA, vol. 147, pp. 123–140. Springer, Cham (2019). https://doi.org/10.1007/978-3-030-16194-1_5

Approximation Guarantees for Deterministic Maximization of Submodular Function with a Matroid Constraint

Xin Sun[1], Dachuan Xu[1], Longkun Guo[2(⊠)], and Min Li[3]

[1] Department of Operations Research and Information Engineering, Beijing University of Technology, Beijing 100124, People's Republic of China
athossun@emails.bjut.edu.cn, xudc@bjut.edu.cn
[2] School of Computer Science and Technology, Qilu University of Technology (Shandong Academy of Sciences), Jinan 250353, People's Republic of China
longkun.guo@gmail.com
[3] School of Mathematics and Statistics, Shandong Normal University, Jinan 250014, People's Republic of China
liminemily@sdnu.edu.cn

Abstract. In the paper, we propose a deterministic approximation algorithm for maximizing a generalized monotone submodular function subject to a matroid constraint. The function is generalized through a curvature parameter $c \in [0,1]$, and essentially reduces to a submodular function when $c = 1$. Our algorithm employs the deterministic approximation devised by Buchbinder et al. [3] for the $c = 1$ case of the problem as a building block, and eventually attains an approximation ratio of $\frac{1+g_c(x)+\Delta \cdot [3+c-(2+c)x-(1+c)g_c(x)]}{2+c+(1+c)(1-x)}$ for the curvature parameter $c \in [0,1]$ and for a calibrating parameter that is any $x \in [0,1]$. For $c = 1$, the ratio attains 0.5008 by setting $x = 0.9$, coinciding with the renowned performance guarantee of the problem. Moreover, when the submodular set function degenerates to a linear function, our generalized algorithm always produces optimum solutions and thus achieves an approximation ratio 1.

Keywords: Submodular optimization · Matroid constraint · Curvature · Deterministic algorithm

1 Introduction

As a classical problem in submodular optimization, maximization of a monotone submodular function f subject to a single matroid constraint has broad applications in industry and data science, such as Submodular Welfare Problem [13] which is essentially a submodular maximization problem subject to a partition matroid. In the problem, typically we are given $\mathcal{S}$ a ground set of n elements and an oracle that accesses to a non-negative monotone submodular

© Springer Nature Switzerland AG 2020
J. Chen et al. (Eds.): TAMC 2020, LNCS 12337, pp. 205–214, 2020.
https://doi.org/10.1007/978-3-030-59267-7_18

function $f : 2^S \rightarrow R_{\geq 0}$. The aim is to find the a feasible set $B \subseteq S$ satisfying the constraint over a matroid $\mathcal{M}$ to maximize f.

Recently, a well-studied generalization of submodular maximization is renewed attracting research interest [2,7,11,12,16,18]. The generalization considers the so-called total curvature of f introduced by Conforti and Cornuéjols [7]. By definition, curvature is the minimum $c \in [0,1]$, for which for any subset $U \subset S$ and element $j \in S\backslash U$, $f(U \cup \{j\}) - f(U) \geq (1 - c)f(\{j\})$ holds. The concept indicates how far a function f deviates from linearity.

It is well-known that the standard greedy algorithm gave a $1/2$-approximation ratio for the problem of monotone submodular maximization over a matroid constraint [10], and $1/(1 + c)$-approximation when f has curvature c [7]. In previous works, Feige [8] showed that there exists no polynomial-time algorithm with an approximation ratio better than $(1 - 1/e)$. Also, Nemhauser and Wolsely [14] showed that any improvement over $(1 - 1/e)$ has to make sacrifice for an exponential number of queries to the value oracle. In recent years, Calinescu et al. [5,17] presented a randomized $(1 - e^{-c})/c$-approximation algorithm for the problem. This is a breakthrough result that perfectly matches the conjecture given by Feige [8]. Their technique is continuous greedy with pipage rounding [1], as well as swap rounding [6] which is a rapid method running in continuous time. Another excellent work to this problem is done by Filmus and Ward [9], who gave a non-oblivious local search algorithm with an approximation ratio of $(1 - e^{-c})/c$. The randomized algorithm is combinatorial and based on local search. Notably, the two algorithms above are not easy to derandomize because they are intrinsically randomized. Therefore, Calinescu et al. [5] propose the still open problem "whether a $(1 - e^{-c})/c$-approximation can be obtained using a deterministic algorithm".

For the open problem, Buchbinder et al. [3] lately made a step forward by proposing a deterministic 0.5008-approximation algorithm. Observing the difficulty of applying derandomization method to submodular maximization problems, their algorithm consists of two simple components. One is a deterministic greedy-like split algorithm; The other is a randomized algorithm which has a known deterministic version. The latter algorithm, known as Residual Random Greedy (RRGreedy) algorithm and originally described in [4], is a simple greedy-like randomized algorithm but not difficult to derandomize. The main algorithm unites these two greedy algorithms in the following way: it first obtains two disjoint sets whose union is a base through the split algorithm, then finds the other parts for each one of them to make them a base on appropriate contracted matroids; at last, outputs the better one within these two bases. Buchbinder et al. devised a novel method to select the parameters to ensure the approximation ratio is strictly better than 0.5.

Contribution. In this paper, we present an approximation algorithm for the submodular maximization problem with a curvature parameter c by generalizing the deterministic 0.5008-approximation by Buchbinder [3]. As a by-product, we first show the algorithm achieves an identical ratio of 0.5008 when curvature

$c = 1$. Then, we generalize the analysis to arbitrary curvature $c \in [0, 1]$ and argue that the algorithm achieves the following performance guarantee:

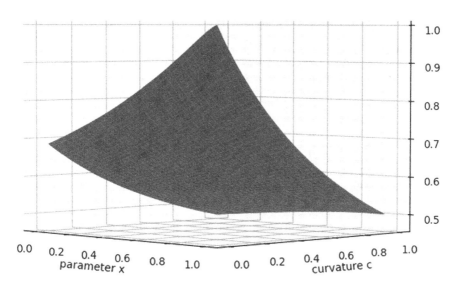

Fig. 1. The relationship between the approximation ratios, curvature c and the calibrating parameter x.

Theorem 1.1. *For curvature $c \in [0, 1]$ and for any $x \in [0, 1]$, there exists a deterministic algorithm which yields an approximation ratio*

$$\frac{1 + g_c(x) + \Delta \cdot [3 + c - (2 + c)x - (1 + c)g_c(x)]}{2 + c + (1 + c)(1 - x)},$$

where $g_c(x) \doteq \frac{1 - (1 - x)^{1+c}}{1+c}$.

The relationship between the approximation ratios, curvature c and parameter x is demonstrated as in Fig. 1 and Table 1. Note that both the total curvature c and the calibrating parameter x are in $[0, 1]$. Comparing to the case in [3] when curvature c equals to 1, the above ratio also attains 0.5008 by setting $x = 0.9$. Moreover, for other values of c, the ratio is strictly better than 0.5008 as depicted in the figure. In particular, the approximation ratio attains 1 when the curvature of the submodular function is 0 so that the set function degenerates to linear function.

Organization. The remainder of the paper is organized as below: Sect. 2 gives preliminary definitions of the paper; Sect. 3 presents the algorithms and all analysis; Sect. 4 finally concludes the paper. In addition, the formal proofs are deferred to the journal version.

Table 1. The approximation ratios of our algorithm respectively against a list of fixed curvature $c \in [0,1]$ obtained through a list of carefully chosen x.

Cur. c	Best x	Ratio	Cur. c	Best x	Ratio
1.0	**0.9**	**0.50087**	0.48	1.0	0.67568
0.98	0.9	0.50527	0.46	1.0	0.68493
0.96	1.0	0.5102	0.44	1.0	0.69444
0.94	1.0	0.51546	0.42	1.0	0.70423
0.92	1.0	0.52083	0.4	1.0	0.71429
0.9	1.0	0.52632	0.38	1.0	0.72464
0.88	1.0	0.53191	0.36	1.0	0.73529
0.86	1.0	0.53763	0.34	1.0	0.74627
0.84	1.0	0.54348	0.32	1.0	0.75758
0.82	1.0	0.54945	0.3	1.0	0.76923
0.8	1.0	0.55556	0.28	1.0	0.78125
0.78	1.0	0.5618	0.26	1.0	0.79365
0.76	1.0	0.56818	0.24	1.0	0.80645
0.74	1.0	0.57471	0.22	1.0	0.81967
0.72	1.0	0.5814	0.2	1.0	0.83333
0.7	1.0	0.58824	0.18	1.0	0.84746
0.68	1.0	0.59524	0.16	1.0	0.86207
0.66	1.0	0.60241	0.14	1.0	0.87719
0.64	1.0	0.60976	0.12	1.0	0.89286
0.62	1.0	0.61728	0.1	1.0	0.90909
0.6	1.0	0.625	0.08	1.0	0.92593
0.58	1.0	0.63291	0.06	1.0	0.9434
0.56	1.0	0.64103	0.04	1.0	0.96154
0.54	1.0	0.64935	0.02	1.0	0.98039
0.52	1.0	0.65789	**0.0**	**1.0**	**1.0**
0.5	1.0	0.66667			

2 Preliminaries

This section gives the formal definition of the terms and notations used in the paper. A set function $f : 2^S \rightarrow \mathbb{R}$ defined on a *ground set* S of size n is *submodular* if and only if for any $U, V \subseteq S$, $f(U \cup V) + f(U \cap V) \leq f(U) + f(V)$. Moreover, we say f is *monotone* if $f(U) \leq f(V)$ whenever $U \subseteq V$. Given a submodular function, the *marginal profit* of adding an element $j \in S$ to U is defined by $f(\{j\}|U) \doteq f(U \cup \{j\}) - f(U)$. The *total curvature* of a monotone submodular function f is the minimum $c \in [0,1]$ such that for any $U \subset S$ and $j \in S \backslash U$,

$f(U \cup j) - f(U) \geq (1-c)f(j)$. For simplicity, we abbreviate $f(\{j\}|U), U \cup \{j\}$ and $U \backslash \{j\}$ as $f(j|U), U+j$ and $U-j$.

Algorithm 1. Main Algorithm

Input: function f, matroid $\mathcal{M}$

Output: The better solution out of $U = (U_1 \cup U_2)$ and $V = (V_1 \cup V_2)$.

  1: $(U_1, V_1) \leftarrow$ Partition$(f, \mathcal{M}, q)$.

  2: $U_2 \leftarrow$ RRG$(f(\cdot|U_1), \mathcal{M}/U_1)$.

  3: $V_2 \leftarrow$ RRG$(f(\cdot|V_1), \mathcal{M}/V_1)$.

A pair $\mathcal{M} = (\mathcal{S}, \mathcal{I})$ is called a *matroid* w.r.t. a ground set $\mathcal{S}$, if and only if the independence system $\mathcal{I}$ is a nonempty collection of subsets of $\mathcal{S}$ satisfying the following properties:

(i) If $U \subseteq V \subseteq \mathcal{S}$ and $V \in \mathcal{I}$, then $U \in \mathcal{I}$;
(ii) If $U, V \in \mathcal{I}$ and $|U| < |V|$, then there is an element $j \in V \backslash U$ such that $U + j \in \mathcal{I}$.

For a matroid $\mathcal{M} = (\mathcal{S}, \mathcal{I})$, a subset U of $\mathcal{S}$ is called *independent* if and only if U belongs to $\mathcal{I}$. A set B is called the *base* of the matroid $\mathcal{M}$ if $B \in \mathcal{S}$ is a maximal independent subset of $\mathcal{S}$. The common size of all bases is called the *rank* of $\mathcal{M}$ and denoted by r. We assume $r \geq 2$ throughout this paper. Moreover, for an independent set U of $\mathcal{M}$, we denote by $\mathcal{M}/U$ the matroid obtained from $\mathcal{M}$ by *contracting* U. We recommend [15] to the readers for more information on matroid theory. Similar to the case of submodular functions, the size of the description of a matroid can be exponential in the size of its ground set. Hence, we assume that the algorithms have access to matroids only through an independence oracle that for a given set $U \subseteq \mathcal{S}$ answers whether U is independent or not.

We are interested in the problem of maximizing a non-negative monotone function $f : 2^{\mathcal{S}} \to \mathbb{R}_{\geq 0}$ subject to curvature $c \in [0,1]$ and a matroid $\mathcal{M} = (\mathcal{S}, \mathcal{I})$ constraint. We use OPT to denote the optimal solution for this problem. For the case of $r = 1$, the above problem is apparently polynomial solvable.

3 Generalizing the Deterministic 0.5008-Approximation Algorithm

In the section, we shall present our main algorithm, which uses two simple greedy-like algorithms as building blocks. The first is the Partition algorithm for generating the partitions and the second is the Residual Random Greedy algorithm that was introduced in 2014 [4]. The formal layout of the main algorithm is as stated in Algorithm 1.

Our analysis starts from the Partition algorithm, which is a greedy-like algorithm and takes the set function f, matroid $\mathcal{M}$ and a parameter $q \in [0,1]$ as inputs. At each iteration, it selects the element that has maximal marginal profit in a certain contracted matroid. Formally, the Partition algorithm proceeds as in Algorithm 2.

Algorithm 2. Partition

Input: function f, matroid $\mathcal{M}$, parameter q
Output: (U_r, V_r)
1: Initialize: $U_0 \leftarrow \emptyset$, $V_0 \leftarrow \emptyset$.
2: **for** $k = 1$ to r **do**
3:      $i_k^U = \arg\max_{i \in \mathcal{M}/(U_{k-1} \cup V_{k-1})} \{f(i|U_{k-1})\}$.
4:      $i_k^V = \arg\max_{i \in \mathcal{M}/(U_{k-1} \cup V_{k-1})} \{f(i|V_{k-1})\}$.
5:      **if** $q \cdot f(i_k^U|U_{k-1}) \geq (1-q) \cdot f(i_k^V|V_{k-1})$ **then**
6:          $U_k \leftarrow U_{k-1} + i_k^U$ and $V_k \leftarrow V_{k-1}$.
7:      **else**
8:          $V_k \leftarrow V_{k-1} + i_k^V$ and $U_k \leftarrow U_{k-1}$.
9:      **end if**
10: **end for**

The algorithm outputs a partition of a base B with a lower bound as stated in the following lemma:

Lemma 3.1. *For each base B of matroid $\mathcal{M}$, set function f with curvature c, and parameters $\alpha \in [0,1]$, $q \in [0,1]$, the Partition algorithm satisfies*

$$\theta f(U_k) + (1-\theta)f(V_k) \geq \Delta \cdot f(B) \geq \frac{1}{3} \cdot f(B),$$

where

$$\theta = \frac{q\left[(1+\alpha) - \alpha(1-c)\right]}{q\left[(1+\alpha) - \alpha(1-c)\right] + (1-q)\left[(2-\alpha) - (1-\alpha)(1-c)\right]}$$

and

$$\Delta = \frac{\left[\alpha q + (1-\alpha)(1-q)\right]}{q\left[(1+\alpha) - \alpha(1-c)\right] + (1-q)\left[(2-\alpha) - (1-\alpha)(1-c)\right]}.$$

Moreover, the following lemma shows the output of Algorithm 2 is always a partition for a base B of $\mathcal{M}$ with guaranteed properties as follows.

Lemma 3.2. *For every base B of $\mathcal{M}$, there exists a partition $B = B_U \cup B_V$ such that the following two characteristics are true*

(1) $U_r \cup B_U$ and $V_r \cup B_V$ are bases of $\mathcal{M}$.
(2) $c \cdot f(U_r) + f(U_r \cup B_U) \geq f(B)$ and $c \cdot f(V_r) + f(V_r \cup B_V) \geq f(B)$.

After obtaining the properties of the partition as in Algorithm 2, we adopt the Residual Random Greedy (RRG) algorithm given by [4] to eventually produce our desired subset. The detailed algorithm is illustrated in Algorithm 3.

Algorithm 3. Residual Random Greedy Algorithm

Input: function f, matroid $\mathcal{M}$
Output: U_r
1: Initialize: $U_0 \leftarrow \emptyset$.
2: **for** $k = 1$ to r **do**
3:    X_k be a base of $\mathcal{M}/U_{k-1}$ maximizing $\sum_{u \in X_k} f(i|U_{k-1})$.
4:    $U_k \leftarrow U_{k-1} + i_k$, where i_k is a uniformly random element from X_k.
5: **end for**

Apparently, Algorithm 3 is again a greedy-like algorithm. We will show the algorithm has an approximation ratio of $1/(1+c)$. Before presenting the analysis of this algorithm, we construct for every integer $k \in \{0, \ldots, r\}$ a set B_k from an arbitrary base B of $\mathcal{M}$, which is a base of the contracted matroid $\mathcal{M}/U_k$ and a bijection mapping $\omega_k : X_k \rightarrow B_{k-1}$ as in [3]. This maps every element $i \in X_k$ to an element of B_{k-1} in a way that $(B_{k-1} - \omega_k(i)) + i_k$ is a base of $\mathcal{M}/U_{k-1}$. The following lemma is the beginning for showing the main result regarding Algorithm 3.

Lemma 3.3. *For every integer $k \in [r]$ and a possibly random set $T \subset \mathcal{S}$,*

$$E\left[f(U_k) + f(U_k \cup B_k \cup T)\right] \geq E\left[f(U_{k-1}) + f(U_{k-1} \cup B_{k-1} \cup T) + (1-c) \cdot f(i_k)\right].$$

Naturally, we can generalize the result of the above lemma to a conclusive result related to the matroid rank r.

Corollary 3.1. *For every $1 \leq k \leq r$ and a possibly random set $T \subset \mathcal{S}$,*

$$E\left[f(U_r) + f(U_r \cup T)\right] \geq E\left[f(U_k) + f(U_k \cup B_k \cup T) + (1-c) \cdot \sum_{l=k+1}^{r} f(i_l)\right]$$

$$\geq E\left[f(B \cup T) + (1-c) \cdot f(U_r)\right].$$

By setting $T = \emptyset$ in the above corollary, we immediately obtain the approximation ratio $1/2$. Then we use the next lemma to show the property of $E[f(U_k)]$, which is intuitively the expected value of U_k in each round.

Lemma 3.4. *For every $0 \leq k \leq r$, we have the following result*

$$E\left[f(U_k)\right] \geq \left[g_c\left(\frac{k}{r}\right) + \delta_c\right] \cdot f(B),$$

where

$$g_c(x) = \frac{1 - (1-x)^{1+c}}{1+c}$$

and

$$\delta_c = \begin{cases} \frac{c}{(1+c)k^{1+c}} & 0 < k < r \\ 0 & otherwise. \end{cases}$$

Furthermore, we can actually extend Lemma 3.4 to non-integer values, through the following lemma:

Lemma 3.5. *For every* $0 \leq x < 1$, $\alpha = \lfloor rx + 1 \rfloor - rx$ *and* $k = \lfloor rx \rfloor$, *then*

$$\alpha \cdot E[f(U_k)] + (1 - \alpha) \cdot E[f(U_{k+1})] \geq g_c(x) \cdot f(B).$$

Moreover, the next lemma gives another lower bound to Algorithm 3, which helps analyzing the main algorithm.

Lemma 3.6. *For every* $0 \leq x \leq 1$ *and an arbitrary base* B' *of* $\mathcal{M}$,

$$(2 + c) \cdot E[f(U_r)] \geq (1 + g_c(x)) \cdot f(B') + (1 - x) \cdot f(B|B').$$

Recall the main algorithm produces two sets U_1 and V_1 in the first step. According to Lemma 3.2, the optimal solution OPT, which is certainly a base of $\mathcal{M}$ due to the assumption of this problem, can be separated into OPT_U and OPT_V. Those two sets can be supplementary set to U_1 and V_1, respectively. This indicates that the unions of $U_1 \cup OPT_U$ and $V_1 \cup OPT_V$ are two bases of $\mathcal{M}$. Consequently, we begin the analysis of the main algorithm with the following lemma on the property of the two complementary sets V_1 and OPT_U of U_1.

Lemma 3.7. *The union of the two complementary sets* V_1 *and* OPT_U *of* U_1 *satisfies the following inequality*

$$f(V_1 \cup OPT_U) \geq f(OPT) - (1 + c) \cdot E\left[f(V|V_1)\right].$$

Therefore, we have a lower bound on the linear combination of the outputs produced by the main algorithm.

Lemma 3.8. *For every* $0 \leq x \leq 1$,

$$(2 + c) \cdot E[f(U_2|U_1)] + (1 + c)(1 - x) \cdot E[f(V)]$$
$$\geq (1 + g(x)) \cdot f(OPT) + (2 - x - (1 + c)g_c(x)) \cdot f(U_1) + (1 + c)(1 - x) \cdot f(V_1).$$

Combining the above lemmas, we can eventually prove the main result as stated in Theorem 1.1.

Proof of Theorem 1.1. According to the result of Lemma 3.8, we can let

$$\theta = \frac{2 - x - (1 + c)g_c(x)}{2 - x - (1 + c)g_c(x) + (1 + c)(1 - x)}$$
$$= \frac{2 - x - (1 + c)g_c(x)}{3 + c - (2 + c)x - (1 + c)g_c(x)}.$$

Then we get the followings from Lemma 3.1 by setting base $B = OPT$

$$(2 - x - (1 + c)g_c(x)) \cdot f(U_1) + (1 + c)(1 - x) \cdot f(V_1)$$
$$\geq \Delta \cdot (3 + c - (2 + c)x - (1 + c)g_c(x)) \cdot f(OPT).$$

Right now we pay attention to the two outputs U and V of the main algorithm. We have

$$
\begin{aligned}
&\max\left\{E[f(U)], E[f(V)]\right\}\\
\geq\ & \frac{(2+c)\cdot E[f(U)] + (1+c)(1-x)\cdot E[f(V)]}{2+c+(1+c)(1-x)}\\
\geq\ & \frac{(1+g_c(x))\cdot f(OPT) + ((2-x)-(1+c)g_c(x))\cdot f(U_1) + (1+c)(1-x)\cdot f(V_1)}{2+c+(1+c)(1-x)}\\
\geq\ & \frac{1+g_c(x) + \Delta\cdot[3+c-(2+c)x-(1+c)g_c(x)]}{2+c+(1+c)(1-x)}\cdot f(OPT),
\end{aligned}
$$

where the second inequality is given by Lemma 3.8 and the final one follows from Lemma 3.1.

We show the relationship between the approximation ratios, curvature c and parameter x in Fig. 1. Note that the total curvature $c \in [0,1]$ and $x \in [0,1]$. And we have the same condition comparing to [3] if curvature c equals to 1. In this case, the above ratio attains 0.5008 via setting $x = 0.9$. In other cases of c, we can verify it is strictly better than 0.5008 through a numerical experiment Table 1. In particular, the ratio achieves 1 when the curvature of the submodular function equals to 0 that the set function degenerates to linear function.   □

Note that the main algorithm is still a randomized algorithm because of the randomization procession in the RRG algorithm. Anyhow, it can be derandomized in a way following a similar line as the derandomization of the RRG algorithm in [3]. In fact, we can adopt their derandomization method against our algorithm without compromising the lemmas above. Thus, the derandomized algorithm derived from our algorithm deserves the same approximation ratio as the randomized algorithm.

4   Conclusions

In this paper, we present an algorithm for the problem of maximizing a monotone submodular function with a total curvature $c \in [0,1]$ and a general matroid constraint. We show the algorithm achieves an approximation ratio of $\frac{1+g_c(x)+\Delta\cdot[3+c-(2+c)x-(1+c)g_c(x)]}{2+c+(1+c)(1-x)}$ for calibrating parameter $x \in [0,1]$. Our algorithm essentially generalizes the 0.5008-approximation algorithm due to Buchbinder et al. [3], which solves the special case of the problem when $c = 1$ and is renown for being the first polynomial time deterministic algorithm with an approximation guarantee better than 1/2. Moreover, our algorithm achieves an approximation guarantee 1 when the set function degenerates to a linear function.

Acknowledgements. The first two authors are supported by Natural Science Foundation of China (No. 11871081). The third author is supported by Natural Science Foundation of China (No. 61772005) and Natural Science Foundation of Fujian Province (No. 2017J01753). The fourth author is supported by Higher Educational Science and Technology Program of Shandong Province (No. J17KA171) and Natural Science Foundation of Shandong Province (No. ZR2019MA032) of China.

References

1. Ageev, A.A., Sviridenko, M.I.: Pipage rounding: a new method of constructing algorithms with proven performance guarantee. J. Comb. Optim. **8**, 307–328 (2004)
2. Balkanski, E., Rubinstein, A., Singer, Y.: The power of optimization from samples. In: Proceedings of NIPS, pp. 4017–4025 (2016)
3. Buchbinder, N., Feldman, M., Garg, M.: Deterministic $(1/2+\varepsilon)$-approximation for submodular maximization over a matroid. In: Proceedings of SODA, pp. 241–254 (2019)
4. Buchbinder, N., Feldman, M., Naor, J.S., Schwartz, R.: Submodular maximization with cardinality constraints. In: Proceedings of SODA, pp. 1433–1452 (2014)
5. Călinescu, G., Chekuri, C., Pál, M., Vondrák, J.: Maximizing a monotone submodular function subject to a matroid constraint. SIAM J. Comput. **40**, 1740–1766 (2011)
6. Chekuri, C., Vondrák, J., Zenklusen, R.: Dependent randomized rounding via exchange properties of combinatorial structures. In: Proceedings of FOCS, pp. 575–584 (2010)
7. Conforti, M., Cornuéjols, G.: Submodular set functions, matroids and the greedy algorithm: tight worst-case bounds and some generalizations of the Rado-Edmonds theorem. Discrete Appl. Math. **7**, 251–274 (1984)
8. Feige, U.: A threshold of $\ln n$ for approximating set cover. J. ACM **45**, 634–652 (1998)
9. Filmus, Y., Ward, J.: Monotone submodular maximization over a matroid via non-oblivious local search. SIAM J. Comput. **43**, 514–542 (2014)
10. Fisher, M.L., Nemhauser, G.L., Wolsey, L.A.: An analysis of approximations for maximizing submodular set functions-II. Math. Program. Study **8**, 73–87 (1978)
11. Iyer, R.K., Bilmes, J.A.: Submodular optimization with submodular cover and submodular knapsack constraints. In: Proceedings of NIPS, pp. 2436–2444 (2013)
12. Iyer, R.K., Jegelka, S., Bilmes, J.A.: Curvature and optimal algorithms for learning and minimizing submodular functions. In: Proceedings of NIPS, pp. 2742–2750 (2013)
13. Lehmann, B., Lehmann, D.J., Nisan, N.: Combinatorial auctions with decreasing marginal utilities. Games Econ. Behav. **55**, 270–296 (2006)
14. Nemhauser, G.L., Wolsey, L.A.: Best algorithms for approximating the maximum of a submodular set function. Math. Oper. Res. **3**, 177–188 (1978)
15. Schrijver, A.: Combinatorial Optimization: Polyhedra and Efficiency. Springer, Heidelberg (2003)
16. Sviridenko, M., Vondrák, J., Ward, J.: Optimal approximation for submodular and supermodular optimization with bounded curvature. Math. Oper. Res. **42**, 1197–1218 (2017)
17. Vondrák, J.: Optimal approximation for the submodular welfare problem in the value oracle model. In: Proceedings of STOC, pp. 67–74 (2008)
18. Vondrák, J.: Submodularity and curvature: the optimal algorithm. RIMS Kokyuroku Bessatsu **B23**, 253–266 (2010)

A Novel Initialization Algorithm for Fuzzy C-means Problem

Qian Liu, Jianxin Liu, Min Li, and Yang Zhou[✉]

School of Mathematics and Statistics, Shandong Normal University,
Jinan 250014, People's Republic of China
lq_qsh@163.com, 3286330436@qq.com, liminEmily@sdnu.edu.cn, zhyg1212@163.com

Abstract. The fuzzy C-means problem belongs to soft clustering problem, where each given point has relationship to every center point. This problem is different from the k-means problem, where each point should belong to only one cluster. In this paper, we design one seeding algorithm for fuzzy C-means problem and obtain performance ratio $O(k\ln k)$. We also give the performance guarantee $O(k^2\ln k)$ of the seeding algorithm based on k-means++ for fuzzy C-means problem. At last, we present our numerical experiment to show the validity of the algorithms.

Keywords: Fuzzy C-means problem · Approximation algorithm · Seeding algorithm

1 Introduction

As a classical problem, data clustering may arise in plenty of areas such as machine learning, pattern recognition, facility location, etc. Among different models for clustering, k-means problem is known as the most widely used model, in which given an integer k and a set of n data points, the goal is to choose k centers so as to minimize the sum of the squared distances between each point and its closest center. The data samples can then be assigned into k clusters according to their distances to the centers. This problem is proved to be NP-hard [2,5]. The first constant approximation algorithm and the best one are given with performance guarantees 108 [10] and 6.357 [1], respectively.

As a heuristic algorithm, the Lloyd's method [14] is proposed and widely used for this problem for its easy implementation and speediness. In Lloyd's method, one begins with k randomly chosen centers from the data points. Each point is

Supported by Higher Educational Science and Technology Program of Shandong Province (No. J17KA171), Natural Science Foundation of Shandong Province (Nos. ZR2019MA032, ZR2019PA004) of China.

Electronic supplementary material The online version of this chapter (https://doi.org/10.1007/978-3-030-59267-7_19) contains supplementary material, which is available to authorized users.

© Springer Nature Switzerland AG 2020
J. Chen et al. (Eds.): TAMC 2020, LNCS 12337, pp. 215–225, 2020.
https://doi.org/10.1007/978-3-030-59267-7_19

then assigned to the nearest center, and each center is recomputed as the center of mass of all points assigned to it. These two steps (assignment and center calculation) are repeated until the process stabilizes. To improve the accuracy of Lloyd's method, a careful seeding technic is added called k-means++ algorithm is proposed and it is proved that this algorithm is $O(\log k)$-competitive with the optimal clustering [3]. The underlying idea of the seeding algorithm is picking the first centers uniformly from the dataset, and then picking the rest points one by one from the dataset with probabilities proportional to their contributions to the potential function. The idea of seeding algorithm has been applied to many variants of k-means problem, such as spherical k-means problem [13], k-means problem with penalties [12], functional k-means problem [11] and so on. For more study of approximation algorithm about k-means problem, one can refer to [8, 15, 21, 22].

However, in many of real applications there are no strict boundaries among different categories. One object may belong to several clusters in different extents. In those cases, instead of hard clustering, it is more natural to consider the soft clustering strategy. For example, we consider the location problem of k super markets around a district. When we assume that people prefer to go to the nearest one, then the location of markets can be formulated as a k-means problem. However, people do not go to only one market since there are plenty of factors which will affect their preference, which consequently leads to a fuzzy C-means model [6, 19]. And there are some work about the comparison between k-means problem and fuzzy C-means problem [9, 20]. For more details of applications of fuzzy C-means, see [16, 18].

In fuzzy C-means problem, instead of determine the affiliation of objects to clusters, degrees of membership (between 0 and 1) of each object to each representative are defined to describe the level of closeness between them. Fuzzy C-means problem can also be seen as a generalization of k-means problem. In k-means problem, the degrees of membership are either 0 or 1. When k is fixed, Blömer et al. present a PTAS for fuzzy C-means problem [7].

In 2015, Stetco et al. propose the fuzzy C-means++ algorithm (FCM++, Algorithm 2) to solve fuzzy C-means problem, in which the seeding strategy of k-means++ is utilized to improve the effectiveness and speed of the classical algorithms [17]. Numerical experiments show that this algorithm have significantly improved both the computation time and final cost function values, whereas theoretical analysis on FCM++ stays unsolved.

The motivation of this paper is as following. It is well known that for the standard FCM algorithm to solve fuzzy C-means problem given in Algorithm 1 [14], the initialization has a great influence on its effectiveness. Thus in [17], the algorithm is improved by utilizing the seeding mechanism of the k-means++ algorithm, in which initial centers are sampled by special probability as in Algorithm 2. However, the probability only depends on the contribution of the given point to the potential in k-means problem, whereas any information related to the membership degree is not used. Moreover, the performance ratio is $O(\ln k)$ when Algorithm 2 is applied to k-means problem whereas no performance guarantee are given for fuzzy C-means problem. Therefore, it is natural to consider that the initial centers are chosen according to a new probability which depends on the contribution of the given point to the potential in fuzzy C-means problem.

The contribution of this paper is twofold. Firstly, noting that the seeding strategy in FCM++ is based on the contribution of data points to the potential of k-means problem, we propose a novel seeding algorithm to fuzzy C-means problem. In this algorithm the centers (except the first one) are also chosen randomly according to a distribution, which is constructed according to their contribution to the potential function of the fuzzy C-means problem. We show that an $O(k\ln k)$-competitive solution can be obtained by the algorithm when the fuzzifier parameter is selected as 2, denoted by m in this paper. For a more general case that $m \in \mathbb{N}_{++}$, which denotes the set of positive integers, the performance ratio can be extended as $O(k^{m-1}\ln k)$. Secondly, we show that the performance ratio of FCM++ algorithm is $O(k^{2m-2}\ln k)$ for $m \in \mathbb{N}_{++}$. Numerical results show that the new algorithm performs well in solving the fuzzy C-means problem.

The rest of this paper is organized as follows. In Sect. 2, we present the fuzzy C-means problem and some basic notations. We introduce the seeding algorithms and the main results for fuzzy C-means problem in Sect. 3. In Sect. 4, the proof to show the correctness of the algorithm is given. In Sect. 5, the numerical experiment about the seeding algorithm for fuzzy C-means problem is presented. The final remarks are concluded in Sect. 6.

2   Preliminaries

In this section, the definition of fuzzy C-means problem, some symbols and notations, as well as some important results are mainly introduced.

Given a set $X = \{x_1, x_2, \ldots, x_n\}$ and $C = \{c_1, c_2, \ldots, c_k\}$ in $\mathbb{R}^d$, $\mu_{ij} \in [0,1]$ $(i = 1, 2, \ldots, n; j = 1, 2, \ldots, k)$ such that $\sum_{j=1}^{k} \mu_{ij} = 1$ for $i = 1, 2, \ldots, n$, and $m \geq 2$, we can define the loss function or potential function of X over C as follows.

$$\phi(X, C, m) = \sum_{i=1}^{n} \sum_{j=1}^{k} \mu_{ij}^{m} ||x_i - c_j||^2.$$

In general, we call m fuzzifier parameter and $\mu = (\mu_{ij})_{n \times k}$ membership degree. The fuzzy C-means problem is to find a clustering C and membership degree μ minimizing the loss function $\phi(X, C, m)$. We use $(C^*(m), \mu^*(m))$ to denote an optimal solution and $\phi^*(m)$ to denote the corresponding objective value. If $m = 1$, it is easy to see that fuzzy C-means problem is reduced to k-means problem. However, there is an example showing that fuzzy C-means problem and k-means problem have different optimal centers when $m = 2$.

Example 1 (A problem with different optimal centers under k-means problem and fuzzy C-means model). Let $X = \{0.0153, 0.7353, 0.4143, 0.2110\} \subseteq \mathbb{R}$ and $k = 2$. The k-means optimal solution of this problem is $\{0.5748, 0.1132\}$, which are actually two means of every two data points. For the fuzzy C-means problem, the objective value of $\{0.5748, 0.1132\}$ is 0.0624. And there is a better center set $\{0.1414, 0.6533\}$ with potential function value as 0.058955. Then it can be concluded that the optimal centers of k-means and fuzzy C-means problem with the same data sets may still vary.

In the following discussion, we assume that $m = 2$. The case that $m \in \mathbb{N}_{++}$ will be presented in the later journal version of the paper. Then the loss function is defined as follows.

$$\phi(X, C) = \phi(X, C, 2) = \sum_{i=1}^{n} \sum_{j=1}^{k} \mu_{ij}^2 \|x_i - c_j\|^2.$$

We use (C^*, μ^*) to denote an optimal solution and ϕ^* to denote the corresponding objective value. Given a set $A \subseteq X$, we denote

$$\phi(A, C) = \sum_{x_i \in A} \sum_{j=1}^{k} \mu_{ij}^2 \|x_i - c_j\|^2,$$

and

$$\phi^*(A) = \sum_{x_i \in A} \sum_{c_j \in C^*} \mu_{ij}^{*2} \|x_i - c_j^*\|^2.$$

Specially, when $A = \{a\}$, we use $\phi(a, C)$ to denote the loss function of A over C for short.

Remark 1. Given any set $C = \{c_1, c_2, \ldots, c_k\}$, we can get the optimal membership degrees as follows.

$$\mu_{ij} = \frac{1}{\sum_{l=1}^{k} \left(\frac{\|x_i - c_j\|}{\|x_i - c_l\|} \right)^2}, i = 1, 2, \ldots, n; j = 1, 2, \ldots, k.$$

Moreover, the loss function of any point $x_i \in X$ without membership degrees is

$$\phi(x_i, C) = \sum_{j=1}^{k} \mu_{ij}^2 \|x_i - c_j\|^2$$

$$= \frac{1}{\sum_{l=1}^{k} \frac{1}{\|x_i - c_l\|^2}}. \tag{1}$$

Inversely, given any μ with $\sum_{j=1}^{k} \mu_{ij} = 1$, $i = 1, 2, \ldots, n$, we can obtain the optimal center points corresponding to μ:

$$c_j = \frac{\sum_{i=1}^{n} \mu_{ij}^2 x_i}{\sum_{i=1}^{n} \mu_{ij}^2}, j = 1, 2, \ldots, k.$$

3 The Seeding Algorithms and Our Main Result

In this section, we will mainly present the seeding algorithms for the fuzzy C-means problem, which are based on the FCM algorithm for fuzzy C-means problem given in Algorithm 1 [14]. The clustering centers and the membership degree are updated in each iteration because of their special relationship. And this algorithm makes enough improvement in each cycle until the clustering or membership degree is no longer changed. This algorithm is improved in [17], which samples the initial centers by special probability as in Algorithm 2. In fact, Algorithm 2 is the seeding section in k-means++ for k-means problem. The probability only depends on the contribution of the given point to the potential in k-means problem, whereas any information related to the membership degree is not used. Moreover, the performance ratio is $O(\ln k)$ when Algorithm 2 is applied to k-means problem. But there is no performance guarantee given when this algorithm is used to solve fuzzy C-means problem. In Algorithm 3, we use a new probability to choose the initial centers, which depends on the contribution of the given point to the potential in fuzzy C-means problem. The motivation of this chosen method comes from the loss function of the observation point without computing membership degrees, which is presented in (1). From now on, the probability used in Algorithm 2 is called D^2-weighting, and the one in Algorithm 3 is called μ^2-weighting.

Algorithm 1. FCM for fuzzy C-means problem

Input: A set of n data points $X = \{x_i\}_{i=1}^n$, the clusters number k, the initial centers $C = \{c_i\}_{i=1}^k$ with k points, and the intial matrix of membership degree $\mu_{n \times k} = 0$.
Output: A fuzzy C-means C of X and the matrix of membership degree $\mu = (\mu_{ij})_{n \times k}$.

1: **for** i from 1 to n and j from 1 to k **do**

2:    Update the membership degree $\mu_{ij} := \dfrac{\frac{1}{\|x_i - c_j\|^2}}{\sum_{l=1}^k \frac{1}{\|x_i - c_l\|^2}}$;

3: **end for**
4: **for** i from 1 to k **do**

5:    Update the centers $c_j := \dfrac{\sum_{i=1}^n \mu_{ij}^2 x_i}{\sum_{i=1}^n \mu_{ij}^2}$;

6: **end for**
7: Repeat Step 1 to Step 6 until C or μ no longer changes;
8: Return C and μ.

Then, we present our main result as follows.

Theorem 1. *Suppose that C is constructed in Algorithm 3 for $X \subseteq \mathbb{R}^d$, then the corresponding cost function satisfies* $\mathrm{E}[\phi(X, C)] \le 16k(\ln k + 2)\phi^*$.

Theorem 2. *Suppose that C is constructed in Algorithm 2 for $X \subseteq \mathbb{R}^d$, then the corresponding cost function satisfies* $\mathrm{E}[\phi(X, C)] \le 16k^2(\ln k + 2)\phi^*$.

Algorithm 2. The seeding algorithm for k-means problem

Input: A set of n data points $X = \{x_i\}_{i=1}^n$, the clusters number k and set $C := \emptyset$.

Output: A clustering C of X.

1: Choose the first center c_1 uniformly at random from X, then set $C := C \cup \{c_1\}$;

2: **for** i from 2 to k **do**

3:    Choose the i-th center c_i from $X \setminus C$ with probability $\frac{d^2(c_i,C)}{\sum_{x_i \in X \setminus C} d^2(x_i,C)}$;

4:    Set $C := C \cup \{c_i\}$;

5: **end for**

6: Return C.

Algorithm 3. The seeding algorithm for fuzzy C-means problem

Input: A set of n data points $X = \{x_i\}_{i=1}^n$, the clusters number k and initial center set $C := \emptyset$.

Output: A set of centers C of X.

1: Choose the first center c_1 uniformly at random from X, then set $C := C \cup \{c_1\}$;

2: **for** i from 2 to k **do**

3:    Choose the i-th center c_i from $X \setminus C$ with probability $\frac{\frac{1}{\sum_{c_j \in C} \|c_i - c_j\|^{-2}}}{\sum_{x_i \in X \setminus C} \frac{1}{\sum_{c_j \in C} \|x_i - c_j\|^{-2}}}$;

4:    Set $C := C \cup \{c_i\}$;

5: **end for**

6: Return C.

---

4   Proof of Correctness

Given an optimal solution (C^*, μ^*), where $C^* = \{c_1^*, c_2^*, \ldots, c_k^*\}$, the set X can be partitioned into k optimal clusters as follows.

$$X_j = \{x \in X | \|x - c_j^*\| \leq \|x - c_l^*\|, \forall l \neq j\}, j = 1, 2, \ldots, k.$$

Lemma 1. *Let A be an optimal cluster and $c(A)^*$ be the center of A. Then*

$$\sum_{x_i \in A} \|x_i - c(A)^*\|^2 \leq k\phi^*(A).$$

Proof.

$$\sum_{x_i \in A} \|x_i - c(A)^*\|^2 = \sum_{x_i \in A} (\mu_{i1}^* + \mu_{i2}^* + \cdots + \mu_{ik}^*)^2 \|x_i - c(A)^*\|^2$$

$$\leq k \sum_{x_i \in A} \sum_{l=1}^{k} {\mu_{il}^*}^2 \|x_i - c(A)^*\|^2$$

$$\leq k \sum_{x_i \in A} \sum_{l=1}^{k} {\mu_{il}^*}^2 \|x_i - c_l^*\|^2$$

$$= k\phi^*(A).$$

Remark 2. The first inequality can be tight when each $\mu^*_{ij} = \frac{1}{k}$, $j = 1, 2, \ldots, k$. In fact, if there is an observation point with the same distance to each center point, the same components of μ can be obtained. For k-means problem, Lemma 1 is trivial since $\sum_{x_i \in A} \|x_i - c(A)^*\|^2 = \phi^*(A)$. Therefore, this lemma shows why the performance guarantee of fuzzy C-means problem is $O(k \ln k)$ and the one of k-means problem is $O(\ln k)$.

When there is only one center point chosen, one can obtain the following bound by using the triangle inequality.

Lemma 2. *Let A be an optimal cluster, and let C be the clustering with just one center, which is chosen uniformly at random from A. Then,*

$$E[\phi(A, C)] \leq 4k\phi^*(A).$$

The following two lemmas can be proved easily by induction and will be used to prove Lemma 5.

Lemma 3. *For any $A, B > 0$ and $a \geq 0$, we have*

$$\frac{1}{\frac{1}{A} + a} + \frac{1}{\frac{1}{B} + a} \geq \frac{1}{\frac{1}{A+B} + a}.$$

Lemma 4. *For any positive integer n, b_i are positive constants for $i = 1, \cdots, n$, and $a \geq 0$. The following inequality holds*

$$\frac{1}{\frac{1}{b_1 + a} + \cdots + \frac{1}{b_n + a}} \leq \frac{1}{\frac{1}{b_1} + \cdots + \frac{1}{b_n}} + a.$$

If another center point is sampled, we can know the influence to the potential function from the following lemma.

Lemma 5. *Let A be an optimal cluster, and let C be an arbitrary representative $C = \{c_1, \cdots, c_t\}$. If we add a random point to C from A, chosen with μ^2-weighting (or D^2-weighting), then*

$$E[\phi(A, C')] \leq 16k\phi^*(A),$$

where C' is the new representative.

When several center points are added, we can know the influence to the potential function from the following lemma.

Lemma 6. *Let C be an arbitrary representative, $u > 0$ be the number of "uncovered" optimal clusters, and X_u denote the set of points in these clusters. Also let $X_c = X - X_u$. Now suppose we add $t \leq u$ random points to C. Let C' denote the resulting set of representative.*
(i) If the new points are chosen with μ^2-weighting, the $E[\phi(X, C')]$ is at most,

$$[\phi(X_c, C) + 16k\phi^*(X_u)](1 + H_t) + \frac{u - t}{u}\phi(X_u, C);$$

(ii) If the new points are chosen with D^2-weighting, the $E[\phi(X, C')]$ is at most,

$$k[\phi(X_c, C) + 16k\phi^*(X_u)](1 + H_t) + \frac{u-t}{u}\phi(X_u, C),$$

where $H_t = 1 + \frac{1}{2} + \cdots + \frac{1}{t}$.

For the proof details of Lemma 2–6, please see the journal version of the paper. Now, we can present the proofs of our main results as follows.

Proof of Theorem 1

Applying Lemma 2 and Lemma 6 (i) with $t = u = k - 1$ and with A being the only covered cluster, where the chosen point is $a \in A$, we have,

$$\begin{aligned}
E[\phi(X, C)] &\le [\phi(A, \{a\}) + 16k\phi^* - 16k\phi^*(A)](1 + H_{k-1}) \\
&\le (2 + \ln k)[4k\phi^*(A) + 16k\phi^* - 16k\phi^*(A)] \\
&\le 16k(2 + \ln k)\phi^*.
\end{aligned}$$

Proof of Theorem 2

Applying Lemma 2 and Lemma 6 (ii) with $t = u = k - 1$ and with A being the only covered cluster, where the chosen point is $a \in A$, we have,

$$\begin{aligned}
E[\phi(X, C)] &\le k[\phi(A, \{a\}) + 16k\phi^* - 16k\phi^*(A)](1 + H_{k-1}) \\
&\le k(2 + \ln k)[4k\phi^*(A) + 16k\phi^* - 16k\phi^*(A)] \\
&\le 16k^2(2 + \ln k)\phi^*.
\end{aligned}$$

5 Numerical Results

In this section we report the numerical results of to test the accuracy and efficiency of the proposed algorithm. The algorithm is implemented in Matlab 2016b and executed on a 2.5 GHz Intel Core i-6500U machine with 16 GB memory.

First, we give a report on solving Example 1 (in Sect. 2) by k-means++, FCM++ and FCM with initialization by Algorithm 2 (NFCM for short). In solving this problem, it is found that there are about a half chance that k-means++ obtains a wrong solution as $\{0.2135, 0.7353\}$, whereas FCM++ and NFCM are more stable for this problem.

Secondly, we test the two algorithms for fuzzy C-means problem with some real data. Two public datasets, SPAM and IRIS, from UC-Irvine Machine Learning Repository [4] is used in this section for the test. The Iris dataset contains three classes of Iris plants with 50 instances each: setosa, versicolor or virginica. For each instance a real positive vector with dimension 4 is recorded. The SPAM dataset contains 4601 instances which describes the characteristics of two categories of emails (solicited and unsolicited). For each instance a dimension 56 real valued vector is recorded to describe frequencies of certain words, capital letters, etc. On account of the randomness of the two algorithms, for each $k = 2, 3, \ldots, 10$

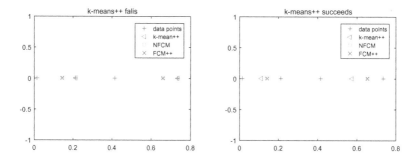

Fig. 1. The distribution of a 2-means problem with its solution computed by k-means++, FCM++ and NFCM algorithm. On the k-means++ algorithm there are chances that it fails to obtain the optimal solution as shown in the second figure.

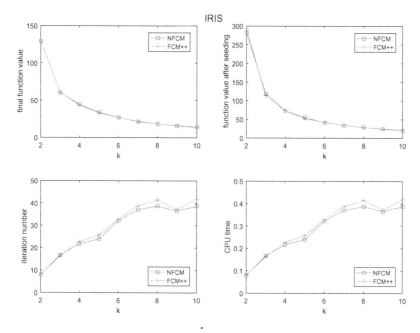

Fig. 2. Numerical results on solving fuzzy C-means problem using FCM++ and NFCM with dataset as IRIS.

and algorithm we run 100 trials and record the means of function values (final and seeding), iteration numbers and CPU times (Fig. 1).

Numerical results of testing FCM++ and NFCM with the two data sets are given in Fig. 2 and 3 respectively. Seen from the figures, the two algorithms perform almost similar to each other. For the IRIS example, NFCM outperforms slightly on iterations and CPU times for certain k. From this point of view it is also natural to conjecture that there holds similar theoretical results on FCM++.

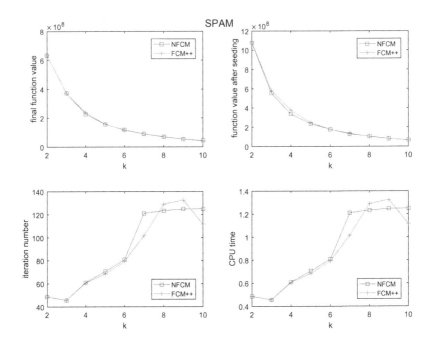

Fig. 3. Numerical results on solving fuzzy C-means problem using FCM++ and NFCM with dataset as SPAM.

6   Conclusions

In this paper, a novel seeding algorithm for fuzzy C-means problem is proposed with its performance ratio proved to be $O(k\ln k)$. The guarantee of the seeding algorithm based on k-means++ for fuzzy C-means problem, as FCM++ [17] is also given to be $O(k^2\ln k)$. Numerical experiment shows the competitiveness of this algorithm to state-of-art methods for fuzzy C-means problems.

References

1. Ahmadian, S., Norouzi-Fard, A., Svensson, O., Ward, J.: Better guarantees for k-means and Euclidean k-median by primal-dual algorithms. SIAM J. Comput. (2019). https://doi.org/10.1137/18M1171321
2. Aloise, D., Deshpande, A., Hansen, P., Popat, P.: NP-hardness of Euclidean sum-of-squares clustering. Mach. Learn. **75**(2), 245–248 (2009)
3. Arthur, D., Vassilvitskii, S.: k-means++: the advantages of careful seeding. In: Proceedings of the Eighteenth Annual ACM-SIAM Symposium on Discrete Algorithms (SODA), pp. 1027–1035 (2007)
4. Asuncion, A., Newman, D.J.: UCI machine learning repository. University of California Irvine School of Information (2007)
5. Awasthi, P., Charikar, M., Krishnaswamy, R., Sinop, A.K.: The hardness of approximation of Euclidean k-means. In: Proceedings of the 31st International Symposium on Computational Geometry (SoCG), pp. 754–767 (2015)

6. Bezdek, J.C.: Pattern Recognition with Fuzzy Objective Function Algorithms. AAPR. Springer, Boston, MA (1981). https://doi.org/10.1007/978-1-4757-0450-1
7. Blömer, J., Brauer, S., Bujna, K.: A theoretical analysis of the fuzzy k-means problem. In: 2016 IEEE 16th International Conference on Data Mining (ICDM), pp. 805–810 (2016)
8. Feng, Q., Zhang, Z., Shi, F., Wang, J.: An improved approximation algorithm for the k-means problem with penalties. In: Chen, Y., Deng, X., Lu, M. (eds.) FAW 2019. LNCS, vol. 11458, pp. 170–181. Springer, Cham (2019). https://doi.org/10.1007/978-3-030-18126-0_15
9. Gafar, A.F.O., Tahyudin, I., et al.: Comparison between k-means and fuzzy C-means clustering in network traffic activities. In: Xu, J., Gen, M., Hajiyev, A., Cooke, F. (eds.) International Conference on Management Science and Engineering Management (ICMSEM), pp. 300–310. Springer, Cham (2017)
10. Jain, K., Vazirani, V.V.: Approximation algorithms for metric facility location and k-median problems using the primal-dual schema and Lagrangian relaxation. J. ACM **48**(2), 274–296 (2001)
11. Li, M., Wang, Y., Xu, D., Zhang, D.: The seeding algorithm for functional k-means problem. In: International Computing and Combinatorics Conference, pp. 387–396 (2019)
12. Li, M., Xu, D., Yue, J., Zhang, D., Zhang, P.: The seeding algorithm for k-means problem with penalties. J. Comb. Optim. **39**(1), 15–32 (2020)
13. Li, M., Xu, D., Zhang, D., Zou, J.: The seeding algorithms for spherical k-means clustering. J. Glob. Optim. **76**(4), 695–708 (2019). https://doi.org/10.1007/s10898-019-00779-w
14. Lloyd, S.: Least squares quantization in PCM. IEEE Trans. Inf. Theory **28**(2), 129–137 (1982)
15. Peng, J., Wei, Y.: Approximating k-means-type clustering via semidefinite programming. SIAM J. Optim. **18**(1), 186–205 (2007)
16. Soomro, S., Munir, A., Choi, K.N.: Fuzzy C-means clustering based active contour model driven by edge scaled region information. Expert Syst. Appl. **120**, 387–396 (2019)
17. Stetco, A., Zeng, X.J., Keane, J.: Fuzzy C-means++: fuzzy C-means with effective seeding initialization. Expert Syst. Appl. **42**(21), 7541–7548 (2015)
18. Tomar, N., Manjhvar, A.K.: Role of clustering in crime detection: application of fuzzy k-means. In: Advances in Computer and Computational Sciences, pp. 591–599 (2018)
19. Wang, P.: Pattern recognition with fuzzy objective function algorithms (James C. Bezdek). SIAM Rev. **25**(3), 442–442 (1983)
20. Wang, S., Zhang, X., Cheng, Y., Jiang, F., Yu, W., Peng, J.: A fast content-based spam filtering algorithm with fuzzy- SVM and k-means. In: 2018 IEEE International Conference on Big Data and Smart Computing (BigComp), pp. 301–307 (2018)
21. Xu, D., Xu, Y., Zhang, D.: A survey on algorithms for k-means problem and its variants. Oper. Res. Trans. **21**(2), 101–109 (2017)
22. Xu, D., Xu, Y., Zhang, D.: A survey on the initialization methods for the k-means algorithm. Oper. Res. Trans. **22**(2), 31–40 (2018)

On the Parameterized Complexity
of *d*-Restricted Boolean Net Synthesis

Ronny Tredup[1][✉] and Evgeny Erofeev[2]

[1] Institut für Informatik, Theoretische Informatik,
Universität Rostock, Albert-Einstein-Straße 22, 18059 Rostock, Germany
`ronny.tredup@uni-rostock.de`
[2] Department of Computing Science, Carl von Ossietzky Universität Oldenburg,
26111 Oldenburg, Germany
`evgeny.erofeev@informatik.uni-oldenburg.de`

Abstract. In this paper, we investigate the parameterized complexity of *d-restricted τ-synthesis* (*d*RτS) parameterized by *d* for a range of Boolean types of nets τ. We show that *d*RτS is *W*[1]-hard for 64 of 128 possible Boolean types that allow places and transitions to be independent.

Keywords: Synthesis · Parameterized complexity · Boolean Petri net

1 Introduction

Boolean Petri nets are one of the most well-known and used families of Petri nets, see [2, pp. 139–152] (and references therein). For Boolean nets, a place *p* contains at most one token in every reachable marking. Thus, *p* is considered as a Boolean condition which is *true* if *p* is marked and false otherwise. In a Boolean Petri net, a place *p* and a transition *t* are related by one of the Boolean *interactions*: *no operation* (nop), *input* (inp), *output* (out), *unconditionally set to true* (set), *unconditionally reset to false* (res), *inverting* (swap), *test if true* (used), and *test if false* (free). These interactions define in which way *p* and *t* influence each other: The interaction inp (out) defines that *p* must be *true* (*false*) before and *false* (*true*) after *t*'s firing; free (used) implies that *t*'s firing proves that *p* is *false* (*true*); nop means that *p* and *t* do not affect each other at all; res (set) implies that *p* may initially be both *false* or *true* but after *t*'s firing it is *false* (*true*); swap means that *t* inverts *p*'s current Boolean value.

A set τ of Boolean interactions is called a *type of net*. Since we have eight interactions to choose from, there are a total of 256 different types. A Boolean Petri net *N* is of type τ (a τ-net) if it applies at most the interactions of τ. For a type τ, the τ-synthesis problem consists in deciding whether a given directed labelled graph *A*, also called *transition system*, is isomorphic to the reachability graph of some τ-net *N*, and in constructing *N* if it exists.

Supported by DFG through grant Be 1267/16-1 `ASYST`.

© Springer Nature Switzerland AG 2020
J. Chen et al. (Eds.): TAMC 2020, LNCS 12337, pp. 226–238, 2020.
https://doi.org/10.1007/978-3-030-59267-7_20

Badouel et al. [1] and Schmitt [4] investigated the computational complexity of τ-synthesis for elementary net systems ($\{\mathsf{nop}, \mathsf{inp}, \mathsf{out}\}$) and flip-flop nets ($\{\mathsf{nop}, \mathsf{inp}, \mathsf{out}, \mathsf{swap}\}$), respectively; while synthesis is NP-complete for the former, it is polynomial for the latter. In [5], the complexity of τ-synthesis restricted to g-bounded inputs (every state of A has at most g incoming and g outgoing arcs) has been completely characterized for the types that contain nop and, thus, allow places and transitions to be independent. For 84 of these 128 types it turned out that synthesis is NP-complete, even for small fixed $g \leq 3$. As a result, τ-synthesis parameterized by g is certainly not *fixed parameter tractable* (FPT).

This paper addresses the computational complexity of a different instance of τ-synthesis, namely *d-restricted τ-synthesis* ($d\mathrm{R}\tau\mathrm{S}$), imposing a limitation for the synthesis output: The d-restricted synthesis targets to those τ-nets in which every place must be in relation nop with all but d transitions of the net, while the synthesis input is no longer confined. This formulation of the synthesis problem is motivated at least twofold. On the one hand, in applications, places are usually meant as resources while transitions are meant as agents. Hence such a restriction ensures that a certain resource binds only few agents. On the other hand, $d\mathrm{R}\tau\mathrm{S}$ is of a particular interest from the theoretical point of view, since, parameterized by d, it belongs to the complexity class XP [5, p. 25]. Consequently, the question for the existence of FPT-algorithms arises.

In this paper, we enhance our understanding of $d\mathrm{R}\tau\mathrm{S}$ from a parameterized complexity point of view and show $W[1]$-hardness for the following types of nets:

1. $\{\mathsf{nop}, \mathsf{inp}, \mathsf{free}\}$, $\{\mathsf{nop}, \mathsf{inp}, \mathsf{free}, \mathsf{used}\}$, $\{\mathsf{nop}, \mathsf{out}, \mathsf{used}\}$, $\{\mathsf{nop}, \mathsf{out}, \mathsf{free}, \mathsf{used}\}$,
2. $\tau = \{\mathsf{nop}, \mathsf{swap}\} \cup \omega$ such that $\omega \subseteq \{\mathsf{inp}, \mathsf{out}, \mathsf{res}, \mathsf{set}, \mathsf{free}, \mathsf{used}\}$ and $\omega \cap \{\mathsf{inp}, \mathsf{out}, \mathsf{free}, \mathsf{used}\} \neq \emptyset$

Our proofs base on parameterized reductions of the well-known $W[1]$-complete problems *Regular Independent Set* and *Odd Set* [3]. While all types of (2) have been shown to be NP-complete [5], the types covered by (1) that does not contain any of $\mathsf{res}, \mathsf{set}$ have been shown to be polynomial [4,6]. However, since our parameterized reductions are actually polynomial-time reductions, here we show NP-completeness and $W[1]$-hardness for these types at the same time.

The paper is organized as follows. After introducing of the necessary definitions in Sect. 2, the main contribution is presented in Sect. 3. Section 4 suggests an outlook of the further research directions.

2   Preliminaries

We assume that the reader is familiar with the concepts relating to fixed-parameter tractability. Due to space restrictions, some formal definitions and some proofs are omitted. See [3] for the definitions of relevant notions in parameterized complexity theory.

Transition Systems. A (deterministic) *transition system* (TS, for short) $A = (S, E, \delta)$ is a directed labeled graph with states S, events E and partial *transition*

x	nop(x)	inp(x)	out(x)	set(x)	res(x)	swap(x)	used(x)	free(x)
0	0		1	1	0	1		0
1	1	0		1	0	0	1	

Fig. 1. All Interactions i of I. If a cell is empty, then i is undefined on the respective x.

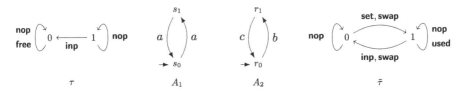

$$\tau \qquad A_1 \qquad A_2 \qquad \tilde{\tau}$$

Fig. 2. Left: $\tau = \{\mathsf{nop}, \mathsf{inp}, \mathsf{free}\}$. Right: $\tilde{\tau} = \{\mathsf{nop}, \mathsf{swap}, \mathsf{used}, \mathsf{set}\}$. The TS A_1 has no ESSP atoms. Hence, it has the τ-ESSP and $\tilde{\tau}$-ESSP. The only SSP atom of A_1 is (s_0, s_1). It is $\tilde{\tau}$-solvable by $R_1 = (sup_1, sig_1)$ with $sup_1(s_0) = 0$, $sup_1(s_1) = 1$, $sig_1(a) = \mathsf{swap}$. Thus, A_1 has the τ-admissible set $\mathcal{R} = \{R_1\}$, and the τ-net $N_A^{\mathcal{R}} = (\{R_1\}, \{a\}, M_0, f)$ with $M_0(R_1) = sup_1(R_1)$ and $f(R_1, a) = sig_1(R_1)$ solves A_1. The SSP atom (s_0, s_1) is not τ-solvable, thus, neither is A_1. TS A_2 has ESSP atoms (b, r_1) and (c, r_0), which are both $\tilde{\tau}$-unsolvable. The only SSP atom (r_0, r_1) in A_2 can be solved by $\tilde{\tau}$-region $R_2 = (sup_2, sig_2)$ with $sup_2(r_0) = 0$, $sup_2(r_1) = 1$, $sig_2(b) = \mathsf{set}$, $sig_2(c) = \mathsf{swap}$. Thus, A_2 has the $\tilde{\tau}$-SSP, but not the $\tilde{\tau}$-ESSP. None of the (E)SSP atoms of A_2 can be solved by any τ-region. Notice that the $\tilde{\tau}$-region R_2 maps two events to a signature different from nop. Thus, in case of d-restricted $\tilde{\tau}$-synthesis, R_2 would be not valid for $d = 1$.

function $\delta : S \times E \longrightarrow S$, where $\delta(s, e) = s'$ is interpreted as $s \xrightarrow{e} s'$. For $s \xrightarrow{e} s'$ we say s is a *source* and s' is a *sink* of e, respectively. An event e *occurs* at a state s, denoted by $s \xrightarrow{e}$, if $\delta(s, e)$ is defined. An *initialized* TS $A = (S, E, \delta, s_0)$ is a TS with a distinct state $s_0 \in S$ where every state $s \in S$ is *reachable* from s_0 by a directed labeled path.

Boolean Types of Nets [2]. The following notion of Boolean types of nets allows to capture *all* Boolean Petri nets in *one* uniform way. A *Boolean type of net* $\tau = (\{0, 1\}, E_\tau, \delta_\tau)$ is a TS such that E_τ is a subset of the Boolean interactions: $E_\tau \subseteq I = \{\mathsf{nop}, \mathsf{inp}, \mathsf{out}, \mathsf{set}, \mathsf{res}, \mathsf{swap}, \mathsf{used}, \mathsf{free}\}$. The interactions $i \in I$ are binary partial functions $i : \{0, 1\} \to \{0, 1\}$ as defined in Fig. 1. For all $x \in \{0, 1\}$ and all $i \in E_\tau$ the transition function of τ is defined by $\delta_\tau(x, i) = i(x)$. By definition, a Boolean type τ is completely determined by its event set E_τ. Hence, in the following we identify τ with E_τ, cf. Fig. 2.

τ-**Nets.** Let $\tau \subseteq I$. A Boolean Petri net $N = (P, T, M_0, f)$ of type τ, (τ-net, for short) is given by finite and disjoint sets P of places and T of transitions, an initial marking $M_0 : P \longrightarrow \{0, 1\}$, and a (total) flow function $f : P \times T \to \tau$. For a natural number d, a τ-net is called d-*restricted* if for every $p \in P$: $|\{t \in T \mid f(p, t) \neq \mathsf{nop}\}| \leq d$. A τ-net realizes a certain behavior by firing sequences of transitions: A transition $t \in T$ can fire in a marking $M : P \longrightarrow \{0, 1\}$ if $\delta_\tau(M(p), f(p, t))$ is defined for all $p \in P$. By firing, t produces the next marking

$M' : P \longrightarrow \{0,1\}$ where $M'(p) = \delta_\tau(M(p), f(p,t))$ for all $p \in P$. This is denoted by $M \xrightarrow{t} M'$. Given a τ-net $N = (P, T, M_0, f)$, its behavior is captured by a transition system A_N, called the reachability graph of N. The state set of A_N is the reachability set $RS(N)$, that is, the set of all markings that, starting from initial state M_0, are reachable by firing a sequence of transitions. For every reachable marking M and transition $t \in T$ with $M \xrightarrow{t} M'$ the state transition function δ of A is defined as $\delta(M, t) = M'$.

τ-**Regions.** Let $\tau \subseteq I$. If an input A of τ-synthesis allows a positive decision then we want to construct a corresponding τ-net N purely from A. Since A and A_N are isomorphic, N's transitions correspond to A's events. However, the notion of a place is unknown for TSs. So called regions mimic places of nets: A τ-region of a given $A = (S, E, \delta, s_0)$ is a pair (sup, sig) of support $sup : S \to S_\tau = \{0,1\}$ and signature $sig : E \to E_\tau = \tau$ where every transition $s \xrightarrow{e} s'$ of A leads to a transition $sup(s) \xrightarrow{sig(e)} sup(s')$ of τ. A region (sup, sig) models a place p and the corresponding part of the flow function f. In particular, $sig(e)$ models $f(p, e)$ and $sup(s)$ models $M(p)$ in the marking $M \in RS(N)$ corresponding to $s \in S(A)$. We say that τ-region (sup, sig) respects the parameter d, if $|\{e \in E \mid sig(e) \neq \mathsf{nop}\}| \leq d$. Every set $\mathcal{R}$ of τ-regions of A defines the synthesized τ-net $N_A^{\mathcal{R}} = (\mathcal{R}, E, f, M_0)$ with flow function $f((sup, sig), e) = sig(e)$ and initial marking $M_0((sup, sig)) = sup(s_0)$ for all $(sup, sig) \in \mathcal{R}, e \in E$. It is well-known that $A_{N_A^{\mathcal{R}}}$ and A are isomorphic if and only if $\mathcal{R}$'s regions solve certain separation atoms [2]. A pair (s, s') of distinct states of A defines a state separation atom (SSP atom, for short). A τ-region $R = (sup, sig)$ solves (s, s') if $sup(s) \neq sup(s')$. The meaning of R is to ensure that $N_A^{\mathcal{R}}$ contains at least one place R such that $M(R) \neq M'(R)$ for the markings M and M' corresponding to s and s', respectively. If there is a τ-region that solves (s, s') then s and s' are called τ-solvable. If every SSP atom of A is τ-solvable then A has the τ-state separation property (τ-SSP, for short). A pair (e, s) of event $e \in E$ and state $s \in S$ where e does not occur at s, that is $\neg s \xrightarrow{e}$, defines an event state separation atom (ESSP atom, for short). A τ-region $R = (sup, sig)$ solves (e, s) if $sig(e)$ is not defined on $sup(s)$ in τ, that is, $\neg \delta_\tau(sup(s), sig(e))$. The meaning of R is to ensure that there is at least one place R in $N_A^{\mathcal{R}}$ such that $\neg M \xrightarrow{e}$ for the marking M corresponding to s. If there is a τ-region that solves (e, s) then e and s are called τ-solvable. If every ESSP atom of A is τ-solvable then A has the τ-event state separation property (τ-ESSP, for short). A set $\mathcal{R}$ of τ-regions of A is called τ-admissible if for every of A's (E)SSP atoms there is a τ-region R in $\mathcal{R}$ that solves it. The following lemma, borrowed from [2, p.163], summarizes the already implied connection between the existence of τ-admissible sets of A and (the solvability of) τ-synthesis:

Lemma 1 ([2]). *A TS A is isomorphic to the reachability graph of a τ-net N if and only if there is a τ-admissible set $\mathcal{R}$ of A such that $N = N_A^{\mathcal{R}}$.*

In this paper, we investigate the following parameterized problem: d-**Restricted τ-Rynthesis** (dRτS). The input (A, d) consists of a TS A and a natural number $d \in \mathbb{N}$. The parameter is d. The question to answer is, if there is a τ-admissible set $\mathcal{R}$ of A such that $|\{e \in E(A) \mid sig(e) \neq \mathsf{nop}\}| \leq d$ is true for all $R \in \mathcal{R}$.

3 $W[1]$-Hardness of d-Restricted τ-Synthesis

Theorem 1. *The problem d-restricted τ-synthesis is $W[1]$-hard if*

1. $\tau = \{\mathsf{nop}, \mathsf{inp}, \mathsf{free}\}$ *or* $\tau = \{\mathsf{nop}, \mathsf{inp}, \mathsf{free}, \mathsf{used}\}$ *or* $\tau = \{\mathsf{nop}, \mathsf{out}, \mathsf{used}\}$ *or* $\tau = \{\mathsf{nop}, \mathsf{out}, \mathsf{free}, \mathsf{used}\}$,
2. $\tau = \{\mathsf{nop}, \mathsf{swap}\} \cup \omega$ *such that* $\omega \subseteq \{\mathsf{inp}, \mathsf{out}, \mathsf{res}, \mathsf{set}, \mathsf{free}, \mathsf{used}\}$ *and* $\omega \cap \{\mathsf{inp}, \mathsf{out}, \mathsf{free}, \mathsf{used}\} \neq \emptyset$

The proofs of Theorem 1.1 and Theorem 1.2 base on parameterized reductions of the problems *Regular Independent Set* and *Odd Set*, respectively. Both problems are well-known to be $W[1]$-complete (see e.g. [3]) and are defined as follows:

Regular Independent Set (RIS). The input $(\mathfrak{U}, M, \kappa)$ consists of a finite set $\mathfrak{U}$, a set $M = \{M_0, \ldots, M_{m-1}\}$, $M_i \subseteq \mathfrak{U}$ and $|M_i| = 2$ for all $i \in \{0, \ldots, m-1\}$, and $\kappa \in \mathbb{N}$. The parameter is κ. Moreover, there is $r \in \mathbb{N}$ for all $X \in \mathfrak{U}$ such that $|\{a \in M \mid X \in a\}| = r$. The question is whether there is an independent set $S \subseteq \mathfrak{U}$, that is, $\{X, X'\} \notin M$ for all $X, X' \in S$, such that $|S| \geq \kappa$.

Odd Set (OD). The input $(\mathfrak{U}, M, \kappa)$ consists of a finite set $\mathfrak{U}$, a set $M = \{M_0, \ldots, M_{m-1}\}$ of subsets of $\mathfrak{U}$ and a natural number κ. The parameter is κ. The question to answer is whether there is a set $S \subseteq \mathfrak{U}$ of size at most κ such that $|S \cap M_i|$ is odd for every $i \in \{0, \ldots, m-1\}$.

The General Reduction Idea. An input $I = (\mathfrak{U}, M, \kappa)$ (of RIS or OD), where $M = \{M_0, \ldots, M_{m-1}\}$, is reduced to an instance (A_I^τ, d) with TS A_I^τ and $d = f(\kappa)$ (f being a polynomial time computable function) as follows: For every $i \in \{0, \ldots, m-1\}$, the TS A_I^τ has for the set $M_i = \{X_{i_0}, \ldots, X_{i_{m_i}-1}\}$ a directed labelled path $P_i = s_{i,0} \xrightarrow{X_{i_0}} \ldots \xrightarrow{X_{i_{m_i}-1}} s_{i,m_i}$ that represents M_i and uses its elements as events. The TS A_I^τ has an ESSP atom α such that if $R = (sup, sig)$ is a τ-region that solves α and respects d, then there are indices $i_0, \ldots, i_j \in \{0, \ldots, m-1\}$ such that $sup(s_{i_\ell,0}) \neq sup(s_{i_\ell,m_{i_\ell}})$ for all $\ell \in \{0, \ldots, j\}$. Since the image of P_{i_ℓ} is a directed path in τ, by $sup(s_{i_\ell,0}) \neq sup(s_{i_\ell,m_{i_\ell}})$, there has to be an element $X \in M_{i_\ell}$ such that $s \xrightarrow{X} s' \in P_{i_\ell}$ implies $sup(s) \neq sup(s')$. That is, X causes a state change in τ. This is simultaneously true for all $P_{i_0}, \ldots, P_{i_j}$.

The reduction ensures that $S = \{X \in \mathfrak{U} \mid s \xrightarrow{X} s' \Rightarrow sup(s) \neq sup(s')\}$ defines a searched independent set or a searched odd set, depending on the actually reduced problem. Thus, if (A_I^τ, d) is a yes-instance, implying the solvability of α, then $I = (\mathfrak{U}, M, \kappa)$ is, too.

Reversely, if $I = (\mathfrak{U}, M, \kappa)$ is a yes-instance, then there is a fitting τ-region of A_I^τ that solves α. The reduction ensures that the τ-solvability of α implies that all (E)SSP atoms of A_I^τ are solvable by τ-regions respecting d. Thus, (A_I^τ, d) is a yes-instance, too.

In what follows, we present the corresponding reductions, show that the solvability of α implies a searched (independent or odd) set and argue that the existence of a searched set implies the solvability of α.

The Proof of Theorem 1.1. Let $\tau \in \{\{\mathsf{nop}, \mathsf{inp}, \mathsf{free}\}, \{\mathsf{nop}, \mathsf{inp}, \mathsf{free}, \mathsf{used}\}\}$. We prove the claim for τ, by symmetry, the proof for the other types is similar.

Let $I = (\mathfrak{U}, M, \kappa)$ be an instance of RIS, where $M = \{M_0, \ldots, M_{m-1}\}$ such that $M_i = \{X_{i_0}, X_{i_1}\}$ and (without loss of generality we assume that) $i_0 < i_1$ for all $i \in \{0, \ldots, m-1\}$. Let $r \in \mathbb{N}$ such that $|\{a \in M \mid X \in a\}| = r$ for all $X \in \mathfrak{U}$.

For a start, we define $d = \kappa \cdot (r+1) + 2$. The TS A_I^τ has the following gadget H with events k_0 and k_1 that provides the atom $\alpha = (k_1, h_0)$:

$$h_0 \xrightarrow{\;k_0\;} h_1 \xrightarrow{\;k_1\;} h_2$$

Moreover, for every $i \in \{0, \ldots, m-1\}$, the TS A_I^τ has the following gadget T_i that represents $M_i = \{X_{i_0}, X_{i_1}\}$:

The gadget T_i uses M_i's elements X_{i_0} and X_{i_1} as events. Moreover, it has exactly $r\kappa$ events $a_i^0, \ldots, a_i^{r\kappa-1}$ that occur consecutively on a path. If $i \in \{0, \ldots, m-1\}$ and $j \in \{0, \ldots, r\kappa - 1\}$, then we say a_i^j is the j-th event of (the set) M_i. For every $j \in \{0, \ldots, r\kappa - 1\}$, the TS A_I^τ has the following gadget G_j:

For all $i \in \{0, \ldots, m-1\}$, the gadget G_i applies the j-th event of M_i, and the events $a_0^j, \ldots, a_{m-1}^j$ occur consecutively in a row.

The initial state of A_I^τ is $\perp_{m-1,0}$. Fresh events $\ominus_0, \ldots, \ominus_{m-1}$ and $\odot_0, \ldots, \odot_{r\kappa-1}$ join the introduced gadgets H, $T_0, \ldots, T_{m-1}$ and $G_0, \ldots, G_{r\kappa-1}$ into the TS A_I^τ and make all states reachable from $\perp_{m-1,0}$. More exactly, for all $i \in \{1, \ldots, m-1\}$, the TS A_I^τ has the edge $\perp_{i,0} \xrightarrow{\ominus_i} \perp_{i-1,0}$, and it has the edge $\perp_{0,0} \xrightarrow{\ominus_0} h_0$. Moreover, A_I^τ has the edge $\perp_{m-1,0} \xrightarrow{\odot_0} T_0$ and, for all $j \in \{0, \ldots, r\kappa - 2\}$, the edge $T_j \xrightarrow{\odot_{j+1}} T_{j+1}$. The resulting TS is A_I^τ, and it is easy to see that (A_I^τ, d) is obtained by a parameterized reduction.

Let (A_I^τ, d) be a yes-instance. We argue that $(\mathfrak{U}, M, \kappa)$ has an independence set of size κ. Since (A_I^τ, d) is a yes-instance, there is a τ-region $R = (sup, sig)$ that solves α and respects the parameter d, that is, $|\{e \in E(A_I^\tau) \mid sig(e) \neq \mathsf{nop}\}| \leq d$. In the following, we argue that $S = \{X \in \mathfrak{U} \mid sig(X) = \mathsf{inp}\}$ is a searched set. The general idea is as follows: The region R selects exactly $r\kappa$ gadgets $T_{i_0}, \ldots T_{i_{r\kappa-1}}$, representing the sets $M_{i_0}, \ldots, M_{i_{r\kappa-1}}$, such that $sup(t_{i_j,0}) = 1$ and $sup(t_{i_j,2}) = 0$ for all $j \in \{0, \ldots, r\kappa - 1\}$. In particular, for all $j \in \{0, \ldots, r\kappa - 1\}$, that makes $sup(t_{i_j,0}) \xrightarrow{X_{i_{j_0}}} sup(t_{i_j,1}) \xrightarrow{X_{i_{j_1}}} sup(t_{i_j,2})$ a path from 1 to 0 in τ. Consequently, for every $j \in \{0, \ldots, r\kappa-1\}$, there is exactly one event $e \in \{X_{i_{j_0}}, X_{i_{j_1}}\}$ with $sig(e) = \mathsf{inp}$. The reduction ensures that there are exactly κ elements $X_{i_0}, \ldots, X_{i_{\kappa-1}} \in \mathfrak{U}$ such that $sig(X_{i_j}) = \mathsf{inp}$ for all $j \in \{0, \ldots, \kappa - 1\}$. Moreover, it also ensures $sig(e) = \mathsf{nop}$ for all $e \in \mathfrak{U} \setminus \{X \in \mathfrak{U} \mid sig(X) = \mathsf{inp}\}$. As a result, $r\kappa$ sets are "covered" by κ elements. Since every elements is a member of exactly r sets, $S = \{X \in \mathfrak{U} \mid sig(X) = \mathsf{inp}\}$ is an independent set of size κ of $(\mathfrak{U}, M)$.

Let us formally argue that the reduction correctly converts this general idea. By definition of τ, one easily finds out that $sig(k_1) = \mathsf{free}$, $sup(h_0) = 1$ and $sig(k_0) = \mathsf{inp}$. By $t_{i,0} \xrightarrow{k_0}$, this implies $sup(t_{i,0}) = 1$ for all $i \in \{0, \ldots, m - 1\}$. Moreover, since R respects d, there are at most $\kappa \cdot (r+1)$ other events left whose signature is different from nop.

Let $j, j' \in \{0, \ldots, r\kappa - 1\}$ such that $j \neq j'$. By $sig(k_0) = \mathsf{inp}$ and $g_{j,0} \xrightarrow{k_0}$, we have $sup(g_{j,0}) = 1$; by $sig(k_1) = \mathsf{free}$ and $g_{j,m} \xrightarrow{k_1}$, we have $sup(g_{j,m}) = 0$. Consequently, $sup(g_{j,0}) \xrightarrow{sig(a_0^j)} \ldots \xrightarrow{sig(a_{m-1}^j)} sup(g_{j,m})$ is a path from 1 to 0 in τ. Since there is no path in τ on which inp occurs twice, there is exactly one $i \in \{0, \ldots, m - 1\}$ such that $sig(a_i^j) = \mathsf{inp}$. Similarly, there is exactly one $i' \in \{0, \ldots, m - 1\}$ such that $sig(a_{i'}^{j'}) = \mathsf{inp}$. For all $i \in \{0, \ldots, m - 1\}$, the events $a_i^0, \ldots, a_i^{r\kappa-1}$ occur consecutively on a path in T_i, and inp never occurs twice on a path in τ. Thus, by $j \neq j'$, we have $i \neq i'$, that is, never the j-th and the j'-th event of the same set M_i are selected. Consequently, by the arbitrariness of j and j', there are exactly $r\kappa$ events $a_{i_0}^{j_0}, \ldots, a_{i_{r\kappa-1}}^{j_{r\kappa-1}}$ such that $sig(a_{i_0}^{j_0}) = \cdots = sig(a_{i_{r\kappa-1}}^{j_{r\kappa-1}}) = \mathsf{inp}$, and all $i_0, \ldots, i_{r\kappa-1} \in \{0, \ldots, m-1\}$ are pairwise distinct. On the one hand, this shows that there are $r\kappa$ gadgets $T_{i_0}, \ldots, T_{i_{r\kappa-1}}$ (representing the sets $M_{i_0}, \ldots, M_{i_{r\kappa-1}}$) such that $sup(t_{i_j,0}) = 1$ and $sup(t_{i_j,2}) = 0$ for all $j \in \{0, \ldots, r\kappa - 1\}$. Thus, for every $j \in \{0, \ldots, r\kappa - 1\}$ there is an event $X \in \{X_{i_{j_0}}, X_{i_{j_1}}\}$ with $sig(X) = \mathsf{inp}$. On the other hand, since R respects d and $|\{k_0, k_1, a_{i_0}^{j_0}, \ldots, a_{i_{r\kappa-1}}^{j_{r\kappa-1}}\}| = r\kappa + 2$, there are at most κ events $X_{i_0}, \ldots, X_{i_{\kappa-1}} \in \mathfrak{U}$ whose signature is different from nop. Thus, $r\kappa$ sets are "covered" by at most κ elements. Since every element is a member of exactly r sets, this is only possible if $S = \{X \in \mathfrak{U} \mid sig(X) = \mathsf{inp}\} = \{X \in \mathfrak{U} \mid s \xrightarrow{X} s' \Rightarrow sup(s) \neq sup(s')\}$ defines an independent set of size κ.

Let $(\mathfrak{U}, M, \kappa)$ be a yes-instance of RIS. In the following we argue that α is solvable by a τ-region that respects the parameter. Let S be an independent set of size κ. Every element of $\mathfrak{U}$ occurs in exactly r sets. Thus, there are exactly

$r\kappa$ sets $M_{i_0}, \ldots, M_{i_{r\kappa-1}} \in M$ such that $S \cap M_{i_j} \neq \emptyset$ for all $j \in \{0, \ldots, r\kappa - 1\}$. We define $R = (sup, sig)$ as follows: $sup(\bot_{m-1,0}) = 1$; for all $e \in E(A_I^\tau)$, if $e \in \{k_0\} \cup S$, then $sig(e) = \mathsf{inp}$; if $e = k_1$, then $sig(k_1) = \mathsf{free}$; if $e = a_{i_j}^j$ and $j \in \{0, \ldots, r\kappa - 1\}$, then $sig(a_{i_j}^j) = \mathsf{inp}$; else $sig(e) = \mathsf{nop}$.

For all $s \in S(A_I^\tau) \setminus \{\bot_{m-1,0}\}$, there is a path $\bot_{m-1,0} = s_0 \xrightarrow{e_1} s_1 \ldots \xrightarrow{e_n} s_n = s$. By inductive defining $sup(s_{i+1}) = \delta_\tau(s_i, sig(e_{i+1}))$ for all $i \in \{0, \ldots, n-1\}$, we obtain sup. One easily verifies that (sup, sig) is a fitting region that solves α. $\qquad\square$

The Proof of Theorem 1.2 for $\tau \cap \{\mathsf{used}, \mathsf{free}\} = \emptyset$. Let $I = (\mathfrak{U}, M, \kappa)$ be an instance of OD, that is, $\mathfrak{U} = \{X_0, \ldots, X_{n-1}\}$, $M = \{M_0, \ldots, M_{m-1}\}$ and $M_i = \{X_{i_0}, \ldots, X_{i_{m_i-1}}\} \subseteq \mathfrak{U}$ for all $i \in \{0, \ldots, m-1\}$. Without loss of generality, we assume $i_0 < i_1 < \cdots < i_{m_i-2} < i_{m_i-1}$ for all $i \in \{0, \ldots, m-1\}$.

For a start, we define $d = 2\kappa + 2$. The TS A_I^τ has the following gadget H that applies the events k, z, o and w_m and provides the atom $\alpha = (k, h_2)$:

$$\mathsf{T}_m \xrightarrow{w_m} h_0 \xrightarrow{k} h_1 \xrightarrow{z} h_2 \xrightarrow{o} h_3 \xrightarrow{k} h_4$$

Next, we introduce A_I^τ's gadgets using the elements of $\mathfrak{U} = \{X_0, \ldots, X_{n-1}\}$ as events. Moreover, these gadgets use also the events of $\mathfrak{u} = \{x_0, \ldots, x_{n-1}\}$, and $\mathfrak{U}$ and $\mathfrak{u}$ are connected as follows: For every $i \in \{0, \ldots, n-1\}$, the event X_i is associated with the event x_i such that $s \xrightarrow{X_i} s'$ is an edge in A_I^τ if and only if $s \xleftarrow{x_i} s'$ is an edge in A_I^τ. In particular, for all $i \in \{0, \ldots, m-1\}$, the TS A_I^τ has for the set $M_i = \{X_{i_0}, \ldots, X_{i_{m_i-1}}\}$ the following gadget T_i that uses the elements of M_i (and their associated events of $\mathfrak{u}$) as events:

$$t_{i,0} \xrightarrow{k} t_{i,1} \underset{x_{i_0}}{\overset{X_{i_0}}{\rightleftarrows}} t_{i,2} \underset{x_{i_1}}{\overset{X_{i_1}}{\rightleftarrows}} t_{i,3} \cdots t_{i,m_i} \underset{x_{i_{m_i-1}}}{\overset{X_{i_{m_i-1}}}{\rightleftarrows}} t_{i,m_i+1} \xrightarrow{z} t_{i,m_i+2} \xrightarrow{k} t_{i,m_i+3}$$

We postpone the actual joining of $H, T_0, \ldots, T_{m-1}$ and argue first that a d-restricted τ-region $R = (sup, sig)$ solving α implies a searched odd set S.

Since R solves α and $\tau \cap \{\mathsf{free}, \mathsf{used}\} = \emptyset$, we have $sig(k) \in \{\mathsf{inp}, \mathsf{out}\}$. In what follows, we assume $sig(k) = \mathsf{inp}$ and argue that $S = \{X \in \mathfrak{U} \mid s \xrightarrow{X} s' \Rightarrow sup(s) \neq sup(s')\}$ defines a fitting odd set of size at most κ. By symmetry, the case $sig(k) = \mathsf{out}$ is similar.

Since R solves α and $sig(k) = \mathsf{inp}$, we have $sup(h_2) = 0$. Moreover, for all $s \in S(A_I^\tau)$, if $\xrightarrow{k} s$, then $sup(s) = 0$, and if $s \xrightarrow{k}$, then $sup(s) = 1$. By $h_1 \xrightarrow{z} h_2$ and $\xrightarrow{z} t_{0,m_0+2}$, this implies $sig(z) = \mathsf{nop}$; by $h_2 \xrightarrow{o} h_3$, this implies $sig(o) \neq \mathsf{nop}$. Let $i \in \{0, \ldots, m-1\}$ be arbitrary but fixed. By $sig(z) = \mathsf{nop}$ and $\xrightarrow{k} t_{i,1}$ and $t_{i,m_i+1} \xrightarrow{z} t_{i,m_i+2} \xrightarrow{k}$ we obtain $sup(t_{i,1}) = 0$ and $sup(t_{i,m_i+1}) = 1$. Thus, the path $sup(t_{i,1}) \xrightarrow{sig(X_{i_1})} \cdots \xrightarrow{sig(X_{i_{m_i-1}})} sup(t_{i,m_i+1})$ is a path from 0 to

1 in τ. In particular, the number of state changes between 0 and 1 on this path is odd. Consequently, since every $X \in M_i$ occurs once in T_i, the number $|\{X \in M_i | s \xrightarrow{X} s' \in T_i \text{ and } sup(s) \neq sup(s')\}|$ is odd. Since i was arbitrary, this is simultaneously true for all gadgets $T_0, \ldots, T_{m-1}$. In the following, we show that $|S \cap M_i|$ is odd for all $i \in \{0, \ldots, m-1\}$. To do so, we argue that for all $X \in S$ and $T_i \neq T_j$, $i, j \in \{0, \ldots, m-1\}$, with $s \xrightarrow{X} s' \in T_i$ and $q \xrightarrow{X} q' \in T_j$ the following is true: If $sup(s) \neq sup(s')$, then $sup(q) \neq sup(q')$. Intuitively, there is no X contributing to a state change in T_i but not in T_j. Since X always occurs with its associated event x, both $s \xleftarrow{x} s'$ and $q \xleftarrow{x} q'$ are present. Thus, if $sup(s) = 0$ and $sup(s') = 1$, then $sig(X) \in \{\text{out}, \text{set}, \text{swap}\}$ and $sig(x) \in \{\text{inp}, \text{res}, \text{swap}\}$. Clearly, if $sig(X) = \text{swap}$ or $sig(x) = \text{swap}$, then $sup(q) \neq sup(q')$. Otherwise, if $sig(X) \in \{\text{out}, \text{set}\}$ and $sig(x) \in \{\text{inp}, \text{res}\}$, then $q \xrightarrow{X} q'$ and $q \xleftarrow{x} q'$ imply $sup(q) = 0 \neq sup(q') = 1$. Similarly, if $sup(s) = 1$ and $sig(s') = 0$, then also $sup(q) \neq sup(q')$. Consequently, $|S \cap M_i|$ is odd for all $i \in \{0, \ldots, m-1\}$.

We argue that $|S| \leq \kappa$. Every $X \in S$ occurs always with its associated event $x \in \mathfrak{u}$: if $s \xrightarrow{X} s'$, then $s \xleftarrow{x} s'$. Moreover, $X \in S$ implies $sup(s) \neq sup(s')$ and, thus, $sig(X) \neq \text{nop}$ and $sig(x) \neq \text{nop}$. Recall that $sig(k), sig(o) \notin \{\text{nop}\}$. Consequently, if $|S| \geq \kappa + 1$, then $|\{e \in E(A_I^\tau) \mid sig(e) \neq \text{nop}\}| \geq 2\kappa + 4$, a contradiction. This proves $|S| \leq \kappa$. In particular, S defines a searched odd set.

In the following, we complete the construction of A_I^τ. In order to do that, for all $i \in \{0, \ldots, m-1\}$, we enhance T_i to a (path) gadget $G_i = T_i \rightsquigarrow T_i$ with starting state $\top_i$. This extension of T_i is necessary to ensure that if α is solvable by a τ-region that respects d, then all of A_I^τ's (E)SSP atoms are too. To finally obtain A_I^τ, we use fresh events $\ominus_1, \ldots, \ominus_m$ and thread $G_0, \ldots, G_{m-1}$ and H on a chain, that is, $\top_0 \xrightarrow{\ominus_1} \ldots \xrightarrow{\ominus_{m-1}} \top_{m-1} \xrightarrow{\ominus_m} \top_m$.

Let $j \in \{0, \ldots, m-1\}$ and $\ell \in \{0, \ldots, m_j\}$. We define the set $V_{j,\ell}$ as follows:

$$
V_{j,\ell} = \begin{cases} \{X_{j_0}\}, & \text{if } \ell = 0 \\ \{X_{j_{\ell-1}}, X_{j_\ell}\}, & \text{if } 1 \leq \ell \leq m_j - 1 \\ \{X_{j_{m_j-1}}\}, & \text{if } \ell = m_j \end{cases}
$$

Let $i \in \{0, \ldots, m-1\}$ and $j \in \{0, \ldots, i-1, i+1, \ldots, m-1\}$. The number $\sigma_{i,j}$ of elements of $V_j = \{V_{j,0}, \ldots, V_{j,m_j}\}$ that are subsets of M_i is defined by $\sigma_{i,j} = |\{V \in V_j \mid V \subseteq M_i\}|$. Let $\ell_0, \ldots, \ell_{\sigma_{i,j}-1} \in \{0, \ldots, m_j-1\}$ be the pairwise distinct indices (in increasing order) such that $V_{j,\ell_k} \subseteq M_i$ for all $k \in \{0, \ldots, \sigma_{i,j} - 1\}$. The gadget G_i implements events $u_{\ell_0}^{i,j}, v_{\ell_0}^j, \ldots, u_{\ell_{\sigma_{i,j}-1}}^{j,i}, v_{\ell_{\sigma_{i,j}-1}}^j$ consecutively on the following path $P_i^j =$

$$
\bot_{i,j,0} \xrightarrow{u_{\ell_0}^{j,i}} \bot_{i,j,1} \xrightarrow{v_{\ell_0}^j} \bot_{i,j,2} \xrightarrow{u_{\ell_1}^{j,i}} \bot_{i,j,3} \xrightarrow{v_{\ell_1}^j} \ldots \xrightarrow{u_{\ell_{\sigma_{i,j}-1}}^{j,i}} \bot_{i,j,2\sigma_{i,j}+1} \xrightarrow{v_{\ell_{\sigma_{i,j}-1}}^j} \bot_{i,j,2\sigma_{i,j}+2}
$$

Notice that the events $v_{\ell_0}^j, \ldots, v_{\ell_{\sigma_{i,j}-1}}^j$ might occur on different paths of A_I^τ, that is, P_i^j and $P_{i'}^j$ where $i \neq i'$. On the other hand, the events $u_{\ell_0}^{j,i}, \ldots, u_{\ell_{\sigma_{i,j}-1}}^{j,i}$ occur

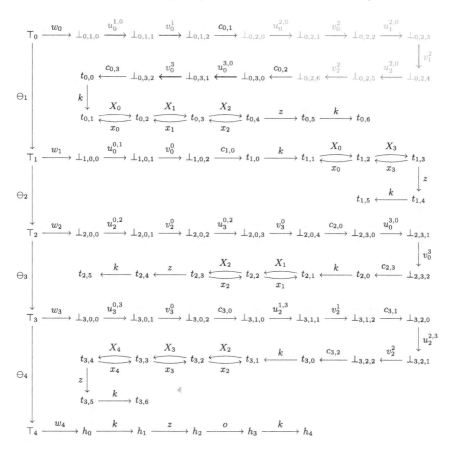

Fig. 3. The TS $A_{I_0}^\tau$ that origins from I_0, defined by Example 1. Top: P_0^1 (red), P_0^2 (olive), P_0^3 (blue). The red colored circles sketch the states mapped to 1 by the region R that bases on S_0, solves (k, h_2) and respects $d = 5$. (Color figure online)

exactly once in A_I^τ (on the path P_i^j). The gadget G_i is finally built as follows. If $\sigma_{i,j} = 0$ for all $j \in \{0, \ldots, i-1, i+1, \ldots, m-1\}$, then $G_i = \top_i \xrightarrow{w_i} \top_i$. That is, we extend T_i simply by the edge $\top_i \xrightarrow{w_i} t_{i,0}$. Otherwise, G_i is given by

$$G_i = \top_i \xrightarrow{w_i} P_{i,j_0} \xrightarrow{c_{i,j_0}} P_{i,j_1} \xrightarrow{c_{i,j_1}} \cdots \xrightarrow{c_{i,j_{\ell-1}}} P_{i,j_\ell} \xrightarrow{c_{i,j_\ell}} T_i$$

where $j_0, \ldots, j_\ell \in \{0, \ldots, i-1, i+1, \ldots, m-1\}$, $j_0 < \cdots < j_\ell$, are exactly the indices such that $\sigma_{i,j_k} > 0$ for all $k \in \{0, \ldots, \ell\}$. This finally results in A_I^τ, and it is easy to see that this is a parameterized (and even polynomial) reduction.

Example 1. Let $I_0 = (\mathfrak{U}, M, \kappa)$ be (the yes-instance) defined by $\mathfrak{U} = \{X_0, \ldots, X_4\}$, $M = \{M_0, \ldots, M_3\}$ with $M_0 = \{X_0, X_1, X_2\}$, $M_1 = \{X_0, X_3\}$, $M_2 = \{X_1, X_2\}$ and $M_3 = \{X_2, X_3, X_4\}$ and $\kappa = 3$. The set $S_0 = \{X_2, X_3, X_4\}$

is a fitting odd set of size 3. By definition, $V_{0,0} = \{X_0\}$, $V_{0,1} = \{X_0, X_1\}$, $V_{0,2} = \{X_1, X_2\}$ and $V_{0,3} = \{X_2\}$; $V_{1,0} = \{X_0\}$, $V_{1,1} = \{X_0, X_3\}$ and $V_{1,2} = \{X_3\}$; $V_{2,0} = \{X_1\}$, $V_{2,1} = \{X_1, X_2\}$ and $V_{2,2} = \{X_2\}$; $V_{3,0} = \{X_2\}$, $V_{3,1} = \{X_2, X_3\}$, $V_{3,2} = \{X_3, X_4\}$ and $V_{3,3} = \{X_4\}$.

For G_0, we have $V_{1,0} \subseteq M_0$, $V_{1,1} \not\subseteq M_0$ and $V_{1,2} \not\subseteq M_0$. Thus, $\sigma_{0,1} = 1$. By $V_{2,0} \subseteq M_0$, $V_{2,1} \subseteq M_0$ and $V_{2,2} \subseteq M_0$, we have $\sigma_{0,2} = 3$. Finally, only $V_{3,0}$ is a subset of (interest of) M_0, thus, $\sigma_{0,3} = 1$. The red, olive and blue colored paths of Fig. 3 show $P_{0,1}, P_{0,2}$ and $P_{0,3}$, respectively.

For G_1, the only set of interest is $V_{0,0} \subseteq M_1$, thus $\sigma_{1,0} = 0$ and $\sigma_{1,2} = \sigma_{1,3} = 0$. For G_2, we have $V_{0,2}, V_{0,3}, V_{3,0} \subseteq M_2$, thus, $\sigma_{2,0} = 2$, $\sigma_{2,1} = 0$ and $\sigma_{2,3} = 1$. For G_3, we observe $V_{0,3}, V_{1,2}, V_{2,2} \subseteq M_3$, thus, $\sigma_{3,0} = \sigma_{3,1} = \sigma_{3,2} = 1$. Figure 3 finally shows the joining of $G_0, \ldots, G_3$ and H into $A_{I_0}^\tau$.

So far, we have argued that if (A_I^τ, d) is a yes-instance, then $(\mathfrak{U}, M, \kappa)$ is too. In the following, we argue that if S is a fitting odd set S of $(\mathfrak{U}, M, \kappa)$, then α is solvable by a τ-region $R = (sup, sig)$ that respects d: $sup(\top_0) = 1$; for all $e \in E(A_I^\tau)$, if $e = k$, then $sig(k) = \mathsf{inp}$; if $e \in \{o\} \cup S \cup \{x \in \mathfrak{u} \mid X \in S\}$, then $sig(e) = \mathsf{swap}$; otherwise, $sig(e) = \mathsf{nop}$. By A_I^τ's reachability, one easily finds that this properly defines R. Figure 3 sketches R for the odd set $S = \{X_2, X_3, X_4\}$.   □

If τ is a type of Theorem 1.2 such that $\tau \cap \{\mathsf{used}, \mathsf{free}\} \neq \emptyset$, then the former reduction generally does not fit. For example, if $\tau = \{\mathsf{nop}, \mathsf{swap}, \mathsf{used}\}$, then a τ-solvable TS A satisfies that $s \xrightarrow{e} s' \in A$ implies $s' \xrightarrow{e}$. Since used is the only interaction of τ that ever allows τ-solvability of ESSP atoms, (e, s') would be unsolvable otherwise. Thus, for $\tau = \{\mathsf{nop}, \mathsf{swap}, \mathsf{used}\}$, the previous reduction yields always no-instances. However, if τ is a type of Theorem 1.2 such that $\tau \cap \{\mathsf{used}, \mathsf{free}\}$, then the reduction of the following proof fits for τ.

The Proof of Theorem 1.2 for $\tau \cap \{\mathsf{used}, \mathsf{free}\} \neq \emptyset$. For a start, we define $d = \kappa + 4$. The TS A_I^τ has the following gadgets H_0, H_1 with events k, z_0, z_1, o_0 and o_1 that provide the atom $\alpha = (k, h_{0,2})$:

$$H_0 = \top_m \xleftarrow{w_m} h_{0,0} \xleftarrow{k} h_{0,1} \xleftarrow{o_0} h_{0,2} \xleftarrow{o_1} h_{0,3} \xleftarrow{k} h_{0,4}$$

$$H_1 = \top_{m+1} \xleftarrow{w_{m+1}} h_{1,0} \xleftarrow{k} h_{1,1} \xleftarrow{z_0} h_{1,2} \xleftarrow{o_0} h_{1,3} \xleftarrow{z_1} h_{1,4} \xleftarrow{k} h_{1,5}$$

Moreover, for every set $M_i = \{X_{i_0}, \ldots, X_{m_i-1}\}$, $i \in \{0, \ldots, m-1\}$, the TS A_I^τ has the following gadget T_i that uses the elements of M_i as events:

$$t_{i,0} \xleftarrow{k} t_{i,1} \xleftarrow{z_0} t_{i,2} \xleftarrow{X_{i_0}} \cdots \xleftarrow{X_{i_{m_i-1}}} t_{i,m_i+2} \xleftarrow{z_1} t_{i,m_i+3} \xleftarrow{k} t_{i,m_i+4}$$

Moreover, we extend T_i to a gadget $G_i = \top_1 \rightsquigarrow T_i$ in exactly the same way like the previous reduction for $\tau \cap \{\mathsf{used}, \mathsf{free}\} = \emptyset$. Finally, for all $i \in \{0, \ldots, m-1\}$, we use the fresh events $\ominus_1, \ldots, \ominus_{m+1}$ and apply the edges $\top_0 \xrightarrow{\ominus_1}, \ldots \xrightarrow{\ominus_{m+1}} \top_{m+1}$.

Let $R = (sup, sig)$ be a τ-region that solves α and respects d. Let $e \in E(A_I^\tau)$ be arbitrary. Since $s\xrightarrow{e}s'$ implies $s'\xrightarrow{e}s$, $sig(e) \notin \{\mathsf{inp}, \mathsf{out}\}$ is true. In particular, since R solves α, $sig(k) \in \{\mathsf{used}, \mathsf{free}\}$. Moreover, if $sup(s) \neq sup(s')$, then $sig(e) = \mathsf{swap}$.

In what follows, we assume $sig(k) = \mathsf{used}$, which implies $sup(h_{0,0}) = 0$ and, thus, $sig(o_0) = sig(o_1) = \mathsf{swap}$. By symmetry, the case $sig(k) = \mathsf{free}$ is similar. By $sig(k) = \mathsf{used}$, $sup(h_{1,1}) \xrightarrow{sig(z_0)} \ldots \xrightarrow{sig(z_1)} sup(h_{1,4})$ is a path from 1 to 1 in τ. In particular, the number $|\{e \in \{z_0, z_1, o_0\} \mid sig(e) = \mathsf{swap}\}|$ is *even*. Since $sig(o_0) = \mathsf{swap}$, there is exactly one event $e \in \{z_0, z_1\}$ such that $sig(e) = \mathsf{swap}$. In the following, we assume $sig(z_0) = \mathsf{swap}$ implying $sig(z_1) \in \{\mathsf{nop}, \mathsf{used}\}$. By symmetry, the case $sig(z_1) = \mathsf{swap}$ is similar. Since R respects d, there are at most κ events left whose signature is different from nop. Let $i \in \{0, \ldots, m-1\}$ be arbitrary but fixed. By $sig(k) = \mathsf{inp}$, $sig(z_0) = \mathsf{swap}$ and $sig(z_1) \in \{\mathsf{nop}, \mathsf{used}\}$, the path $sup(t_{i,2}) \xrightarrow{sig(X_{i_0})} \ldots \xrightarrow{sig(X_{i_{m_i-1}})} sup(t_{i,m_i+2})$ is a path from 0 to 1 in τ. Similar to the case $\tau \cap \{\mathsf{used}, \mathsf{free}\} = \emptyset$, the set $S = \{X \in \mathfrak{U} \mid s\xrightarrow{X}s' \Rightarrow sup(s) \neq sup(s')\}$ of elements of $\mathfrak{U}$ mapped to swap implies a searched odd set of M.

For the reverse direction, let $S \subseteq \mathfrak{U}$ be an odd size of size at most κ of M. We obtain a τ-region $R = (sup, sig)$ that solves $(k, h_{0,2})$ an respects d as follows: For a start, we let $sup(\top_0) = 1$. Moreover, for all $e \in E(A_I^\tau)$, if $e = k$, then $sig(e) = \mathsf{used}$; if $e \in \{o_0, o_1, z_0\} \cup S$, then $sig(e) = \mathsf{swap}$; otherwise $sig(e) = \mathsf{nop}$. This implicitly defines a fitting region that solves α. $\square$

4   Conclusion

In this paper, we investigate the parameterized complexity of $d\mathrm{R}\tau\mathrm{S}$ parameterized by d and show $W[1]$-completeness for a range of Boolean types. As a result, d is ruled out for fpt-approaches for the considered types of nets. As future work, one may investigate the parameterized complexity of $d\mathrm{R}\tau\mathrm{S}$ for other boolean types [5]. Moreover, one may look for other more promising parameters: If $N = (P, T, M_0, f)$ is a Boolean net, $p \in P$ and if the *occupation number* o_p of p is defined by $o_p = |\{M \in RS(N) \mid M(p) = 1\}|$ then the *occupation number* o_N of N is defined by $o_N = \max\{o_p \mid p \in P\}$. If $\mathcal{R}$ is a τ-admissible set (of a TS A) and $R \in \mathcal{R}$, then the support of R determines the number of markings of $N_A^{\mathcal{R}}$ that occupy R, that, is, $o_R = |\{s \in S(A) \mid sup(s) = 1\}|$. Thus, searching for a τ-net where $o_N \leq n$, $n \in \mathbb{N}$, corresponds to searching for a τ-admissible set $\mathcal{R}$ such that $|\{s \in S(A) \mid sup(s) = 1\}| \leq n$ for all $R \in \mathcal{R}$. As a result, for each (E)SSP atom α there are at most $\mathcal{O}(\binom{|S|}{o_N})$ fitting supports for τ-regions solving α. Thus, the corresponding problem o_N-*restricted τ-synthesis* parameterized by o_N is in XP if, in a certain sense, τ-regions are fully determined by a given support sup.

References

1. Badouel, E., Bernardinello, L., Darondeau, P.: The synthesis problem for elementary net systems is NP-complete. Theor. Comput. Sci. **186**(1–2), 107–134 (1997). https://doi.org/10.1016/S0304-3975(96)00219-8
2. Badouel, E., Bernardinello, L., Darondeau, P.: Petri Net Synthesis. TTCSAES. Springer, Heidelberg (2015). https://doi.org/10.1007/978-3-662-47967-4
3. Cygan, M., et al.: Parameterized Algorithms. Springer, Cham (2015). https://doi.org/10.1007/978-3-319-21275-3
4. Schmitt, V.: Flip-flop nets. In: Puech, C., Reischuk, R. (eds.) STACS 1996. LNCS, vol. 1046, pp. 515–528. Springer, Heidelberg (1996). https://doi.org/10.1007/3-540-60922-9_42
5. Tredup, R.: The complexity of synthesizing NOP-equipped Boolean nets from G-bounded inputs (technical report), submitted for topnoc 2020 (2019)
6. Tredup, R., Rosenke, C.: The complexity of synthesis for 43 Boolean Petri net types. In: Gopal, T.V., Watada, J. (eds.) TAMC 2019. LNCS, vol. 11436, pp. 615–634. Springer, Cham (2019). https://doi.org/10.1007/978-3-030-14812-6_38

Approximate #Knapsack Computations to Count Semi-fair Allocations

Theofilos Triommatis[1(✉)] and Aris Pagourtzis[2]

[1] School of Electrical Engineering, Electronics and Computer Science,
University of Liverpool, Liverpool L69-3BX, UK
`Theofilos.Triommatis@liverpool.ac.uk`
[2] School of Electrical and Computer Engineering, National Technical
University of Athens, Polytechnioupoli, 15780 Zografou, Athens, Greece
`pagour@cs.ntua.gr`

Abstract. In this paper, we study the problem of counting the number of different knapsack solutions with a prescribed cardinality. We present an FPTAS for this problem, based on dynamic programming. We also introduce two different types of semi-fair allocations of indivisible goods between two players. By semi-fair allocations, we mean allocations that ensure that at least one of the two players will be free of envy. We study the problem of counting such allocations and we provide FPTASs for both types, by employing our FPTAS for the prescribed cardinality knapsack problem.

Keywords: Knapsack problems · Counting problems · Fptas · Fair allocations · Envy-freeness

1 Introduction

We define and study three counting problems. The first of them concerns knapsack solutions with a prescribed number of items allowed in the knapsack, while the other two concern two new notions of allocations of indivisible goods among two players. We show that both our allocation notions imply a *semi-fairness* property, namely that at least one of the two players is envy-free. From a computational point of view both types of allocations are shown to be easy to satisfy, however the corresponding counting problems seem to be hard. We provide fully polynomial-time approximation schemes for all three problems that we study. Along the way we compare our new notions of allocations to the standard notion of envy-freeness (EF) [5] and show that while one of them is incomparable to EF, the other one includes all EF allocations. Note that the problem of approximate

Theofilos Triommatis was supported in part for this work by EPSRC grant EP/S023445/1 EPSRC Centre for Doctoral Training in Distributed Algorithms: the what, how and where of next-generation data science, https://gow.epsrc.ukri.org/NGBOViewGrant.aspx?GrantRef=EP/S023445/1.

© Springer Nature Switzerland AG 2020
J. Chen et al. (Eds.): TAMC 2020, LNCS 12337, pp. 239–250, 2020.
https://doi.org/10.1007/978-3-030-59267-7_21

counting allocations, apart from its own interest, may serve as a basis for solving problems under *uncertainty* [2].

In the counting version of a decision problem that asks for the existence of a solution to the given instance we are interested in counting the number of solutions to the instance. The complexity class that characterizes counting problems with polynomial-time verifiable solutions is the well-known class #P [12] and it is known that it contains some hard counting problems. It is also known that counting problems that have NP-complete existence versions are not approximable unless NP = RP [4]. In contrast, the class #PE [6,9], consisting of counting problems in #P that have an easy existence version, contains many approximable counting problems. Moreover, several well-known approximable counting problems belong to a subclass of #PE, called TotP [9]; in [9] it is proven that TotP is the class that contains all the functions of #PE that are self-reducible. Such a problem is #Knapsack, which admits an FPTAS [10]. Here we show, among others, that our counting problems share the property of having easy existence version, thus providing the first evidence that they admit an FPTAS.

The first problem that we study is #EXACT M-ITEMS KNAPSACK, a problem the optimization version of which has recently been studied [7]. We will first present an FPTAS for this problem, by extending the FPTAS of Stefankovic et al. [10], and then use it to obtain FPTASs for the allocation counting problems that we define in this paper. This connection could be of further interest as only few variants of KNAPSACK have been associated to allocation problems; such an example is the NON-LINEAR FRACTIONAL EQUALITY KNAPSACK [13].

Due to space limitations, some proofs will be omitted; they can be found in the full version of the paper [11].

2   The #Exact M-Items KnapsackProblem

In this section, we define #EXACT M-ITEMS KNAPSACKand provide an FPTAS for it. Our algorithm uses dynamic programming and builds on techniques developed in [3] and [10]. More specifically we adapt the technique of [10] by defining an extra parameter that takes into account the number of knapsack items, restricting the set of feasible solutions accordingly.

We will first define the decision version of #EXACT M-ITEMS KNAP-SACKwhich is very similar to the standard KNAPSACK problem with the additional restriction that a specific number of objects should be put in the knapsack. Note that we ignore objects' values as we are interested in all feasible solutions, that is, solutions in which the sum of weights does not exceed the capacity of the knapsack.

Definition 2.1 (Exact M-Items Knapsack). *Given the weights $\{w_1, \dots, w_n\}$ of n objects, an integer $M \in \{1, \dots, n\}$ and a capacity C, is there a subset S of $\{1, \dots, n\}$ such that*

$$\sum_{i \in S} w_i \leq C \text{ and } |S| = M \tag{1}$$

To describe the set of feasible solutions of EXACT M-ITEMS KNAPSACK in any sub-problem we examine, we will use a function $f : \{1, \ldots, n\} \times \{1, \ldots, M\} \times \mathbb{R}^+ \to \mathscr{P}(\mathscr{P}(S))$ with

$$f(i, m, c) = \left\{ S \subseteq \{1, \ldots, i\} : \sum_{j \in S} w_j \leq c \text{ and } |S| = m \right\} \tag{2}$$

where $\mathscr{P}(A)$ denotes the power set of A.

Thus, $f(i, m, c)$ is the set of feasible knapsack solutions that use only the first i objects and have exactly m objects in the knapsack and total weight at most c. Clearly, the set of solutions to the EXACT M-ITEMS KNAPSACK problem is given by $f(n, M, C)$.

Let us now define the counting version of EXACT M-ITEMS KNAPSACK.

Definition 2.2 (#Exact M-Items Knapsack). *Given the weights $\{w_1, \ldots, w_n\}$ of n objects, an integer $M \in \{1, \ldots, n\}$ and a capacity C, count how many subsets S of $\{1, \ldots, n\}$ are there such that*

$$\sum_{i \in S} w_i \leq C \text{ and } |S| = M \tag{3}$$

Remark 2.1. Note that the solution to an instance of #EXACT M-ITEMS KNAPSACK is the cardinality of $f(n, M, C)$, i.e. $|f(n, M, C)|$.

Remark 2.2. If the values n and M are fixed and $c, c' \in \mathbb{R}^+$ with $c \leq c'$ then

$$|f(n, M, c)| \leq |f(n, M, c')|$$

This means that f is monotone w.r.t. the capacity.

The #EXACT M-ITEMS KNAPSACKproblem is #P-hard, since #KNAPSACK can be easily reduced to it. We therefore aim at approximating it. Following ideas of Stefankovic *et al.* [10] we will define a function τ in order to approximate the solution of #EXACT M-ITEMS KNAPSACK.

Definition 2.3. *We define $\tau : \{0, \ldots, M\} \times \{0, \ldots, n\} \times \mathbb{R}^+ \longrightarrow \overline{\mathbb{R}}$ with*

$$\tau(m, i, a) = \begin{cases} +\infty & \text{if } a = 0 \text{ or } m > i, \\ \min \{c \in \mathbb{R} : |f(i, m, c)| \geq a\} & \text{if } a \leq \binom{i}{m} \text{ and } m \leq i, \\ +\infty & \text{otherwise} \end{cases} \tag{4}$$

Remark 2.3. Note that we consider as a feasible solution the one that leaves the knapsack empty, hence $\tau(0, i, 1) = 0$.

So $\tau(m, i, a)$ represents the minimum capacity such that the number of solutions of EXACT M-ITEMS KNAPSACKwith exactly m items from $\{1, \ldots, i\}$ is at least a.

We also note that a should be a non negative integer, more precisely $a \in \{0, 1, \ldots, 2^n\}$, but instead in the above definition we let $a \in \mathbb{R}^+$. This happens because we will approximate the number of solutions of EXACT M-ITEMS KNAPSACK.

Notice that with the help of function τ we can redefine the solution to an instance of #EXACT M-ITEMS KNAPSACKas follows:

$$|f(n, M, C)| = \max\{a \in \{0, 1, \ldots, 2^n\} : \tau(M, n, a) \leq C\}$$

Remark 2.4. From Remark 2.2 and the definition of τ it is easy to see that for fixed $0 \leq i \leq n$, $0 \leq m \leq i$ and $a \leq a'$ we have that

$$\tau(m, i, a) \leq \tau(m, i, a')$$

This means that τ is non-decreasing w.r.t. a.

Lemma 2.1. *For every $i \in \{1, \ldots, n\}$, $m \in \{1, \ldots, M\}$ and $a \in \mathbb{R}^+$, τ satisfies the following recursion*

$$\tau(m, i, a) = \min_{k \in [0,1]} \max \begin{cases} \tau(m-1, i-1, ka) + w_i \\ \tau(m, i-1, (1-k)a) \end{cases} \tag{5}$$

Note that in the i-th step of the recursion, there are $(1-k)a$ solutions that do not contain w_i and ka solutions that contain it. Furthermore we can calculate the minimum in each step if we consider every

$$k = \frac{r}{a}, \text{ where } r \in \mathbb{Z} \text{ and } 0 \leq r \leq a$$

By Definition 2.3 the domain of τ is $Dom(\tau) = \{1, \ldots, M\} \times \{1, \ldots, n\} \times \mathbb{R}^+$. In order to compute the exact minimum in every step of the recursion we would have to check every possible value of r, $0 \leq r \leq a \leq \binom{i}{m}$, thus needing in the end $\mathcal{O}(2^n)$ evaluations. We can approximate the minimum efficiently by restricting τ in Ω where

$$\Omega = \{1, \ldots, M\} \times \{1, \ldots, n\} \times \left\{0, 1, \ldots, \lceil n \log_{Q(\varepsilon)} 2 \rceil\right\} \text{ and } Q(\varepsilon) = 1 + \frac{\varepsilon}{n+1}$$

Let $s = \lceil n \log_Q 2 \rceil$ and $T = \tau|_\Omega$, the restriction of τ in Ω. As T is a restriction of τ it must satisfy recursion 5, and in particular:

$$T(m, i, j) = \min_{k \in [0,1]} \max \begin{cases} T\left(m-1, i-1, \lfloor j + \log_Q k \rfloor\right) + w_i \\ T\left(m, i-1, \lfloor j + \log_Q(1-k) \rfloor\right) \end{cases} \tag{6}$$

Now with the following algorithm we can compute T efficiently and as a result we get an approximation of its optimal solution.

Algorithm 1. Count EXACT M-ITEMS KNAPSACK

Require: Integers $w_1, \ldots, w_n, C, M$ and $\varepsilon > 0$
Ensure: $(1 + \varepsilon)$-approximation of #EXACT M-ITEMS KNAPSACK
1: Set $T[0, i, 1] = 0$ for $i \geq 0$ and $T[0, i, 0] = \infty$ for $i \geq 0$
2: Set $T[1, i, 0] = \infty$ for $i \geq 0$ and $T[1, 0, j] = \infty$ for $j \geq 0$
3: Set $T[0, i, j] = \infty$ for $i, j \geq 0$
4: Set $Q = 1 + \frac{\varepsilon}{n+1}$
5: **for** $m = 1$ to M **do**
6:     **for** $i = 1$ to n **do**
7:         **for** $j = 1$ to s **do**
8:             **if** $\left(m > i \text{ or } j > \binom{i}{m}\right)$ **then**
9:                 $T[m, i, j] = \infty$
10:            **else**
11:                $T[m, i, j] = \min_{k \in [0,1]} \max \begin{cases} T\left[m - 1, i - 1, \lfloor j + \log_Q k \rfloor\right] + w_i \\ T\left[m, i - 1, \lfloor j + \log_Q(1 - k) \rfloor\right] \end{cases}$
12: Set $j' = \max\{j : T[M, n, j] \leq C\}$
13: **Return:** $Z' = Q^{j'+1}$

Now we will prove that T approximates τ in the following manner

Lemma 2.2. *Let $i \geq 1$, $0 \leq m \leq i$. Assume that for every $j \in \{0, \ldots, s\}$, $T[m, i - 1, j]$ satisfies*

$$\tau\left(m, i - 1, Q^{j-i+1}\right) \leq T[m, i - 1, j] \leq \tau\left(m, i - 1, Q^j\right)$$

Then for all $j \in \{0, \ldots, s\}$ we have that $T[m, i, j]$ computed using 6 satisfies:

$$\tau\left(m, i, Q^{j-i}\right) \leq T[m, i, j] \leq \tau\left(m, i, Q^j\right)$$

Now we are ready to prove that the output Z' of Algorithm 1 is a $(1 + \varepsilon)$ approximation of the solution of #EXACT M-ITEMS KNAPSACK.

Theorem 2.1. *Let Z be the solution of #EXACT M-ITEMS KNAPSACK problem on an instance with n items. Then for every $\varepsilon \in (0, 1)$, Algorithm 1 outputs Z' such that*

$$(1 - \varepsilon)Z \leq Z' \leq (1 + \varepsilon)Z, \quad \text{and the algorithm runs in time} \quad \mathcal{O}\left(\frac{n^4}{\varepsilon} \log \frac{n}{\varepsilon}\right)$$

Proof. By Lemma 2.2 we have for $j' = \max\{j : T[M, n, j] \leq C\}$ that the approximation Z' does not underestimates Z because

$$C \leq T[M, n, j' + 1] \leq \tau\left(M, n, Q^{j'+1}\right)$$

Moreover we have at least $Q^{(j'-n)}$ solutions of EXACT M-ITEMS KNAPSACK because

$$\tau\left(M, n, Q^{(j'-n)}\right) \leq T[M, n, j'] \leq C$$

hence,

$$\frac{Z'}{Z} \leq \frac{Q^{j'+1}}{Q^{j'-n}} = Q^{n+1} = \left(1 + \frac{\varepsilon}{n+1}\right)^{n+1} \leq e^{\varepsilon}$$

This proves that the output of the algorithm satisfies the statement of the theorem. All that is left to prove is the running time.

The algorithm fills up a $(n \times m \times s)$ matrix with $m = \mathcal{O}(n)$. Also we have discussed above that in order to compute the minimum in recursion (6) we must search all the values of a finite and discrete set S. More particular for every $j \in \{0, 1, \ldots, s\}$, we have that $S = S_1 \cup S_2$ where $S_1 = \{Q^{-j}, \ldots, Q^0\}$ and $S_2 = \{1 - Q^0, \ldots, 1 - Q^{-j}\}$. So it will take time $\mathcal{O}(s)$ to calculate the $T[m, i, j]$ cell of the matrix. Therefore it will take time $\mathcal{O}(nms^2)$ to fill up the matrix.

Moreover we have that $s = \lceil n \log_Q 2 \rceil = \mathcal{O}\left(\frac{n^2}{\varepsilon}\right)$. So if the algorithm searches all the values of S in each step in order to compute the minimum of recursion (6) it will take time $\mathcal{O}\left(\frac{n^6}{\varepsilon^2}\right)$.

But from Remark 2.4, we know that τ is increasing, so as $k \in [0, 1]$ increases, $T[m-1, i-1, \lfloor j + \log_Q k \rfloor] + w_i$ increases and $T[m, i-1, \lfloor j + \log_Q (1-k)\rfloor]$ decreases.

Now the minimum of the maximum, will be achieved for $k \in [0, 1]$ with the following property: Either $k \in \{0, 1\}$ or for every $k' < k$ we have

$$T[m, i-1, \lfloor j + \log_Q (1-k')\rfloor] < T[m-1, i-1, \lfloor j + \log_Q k'\rfloor] + w_i$$

and for every $k' > k$

$$T[m, i-1, \lfloor j + \log_Q (1-k')\rfloor] \geq T[m-1, i-1, \lfloor j + \log_Q k'\rfloor] + w_i$$

Unfortunately we can't have S sorted, but we can compute S_1 and S_2 in such a way that their elements will already be in order. If we apply binary search to S_1 then we can find $k_1 \in [0, 1]$ that satisfies the property to be the minimum of the maximum of T. Accordingly by applying binary search to S_2 we will find $k_2 \in [0, 1]$ that satisfies the above property. So with this technique it takes time $\mathcal{O}(\log s)$ to compute $T[m, i, j]$ and finally the running time of the algorithm is $\mathcal{O}(nms \log s) = \mathcal{O}\left(\frac{n^4}{\varepsilon} \log \frac{n}{\varepsilon}\right)$, concluding the proof. $\qquad\square$

3   Allocations Where Players Value Their Bundle More Than Others Do

In this section we will define the problem of allocating n goods between two players A and B in such a way that each player values its bundle more than the other player. We will assume that the i-th goods has value a_i for A and b_i for B. More formally we have

Definition 3.1 (Larger-than-swap-Player-Valuation (LPV) allocation).
Given two sets $A = \{a_i \in \mathbb{Z}^+ : 1 \leq i \leq n\}$ *and* $B = \{b_i \in \mathbb{Z}^+ :$

$1 \le i \le n\}$, where $n \in \mathbb{N}$, the goal is to find a partition of $S = \{1, \ldots, n\}$ into two sets S_A and S_B, such that

$$\sum_{S_A} a_i \ge \sum_{S_A} b_i \text{ and } \sum_{S_B} b_i \ge \sum_{S_B} a_i \qquad (7)$$

In other words, the LPV allocation is a pair of bundles (S_A, S_B) such that bundle S_A is more valuable to A than to B and bundle S_B is more valuable to B than to A.

Usually in this type of problems we are interested in fair solutions, but the interesting part is that there are many definitions of fairness. The most common notion of a fair solution is that the players should be envy free as was introduced in [5] and as the name suggests the goal is, A not to envy the bundle of B and vice versa.

Definition 3.2 (Envy-Free (EF) allocation). *An allocation (S_A, S_B) of n goods among two players A and B, where the i-th good has value a_i for A and b_i for B, is called* Envy Free *if*

$$\sum_{S_A} a_i \ge \sum_{S_B} a_i \text{ and } \sum_{S_B} b_i \ge \sum_{S_A} b_i \qquad (8)$$

Definition 3.3 (semi-Envy-Free (sEF) allocation). *For an allocation (S_A, S_B) of n goods among two players A and B wi will say that A **doesn't envy** B **or** A **is free of envy**, and we will denote it with sEF(A), if*

$$\sum_{S_A} a_i \ge \sum_{S_B} a_i \qquad (9)$$

Lemma 3.1. *In an LPV allocation at least one of the two players is free of envy.*

Proof. There are two possible cases either

$$\sum_{S_A} b_i \ge \sum_{S_B} b_i \text{ or } \sum_{S_B} b_i \ge \sum_{S_A} b_i$$

If $\sum_{S_B} b_i \ge \sum_{S_A} b_i$ then, by definition, B is free of envy. If $\sum_{S_A} b_i \ge \sum_{S_B} b_i$ then considering the property of LPV allocation (7) we have

$$\sum_{S_A} a_i \ge \sum_{S_A} b_i \ge \sum_{S_B} b_i \ge \sum_{S_B} a_i \implies \sum_{S_A} a_i \ge \sum_{S_B} a_i$$

hence player A is free of envy. $\qquad \square$

Remark 3.1. It is easy to find an LPV allocation. We can look at all values a_i and b_i, if $a_i \ge b_i$ then $i \in S_A$ else $i \in S_B$. Note that there is only one LPV allocation if $a_i > b_i$ for every $1 \le i \le n$, namely $S_A = S, S_B = \emptyset$ (and similarly if $b_i > a_i$ for every $1 \le i \le n$). This means that the problem of counting LPV allocations belongs to #PE as mentioned earlier (cf. [6,9]).

Definition 3.4 (#LPV Allocations Problem). *Given two sets* $A = \{a_i \in \mathbb{Z}^+ : 1 \leq i \leq n\}$ *and* $B = \{b_i \in \mathbb{Z}^+ : 1 \leq i \leq n\}$, *where* $n \in \mathbb{N}$, *find the number of partitions of* $S = \{1, \ldots, n\}$ *into two sets* S_A *and* S_B, *such that*

$$\sum_{S_A} a_i \geq \sum_{S_A} b_i \text{ and } \sum_{S_B} b_i \geq \sum_{S_B} a_i \tag{10}$$

We will now give a reduction of #LPV Allocations problem to #Exact M-Items Knapsack. This will lead to an FPTAS for the former.

Lemma 3.2. *The solution* Y *of #LPV Allocations on input* $A = \{a_1, \ldots, a_n\}$ *and* $B = \{b_1, \ldots, b_n\}$, *coincides with the sum of solutions* Z_m *of #Exact M-Items Knapsack on input* $w_1, \ldots, w_n$, *where* $w_i = a_i - b_i + b$, *and* $b = \max_{1 \leq i \leq n} b_i$, *capacity* $C = mb$ *and exactly* m *items in the knapsack, for* $m \in \{1, \ldots, n-1\}$ *(assuming w.l.o.g.* $\sum_{i=1}^{n} a_i \geq \sum_{i=1}^{n} b_i$*):*

$$Y = \sum_{m=1}^{n-1} Z_m$$

So an FPTAS algorithm for #LPV Allocations is the following:

Algorithm 2. Count LPV Allocations

Require: Integers $a_1, \ldots, a_n, b_1, \ldots, b_n$ and $\varepsilon > 0$
Ensure: $(1 + \varepsilon)$-approximation of #LPV Allocations
1: $S_a = a_1 + \cdots + a_n$
2: $S_b = b_1 + \cdots + b_n$
3: $Y = 0$
4: **if** $S_a \geq S_b$ **then**
5:     $b = \max(b_1, \ldots, b_n)$
6:     **for** i=1 to n **do**
7:         $w_i = a_i - b_i + b$
8:     **for** $m = 1$ to $n - 1$ **do**
9:         $Y = Y +$ Count Exact M-Items Knapsack$(w_1, \ldots, w_n, mb, m, \varepsilon)$
10: **else**
11:     $b = \max(a_1, \ldots, a_n)$
12:     **for** $i=1$ to n **do**
13:         $w_i = b_i - a_i + b$
14:     **for** $m = 1$ to $n - 1$ **do**
15:         $Y = Y +$ Count Exact M-Items Knapsack$(w_1, \ldots, w_n, mb, m, \varepsilon)$
16: **Return:** Y

Theorem 3.1. *Let* Y *be the solution of #LPV Allocations problem. Then for every* $\varepsilon \in (0, 1)$, *Algorithm 2 outputs* Y' *such that*

$$(1 - \varepsilon)Y \leq Y' \leq (1 + \varepsilon)Y, \quad \text{and the algorithm runs in time } \mathcal{O}\left(\frac{n^5}{\varepsilon} \log \frac{n}{\varepsilon}\right)$$

Proof. Let Z'_m be the output of Algorithm 1 for the #EXACT M-ITEMS KNAP-SACKproblem with m items in the knapsack, capacity $C(m)$ (depends on m) and weights $w_1, w_2, \ldots, w_n$ and Z_m be its exact solution. From Lemma 3.2 we have that Algorithm 2 outputs

$$Y' = \sum_{m=1}^{n-1} Z'_m \leq \sum_{m=1}^{n-1} (1+\varepsilon)Z_m = (1+\varepsilon) \sum_{m=1}^{n-1} Z_m = (1+\varepsilon)Y$$

Accordingly

$$Y' = \sum_{m=1}^{n-1} Z'_m \geq \sum_{m=1}^{n-1} (1-\varepsilon)Z_m = (1-\varepsilon) \sum_{m=1}^{n-1} Z_m = (1-\varepsilon)Y$$

As far as running time is concerned, Algorithm 2 consists of simple operations that take time $\mathcal{O}(n)$ and then executes $(n-1)$ times the algorithm for #EXACT M-ITEMS KNAPSACK. Since Algorithm 1 runs in time $\mathcal{O}\left(\frac{n^4}{\varepsilon} \log \frac{n}{\varepsilon}\right)$, we obtain the claimed running time. $\qquad\square$

4   LTV Allocations

In the previous section we have studied the LPV ALLOCATIONS and we proved that in a solution of the LPV ALLOCATIONS, at least one of the two players will be free of envy. However, the converse is not true, as there exist instances such as $A = \{8, 4, 6, 5\}$ and $B = \{5, 8, 7, 7\}$: if A picks items 1 and 2 and B picks items 3 and 4, i.e. $S_A = \{1, 2\}$ and $S_B = \{3, 4\}$, it is easy to confirm that the couple (S_A, S_B) is an envy free solution but it doesn't satisfy Definition 3.1. The Venn diagram in Fig. 1 visualizes the situation.

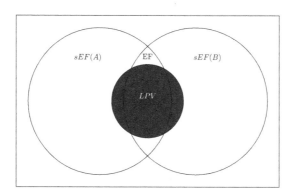

Fig. 1. Relation of LPV to (semi-)envy-free allocations. EF denotes the set of Envy-Free allocations, sEF(A) denotes the set of semi-Envy-Free allocations of A (accordingly for B) and LPV is the set of LPV allocations.

Observe that by counting LPV allocations we may miss several EF or sEF allocations. In order to capture more EF and sEF allocations, we will now define a second notion of allocations.

Definition 4.1 (Larger-than-swap-Total-Valuation (LTV) allocations).
Given two sets $A = \{a_i \in \mathbb{Z}^+ : 1 \leq i \leq n\}$ *and* $B = \{b_i \in \mathbb{Z}^+ : 1 \leq i \leq n\}$ *where* $n \in \mathbb{N}$, *the goal is to find a partition of* $S = \{1, \ldots, n\}$ *into two sets* S_A *and* S_B, *such that* $S_A \cap S_B = \emptyset$ *and* $S_A \cup S_B = S$ *with the following property*

$$\sum_{S_A} (a_i - b_i) \geq \sum_{S_B} (a_i - b_i) \tag{11}$$

Proposition 4.1. *LTV allocations contain all EF and all LPV allocations.*

We can now update the Venn diagram of the allocations to include the LTV allocations, giving a much clearer view of the inclusion relation between the allocations (Fig. 2).

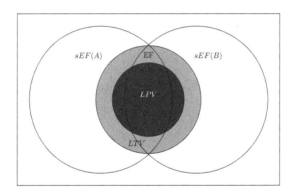

Fig. 2. Relations between LTV, LPV and (semi) Envy Free Allocations. *LTV* denotes the set of LTV allocations.

Remark 4.1. The LTV ALLOCATIONS problem also has some easy-to-find solutions, e.g. the solution that assigns to A all objects that A values more than B and vice versa.

We will now define the corresponding counting problem and study its complexity.

Definition 4.2 (#LTV Allocations). *Given two sets* $A = \{a_i \in \mathbb{Z}^+ : 1 \leq i \leq n\}$ *and* $B = \{b_i \in \mathbb{Z}^+ : 1 \leq i \leq n\}$, *where* $n \in \mathbb{N}$, *the goal is to find how many partitions of* $S = \{1, \ldots, n\}$ *into two sets* (S_A, S_B) *are there, such that* $S_A \cap S_B = \emptyset$ *and* $S_A \cup S_B = S$, *satisfying the following property*

$$\sum_{S_A} (a_i - b_i) \geq \sum_{S_B} (a_i - b_i) \tag{12}$$

Algorithm 3. Count LTV ALLOCATIONS

Require: Integers $a_1, \ldots, a_n, b_1, \ldots, b_n$ and $\varepsilon > 0$
Ensure: $(1 + \varepsilon)$-approximation of #LTV ALLOCATIONS
 1: **for** i=1 to n **do**
 2:     $d_i = a_i - b_i$
 3: $Sum = (d_1 + \ldots + d_n)/2$
 4: Set $b = 1 - \min(d_1, \ldots, d_n)$
 5: **for** i=1 to n **do**
 6:     Set $w_i = a_i - b_i + b$
 7: **for** $m = 1$ to $n - 1$ **do**
 8:     $Y = Y +$ Count EXACT M-ITEMS KNAPSACK$(w_1, \ldots, w_n, Sum + mb, m, \varepsilon)$
 9: **return** Y

By Remark 4.1 it is not unlikely that #LTV ALLOCATIONS be approximable, as it belongs to #PE (cf. [6,9]). Indeed, using similar arguments to those in the proof of Lemma 3.2 we can prove that Algorithm 3 is an FPTAS algorithm for #LTV ALLOCATIONS.

Lemma 4.1. *The solution* Y *of #LTV* ALLOCATIONS *on input* $A = \{a_1, \ldots, a_n\}$ *and* $B = \{b_1, \ldots, b_n\}$, *coincides with the sum of solutions* Z_m *of #EXACT* M-ITEMS KNAPSACK *on input* $w_1, \ldots, w_n$, *capacity* $((\sum_{i=1}^n w_i)/2 + m\beta)$, $\beta = \min\{a_i - b_i : 1 \le i \le n\}$ *and* $w_i = a_i - b_i - \beta + 1$, *and exactly* m *items in the knapsack, for* $m \in \{1, \ldots, n-1\}$, *that is,*

$$Y = \sum_{m=1}^{n-1} Z_m \tag{13}$$

Theorem 4.1. *Let* Y *be the exact solution of #LTV* ALLOCATIONS *problem on some input. Then for every* $\varepsilon \in (0,1)$, *Algorithm 3 outputs* Y' *such that*

$$(1 - \varepsilon)Y \le Y' \le (1 + \varepsilon)Y, \quad \text{and runs in time} \quad \mathcal{O}\left(\frac{n^5}{\varepsilon} \log \frac{n}{\varepsilon}\right)$$

The proof is similar to the proof of Theorem 3.1.

5   Discussion

We presented an FPTAS for the problem of counting feasible knapsack solutions with a specific (given) number of items; to the best of our knowledge no FTPAS has been proposed for this problem so far, despite its evident importance. We built on Dyer's dynamic programming algorithm [3]. An interesting future work would be to improve the complexity of the FPTAS by exploring dimension reduction techniques (see e.g. [8]).

We also defined two new notions of allocations of indivisible goods and provided FPTASs for the counting problems associated with them by employing the above mentioned FPTAS. We leave as an open question whether our results can be extended to more than two players.

Different notions of fair allocation are examined in various papers (see, e.g., [1] and references therein); it would be interesting to compare these notions to our notions of LPV and LTV allocations. Moreover, we would like to see which of our techniques might be applicable to counting versions of other fair allocation problems.

Finally, we would like to settle the complexity of counting LPV and LTV allocations either by proving #P-hardness (as we believe is the case) or by providing polynomial-time algorithms.

References

1. Amanatidis, G., Birmpas, G., Markakis, V.: Comparing approximate relaxations of envy-freeness. In: Proceedings of IJCAI 2018, pp. 42–48 (2018). ijcai.org
2. Buhmann, J.M., Gronskiy, A., Mihalák, M., Pröger, T., Srámek, R., Widmayer, P.: Robust optimization in the presence of uncertainty: a generic approach. J. Comput. Syst. Sci. **94**, 135–166 (2018)
3. Dyer, M.E.: Approximate counting by dynamic programming. In: Proceedings of the 35th Annual ACM Symposium on Theory of Computing, STOC 2003, pp. 693–699. ACM (2003)
4. Dyer, M.E., Goldberg, L.A., Greenhill, C.S., Jerrum, M.: The relative complexity of approximate counting problems. Algorithmica **38**(3), 471–500 (2004)
5. Foley, D.K.: Resource allocation and the public sector. Yale Econ Essays **7**, 45–98 (1967)
6. Kiayias, A., Pagourtzis, A., Sharma, K., Zachos, S.: Acceptor-definable counting classes. In: Manolopoulos, Y., Evripidou, S., Kakas, A.C. (eds.) PCI 2001. LNCS, vol. 2563, pp. 453–463. Springer, Heidelberg (2003). https://doi.org/10.1007/3-540-38076-0_29
7. Li, W., Lee, J., Shroff, N.B.: A faster FPTAS for knapsack problem with cardinality constraint. CoRR, abs/1902.00919 (2019)
8. Melissinos, N., Pagourtzis, A.: A faster FPTAS for the subset-sums ratio problem. In: Wang, L., Zhu, D. (eds.) COCOON 2018. LNCS, vol. 10976, pp. 602–614. Springer, Cham (2018). https://doi.org/10.1007/978-3-319-94776-1_50
9. Pagourtzis, A., Zachos, S.: The complexity of counting functions with easy decision version. In: Královič, R., Urzyczyn, P. (eds.) MFCS 2006. LNCS, vol. 4162, pp. 741–752. Springer, Heidelberg (2006). https://doi.org/10.1007/11821069_64
10. Stefankovic, D., Vempala, S., Vigoda, E.: A deterministic polynomial-time approximation scheme for counting knapsack solutions. SIAM J. Comput. **41**(2), 356–366 (2012)
11. Triommatis, T., Pagourtzis, A.: Approximate #knapsack computations to count semi-fair allocations. CoRR, abs/1912.12430 (2019)
12. Valiant, L.G.: The complexity of computing the permanent. Theoret. Comput. Sci. **8**, 189–201 (1979)
13. Yazidi, A., Jonassen, T.M., Herrera-Viedma, E.: An aggregation approach for solving the non-linear fractional equality knapsack problem. Expert Syst. Appl. **110**, 323–334 (2018)

Characterizations and Approximability of Hard Counting Classes Below #P

Eleni Bakali, Aggeliki Chalki$^{(\boxtimes)}$, and Aris Pagourtzis

School of Electrical and Computer Engineering,
National Technical University of Athens, 15780 Athens, Greece
{mpakali,achalki}@corelab.ntua.gr, pagour@cs.ntua.gr

Abstract. An important objective of research in counting complexity is to understand which counting problems are approximable. In this quest, the complexity class TotP, a hard subclass of #P, is of key importance, as it contains self-reducible counting problems with easy decision version, thus eligible to be approximable. Indeed, most problems known so far to admit an fpras fall into this class.

An open question raised recently by the community of descriptive complexity is to find a logical characterization of TotP and of *robust* subclasses of TotP. In this work we define two subclasses of TotP, in terms of descriptive complexity, both of which are robust in the sense that they have natural complete problems, which are defined in terms of satisfiability of Boolean formulae.

We then explore the relationship between the class of approximable counting problems and TotP. We prove that TotP $\not\subseteq$ FPRAS if and only if NP $\neq$ RP and FPRAS $\not\subseteq$ TotP unless RP = P. To this end we introduce two ancillary classes that can both be seen as counting versions of RP. We further show that FPRAS lies between one of these classes and a counting version of BPP.

Finally, we provide a complete picture of inclusions among all the classes defined or discussed in this paper with respect to different conjectures about the NP vs. RP vs. P questions.

1 Introduction

The class #P [21] is the class of functions that count the number of solutions to problems in NP, e.g. #SAT is the function that on input a formula ϕ returns the number of satisfying assignments of ϕ. Equivalently, functions in #P count accepting paths of non-deterministic polynomial time Turing machines (NPTMs).

NP-complete problems are hard to count, but it is not the case that problems in P are easy to count as well. When we consider counting, non-trivial facts hold. First of all there exist #P-complete problems, that have decision version in P, e.g. #DNF. Moreover, some of them can be approximated, e.g. the Permanent [13] and #DNF [14], while others cannot, e.g. #IS [8]. The class of problems in #P with decision version in P is called #PE, and a subclass of #PE is TotP,

© Springer Nature Switzerland AG 2020
J. Chen et al. (Eds.): TAMC 2020, LNCS 12337, pp. 251–262, 2020.
https://doi.org/10.1007/978-3-030-59267-7_22

which contains all self-reducible problems in #PE [17]. Their significance will be apparent in what follows.

Since many counting problems cannot be exactly computed in polynomial time, the interest of the community has turned to the complexity of approximating them. On one side, there is an enormous literature on approximation algorithms and inapproximability results for individual problems in #P [8,10,13,14,21]. On the other hand, there have been attempts to classify counting problems with respect to their approximability [2,3,9,19].

Related Work. From a unifying point of view, the most important results regarding approximability are the following. Every function in #P either admits an fpras, or does not admit any polynomial approximation ratio [20]; we will therefore call the latter *inapproximable*. For self-reducible problems in #P, fpras is equivalent to almost uniform sampling [20]. With respect to approximation preserving reductions, there are three main classes of functions in #P [9]: (a) the class of functions that admit an fpras, (b) the class of functions that are interreducible with #SAT, and (c) the class #RHΠ_1 of problems that are interreducible with #BIS. Problems in the second class do not admit an fpras unless NP = RP, while the approximability status of problems in the third class is unknown and the conjecture is that they are neither interreducible with #SAT, nor they admit an fpras. We will denote FPRAS the class of #P problems that admit an fpras.

Several works have attempted to provide a structural characterization that exactly captures FPRAS, in terms of path counting [4,17], interval size functions [5], or descriptive complexity [3]. Since counting problems with NP-complete decision version are inapproximable unless NP = RP [9], those that admit fpras should be found among those with easy decision version (i.e., in BPP or even in P). Even more specifically, in search of a logical characterization of a class that exactly captures FPRAS, Arenas et al. [3] show that subclasses of FPRAS are contained in TotP, and they implicitly propose to study subclasses of TotP with certain additional properties in order to come up with approximable problems. Notably, most problems proven so far to admit an fpras belong to TotP, and several counting complexity classes proven to admit an fpras, namely #Σ_1, #RΣ_2 [19], ΣQSO(Σ_1), ΣQSO(Σ_1[FO]) [3] and spanL [2], are subclasses of TotP.

Counting problems in #P have also been studied in terms of descriptive complexity [3,6,7,9,19]. Arenas et al. [3] raised the question of defining classes in terms of descriptive complexity that capture either TotP or robust subclasses of TotP, as one of the most important open questions in the area. A robust class of counting problems needs either to have a natural complete problem or to be closed under addition, multiplication and subtraction by one [3]. In particular, TotP satisfies both of the above properties [3,4].

Our Contribution. In the first part of the paper we focus on the exploration of the structure of #P through descriptive complexity. In particular, we define two subclasses of TotP, namely ΣQSO(Σ_2-2SAT) and #Π_2-1VAR, via logical characterizations; for both these classes we show robustness by providing natural complete problems for them. Namely, we prove that the problem #DISJ2SAT of computing the number of satisfying assignments to disjunctions of 2SAT formulae is complete for $\Sigma QSO(\Sigma_2\text{-}2SAT)$ under parsimonious reductions. This reveals that problems hard for $\Sigma QSO(\Sigma_2\text{-}2SAT)$ under parsimonious reductions cannot admit an fpras unless NP = RP. We also prove that #MONOTONESAT is complete for $\#\Pi_2\text{-}1VAR$ under product reductions. Our result is the first completeness result for #MONOTONESAT under reductions stronger than Turing. Notably, the complexity of #MONOTONESAT has been investigated in [5,12] and it is still open whether it is complete for TotP, or for a subclass of TotP under reductions for which the class is downwards closed. Although, $\#\Pi_2\text{-}1VAR$ is not known to be downwards closed under product reductions, our result is a step towards understanding the exact complexity of #MONOTONESAT.

In the second part of this paper we examine the relationship between the class TotP and FPRAS. As we already mentioned, most (if not all) problems proven so far to admit fpras belong to TotP, so we would like to examine whether FPRAS $\subseteq$ TotP. Of course, problems in FPRAS have decision version in BPP [11], so if we assume P $\neq$ BPP this is probably not the case. Therefore, a more realistic goal is to determine assumptions under which the conjecture FPRAS $\subseteq$ TotP might be true. The world so far is depicted in Fig. 1, where #BPP denotes the class of problems in #P with decision version in BPP.

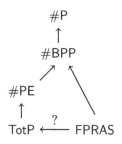

Fig. 1. Relation of FPRAS to counting classes below #P.

In this work we refine this picture by proving that (a) FPRAS $\not\subseteq$ TotP unless RP = P, which means that proving FPRAS $\subseteq$ TotP would be at least as hard as proving RP = P, (b) TotP $\not\subseteq$ FPRAS if and only if NP $\neq$ RP, (c) FPRAS lies between two classes that can be seen as counting versions of RP and BPP, and (d) FPRAS′, which is the subclass of FPRAS with zero error probability when the function value is zero, lies between two classes that we introduce here, that can both be seen as counting versions of RP, and which surprisingly do not coincide unless RP = NP. Finally, we give a complete picture of inclusions among all the classes defined or discussed in this paper with respect to different conjectures about the NP vs. RP vs. P questions.

2 Two Robust Subclasses of TotP

In this section we give the logical characterization of two robust subclasses of TotP. Each one of them has a natural complete problem. Two kinds of reductions will be used for the completeness results; parsimonious and product reductions. Note that both of them preserve approximations of multiplicative error [9,19].

Definition 1. *Let f, $g : \Sigma^* \to \mathbb{N}$ be two counting functions.*

(a) We say that there is a parsimonious (or Karp) reduction from f to g, symb. $f \leq_m^p g$, if there is a polynomial-time computable function h, such that for every $x \in \Sigma^*$ it holds that $f(x) = g(h(x))$.

(b) We say that there is a product reduction from f to g, symb. $f \leq_{pr} g$, if there are polynomial-time computable functions h_1, h_2 such that for every $x \in \Sigma^*$ it holds that $f(x) = g(h_1(x)) \cdot h_2(x)$.

Formal definitions of the classes #PE and TotP can be found in [17].

Some proofs are omitted due to space limitations. They will appear in the full version of the paper.

2.1   The Class ΣQSO(Σ_2-2SAT)

In order to define the first class we make use of the framework of Quantitative Second-Order Logics (QSO) defined in [3].

Given a relational vocabulary σ, the set of First-Order logic formulae over σ is given by the grammar:

$$\phi := x = y \mid R(\overrightarrow{x}) \mid \neg\phi \mid \phi \vee \phi \mid \exists x\phi \mid \top \mid \bot$$

where x, y are first-order variables, $R \in \sigma$, $\overrightarrow{x}$ is a tuple of first order variables, $\top$ represents a tautology, and $\bot$ represents the negation of a tautology.

We define a literal to be either of the form $X(\overrightarrow{x})$ or $\neg X(\overrightarrow{x})$, where X is a second-order variable and $\overrightarrow{x}$ is a tuple of first-order variables. A 2SAT clause over σ is a formula of the form $\phi_1 \vee \phi_2 \vee \phi_3$, where each of the ϕ_i's, $1 \leq i \leq 3$, can be either a literal or a first-order formula over σ. In addition, at least one of them is a first-order formula. The set of Σ_2-2SAT formulae over σ are given by:

$$\psi := \exists\overrightarrow{x}\forall\overrightarrow{y} \bigwedge_{j=1}^{k} C_j(\overrightarrow{x}, \overrightarrow{y})$$

where $\overrightarrow{x}, \overrightarrow{y}$ are tuples of first-order variables, $k \in \mathbb{N}$ and C_j are 2SAT clauses for every $1 \leq j \leq k$.

The set of $\Sigma QSO(\Sigma_2\text{-}2SAT)$ formulae over σ is given by the following grammar:

$$\alpha := \phi \mid s \mid (\alpha + \alpha) \mid \Sigma x.\alpha \mid \Sigma X.\alpha$$

where ϕ is a Σ_2-2SAT formula, $s \in \mathbb{N}$, x is a first-order variable and X is a second-order variable. The syntax of $\Sigma QSO(\Sigma_2\text{-}2SAT)$ formulae includes the counting operators of addition $+$, Σx, ΣX. Specifically, Σx, ΣX are called first-order and second-order quantitative quantifiers respectively.

Let σ be a relational vocabulary, $\mathcal{A}$ a σ-structure with universe A, v a first-order assignment for $\mathcal{A}$ and V a second-order assignment for $\mathcal{A}$. Then the evaluation of a $\Sigma QSO(\Sigma_2\text{-}2SAT)$ formula α over $(\mathcal{A}, V, v)$ is defined as a function $[[\alpha]]$ that on input $(\mathcal{A}, V, v)$ returns a number in $\mathbb{N}$. The function $[[\alpha]]$ is recursively defined in Table 1. A $\Sigma QSO(\Sigma_2\text{-}2SAT)$ formula α is said to be a sentence if it does not have any free variable, that is, every variable in α is under the scope of a usual quantifier ($\exists$, $\forall$) or a quantitative quantifier. It is important to notice that if α is a $\Sigma QSO(\Sigma_2\text{-}2SAT)$ sentence over a vocabulary σ, then for every σ-structure $\mathcal{A}$, first-order assignments v_1, v_2 for $\mathcal{A}$ and second-order assignments

Table 1. The semantics of $\Sigma QSO(\Sigma_2\text{-}2SAT)$ formulae

$$[[\phi]](\mathcal{A}, v, V) = \begin{cases} 1, & \text{if } \mathcal{A} \models \phi \\ 0, & \text{otherwise} \end{cases} \qquad [[s]](\mathcal{A}, v, V) = s$$

$$[[\alpha_1 + \alpha_2]](\mathcal{A}, v, V) = [[\alpha_1]](\mathcal{A}, v, V) + [[\alpha_2]](\mathcal{A}, v, V)$$

$$[[\Sigma x.\alpha]](\mathcal{A}, v, V) = \sum_{a \in A} [[\alpha]](\mathcal{A}, v[a/x], V)$$

$$[[\Sigma X.\alpha]](\mathcal{A}, v, V) = \sum_{B \subseteq A^{arity(X)}} [[\alpha]](\mathcal{A}, v, V[B/X])$$

V_1, V_2 for $\mathcal{A}$, it holds that $[[\alpha]](\mathcal{A}, v_1, V_1) = [[\alpha]](\mathcal{A}, v_2, V_2)$. Thus, in such a case we use the term $[[\alpha]](\mathcal{A})$ to denote $[[\alpha]](\mathcal{A}, v, V)$ for some arbitrary first-order assignment v and some arbitrary second-order assignment V for $\mathcal{A}$.

At this point it is clear that for any $\Sigma QSO(\Sigma_2\text{-}2SAT)$ formula α, a function $[[\alpha]]$ is defined. In the rest of the paper we will use the same notation, namely $\Sigma QSO(\Sigma_2\text{-}2SAT)$, both for the set of formulae and the set of corresponding counting functions.[1]

The following inclusion holds between the class #RHΠ₁ [9] and the class $\Sigma QSO(\Sigma_2\text{-}2SAT)$ defined presently.

Proposition 1. #RHΠ₁ $\subseteq \Sigma QSO(\Sigma_2\text{-}2SAT)$

The class $\Sigma QSO(\Sigma_2\text{-}2SAT)$ contains problems that are tractable, such as #2COL, which is known to be computable in polynomial time [10]. It also contains all the problems in #RHΠ₁, such as #BIS, #1P1NSAT, #DOWNSETS [9]. These three problems are complete for #RHΠ₁ under approximation preserving reductions and are not believed to have an fpras. At last, the problem # IS [9], which is interriducible with #SAT under approximation preserving reductions, belongs to $\Sigma QSO(\Sigma_2\text{-}2SAT)$ as well.

We next show that a generalization of #2SAT, which we will call #DISJ2SAT, is complete for $\Sigma QSO(\Sigma_2\text{-}2SAT)$ under parsimonious reductions.

Membership of #DISJ2SAT in $\Sigma QSO(\Sigma_2\text{-}2SAT)$

In propositional logic, a 2SAT formula is a conjunction of clauses that contain at most two literals. Suppose we are given a propositional formula ϕ, which is a disjunction of 2SAT formulae, then #DISJ2SAT on input ϕ equals the number of satisfying assignments of ϕ.

In this subsection we assume that 2SAT formulae consist of clauses which contain exactly two literals since we can rewrite a clause of the form l as $l \lor l$, for any literal l.

---

[1] Moreover, we will use the terms '(counting) problem' and '(counting) function' interchangeably throughout the paper.

Theorem 1. $\#\text{DISJ2SAT} \in \Sigma\text{QSO}(\Sigma_2\text{-2SAT})$

Proof. Consider the vocabulary $\sigma = \{C_1, C_2, C_3, C_4, D\}$ where C_i, $1 \leq i \leq 4$, are ternary relations and D is a binary relation. This vocabulary can encode any formula which is a disjunction of 2SAT formulae. More precisely, $C_1(c, x, y)$ iff clause c is of the form $x \vee y$, $C_2(c, x, y)$ iff c is $\neg x \vee y$, $C_3(c, x, y)$ iff c is $x \vee \neg y$, $C_4(c, x, y)$ iff c is $\neg x \vee \neg y$ and $D(d, c)$ iff clause c appears in the "disjunct" d.

Let ϕ be an input to $\#\text{DISJ2SAT}$ encoded by an ordered σ-structure $\mathcal{A} = \langle A, C_1, C_2, C_3, C_4, D \rangle$, where the universe A consists of elements representing variables, clauses and "disjuncts". Then, it holds that the number of satisfying assignments of ϕ is equal to $[[\Sigma T.\psi(T)]](\mathcal{A})$, where

$$\psi(T) := \exists d \forall c \forall x \forall y \big((\neg D(d, c) \vee \neg C_1(c, x, y) \vee T(x) \vee T(y)) \wedge$$
$$(\neg D(d, c) \vee \neg C_2(c, x, y) \vee \neg T(x) \vee T(y)) \wedge$$
$$(\neg D(d, c) \vee \neg C_3(c, x, y) \vee T(x) \vee \neg T(y)) \wedge$$
$$(\neg D(d, c) \vee \neg C_4(c, x, y) \vee \neg T(x) \vee \neg T(y)) \big)$$

Thus, $\#\text{DISJ2SAT}$ is defined by $\Sigma T.\psi(T)$ which is in $\Sigma\text{QSO}(\Sigma_2\text{-2SAT})$. $\qquad\square$

Hardness of $\#\text{DISJ2SAT}$

Suppose we have a formula α in $\Sigma\text{QSO}(\Sigma_2\text{-2SAT})$ and an input structure $\mathcal{A}$ over a vocabulary σ. We describe a polynomial-time reduction that given α and $\mathcal{A}$, it returns a propositional formula $\phi_{\alpha_\mathcal{A}}$ which is a disjunction of 2SAT formulae and it holds that $[[\alpha]](\mathcal{A}) = \#\text{DISJ2SAT}(\phi_{\alpha_\mathcal{A}})$. The reduction is a parsimonious reduction, i.e. it preserves the values of the functions involved.

Theorem 2. $\#\text{DISJ2SAT}$ *is hard for* $\Sigma\text{QSO}(\Sigma_2\text{-2SAT})$ *under parsimonious reductions.*

It is known that $\#\text{2SAT}$ has no fpras unless $\text{NP} = \text{RP}$, since it is equivalent to counting all independent sets in a graph [9]. Thus, problems hard for $\Sigma\text{QSO}(\Sigma_2\text{-2SAT})$ under parsimonious reductions also cannot admit an fpras unless $\text{NP} = \text{RP}$.

Inclusion of $\Sigma\text{QSO}(\Sigma_2\text{-2SAT})$ in TotP

Several problems in $\Sigma\text{QSO}(\Sigma_2\text{-2SAT})$, like $\#\text{1P1NSAT}$, $\#\text{IS}$, $\#\text{2COL}$, and $\#\text{2SAT}$, are also in TotP. We next prove that this is not a coincidence.

Theorem 3. $\Sigma\text{QSO}(\Sigma_2\text{-2SAT}) \subseteq \text{TotP}$

Proof. Since TotP is exactly the Karp closure of self-reducible functions of $\#\text{PE}$ [17], it suffices to show that the $\Sigma\text{QSO}(\Sigma_2\text{-2SAT})$-complete problem $\#\text{DISJ2SAT}$ is such a function.

First of all, DISJ2SAT belongs to P. Thus $\#\text{DISJ2SAT} \in \#\text{PE}$.

Secondly, every counting function associated with the problem of counting satisfying assignments for a propositional formula is self-reducible. So $\#\text{DISJ2SAT}$ has this property as well.

Therefore, any $\Sigma\mathrm{QSO}(\Sigma_2\text{-}2\mathrm{SAT})$ formula α defines a function $[[\alpha]]$ that belongs to TotP.   □

Corollary 1. #RHΠ$_1$ ⊆ TotP

2.2   The Class #Π$_2$-1VAR

To define the second class #Π$_2$-1VAR, we make use of the framework presented in [19].

We say that a counting problem #B belongs to the class #Π$_2$-1VAR if for any ordered structure $\mathcal{A}$ over a vocabulary σ, which is an input to #B, it holds that $\#B(\mathcal{A}) = |\{\langle X \rangle : \mathcal{A} \models \forall \overrightarrow{y} \exists \overrightarrow{z} \psi(\overrightarrow{y}, \overrightarrow{z}, X)\}|$. The formula $\psi(\overrightarrow{y}, \overrightarrow{z}, X)$ is of the form $\phi(\overrightarrow{y}, \overrightarrow{z}) \wedge X(\overrightarrow{z})$, where ϕ is a first-order formula over σ and X is a positive appearance of a second-order variable. We call the formula ψ a variable, since it contains only one second-order variable. Moreover, we allow counting only the assignments to the second-order variable X under which the structure $\mathcal{A}$ satisfies $\forall \overrightarrow{y} \exists \overrightarrow{z} \psi(\overrightarrow{y}, \overrightarrow{z}, X)$.

Proposition 2. #$Vc \in$ #Π$_2$-1VAR, *where* #VC *is the problem of counting the vertex covers of all sizes in a graph.*

Completeness of #MONOTONESAT for #Π$_2$-1VAR

Given a propositional formula ϕ in conjunctive normal form, where all the literals are positive, #MONOTONESAT on input ϕ equals the number of satisfying assignments of ϕ.

Theorem 4. #MONOTONESAT $\in$ #Π$_2$-1VAR

Proof. Consider the vocabulary $\sigma = \{C\}$ with the binary relation $C(c, x)$ to indicate that the variable x appears in the clause c. Given a σ-structure $\mathcal{A} = \langle A, C \rangle$ that encodes a formula ϕ, which is an input to #MONOTONESAT, it holds that $\#\mathrm{MONOTONESAT}(\phi) = |\{\langle T \rangle : \mathcal{A} \models \forall c \exists x (C(c, x) \wedge T(x))\}|$.

Therefore, #MONOTONESAT $\in$ #Π$_2$-1VAR.   □

Theorem 5. #MONOTONESAT *is hard for* #Π$_2$-1VAR *under product reductions.*

Inclusion of #Π$_2$-1VAR in TotP

Theorem 6. #Π$_2$-1VAR $\in$ TotP

Proof. It is easy to prove that #MONOTONESAT $\in$ TotP and that TotP is closed under product reductions. Thus, the above results imply that every counting problem in #Π$_2$-1VAR belongs to TotP.   □

3   On TotP vs. FPRAS

In this section we study the relationship between the classes TotP and FPRAS. First of all we give some definitions and facts that will be needed.

Theorem 7. [17] (a) FP $\subseteq$ TotP $\subseteq$ #PE $\subseteq$ #P. The inclusions are proper unless P = NP.
  (b) TotP is the Karp closure of self-reducible #PE functions.

We consider FPRAS to be the class of functions in #P that admit fpras, and we also introduce an ancillary class FPRAS'. Formally:

Definition 2. A function f belongs to FPRAS if $f \in$ #P and there exists a randomized algorithm that on input $x \in \Sigma^*$, $\epsilon > 0$, $\delta > 0$, returns a value $\widehat{f(x)}$ such that
$$\Pr[(1 - \epsilon)f(x) \leq \widehat{f(x)} \leq (1 + \epsilon)f(x)] \geq 1 - \delta$$
in time $poly(|x|, \epsilon^{-1}, \log \delta^{-1})$.
  We further say that a function $f \in$ FPRAS belongs to FPRAS' if whenever $f(x) = 0$ the returned value $\widehat{f(x)}$ equals 0 with probability 1.

We begin with the following observation.[2]

Theorem 8. #P $\subseteq$ FPRAS if and only if NP = RP.

Corollary 2. TotP $\subseteq$ FPRAS if and only if TotP $\subseteq$ FPRAS' if and only if NP = RP.

Now we examine the opposite inclusion, i.e. whether FPRAS is a subset of TotP. To this end we introduce two classes that contain counting problems with decision in RP.
  Recall that if a counting function f admits an fpras, then its decision version, i.e. deciding whether $f(x) = 0$, is in BPP. In a similar way, if a counting function belongs to FPRAS', then its decision version is in RP. So we need to define the subclass of #P with decision in RP. Clearly, if for a problem Π in #P the corresponding counting machine has an RP behavior (i.e., either a majority of paths are accepting or all paths are rejecting) then the decision version is naturally in RP. However, this seems to be a quite restrictive requirement. Therefore we will examine two subclasses of #P.
  For that we need the following definition of the set of Turing Machines associated to problems in RP.

Definition 3. Let M be an NPTM. We denote by p_M the polynomial such that on inputs of size n, M makes $p_M(n)$ non-deterministic choices.
$\mathcal{MR} = \{M \mid M$ is an NPTM and for all $x \in \Sigma^*$ either $acc_M(x) = 0$ or $acc_M > \frac{1}{2} \cdot 2^{p_M(|x|)}\}$.

---

[2] The following theorem is probably well-known among counting complexity researchers. However, since we have not been able to find a proof in the literature we provide one here for the sake of completeness.

Definition 4. $\#RP_1 = \{f \in \#P \mid \exists M \in \mathcal{MR} \, \forall x \in \Sigma^* : f(x) = acc_M(x)\}$.

Definition 5. $\#RP_2 = \{f \in \#P \mid L_f \in RP\}$.

Note that $\#RP_1$, although restrictive, contains counting versions of some of the most representative problems in RP, for which no deterministic algorithms are known. For example consider the polynomial identity testing problem (PIT[3]): Given an arithmetic circuit of degree d that computes a polynomial in a field, determine whether the polynomial is not equal to the zero polynomial. A probabilistic solution to it is to evaluate it on a random point (from a sufficiently large subset S of the field). If the polynomial is zero then all points will be evaluated to 0, else the probability of getting 0 is at most $\frac{d}{|S|}$. A counting analogue of PIT is to count the number of elements in S that evaluate to non-zero values; clearly this problem belongs to $\#RP_1$. Another problem in $\#RP_1$ is to count the number of compositeness witnesses (as defined by the Miller-Rabin primality test) on input an integer $n > 2$; although in this case the decision problem is in P (a prime number has no such witnesses and this can be checked deterministically by AKS algorithm [1]), for a composite number n at least half of the integers in $\mathbb{Z}_n$ are Miller-Rabin witnesses, hence there exists a NPTM $M \in \mathcal{MR}$ that has as many accepting paths as the number of witnesses.

$\#RP_2$ contains natural counting problems as well. Two examples in $\#RP_2$ are #EXACT MATCHINGS and #BLUE-RED MATCHINGS, which are counting versions of EXACT MATCHING [18] and BLUE-RED MATCHING [16], respectively, both of which belong to RP (in fact in RNC) as shown in [15,16]; however, it is still open so far whether they can be solved in polynomial time. Therefore it is also open whether #EXACT MATCHINGS and #BLUE-RED MATCHINGS belong to TotP.

We will now focus on relationships among the aforementioned classes. We start by presenting some unconditional inclusions and then we explore possible inclusions under the condition that either $NP \neq RP \neq P$ or $NP \neq RP = P$ holds.

The results are summarized in Figs. 2 and 3.

3.1   Unconditional Inclusions

Theorem 9. $FP \subseteq \#RP_1 \subseteq \#RP_2 \subseteq \#P$. *Also* $TotP \subseteq \#PE \subseteq \#RP_2$.

Theorem 10. $\#RP_1 \subseteq FPRAS' \subseteq \#RP_2$.

Corollary 3. $\#RP_1 \subseteq FPRAS' \subseteq FPRAS \subseteq \#BPP$.

Corollary 4. *If* $FPRAS \subseteq TotP$ *then* $RP = P$.

Corollary 5. *If* $\#RP_1 = \#RP_2$ *then* $NP = RP$.

Theorems 9 and 10 together with Theorem 7 are summarised in Fig. 2.

---

[3] Determining the computational complexity of polynomial identity testing is considered one of the most important open problems in the mathematical field of Algebraic Computing Complexity.

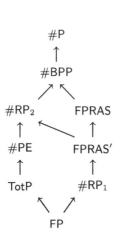

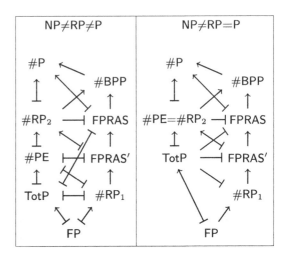

Fig. 2. Unconditional inclusions.

Fig. 3. Conditional inclusions. The following notation is used: $A \rightarrow B$ denotes $A \subseteq B$, $A \dashv B$ denotes $A \not\subseteq B$, and $A \mapsto B$ denotes $A \subsetneq B$.

3.2 Conditional Inclusions/Possible Worlds

Now we will explore further relationships between the above mentioned classes, and we will present two possible worlds inside #P, with respect to NP vs. RP vs. P.

Theorem 11. *The inclusions depicted in Figure 3 hold under the corresponding assumptions on top of each subfigure.*

4   Conclusions and Open Questions

Regarding the question of whether FPRAS is a subset of TotP, Corollary 4 states that if it actually holds, then proving it is at least as difficult as proving RP = P.

A long-sought structural characterization for FPRAS might be obtained by exploring the fact that it lies between $\#RP_1$ and #BPP.

Another open question is whether FPRAS′ is included in $\#RP_1$. It seems that both a negative and a positive answer are compatible with our two possible worlds.

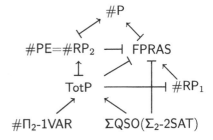

Fig. 4. Inclusions and separations in the case of NP $\neq$ RP = P.

By employing descriptive complexity methods we obtained two new robust subclasses of TotP; the class $\Sigma QSO(\Sigma_2\text{-}2SAT)$ for which the counting problem #DISJ2SAT is complete under parsimonious reductions and the class $\#\Pi_2\text{-}1VAR$

for which #MONOTONESAT is complete under product reductions. We do not expect $\Sigma QSO(\Sigma_2\text{-2SAT})$ to be a subclass of FPRAS, given that #DISJ2SAT does not admit an fpras unless NP = RP. On the other hand, it is an open question whether all problems #Π_2-1VAR admit an fpras. This is equivalent to asking whether #MONOTONESAT admits an fpras.

Although proving #MONOTONESAT complete for #Π_2-1VAR under product reductions, allows a more precise classification of the problem within #P, the question of [12] remains open, i.e. whether #MONOTONESAT is complete for some counting class under reductions under which the class is downwards closed.

Finally, assuming NP $\neq$ RP = P, which is the most widely believed conjecture, the relationships among the classes studied in this paper are given in Fig. 4.

References

1. Agrawal, M., Kayal, N., Saxena, N.: PRIMES is in P. Ann. Math. **160**(2), 781–793 (2004). https://doi.org/10.4007/annals.2004.160.781
2. Arenas, M., Croquevielle, L.A., Jayaram, R., Riveros, C.: Efficient logspace classes for enumeration, counting, and uniform generation. CoRR abs/1906.09226 (2019). https://doi.org/10.1145/3294052.3319704. http://arxiv.org/abs/1906.09226
3. Arenas, M., Muñoz, M., Riveros, C.: Descriptive complexity for counting complexity classes. CoRR abs/1805.02724 (2018). http://arxiv.org/abs/1805.02724
4. Bakali, E., Chalki, A., Pagourtzis, A., Pantavos, P., Zachos, S.: Completeness results for counting problems with easy decision. In: Fotakis, D., Pagourtzis, A., Paschos, V.T. (eds.) CIAC 2017. LNCS, vol. 10236, pp. 55–66. Springer, Cham (2017). https://doi.org/10.1007/978-3-319-57586-5_6
5. Bampas, E., Göbel, A., Pagourtzis, A., Tentes, A.: On the connection between interval size functions and path counting. Comput. Complex. **26**(2), 421–467 (2017). https://doi.org/10.1007/s00037-016-0137-8
6. Bulatov, A., Dalmau, V., Thurley, M.: Descriptive complexity of approximate counting CSPs. In: Rocca, S.R.D. (ed.) Computer Science Logic 2013 (CSL 2013). LIPIcs, vol. 23, pp. 149–164. Schloss Dagstuhl-Leibniz-Zentrum fuer Informatik, Dagstuhl, Germany (2013). https://doi.org/10.4230/LIPIcs.CSL.2013.149. http://drops.dagstuhl.de/opus/volltexte/2013/4195
7. Dalmau, V.: Linear datalog and bounded path duality of relational structures. Log. Methods Comput. Sci. **1**(1) (2005). https://doi.org/10.2168/LMCS-1(1:5)2005. https://lmcs.episciences.org/2275
8. Dyer, M., Frieze, A., Jerrum, M.: On counting independent sets in sparse graphs. SIAM J. Comput. **31**(5), 1527–1541 (2002). https://doi.org/10.1137/S0097539701383844
9. Dyer, M.E., Goldberg, L.A., Greenhill, C.S., Jerrum, M.: The relative complexity of approximate counting problems. Algorithmica **38**(3), 471–500 (2004). https://doi.org/10.1007/s00453-003-1073-y
10. Galanis, A., Goldberg, L.A., Jerrum, M.: A complexity trichotomy for approximately counting list H-colorings. ACM Trans. Comput. Theory **9**(2) (2017). https://doi.org/10.1145/3037381
11. Gill, J.: Computational complexity of probabilistic Turing machines. SIAM J. Comput. **6**(4), 675–695 (1977). https://doi.org/10.1137/020604910.1137/0206049

12. Hemaspaandra, L.A., Homan, C.M., Kosub, S., Wagner, K.W.: The complexity of computing the size of an interval. SIAM J. Comput. **36**(5), 1264–1300 (2007). https://doi.org/10.1137/S0097539705447013

13. Jerrum, M., Sinclair, A.: The Markov Chain Monte Carlo Method: An Approach to Approximate Counting and Integration, pp. 482–520. PWS Publishing Co., USA (1996) https://doi.org/10.1145/261342.571216

14. Karp, R.M., Luby, M., Madras, N.: Monte-Carlo approximation algorithms for enumeration problems. J. Algorithms **10**(3), 429–448 (1989). https://doi.org/10.1016/0196-6774(89)90038-2. http://www.sciencedirect.com/science/article/pii/0196677489900382

15. Mulmuley, K., Vazirani, U.V., Vazirani, V.V.: Matching is as easy as matrix inversion. Combinatorica **7**(1), 105–113 (1987). https://doi.org/10.1007/BF02579206

16. Nomikos, C., Pagourtzis, A., Zachos, S.: Randomized and approximation algorithms for blue-red matching. In: Kučera, L., Kučera, A. (eds.) MFCS 2007. LNCS, vol. 4708, pp. 715–725. Springer, Heidelberg (2007). https://doi.org/10.1007/978-3-540-74456-6_63

17. Pagourtzis, A., Zachos, S.: The complexity of counting functions with easy decision version. In: Královič, R., Urzyczyn, P. (eds.) MFCS 2006. LNCS, vol. 4162, pp. 741–752. Springer, Heidelberg (2006). https://doi.org/10.1007/11821069_64

18. Papadimitriou, C.H., Yannakakis, M.: The complexity of restricted minimum spanning tree problems (extended abstract). In: Maurer, H.A. (ed.) ICALP 1979. LNCS, vol. 71, pp. 460–470. Springer, Heidelberg (1979). https://doi.org/10.1007/3-540-09510-1_36

19. Saluja, S., Subrahmanyam, K., Thakur, M.: Descriptive complexity of #P functions. J. Comput. Syst. Sci. **50**(3), 493–505 (1995). https://doi.org/10.1006/jcss.1995.1039. http://www.sciencedirect.com/science/article/pii/S0022000085710392

20. Sinclair, A., Jerrum, M.: Approximate counting, uniform generation and rapidly mixing Markov chains. Inf. Comput. **82**(1), 93–133 (1989). https://doi.org/10.1016/0890-5401(89)90067-9. http://www.sciencedirect.com/science/article/pii/0890540189900679

21. Valiant, L.: The complexity of computing the permanent. Theor. Comput. Sci. **8**(2), 189–201 (1979). https://doi.org/10.1016/0304-3975(79)90044-6. http://www.sciencedirect.com/science/article/pii/0304397579900446

On Existence of Equilibrium Under Social Coalition Structures

Bugra Caskurlu[1]([✉]), Ozgun Ekici[2], and Fatih Erdem Kizilkaya[1]

[1] TOBB University of Economics and Technology, Ankara, Turkey
{bcaskurlu,f.kizilkaya}@etu.edu.tr
[2] Ozyegin University, Istanbul, Turkey
ozgun.ekici@ozyegin.edu.tr

Abstract. In a strategic form game, a strategy profile is an equilibrium if no viable coalition of agents benefits (in the Pareto sense) from jointly changing their strategies. Weaker or stronger equilibrium notions can be defined by considering various restrictions on coalition formation. In a Nash equilibrium, for instance, the assumption is that viable coalitions are singletons, and in a super strong equilibrium, every coalition is viable. Restrictions on coalition formation can be justified by communication, coordination or institutional constraints. In this paper, inspired by social structures in various real-life scenarios, we introduce certain restrictions on coalition formation, and on their basis, we introduce a number of equilibrium notions. We study our equilibrium notions in resource selection games (RSGs), and we present a complete set of existence and non-existence results for general RSGs and their important special cases.

1 Introduction

In game theory, the centerpiece of analysis is the notion of an equilibrium. An equilibrium is a strategy profile at which certain types of coalitions of agents do not have profitable deviations. The strongest notion that can be defined along this line is a super strong equilibrium: no coalition of agents benefits from jointly changing their strategies. Note that in a game with n agents there are as many as $2^n - 1$ possible coalitions if any coalition is deemed viable. However, deeming every coalition viable and disqualifying strategy profiles as non-equilibrium may be misguided.[1] Indeed, this is the very same idea behind the Nash equilibrium [22] solution concept, wherein only singletons are deemed to be viable coalitions. In this paper, our goal is to fill the gap between the less restrictive Nash

---

[1] In defining our equilibrium notions we use the weak domination relation: a profitable deviation is one that makes the coalition members better off in the Pareto sense. An alternative approach is to define an equilibrium using the strong domination relation, which requires *every* coalition member to be strictly better off. For studies on strong equilibrium, its existence, and some other related work, see [3,4,13–15,20].

This work is supported by The Scientific and Technological Research Council of Turkey (TÜBİTAK) through grant 118E126.

© Springer Nature Switzerland AG 2020
J. Chen et al. (Eds.): TAMC 2020, LNCS 12337, pp. 263–274, 2020.
https://doi.org/10.1007/978-3-030-59267-7_23

equilibrium notion and the very restrictive super strong equilibrium notion. We introduce and study various notions of equilibrium, defined by various restrictions on coalition formation. The restrictions that we consider are motivated by constraints or difficulties that agents may face in real life in coalition formation. We give below a number of examples.

Coordinational Difficulties: A deviation by a coalition requires coalition members to act in unison. However, if coalition members are not familiar with one another, taking coordinated action may not be feasible. Or, it may be that agents are familiar with one another yet they find it difficult to coordinate their actions when the number of coalition members is large.

Communicational Difficulties: The formation of a coalition requires that agents communicate. However, imagine that agents communicate through a network where each agent is located at one of the nodes. If agent i is to offer a deviation to agent j, then every agent along the path from i to j may have to be part of the coalition, too, or else they may block any offer that excludes them and leads to a deterioration in their well-being.

Institutional Constraints: Coalition formation possibilities may also be restricted by institutional constraints. For instance, in global affairs, it is not uncommon that a government feels compelled to act in unison with its allies against a third country. For instance, it may be forced to uphold trade sanctions on a neighboring country, even if doing so may cause much harm to its economy.

A full consideration of what restrictions on coalition formation is reasonable in a specific real-life scenario is beyond the scope of our paper. We focus on restrictions motivated by natural real-life social structures. On the basis of our restrictions, we define new equilibrium notions. We study how they relate to one another, and when they are guaranteed to exist. Indeed, adding social structures to games is a growing trend in recent literature. The following equilibrium notion introduced in an earlier study is related to our study in particular[2]:

Partition Equilibrium: It is assumed that the set of viable coalitions is a partition of the set of agents; see Fig. 1a. This notion, introduced by Feldman and Tennenholtz [10], generalizes the notion of a Nash equilibrium.

Along similar lines, we introduce the following three notions of equilibrium.

Laminar Equilibrium: It is assumed that the set of viable coalitions exhibits a laminar structure; see Fig. 1b. This notion is motivated by institutional constraints as it relates to hierarchical communities in real life. For instance, a military is divided into corps, legions, and brigades; a cabinet is divided into ministries, departments, and directorates; and a company is divided into divisions and departments.

---

[2] For three other related studies see Ashlagi et al. [2], Caskurlu et al. [7] and Hoefer et al. [12].

(a) A Partition Coalition Structure   (b) A Laminar Coalition Structure

Fig. 1. Examples of partition and laminar coalition structures

Contiguous Equilibrium: It is assumed that agents are distributed on a line and each viable coalition consists of some agents that are ordered subsequently; see Fig. 2a. For instance, in politics, political parties are often positioned on a left-right political spectrum, and arguably, coalitions are formed by parties that lie in the same ideological neighborhood on the political spectrum.

Centralized Equilibrium: It is assumed that agents are distributed on a plane and each viable coalition corresponds to a circle on the plane: agents that lie inside the circle are the coalition members, and at the circle's center lies one of the coalition members (perhaps the coalition leader); see Fig. 2b.

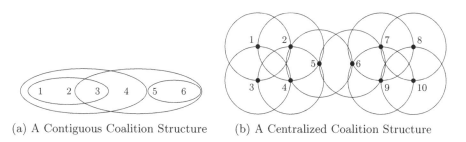

(a) A Contiguous Coalition Structure   (b) A Centralized Coalition Structure

Fig. 2. Examples of contiguous and centralized coalition structures

The number of viable coalitions is $O(n)$ in the case of a partition equilibrium or a laminar equilibrium, and it is $O(n^2)$ in the case of a contiguous equilibrium or a centralized equilibrium. As opposed to a partition equilibrium, viable coalitions w.r.t. our equilibrium notions may be overlapping. A similar solution concept in this regard [12] assumes that the agents are embedded in a social network, and the only viable coalitions are the cliques in the network.[3]

We first show that each equilibrium notion given above generalizes the preceding one by Theorem 1. Then, we study the existence of the above notions of equilibrium in resource selection games (RSGs), for the following reasons:

– RSGs are a subclass of congestion games [24] for which the existence of a Nash equilibrium is guaranteed (see [19,24]). However, a super strong equilibrium

---

[3] There are other settings where overlapping coalitions are considered such as coalition games in interaction graphs [6,21]. The power of agents which are distributed on a line, and where each interval corresponds to a coalition has been studied in the context of voting games [9].

may not exist even in the simplest special cases of RSGs [10]. Since the newly defined solution concepts are generalizations of a Nash equilibrium, and the super strong equilibrium notion is a refinement of them, the existence of equilibria w.r.t. these solution concepts is nontrivial.

- Simple as they may be, RSGs not only have an immense number of applications [11,13,16–18,23] but also capture the essence of more general games.[4]
- RSGs are a benchmark to study the existence of equilibrium w.r.t. newly defined solution concepts.[5]

Our results in RSGs and their relation to the results in the literature are as follows: Feldman and Tennenholtz [10] showed that a partition equilibrium always exists in RSGs under the following restrictions: (i) if the size of a viable coalition is bounded by 2; or (ii) if there are only two resources; or (iii) if the resources are identical. Anshelevich et al. [1] generalized this result by proving that a strategy profile that is both a partition equilibrium and a Nash equilibrium is guaranteed to exist in general RSGs. Caskurlu et al. [7] studied the computational complexity of deciding existence of equilibrium with respect to a given coalition structure in RSGs. They proved that the problem is NP-HARD in general, and identified restricted settings for which polynomial-time algorithms exist.

Our findings are as follows:

- We generalize the results (ii) and (iii) above in [10] to the notion of a laminar equilibrium. We prove that a laminar equilibrium exists if there are only two resources (Theorem 3), or if the resources are identical (Corollary 4).[6]
- We show that an analogous generalization of the result in [1] is not possible. Via an intricate counterexample, we show that a laminar equilibrium may not exist in general RSGs (Theorem 4). Indeed, our counterexample shows that in general RSGs there may not exist a strategy profile that is Pareto efficient, a partition equilibrium, and a Nash equilibrium (Corollary 2).[7]
- We show that a contiguous equilibrium may not exist in an RSG with two resources (Theorem 6), however, contiguous equilibrium exists when resources are identical (Theorem 5). We also show that a centralized equilibrium may not exist even in the setting with two identical resources (Theorem 7).

---

[4] In the routing games literature, they are known as parallel-link networks. For recent literature on routing games on parallel-link networks, see [5] and the references therein.

[5] Not only the immediate previous work [1,10], but also several other newly defined solution concepts [11,12] are studied for RSGs.

[6] Note that RSGs with two resources is interesting in its own right. For instance, the well-known $PoA = 4/3$ result for selfish routing also holds for parallel-link networks with two links [8].

[7] Notice that the main existence result in [1] does not survive a minimal extension of their domain of viable set of coalitions, i.e., when the set of all agents is added to the viable set of coalitions.

Table 1 below summarizes these findings:

Table 1. Existence $(+)$ and Non-existence $(-)$ results

Solution concepts	Resources			
	General	Two	Identical	Two identical
Partition	$+^{**}$	$+^*$	$+^*$	$+^*$
Laminar	$-$	$+$	$+$	$+$
Contiguous	$-$	$-$	$+$	$+$
Centralized	$-$	$-$	$-$	$-$

*due to Feldman and Tennenholtz [10] **due to Anshelevich et al. [1]

2   The Equilibrium Notions

This section introduces our equilibrium notions in the context of a strategic form game and then studies how these notions are related.

Let $\langle N, S, U \rangle$ be a *strategic form game* where N is a finite set of *agents*, $S : (S_j)_{j \in N}$ is the *strategy space* and $U : S \to \mathbb{R}^{|N|}$ is the *payoff function*. Agent j's payoff at strategy profile $s \in S$ is denoted by $U_j(s)$.[8] A *coalition* c is a non-empty subset of agents. Let $\mathcal{P}(N)$ be the power set of N. Then the domain of coalitions is $\mathcal{P}(N) - \{\emptyset\}$. Let $\mathcal{P}_{\geq 1}(N)$ denote this domain. A *coalition structure* C is a set of viable coalitions; i.e., $C \subseteq \mathcal{P}_{\geq 1}(N)$.

Let S_c denote the restriction of the strategy space for coalition c. Let s_c denote the restriction of the strategy profile s for coalition c. That is, $S_c = (S_j)_{j \in c}$ and $s_c = (s_j)_{j \in c}$. Note that the strategy space can be written as $(S_c, S_{N \setminus c})$. The space S_c represents the domain of *deviations* for coalition c. At s if coalition c takes deviation $\tilde{s}_c \in S_c$, the resulting strategy profile is $(\tilde{s}_c, s_{N \setminus c}) \in (S_c, S_{N \setminus c})$. This is a *profitable deviation* for coalition c if for each $j \in c$, $U_j(\tilde{s}_c, s_{N \setminus c}) \geq U_j(s)$, and for some $j \in c$, $U_j(\tilde{s}_c, s_{N \setminus c}) > U_j(s)$. That is, the deviation makes coalition c better off in the Pareto sense. A strategy profile s is called *c-stable* if coalition c has no profitable deviation at s, and *C-stable* if for coalition structure C, s is c-stable for each $c \in C$.

Notice that a strategy profile is a *super strong equilibrium* if it is $\mathcal{P}_{\geq 1}(N)$-stable, and a strategy profile is a *Nash equilibrium* if it is $\mathcal{P}_{=1}(N)$-stable where $\mathcal{P}_{=1}(N) = \{c \subset N| \ |c| = 1\}$. We now define the partition equilibrium which was introduced in the earlier literature, and the three notions of equilibrium which are introduced first in our paper.

Partition Equilibrium: A coalition structure C is a *partition* if for each $j \in N$, there exists a unique coalition $c \in C$ such that $j \in c$. Given a partition coalition structure C, a strategy profile is a *partition equilibrium* if it is C-stable.

---

[8] Throughout, $\subset$ and $\subseteq$ denote the "strict subset of" and the "subset of" relations. For a set X, $|X|$ denotes the cardinality of X. For a number x, $|x|$ denotes the absolute value of x, and $\lfloor x \rfloor$ denotes the greatest integer smaller than x.

Laminar Equilibrium: A coalition structure C is *laminar* if for any two coalitions $c_1, c_2 \in C$ such that $c_1 \cap c_2 \neq \emptyset$, either $c_1 \subseteq c_2$ or $c_2 \subseteq c_1$. Given a laminar coalition structure C, a strategy profile is a *laminar equilibrium* if it is C-stable.

Contiguous Equilibrium: A coalition structure C is *contiguous* if there exists a path $P : j_1 - j_2 - \cdots - j_{|N|}$ (the vertices are agents) in accordance with C in the following sense: for each $c \in C$, the agents in c are subsequently ordered under P. Given a contiguous coalition structure C, a strategy profile is a *contiguous equilibrium* if it is C-stable.

Centralized Equilibrium: A coalition structure C is *centralized* if the agents lie on a plane and a viable coalition consists of agents that lie inside a circle with the restriction that one coalition member lies at the circle's center. Given a centralized coalition structure C, a strategy profile is a *centralized equilibrium* if it is C-stable.

Let $\mathcal{C}^{sse} = \{\mathcal{P}_{\geq 1}(N)\}$. Let $\mathcal{C}^{ne} = \{\mathcal{P}_{=1}(N)\}$. Also, let $\mathcal{C}^{pe}, \mathcal{C}^{le}, \mathcal{C}^{coe}, \mathcal{C}^{cee}$ be, respectively, the domains of coalition structures that are partitions, laminar, contiguous, and centralized. Thus, a strategy profile that is C-stable is a super strong equilibrium if $C \in \mathcal{C}^{sse}$; a Nash equilibrium if $C \in \mathcal{C}^{ne}$; a partition equilibrium if $C \in \mathcal{C}^{pe}$, and so on. Theorem 1 states how $\mathcal{C}^{ne}, \mathcal{C}^{pe}, \mathcal{C}^{le}, \mathcal{C}^{coe}, \mathcal{C}^{cee}$, and $\mathcal{C}^{sse}$ are related.

Theorem 1. *We have* $\mathcal{C}^{ne} \subseteq \mathcal{C}^{pe} \subseteq \mathcal{C}^{le} \subseteq \mathcal{C}^{coe} \subseteq \mathcal{C}^{cee}$. *Also,*

- $\mathcal{C}^{ne} \subset \mathcal{C}^{pe} \subset \mathcal{C}^{le}$ *for* $|N| \geq 2$,
- $\mathcal{C}^{le} \subset \mathcal{C}^{coe}$ *for* $|N| \geq 3$,
- $\mathcal{C}^{coe} \subset \mathcal{C}^{cee}$ *for* $|N| \geq 4$,
- *for each* $C \in \mathcal{C}^{cee}$, $C \subseteq \mathcal{P}_{\geq 1}(N)$,
- *for* $|N| \geq 3$, $\mathcal{P}_{\geq 1}(N) \notin \mathcal{C}^{cee}$.

3   An Application: Resource Selection Games

A *resource selection game* (RSG) is a triplet $\langle N, M, f \rangle$ where $N : \{1, 2, \ldots, n\}$ is the set of *agents*, $M : \{1, 2, \ldots, m\}$ is the set of *resources* and $f : (f_i)_{i=1}^m$ is the profile of strictly monotonic increasing *cost functions* such that $f_i(0) = 0$ for all $i \in \{1, 2, \ldots, m\}$. When q agents use resource i, each incurs a cost equal to $f_i(q)$. Each agent tries to minimize the cost it incurs. In the rest of the paper, we fix the game $\langle N, M, f \rangle$.

An *allocation* is a sequence $a : (a_i)_{i=1}^m$ such that: *i*) For each $i \in M$, we have $a_i \subseteq N$; *ii*) For every $i, i' \in M$ such that $i \neq i'$, we have $a_i \cap a_{i'} = \emptyset$; and *iii*) We have $\bigcup_{i \in M} a_i = N$. Above, a_i denotes the set of agents that are assigned to resource i at allocation a. Thus, at allocation a, each agent in a_i incurs a cost equal to $f_i(|a_i|)$. Let $\mathcal{A}$ be the domain of allocations.

The *maxcost* of an allocation a is the maximum cost incurred by an agent at a. That is, the maxcost of allocation a equals $max_{i \in M} f_i(|a_i|)$. The *minmaxcost* of the RSG, to be denoted by α, is the maxcost of the allocation whose maxcost is smallest. That is, $\alpha = min_{a \in \mathcal{A}} max_{i \in M} f_i(|a_i|)$.

Let $q_i = max_{q \in \mathbb{Z}_{\geq 0}} f_i(q) \leq \alpha$. We refer to q_i as resource i's quota. That is, a resource's quota is the maximum number of agents that can be assigned to it without making its cost exceed α. We distinguish between resources that can and that cannot attain the minmaxcost α. A resource i is a *Type 1 resource* if $f_i(q_i) = \alpha$, and a *Type 2 resource* if $f_i(q_i) < \alpha$. Let T_1 and T_2 denote, respectively, the sets of type 1 and type 2 resources. Since the minmaxcost of the game is α, we have $T_1 \neq \emptyset$. Also, for $i \in T_1$, let $\beta_i = f_i(q_i - 1)$. We refer to β_i as resource i's *beta value*. Note that for a type 1 resource i, its beta value is its cost when the number of agents assigned to it is one less than its quota.

Note that an RSG is a non-cooperative game in the strategic form although its formulation here is different from the formulation of a strategic form game in Sect. 2. Here, agents' payoffs are negative (i.e., they incur costs rather than receive payoffs) and an agent's strategy space is the set M.

In this context, we continue to use the terminology in Sect. 2 in regards to coalitions and coalition structures; i.e., $c, C, \mathcal{P}_{=1}(N), \mathcal{P}_{\geq 1}(N), \mathcal{C}^{sse}, \mathcal{C}^{ne}, \mathcal{C}^{le}$, $\mathcal{C}^{coe}, \mathcal{C}^{cee}$ are as described in Sect. 2. We also use the terminology in Sect. 2 regarding the stability and equilibrium notions but with one exception: Note that in an RSG an allocation fully specifies the strategies of agents. Therefore, in this context, we speak of an "allocation" as a substitute for a strategy profile. Hence, in this context, rather than a strategy profile we speak of an allocation being c-stable or C-stable; or being a laminar equilibrium or a contiguous equilibrium.

Also, in this context, we represent a deviation by a coalition c as a sequence $(c_i)_{i=1}^m$ such that: (i) $c_1 \cup c_2 \cup \cdots \cup c_m = c$; and (ii) for each $i, i' \in M$ and $i \neq i'$, $c_i \cap c_{i'} = \emptyset$. That is, a deviation is an agreement by coalition members on which resources they will use: c_i is the set of coalition members who agree to use resource i. We use $a \circ (c_i)_{i=1}^m$ to denote the allocation that results when coalition c takes deviation $(c_i)_{i=1}^m$ at allocation a: i.e., after the deviation the set of agents that are assigned to resource i is $(a_i setminus c) \cup c_i$. Also, note that a deviation is a profitable deviation if at the resulting allocation each coalition member becomes *weakly better off* (i.e., the cost it incurs does not increase) and at least one of them becomes *better off* (i.e., the cost it incurs decreases).

The notion of a super strong equilibrium is very appealing since it precludes profitable deviations by any coalition of agents. However, in most game forms a super strong equilibrium is not guaranteed to exist. This is also true for RSGs, even in the very restricted setting with three agents and two identical resources [10]. We next present a characterization of Nash equilibrium in RSGs given by [1].

Theorem 2 (due to Anshelevich et al. [1]). *An allocation a of an RSG is a Nash equilibrium if and only if: (i) for each $i \in T_2$, $|a_i| = q_i$; (ii) for each $i \in T_1$, $|a_i| \in \{q_i - 1, q_i\}$; and (iii) for some $i \in T_1$, $|a_i| = q_i$.*

Let allocation a be a Nash equilibrium. We need to designate the set of type 1 resources that are not assigned at a up to their quotas: Let $L(a) = \{i \in T_1 \mid |a_i| = q_i - 1\}$. Also, let $H(a) = M \setminus L(a)$. We refer to the resources in $L(a)$ and in $H(a)$ as *low* and *high* resources at a, respectively. The corollary below immediately follows from the above theorem and it will be useful later on.

Corollary 1. *Let allocation a be a Nash equilibrium. Then, $|L(a)| = \sum_{i \in M} q_i - n$ and $|H(a)| = m - |L(a)|$. Therefore, the number of low and high resources are the same at every Nash equilibrium allocation.*

The rest of this section is divided into two parts. We present our results for the laminar equilibrium notion in Sect. 3.1, and the results for contiguous and centralized equilibrium notions in Sect. 3.2.

3.1   Existence and Non-existence Results for Laminar Equilibrium

In this section, we first present Theorem 3 that states the existence of laminar equilibrium in two-resource RSGs. Existence result for laminar equilibrium in identical-resource RSGs (Corollary 4) follows from the more general result that contiguous equilibrium always exists in identical-resource RSGs, which is presented in Sect. 3.2.

Theorem 3. *In a two-resource RSG, for any laminar coalition structure $C \in \mathcal{C}^{le}$, there exists a C-stable allocation.*

The rest of this section is, thus, devoted to proving that a laminar equilibrium does not necessarily exist in RSGs (Theorem 4). The example that we use to show Theorem 4 is an intricate one, consisting of a large number of agents and resources. We present it below.

Example 1. Consider an RSG as follows:

- There are $n = 14052$ agents and $m = 2001$ resources.
- Every resource is of type 1.
- The set of resources can be written as $M = M_x \cup M_y \cup M_z$ such that:
  - $M_x = \{x\}$ and $q_x = 53$.
  - $M_y = \{y_1, y_2, \ldots, y_{1000}\}$ where each resource in M_y has the same cost function, and $q_y = 8$ for all $y \in M_y$.
  - $M_z = \{z_1, z_2, \ldots, z_{1000}\}$ where each resource in M_z has the same cost function, and $q_z = 7$ for all $z \in M_z$.
  - For all $y \in M_y$ and $z \in M_z$, we have $\beta_x > \beta_y > \beta_z > f_x(q_x - 2)$.   ◇

Theorem 4. *In an RSG, for $C \in \mathcal{C}^{le}$, it may be that no allocation is C-stable. That is, in RSGs a laminar equilibrium is not guaranteed to exist.*

Proof. In Example 1, consider the following coalition structure: $C = \{c_1, \ldots, c_6\} \cup \mathcal{P}_{=1}(N) \cup \{N\}$ where the sets $c_1, \ldots, c_6$ are disjoint and each has a cardinality of $14052/6 = 2342$. Note that C is laminar. We prove the theorem by showing that no C-stable allocation exists in Example 1. By way of contradiction, suppose that in Example 1 there exists an allocation a which is C-stable.

Note that by Corollary 1: $|L(a)| = 1001 \ (= 53 + 8 \times 1000 + 7 \times 1000 - 14052)$. And $|H(a)| = 2001 - 1001 = 1000$. Since $\mathcal{P}_{=1}(N) \subset C$, a is a Nash equilibrium.

Therefore, using Corollary 1, at allocation a there are 1001 low resources and 1000 high resources.

We divide the proof into six parts:

(1) We show that $x \in L(a)$.

By way of contradiction, suppose that $x \in H(a)$. Then, in $M_y \cup M_z$, there are 1001 resources that are low. Let i, i' be two of them ($i \neq i'$). Consider the agents $a_i \cup a_{i'}$. Note that $|a_i \cup a_{i'}| = q_i + q_{i'} - 2 \leq 14$. Let $N_1 \cup N_2 \subset a_x$ be such that N_1 and N_2 are disjoint, $|N_1| = q_i$, and $|N_2| = q_{i'}$. We define allocation a' from a as follows.

– Remove the agents in $N_1 \cup N_2 \cup a_i \cup a_{i'}$ from their assigned resources.
– Assign agents in N_1 to resource i, assign agents in N_2 to resource i', and assign agents in $a_i \cup a_{i'}$ to resource x.
(The assignments of remaining agents are the same as before.)

At allocation a', the agents assigned to resource x are now better off (since x is now assigned $q_x - 2$ agents). All other agents are equally well off at the two allocations. But then a is not N-stable, a contradiction. Thus, $x \in L(a)$.

(2) We show that $|H(a) \cap M_y| \leq 7$. (Hence, $|H(a) \cap M_z| \geq 993$.)

By (1), we know that $x \in L(a)$. Then, at a, in $M_y \cup M_z$ there are 1000 high resources and 1000 low resources. By way of contradiction, suppose that $|H(a) \cap M_y| \geq 8$. This implies that $|L(a) \cap M_z| \geq 8$. We define allocation a' from a as follows. We pick 7 high resources in M_y: Wlog., let $y_1, \cdots, y_7 \in H(a) \cap M_y$. We pick 8 low resources in M_z: Wlog., let $z_1, \cdots, z_8 \in L(a) \cap M_z$. We pick 49 agents assigned to x at a: Let $N_x \subset a_x$ be such that $|N_x| = 49$. Then:

– Remove the agents in $N_x \cup a_{y_1} \cup \cdots \cup a_{y_7} \cup a_{z_1} \cup \cdots \cup a_{z_8}$ from their assigned resources.
– Assign the 49 agents in N_x to resources $y_1, \cdots, y_7$ such that each resource is assigned 7 agents.
– Assign the 56 agents in $a_{y_1} \cup \cdots \cup a_{y_7}$ to resources $z_1, \cdots, z_8$ such that each resource is assigned 7 agents.
– Assign the 48 agents in $a_{z_1} \cup \cdots \cup a_{z_8}$ to resource x.
(The assignments of remaining agents are the same as before.)

At allocation a', the agents assigned to resource x are now better off (since x is now assigned $q_x - 2$ agents). The agents assigned to resources $y_1, \cdots, y_7$ are also better off (because they are now assigned to low resources for which the beta value is smaller). The agents assigned to resources $z_1, \cdots, z_8$ are equally well off (because they are assigned to high resources at both a and a'). The agents assigned to the remaining resources are equally well off. But then a is not N-stable, a contradiction. Thus, $|H(a) \cap M_y| \leq 7$. Hence, we also have $|H(a) \cap M_z| \geq 993$.

(3) We show that there exists $c \in \{c_1, \ldots, c_6\}$ such that there are at least 1159 agents in c which are assigned to resources in $H(a) \cap M_z$ at allocation a.

Above, by (2), at a there are at least 993 high resources in M_z. Since each of them is assigned 7 agents, at a the number of agents assigned to high resources in

M_z is at least $993 \times 7 = 6951$. But then by the generalized pigeonhole principle, there is a coalition $c \in \{c_1, \ldots, c_6\}$ such that at a the number of agents in c that are assigned to high resources in M_z is at least $\lceil \frac{6951}{6} \rceil = 1159$.

(4) Let $c \in \{c_1, \ldots, c_6\}$ be as described in (3). We show that there exists $z \in H(a) \cap M_z$ such that there are at least two agents in c that are assigned to z at allocation a.

Note that at a the number of high resources in M_z is at most 1000 (because $|M_z| = 1000$). By (3) we also know that there are at least 1159 agents in c which are assigned to high resources in M_z at allocation a. But then, by the pigeonhole principle, there exists $z \in H(a) \cap M_z$ such that there are at least two agents in c that are assigned to z at allocation a.

(5) Let $c \in \{c_1, \ldots, c_6\}$ be as described in (3). We show that for each resource $y \in L(a) \cap (M_x \cup M_y)$, there are at least two agents in c that are assigned to y at allocation a.

By (4) there exists $z \in H(a) \cap M_z$ such that there are at least two agents in c that are assigned to z at allocation a. Thus, let $j, j' \in c$ be such that $j \neq j'$ and at a the agents j and j' are assigned to resource z.

By way of contradiction, suppose that there exists $y \in L(a) \cap (M_x \cup M_y)$ such that $|c \cap a_y| \leq 1$.

Suppose that $|c \cap a_y| = 0$. We define allocation a' from a as follows: Agent j is removed from resource z and then assigned to resource y. It is clear that at a' coalition c is better off. But then a is not C-stable, a contradiction. Thus, $|c \cap a_y| \neq 0$.

Suppose that $|c \cap a_y| = 1$. Let $\tilde{j}$ be the agent in $c \cap a_y$. We define allocation a' from a as follows: Agents j and j' are removed from resource z and then assigned to resource y, and agent $\tilde{j}$ is removed from resource y and then assigned to resource z. Note that at a' the agents j and j' are equally well off (they are still assigned to high resources) and the agents in c that are assigned to z ($\tilde{j}$ and perhaps some other agents) are better off (because z is now a low resource, and the beta value for z is smaller than the beta value for y). The remaining agents in c are equally well off. But then a is not C-stable, a contradiction. Thus, $|c \cap a_y| \neq 1$. Therefore, $|c \cap a_y| \geq 2$.

(6) We conclude the proof as follows: Let c be as described in (3). By (1) and (2), there are at least 994 resources in $L(a) \cap (M_x \cup M_y)$. By (5), the number of agents in coalition c that are assigned to resources in $L(a) \cap (M_x \cup M_y)$ is at least $2 \times 994 = 1988$ at allocation a. By (3), there are at least 1159 agents in c which are assigned to resources in $H(a) \cap M_z$ at allocation a. But then we get $|c| \geq 1988 + 1159 = 3147$. This contradicts the fact that $|c| = 2342$. ☐

Corollary 2. *In an RSG, there may not exist a Pareto efficient allocation a that is both a partition equilibrium and a Nash equilibrium.*

Proof. We work with the coalition structure $C = \{c_1, \ldots, c_6\} \cup \mathcal{P}_{=1}(N) \cup \{N\}$ in proving Theorem 4. The part $\{c_1, \ldots, c_6\}$ is a partition of the agents, the part $\mathcal{P}_{=1}(N)$ is the set of singleton coalitions, and the part $\{N\}$ is the grand coalition. ☐

Theorem 1 and Theorem 4 together imply the following result.

Corollary 3. *In an RSG, the existence of a contiguous equilibrium, or of a centralized equilibrium, is also not guaranteed.*

3.2 Existence and Non-existence Results for Contiguous and Centralized Equilibrium

In this section, we first present Theorem 5 that states the existence of contiguous equilibrium in identical-resource RSGs.

Theorem 5. *In an identical-resource RSG, for any given contiguous coalition structure $C \in \mathcal{C}^{coe}$, there exists a C-stable allocation.*

Theorem 1 and Theorem 4 together imply the following result.

Corollary 4. *In an identical-resource RSG, for any given laminar coalition structure $C \in \mathcal{C}^{le}$, there exists a C-stable allocation.*

We next present Theorem 6 which states that in a two-resource RSG a contiguous equilibrium is not guaranteed to exist.

Theorem 6. *In an RSG with two resources, for $C \in \mathcal{C}^{coe}$, it may be that no allocation is C-stable.*

We finally present Theorem 7, which states that a centralized equilibrium may not exist even for the two identical resources case. Note that even though centralized coalition structures contain $O(n^2)$ viable coalitions, equilibrium may not exist in the two-identical resources setting, i.e., it strengthens the non-existence result of super strong equilibrium [10].

Theorem 7. *In an RSG with two identical resources, for $C \in \mathcal{C}^{cee}$, it may be that no allocation is C-stable.*

References

1. Anshelevich, E., Caskurlu, B., Hate, A.: Partition equilibrium always exists in resource selection games. Theory Comput. Syst. **53**(1), 73–85 (2013)
2. Ashlagi, I., Krysta P., Tennenholtz, M.: Social context games. In: International Workshop on Internet and Network Economics (2008)
3. Aumann, R.J.: Acceptable points in general cooperative n-person games. In: Contributions to the Theory of Games, pp. 287–324 (1959)
4. Bernheim, B.D., Peleg, B., Whinston, M.D.: Coalition-proof Nash equilibria concepts. J. Econ. Theory **42**(1), 1–12 (1987)
5. Bhaskar, U., Lolakapuri, P. R.: Equilibrium computation in atomic splittable routing games. In: 26th Annual European Symposium on Algorithms, pp. 1–14 (2018)

6. Bousquet, N., Li, Z., Vetta, A.: Coalition games on interaction graphs: a horticultural perspective. Proceedings of the Sixteenth ACM Conference on Economics and Computation (2015)
7. Caskurlu, B., Ekici, O., Kizilkaya, F.E.: On efficient computation of equilibrium under social coalition structures. Turk. J. Electr. Eng. Comput. Sci. **28**, 1686–1698 (2020)
8. Correa, J.R., Schulz, A.S., Stier-Moses, N.E.: A geometric approach to the price of anarchy in nonatomic congestion games. Games Econ. Behav. **64**(2), 457–469 (2008)
9. Edelman, P.H.: A note on voting. Math. Soc. Sci. **34**(1), 37–50 (1997)
10. Feldman, M., Tennenholtz, M.: Structured coalitions in resource selection games. ACM Trans. Intell. Syst. Technol. **1**(1), 4 (2010)
11. Hayrapetyan, A., Tardos, É., Wexler, T.: The effect of collusion in congestion games. In: Proceedings of the Thirty-Eighth Annual ACM Symposium on Theory of Computing (2006)
12. Hoefer, M., Penn, M., Polukarov, Maria., Skopalik, A., Vöcking, B.: Considerate equilibrium. In: IJCAI (2011)
13. Holzman, R., Law-Yone, N.: Strong equilibrium in congestion games. Games Econ. Behav. **21**(1–2), 85–101 (1997)
14. Konishi, H., Le Breton, N., Weber, S.: Equilibria in a model with partial rivalry. J. Econ. Theory **72**(1), 225–237 (1997)
15. Konishi, H., Le Breton, M., Weber, S.: On coalition-proof Nash equilibria in common agency games. J. Econ. Theory **85**(1), 122–139 (1999)
16. Kuniavsky, S., Smorodinsky, R.: Equilibrium and potential in coalitional congestion games. Theory Decis. **76**(1), 69–79 (2013). https://doi.org/10.1007/s11238-013-9357-4
17. Milchtaich, I.: Congestion games with player-specific payoff functions. Games Econ. Behav. **13**(1), 111–124 (1996)
18. Milinski, M.: An evolutionarily stable feeding strategy in sticklebacks. Zeitschrift für Tierpsychologie **51**(1), 36–40 (1979)
19. Monderer, D., Shapley, L.S.: Potential games. Games Econ. Behav. **14**(1), 124–143 (1996)
20. Moreno, D., Wooders, J.: Coalition-proof equilibrium. Games Econ. Behav. **17**(1), 80–112 (1996)
21. Myerson, R.B.: Graphs and cooperation in games. Math. Oper. Res. **2**, 225–229 (1977)
22. Nash, J.: Non-cooperative games. Ann. Math. **54**, 286–295 (1951)
23. Quint, T., and Martin S.: A model of migration. Cowles Foundation for Research in Economics 1088 (1994)
24. Rosenthal, R.W.: A class of games possessing pure-strategy Nash equilibria. Int. J. Game Theory **2**(1), 65–67 (1973)

Space Complexity of Streaming Algorithms on Universal Quantum Computers

Yanglin Hu, Darya Melnyk[(✉)], Yuyi Wang, and Roger Wattenhofer

ETH Zurich, Zürich, Switzerland
{yahu,dmelnyk,yuwang,wattenhofer}@ethz.ch

Abstract. Universal quantum computers are the only general purpose quantum computers known that can be implemented as of today. These computers consist of a classical memory component which controls the quantum memory. In this paper, the space complexity of some data stream problems, such as PartialMOD and Equality, is investigated on universal quantum computers. The quantum algorithms for these problems are believed to outperform their classical counterparts. Universal quantum computers, however, need classical bits for controlling quantum gates in addition to qubits. Our analysis shows that the number of classical bits used in quantum algorithms is equal to or even larger than that of classical bits used in corresponding classical algorithms. These results suggest that there is no advantage of implementing certain data stream problems on universal quantum computers instead of classical computers when space complexity is considered.

Keywords: Streaming algorithm · Universal quantum computer · Space complexity · Solovay-Kitaev algorithm

1 Introduction

In the past two decades, scientists have made significant progress in the field of quantum computation. Quantum computer protocols based on different physical principles have been constructed and manufactured. Despite this progress, large-scale quantum computers are still not available.

According to the no-programming Theorem [19], a quantum-controlled quantum computer is not better than a classically controlled quantum computer. Therefore, a modern quantum computer consists of a large classical memory controlling a small quantum memory. The limited quantum memory poses great challenges to physicists and computer scientists. In particular, one must decide how to use this limited quantum memory efficiently. One possible way is to build larger-scale quantum computers. Another way is to introduce algorithms that require a small quantum memory, but a large classical memory. In this work, we address the latter case for a special class of problems – the data stream problems.

© Springer Nature Switzerland AG 2020
J. Chen et al. (Eds.): TAMC 2020, LNCS 12337, pp. 275–286, 2020.
https://doi.org/10.1007/978-3-030-59267-7_24

Data stream problems process data streams where the input data comes at a high rate. The massive input data challenges communication, computation, and storage. In particular, one may not be able to transmit, compute and store the whole input. For such problems, classical and quantum algorithms have been proposed with the aim to reduce space complexity. On quantum computers, such algorithms usually use polynomially or even exponentially less quantum memory than their classical counterparts using classical memory.

However, quantum algorithms are generally performed on a universal quantum computer. Note that for some structures quantum gates can change continuously by slowly varying some physical parameters, and it seems that one should use a continuous set of quantum gates to describe them. However, by considering the uncertainty principle, physical parameters can only be measured with errors. Due to these errors, quantum gates with slightly different parameters can therefore often not be distinguished and should be regarded as the same quantum gate. This brings us back to a discrete set of quantum gates and a universal quantum computer. On such a universal quantum computer, only a finite set of quantum gates – the universal quantum gates – can be used directly. Other quantum gates are approximated by quantum gate array to a certain accuracy.

According to the no-programming theorem, universal quantum computers need extra memory, in particular, they need classical memory in order to store the program for the desired quantum gate array. Therefore, the length of the desired quantum gate array would determine the length of the program, which requires extra memory. In this work, we include the extra memory for programs when considering the space complexity, and show that if the extra memory is taken into account, the space complexity of the proposed quantum algorithm for the PartialMOD problem is approximately equal to the space complexity of the respective classical algorithms and that for the Equality problem is even worse. This way, the considered streaming algorithm on universal quantum computers have no advantage over their classical counterparts. Note that our result does not imply that these problems cannot be solved efficiently in a different model. Instead, it suggests that different problems may be solved more efficiently in some particular model, but not in others. We therefore see our result as an inspiration to consider quantum algorithms with respect to the framework in which they can be implemented.

2   Related Work

Classical data stream problems have been first formalized and popularized by Alon et al. [4] in order to estimate the frequency moment of a sequence using as little memory as possible. The PartialMOD [5] and Equality-like problems [23] are well-known examples of problems in this class. For the PartialMOD problem, Ambainis et al. [5] proved a tight bound of $\log p$ bits in the deterministic setting. Ablayev et al. [1,3] proved a tight bound of n bits for the deterministic classical streaming algorithms computing Equality problems.

For the quantum version of data stream problems, Watrous [22] proved the well-known result that the complexity class PrSPACE(s) is equal to the complexity class PrQSPACE(s), which implies that to some extent, quantum algorithms are not better than classical algorithms with respect to their space complexity. For the PartialMOD problem, Ablayev et al. [2] proposed a quantum algorithm that requires only 1 qubit while classical algorithms need $\log p$ bits. Ablayev et al. [3] later proposed quantum streaming algorithms for Equality Boolean functions. Their results show that some problems have both logarithmic or better quantum algorithms, whereas at least a logarithmic number of bits is needed for classical algorithms. Based on both previous results, Khadiev et al. [13–15] proposed quantum stream algorithms with constant space complexity, which is better than classical streaming algorithms that require polylogarithmically many bits. Le Gall [17] also investigated a certain variation of the Equality problem and proposed a quantum algorithm with exponentially lower space complexity (both quantum and classical) than the corresponding classical algorithm.

The field of communication complexity also investigated Equality problems. Buhrman et al. [7] introduced quantum fingerprinting and proposed to use it in communication theory. They chose the Equality problem as an example in their paper. Recently, Guan et al. [10] managed to realize the above progress experimentally.

In our paper, we focus on the space complexity of data stream problems on universal quantum computers. For such computers, the Solovay-Kitaev algorithm [8,11,16] states that any operator can be approximated to an accuracy of ϵ by $\log^c \frac{1}{\epsilon}$ quantum gates from a finite set of gates. Different versions of the Solovay-Kitaev algorithm consider different values of c. In [8], Dawson and Nielsen introduced a version with $c \approx 4$. Kitaev et al. [16] proposed a version with $c \approx 2$, and Harrow et al. [11] finally proved a lower bound of $c = 1$. Moreover, they showed that the corresponding algorithm exists but cannot be given explicitly.

3   Background

We will start by describing the notation for quantum computation we use in this paper. We will introduce fundamental concepts of quantum physics using the Dirac notation, and also present the Bloch sphere model, which is a geometric way to comprehend quantum algorithms. In Sect. 3.2, we will then clarify the Solovay-Kitaev algorithm [8] which gives a way to efficiently approximate any desired operation on a universal quantum computer with a finite set of operations. Finally, in Sect. 3.3, we will explain the quantum no-programming theorem [19]. This theorem points out that we must use orthonormal quantum states to perform different operators with deterministic quantum gate arrays and as such it forms the basis of a classically controlled quantum computer.

3.1 Notation

Let $|i\rangle$ denote the i-th classical state of the (complete orthonormal) computational basis of a Hilbert space. We can write a pure state of a quantum memory as a column vector $|\psi\rangle = (\alpha_1, ..., \alpha_n)^T = \sum_{i=1}^{n} \alpha_i |i\rangle$. Its norm satisfies $\langle \psi | \psi \rangle = \sum_{i=1}^{n} |\alpha_i|^2 = 1$, where $\langle \psi |$ is the conjugate transpose of $|\psi\rangle$.

The evolution of a state can be represented by a unitary operator U. That is, given an initial state $|\psi\rangle$, the final state after applying U is $|\psi'\rangle = U|\psi\rangle$. It is easy to verify that the norm of a state does not change after an evolution.

A single-qubit memory can be represented as a point on the so-called Bloch sphere. Explicitly, a unitary operator

$$U = \begin{pmatrix} \cos(\frac{\theta}{2}) & -e^{-i\phi}\sin(\frac{\theta}{2}) \\ e^{i\phi}\sin(\frac{\theta}{2}) & \cos(\frac{\theta}{2}) \end{pmatrix}.$$

corresponds to the vector $U|0\rangle$, pointing to (θ, ϕ) on the sphere, where θ is the angle between the vector and the z-axis, and ϕ is the angle between the projection of the vector onto xOy plane and the x-axis. Note that the north pole corresponds to all $(0, \phi)$ and the south pole corresponds to all (π, ϕ).

A projective measurement can be represented by a set of orthogonal projectors followed by a normalization. That is, if we apply a measurement $\{P_i\}$ to the initial state $|\psi\rangle$, the final state becomes $|\psi'_i\rangle = P_i|\psi\rangle / \|P_i|\psi\rangle\|$ with probability $|\langle\psi|P|\psi\rangle|$. In particular, if the measurement is $\{|i\rangle\langle i|\}$ and the initial state is $\sum \alpha_i |i\rangle$, the final state is $|i\rangle$ with probability $|\alpha_i|^2$. Note that the total probability of all possible final states is 1.

3.2 Solovay-Kitaev Algorithm

In order to present the Solovay-Kitaev algorithm, we first need to introduce the concept of universality.

Definition 1 (Universal quantum gates [20]). *A set of quantum gates is universal for quantum computation if any unitary operator can be approximated to arbitrary accuracy by a quantum circuit involving only these gates.*

Note that an example of such a set can be found in Chapter 4 of [20].

Based on the well-defined universal set, we can now state the Solovay-Kitaev theorem which talks about how efficient a universal set is:

Theorem 1 (Solovay-Kitaev [8]). *There exist algorithms that can approximate any unitary operator U to an accuracy of $\|U - U_{approx}\|_2 \leq \epsilon$ with $O(\log^c \frac{1}{\epsilon})$ universal quantum gates.*

The proof of the theorem can be found in [8].

According to Harrow [11], $\Omega(\log \frac{1}{\epsilon})$ quantum gates are needed in order to approximate any unitary operator in two dimensions to an accuracy of ϵ. The Solovay-Kitaev algorithm is optimal if we disregard poly-logarithmic differences in the number of quantum gates.

The Solovay-Kitaev theorem does not exclude the possibility that we can approximate some unitary operator with a quantum gate array much shorter than $O(\log \frac{1}{\epsilon})$ to an accuracy of ϵ.

3.3   No-Programming Theorem

The no-programming theorem [19] shows that we cannot use fewer qubits than classical bits for programming if we want to implement a quantum gate array deterministically. We view our quantum computer as a unitary operator G acting on both the quantum program $|P\rangle$ and the memory $|d\rangle$. G acting on a quantum program $|P\rangle$ for unitary U results in $U|d\rangle \otimes |P'\rangle$. After measurement we get $U|d\rangle$ *deterministically*. However, G acting on a superposition of orthogonal quantum programs $\frac{1}{\sqrt{2}}(|P_1\rangle + |P_2\rangle)$ results in a superposition of orthogonal states $\frac{1}{\sqrt{2}}(U_1|d\rangle \otimes |P_1'\rangle + U_2|d\rangle \otimes |P_2'\rangle)$. Therefore, after our measurement, we obtain either $U_1|d\rangle$ or $U_2|d\rangle$ *stochastically*.

Theorem 2. *On a fixed, general purposed quantum computer, if we want to deterministically implement a quantum gate array, quantum programs $|P_1\rangle,...,|P_n\rangle$ performing distinct unitary operator $U_1,...,U_n$ are orthogonal. The program memory is at least N-dimensional, that is, it contains at least $\log(N)$ qubits.*

The theorem shows that, when used for programming a deterministic quantum gate array, a quantum program has no advantage over a classical program, i.e. in this aspect a quantum controlled quantum computer is no better than a classically controlled quantum computer.

When used for programming a probabilistic quantum gate array, there are quantum programs that use exponentially less space but succeed with exponentially smaller probability, which is not practical. In our paper, we thus only consider classical bits for programming.

4   Data Stream Problems

In this section, we present selected examples of data stream problems and study their space complexity. Each section is organized as follows: we first introduce the problem statement and the corresponding proposed algorithm for quantum computers with a continuous set of gates. In practice, quantum computers with a continuous set of gates cannot be realized, which makes such algorithms only of theoretic interest. In the following section, we assume that our universal quantum computer first selects a certain universal set of gates, then it is asked data stream problems with any possible scale and parameter. The quantum computer should answer any possible question using the same universal set of gates. We therefore analyze the space complexity of the respective algorithm on such a universal quantum computer and show that it has no advantage over the space complexity of the best known classical algorithm.

4.1   PartialMOD Problem

In this section, we study the PartialMOD problem as presented in [1,5,15]. In this problem, we receive some unknown bitstring bit by bit of which we know that the number of bits with value 1 is a multiple of a given number. The task is to determine the parity of the multiplier of this number while storing as few bits as possible in the memory.

Definition 2 (PartialMOD problem). *Let $(x_1, ..., x_n)$ be an input sequence of classical bits. Assume that we know in advance that $\#_1$ is a multiple of p, i.e., $\#_1 = v \cdot p$, where $\#_1$ denotes the number of ones in the string. The bits are received one by one by the algorithm. The problem is to determine the parity of v, i.e., to output v mod 2.*

Algorithm with a Continuous Set of Gates. Ambainis and Yakaryilmaz [5] showed that there exists no deterministic or probabilistic algorithm to compute PartialMOD problem with $o(\log p)$ classical bits. In their paper, they also propose a quantum algorithm solving PartialMOD using only one qubit. This algorithm works as follows: There is only one qubit in the quantum memory. Let the initial state of the qubit be $|0\rangle$, which is the north pole of the Bloch sphere, and set $\theta_p = \frac{\pi}{2p}$. Each time we receive a 1 as the next bit, we apply a unitary operator

$$R(\theta_p) = \begin{pmatrix} \cos\theta_p & \sin\theta_p \\ -\sin\theta_p & \cos\theta_p \end{pmatrix}.$$

on the qubit, which is a rotation by $2\theta_p$ around y-axis on the Bloch sphere. After $v \cdot p$ steps, we receive all the input bit and get the state

$$|\psi_f\rangle = \left(\cos(v\frac{\pi}{2}) \quad -\sin(v\frac{\pi}{2}) \right)^{\mathrm{T}}.$$

If $v \bmod 2 = 0$, we return to the north pole of the Bloch sphere and the final state of the qubit is $|0\rangle$. If $v \bmod 2 = 1$, we reach the south pole of the Bloch sphere and the final state is $|1\rangle$. Finally, we can measure the qubit and obtain its state.

With this procedure, we only need one qubit to solve the PartialMOD problem on quantum computers. In contrast, a classical computer requires to use $\log p$ bits, as is shown in [1].

Analysis on Universal Quantum Computers. In the following, we show that the proposed quantum algorithm is not space efficient on universal quantum computers. We suppose that our universal quantum computer is able to solve any specific PartialMOD problem, which requires that we should be able to apply any $R(\theta_p)$ to the demanded accuracy. Observe that there are infinitely many choices of p, and thus infinitely many different $R(\theta_p)$. Since only finitely many gates can be selected in the universal set of a quantum computer, $R(\theta_p)$ have to be approximated by a quantum gate array, where each gate of the array is from the universal set. This leads to possibly wrong outputs. Assume therefore that we approximate $R(\theta_p)$ by $R(\theta_p + \epsilon_p)$, which satisfies

$$\|R(\theta_p) - R(\theta_p + \epsilon_p)\|_2 = 4\sin\frac{\epsilon_p}{4}.$$

Starting with the initial state $|0\rangle$, we reach the state

$$|\psi_f\rangle = \begin{pmatrix} \cos(v\frac{\pi}{2} + vp \cdot \epsilon_p) \\ -\sin(v\frac{\pi}{2} + vp \cdot \epsilon_p) \end{pmatrix}.$$

after vp steps. With probability $\sin^2(vp \cdot \epsilon_p)$ we may get an incorrect output from the measurement. If $vp \cdot \epsilon_p$ is small enough, we can bound the probability of an incorrect output by a positive constant δ as follows

$$\frac{1}{2}vp \cdot 4\sin\left(\frac{\epsilon_p}{4}\right) \leq \sin\left(vp \cdot \epsilon_p\right) \leq \sqrt{\delta}.$$

Therefore, an accuracy of $\frac{2\sqrt{\delta}}{vp}$ must be achieved. Such an accuracy comes at the cost of additional quantum gates.

Intuitively, applying the Solovay-Kitaev algorithm, we need a quantum gate array of $\log\frac{vp}{2\sqrt{\delta}}$. We will show next that a quantum gate array of at least $\Omega(\log(\frac{v}{\sqrt{\delta}}\log p))$ gates must be used in order to approximate $R(\theta_p)$ in the proposed algorithm to an accuracy of $\frac{2\sqrt{\delta}}{vp}$. Note that in this theorem we do not assume the optimality of the Solovay-Kitaev algorithm. Because the optimality of Solovay-Kitaev algorithm is in the sense of polylogarithmic equivalence, and the truly optimal algorithm has not been given, simply assuming this algorithm to be optimal may cause difficulties. However, even without such an assumption, Theorem 3 still shows that the quantum algorithm performs worse in some situations.

Theorem 3. *No algorithm can approximate all $R(\theta_p)$, where $\theta_p = \frac{\epsilon}{2p}$ and $p \leq p_0$, to an accuracy of $\epsilon_p = \frac{\epsilon}{2p}$ using $o(\log(\frac{1}{\epsilon}\log p_0))$ quantum gates on a universal computer, where p_0 is sufficiently large and ϵ sufficiently small. We do not assume the optimality of the Solovay-Kitaev algorithm here.*

We will not present the proof here, but the general idea of the proof is inspired by [11].

Theorem 3 implies that at least $O(\log(\frac{v}{\sqrt{\delta}}\log p_0))$ quantum gates are needed in order to approximate all $R(\theta_p)$, where $p \leq p_0$, to the demanded accuracy of $\epsilon_p = \frac{2\sqrt{\delta}}{vp}$ in order to ensure a success probability of at least $1 - \delta$. Since we have to store the arrangement of the quantum gate array for each $R(\theta_p)$, the number of classical bits required is equal to the number of gates in the quantum gate array, that is, at least $O(\log(\frac{v}{\sqrt{\delta}}\log p))$ classical bits. It is obvious that when v approaches infinity while p remains finite, the quantum algorithm for PartialMOD is not more space-efficient than the corresponding classical algorithm.

Assuming the optimality of the Solovay-Kitaev algorithm, which is discussed in [11], we can also show that in order to obtain such an accuracy, at least $O(\log(\frac{vp}{\sqrt{\delta}}))$ quantum gates must be used by any algorithm.

Theorem 4. *Let p be sufficiently large. No algorithm can approximate all $R(\theta_p)$, where $\theta_p = \frac{\pi}{2p}$ to any accuracy ϵ_p with $o(\log(\frac{1}{\epsilon_p}))$ quantum gates on a universal computer, if the optimality of the Solovay-Kitaev algorithm is assumed.*

This theorem can be proved by contradiction: if one can approximate these operators with $o(\log(\frac{1}{\epsilon_p}))$ quantum gates, then it is possible to construct a better algorithm than the Solovay-Kitaev algorithm.

Theorem 4 shows that there exists some $R(\theta_p)$ for which we need at least $\Omega(\log\frac{1}{\epsilon_p}) = \Omega(\log v + \log p)$ gates in order to approximate it to the demanded accuracy of $\epsilon_p = \frac{2\sqrt{\delta}}{vp}$, assuming the optimality of the Solovay-Kitaev theorem. Since we have to store the arrangement of the quantum gate array, the number of classical bits needed is $\Omega(\log v + \log p)$. When v or p approach infinity, the quantum algorithm is not more space-efficient than the classical algorithm.

Theorem 3 and Theorem 4 are proved under different assumptions. Together they show that the previously proposed algorithm is not more space-efficient than its classical counterpart under certain conditions.

4.2   Equality Problem

In this section, we investigate the so-called Equality problem [3,7,18]. In this problem, two bitstrings are received once one after another bit by bit. The task is to find out whether these two given sequences of bits are equal while storing a minimal amount of information.

Definition 3 (Equality problem). *We are given an input sequence $(x, y) = (x_1, ..., x_n, y_1, ..., y_n)$ of classical bits. The bits are received one by one by the algorithm. We do not receive any bit of y before we have received all bits of x. The output is whether x and y are equal, i.e., $O = \delta(\|x - y\|) = 1, x = y; 0, x \neq y$.*

Algorithm with a Continuous Set of Gates. According to [18,21], there is no classical deterministic algorithm that can compute the equality problem with $o(n)$ classical bits, while there is a randomized algorithm, i.e. Karp-Rabin algorithm, with a space complexity of $O(\log n)$ [12]. There also exists a quantum algorithm that has the same performance. Ablayev et al. [3] applied quantum fingerprinting in a quantum streaming algorithm to solve this problem with $O(\log n)$ qubits on a quantum computer with a continuous set of gates. Their algorithm seems to have the same performance as the Karp-Rabin algorithm.

The quantum memory is divided into two parts. The first part is the first qubit, whose state is in a 2-dimensional space. The second part contains the remaining $\log t$ qubits in a t-dimensional space. The initial state is $|0\rangle \otimes |0\rangle$. The strategy is to first apply Hadamard gates on all qubits of the second part and receive $\frac{1}{\sqrt{t}}|0\rangle \otimes \sum_{j=1}^{t}|j\rangle$. If we receive a 1 for x_i, we apply a unitary operator $U_i = \sum_{j=1}^{t}\{R(\theta_{ij}) \otimes |j\rangle\langle j|\}$, where $R(\theta_{ij})$ is a rotation on the first qubit by $\theta_{ij} = \frac{2\pi m_j}{2^{i+1}}$ and m_j some positive integer. If we receive a 1 for y_i, we replace $R(\theta_{ij})$ with $R(-\theta_{ij})$ in U_i. After receiving all the input bits, the state is

$$\frac{1}{\sqrt{t}}\sum_{j} R\left(\frac{2\pi m_j(x - y)}{2^{n+1}}\right)|0\rangle \otimes |j\rangle.$$

Then we apply Hadamard gates on all qubits in the second part. The final state becomes

$$\frac{1}{t} \sum_j \left(\cos \frac{2\pi m_j (x - y)}{2^{n+1}} \right) |0\rangle \otimes |0\rangle + rest.$$

If $x = y$, we return to the initial state. If $x \neq y$, we reach a non-initial state. We require that the coefficient of $|0\rangle \otimes |0\rangle$ in the final state is approximately a delta function, that is,

$$\left\| \frac{1}{t} \sum_j \cos \frac{2\pi m_j (x - y)}{2^{n+1}} - \delta(x - y) \right\| \leq \sqrt{\epsilon}.$$

Then we can easily verify whether $x = y$ by checking whether we get $|0\rangle \otimes |0\rangle$ after measurement. If $x = y$, we obtain $|0\rangle$ with probability 1. If $x \neq y$, we obtain $|0\rangle \otimes |0\rangle$ with probability less than ϵ.

If we apply discrete Fourier transform to $\delta(g)$, that is, m_js take $t = 2^n$ integers from 0 to 2^n, ϵ is exactly 0. But in that case we need $\log(t) = n$ qubits. It is however possible that if we do not apply discrete Fourier transform, that is, m_js only take $t = O(n \log \frac{1}{\epsilon}) \ll 2^n$ integers from 0 to 2^n, ϵ is also bounded. The next theorem states this fact, its proof can be found in [3].

Theorem 5. *There exists a set of $t > \frac{2}{\epsilon} \ln(2m)$ elements, $\{m_j, \ j = 1, ..., t\}$ such that*

$$\frac{1}{t} \left\| \sum_j \cos \left(\frac{2\pi m_j g}{m} \right) \right\| \leq \sqrt{\epsilon}, \ \forall g \neq 0.$$

Theorem 5 implies that there exists a set of $t = \frac{2}{\epsilon} \ln(2m) + 1$ elements, $\{m_j, j = 1, ..., t\}$, which ensures $\cos \left(\frac{2\pi k_i g}{m} \right)$'s to almost cancel each other. Indeed, if we select integers uniformly at random from 0 to $m - 1$, we are likely to get such m_j. By applying Theorem 5 to the Equality problem, we only need $\log(n) + 1$ qubits on quantum computers with a continuous set of gates, which is exponentially better than n bits deterministic algorithms on computers, as was shown by Babai et al. [6].

Analysis on Universal Quantum Computers. The proposed algorithm to solve the Equality problem is not space-efficient on universal quantum computers. Similar to the PartialMOD problem, we will first bound the accuracy of each operator. Let us denote the probability for the algorithm to accept the input, i.e., in the case where the final state is $|0\rangle \otimes |0\rangle$, as $\Pr(x, y)$. Further, assume that it is possible to approximate the operator $R(\theta_{ij})$ to an accuracy of δ_{ij}. After applying Theorem 5, the partial derivative of $\Pr(x, y)$ becomes

$$\delta \Pr(x, y) \leq \frac{\sqrt{\epsilon}}{t} \left\| \sum_{j=1}^{t} \sum_i \sin \left(\frac{m_j \pi 2^i (x - y)_i}{2^n} \right) \delta_{ij} \right\|.$$

Differently than in the PartialMOD problem, it is challenging to bound the accuracy for the Equality problem precisely. Instead, we simply assume we need $\Omega(1)$ gates for each $R(\theta_{ij})$. The following theorem defines an upper bound on the accuracy needed, and shows our simple assumption is reasonable.

Theorem 6. *Let $|\delta_{ij}| \leq \frac{1}{n}$. Then, there exists a set of $t = \frac{2}{\epsilon}(n+3)$ elements $m_j, j = 1, ..., t$, such that the following two inequalities are satisfied*

$$\frac{1}{t}\left\|\sum_j \cos\left(\frac{\pi m_j g}{2^n}\right)\right\| \leq \sqrt{\epsilon}, \ \forall g \neq 0,$$

and

$$\frac{1}{t}\left\|\sum_{j=1}^{t}\sum_i \sin\left(\frac{m_j \pi 2^i g_i}{2^n}\right)\delta_{ij}\right\| \leq \sqrt{\epsilon}, \ \forall g \neq 0.$$

Here, we prove it via a method similar to that of Theorem 5, shown in [3].

The algorithm for the Equality problem will succeed as long as we reach an accuracy of $\frac{1}{n}$ for a suitably chosen set of $O(n)$ elements. In order to achieve such accuracy, we need at most $O(\log^4 n)$ quantum gates according to the Solovay-Kitaev theorem. Since we need to apply at least one quantum gate in order to be able to implement an operator, it is reasonable to assume that we need at least $\Omega(1)$ quantum gates for each operator to achieve such accuracy.

Now we can analyze the space complexity, for which we also take into account classical bits. When we perform the above algorithm we need to store the set $\{m_j\}$, since the set $\{m_j\}$ is not chosen arbitrarily. There are two natural ways to do so. One way is to store $\{m_j\}$ directly: consider m_j that range from 0 to 2^n, and thus need n classical bits. We have n such integers, and thus at least $\Omega(n^2)$ bits are needed. This strategy requires even more bits than a classical brute force method which saves all $O(n)$ bits of the input. The second way is to store $\{R(\theta_{ij})\}$: note that $R(\theta_{ij})$ need at least $\Omega(1)$ quantum gates for each operator, and thus each need $\Omega(1)$ classical bits. Since we have n^2 such operators in our algorithm, at least $\Omega(n^2)$ bits of storage are needed, which is more than that in the classical deterministic algorithm. In the following theorem, we provide a more rigorous proof.

Theorem 7. *At least $\Omega(n^2)$ bits are needed in order to store a set $\{m_j, j = 1...t\}$ where $m_j \in [0, 2^n - 1]$ and $t = \frac{2}{\epsilon}(n+3)$ without pre-knowledge of the set.*

Proof. We first consider the classical case. The number of possible choices in the classical case is

$$C_t^{2^n} = \frac{2^n!}{t! \cdot (2^n - t)!}.$$

The information entropy of knowing a certain choice from all possible choices with equal possibility is $S = \ln\left(C_t^{2^n}\right)$. Consider when n is sufficiently large, $2^n \gg t = \frac{2}{\epsilon}(n+3)$, use $\ln(1+x) \approx x$ an $\ln(x!) \approx x\ln(x) - x$, we have

$$S = \ln(2^n) + ... + \ln(2^n - t + 1) - \ln(t!) \approx nt - \frac{t^2}{2^n} - t\ln(t) + t = O(n^2).$$

Since the number of bits required is linearly dependent on the information entropy, $O(n^2)$ bits are needed in order to store this set.

We next consider the quantum case. The set $\{m_j, j = 1...t, m_j \in [0, 2^n - 1]\}$ is used to program our quantum computer. Due to the quantum no-programming theorem in Sect. 3.3, quantum programs have no advantage over the classical program with respect to space complexity. Therefore, $\Omega(n^2)$ bits or qubits are needed to store this set.

Therefore, the considered algorithm for the Equality problem has no advantage over the classical deterministic algorithm.

5   Conclusion

Based on the Solovay-Kitaev algorithm, we investigated the space complexity of streaming algorithms on a universal computer when only a finite number of quantum gates are available. We used the PartialMOD problem and the Equality problem to analyze the quantum streaming algorithms in systems where classical bits are used in order to control quantum gates. By applying the Solovay-Kitaev algorithm we concluded that the considered quantum streaming algorithms do not beat their classical counterparts in this system.

Our work shows that not all quantum streaming algorithms can perform well on a universal quantum computer. There are also data stream problems for which quantum algorithms may perform well on a universal quantum computer. One example is the variation of the Equality problem proposed in [17]. In this problem, the input is repeated many times, which is different from the Equality problem discussed in this paper, where we receive the input only once. Another possible candidate is the problem based on the universal (ϵ, l, m)-code of matrices proposed by Sauerhoff et al. in [21] and Gavinsky et al. [9], where the input directly corresponds to a quantum gate array, and one can therefore save space when storing quantum gates for application. By comparing these algorithms, we conclude that a framework can be extremely efficient for a certain set of problems and corresponding algorithms, but not necessarily for all problems. We therefore think that the space complexity of algorithms should be analyzed with respect to the framework of the quantum computer in which they can be implemented.

References

1. Ablayev, F., Gainutdinova, A., Karpinski, M., Moore, C., Pollett, C.: On the computational power of probabilistic and quantum branching program. Inf. Comput. **203**(2), 145–162 (2005)
2. Ablayev, F., Gainutdinova, A., Khadiev, K., Yakaryılmaz, A.: Very narrow quantum OBDDs and width hierarchies for classical OBDDs. In: Jürgensen, H., Karhumäki, J., Okhotin, A. (eds.) DCFS 2014. LNCS, vol. 8614, pp. 53–64. Springer, Cham (2014). https://doi.org/10.1007/978-3-319-09704-6_6
3. Ablayev, F., Khasianov, A., Vasiliev, A.: On complexity of quantum branching programs computing equality-like Boolean functions. Electronic Colloquium on Computational Complexity (2010)

4. Alon, N., Matias, Y., Szegedy, M.: The space complexity of approximating the frequency moments. In: Proceedings of the Twenty-Eighth Annual ACM Symposium on Theory of Computing. STOC (1996)
5. Ambainis, A., Yakaryılmaz, A.: Superiority of exact quantum automata for promise problems. Inf. Process. Lett. **112**(7), 289–291 (2012)
6. Babai, L., Kimmel, P.G.: Randomized simultaneous messages: solution of a problem of Yao in communication complexity. In: Proceedings of the 12th Annual IEEE Conference on Computational Complexity. CCC (1997)
7. Buhrman, H., Cleve, R., Watrous, J., de Wolf, R.: Quantum fingerprinting. Phys. Rev. Lett. **87**, 167902 (2001)
8. Dawson, C.M., Nielsen, M.A.: The Solovay-Kitaev algorithm. Quantum Inf. Comput. **6**(1), 81–95 (2006)
9. Gavinsky, D., Kempe, J., Kerenidis, I., Raz, R., de Wolf, R.: Exponential separations for one-way quantum communication complexity, with applications to cryptography. In: Proceedings of the Thirty-Ninth Annual ACM Symposium on Theory of Computing. STOC (2007)
10. Guan, J.Y., et al.: Observation of quantum fingerprinting beating the classical limit. Phys. Rev. Lett. **116**, 240502 (2016)
11. Harrow, A.W., Recht, B., Chuang, I.L.: Efficient discrete approximations of quantum gates. J. Math. Phys. **43**(9), 4445–4451 (2002)
12. Karp, R.M., Rabin, M.O.: Efficient randomized pattern-matching algorithms. IBM J. Res. Dev. **31**(2), 249–260 (1987)
13. Khadiev, K., Khadieva, A., Kravchenko, D., Rivosh, A.: Quantum versus Classical Online Algorithms with Advice and Logarithmic Space (2017)
14. Khadiev, K., Khadieva, A., Mannapov, I.: Quantum online algorithms with respect to space complexity. Lobachevskii J. Math. **39**, 1377–1387 (2017)
15. Khadiev, K., Ziatdinov, M., Mannapov, I., Khadieva, A., Yamilov, R.: Quantum Online Streaming Algorithms with Constant Number of Advice Bits (2018)
16. Kitaev, A.Y., Shen, A., Vyalyi, M.N.: Classical and Quantum Computation. American Mathematical Society, Boston (2002)
17. Le Gall, F.: Exponential separation of quantum and classical online space complexity. Theory Comput. Syst. **45**, 188–202 (2009)
18. Newman, I., Szegedy, M.: Public vs. private coin flips in one round communication games (extended abstract). In: Proceedings of the Twenty-eighth Annual ACM Symposium on Theory of Computing. STOC (1996)
19. Nielsen, M.A., Chuang, I.L.: Programmable quantum gate arrays. Phys. Rev. Lett. **79**, 321 (1997)
20. Nielsen, M.A., Chuang, I.L.: Quantum Computation and Quantum Information: 10th, Anniversary Edition. Cambridge University Press, Cambridge (2010)
21. Sauerhoff, M., Sieling, D.: Quantum branching programs and space-bounded nonuniform quantum complexity. Theoret. Comput. Sci. **334**(1), 177–225 (2005)
22. Watrous, J.H.: Space-bounded quantum computation. Ph.D. thesis, The University of Wisconsin, Madison (1998)
23. Yao, A.C.C.: Some complexity questions related to distributive computing (preliminary report). In: Proceedings of the Eleventh Annual ACM Symposium on Theory of Computing. STOC (1979)

On Coresets for Support Vector Machines

Murad Tukan[1]([✉]), Cenk Baykal[2]([✉]), Dan Feldman[1]([✉]), and Daniela Rus[2]([✉])

[1] Computer Science Department, University of Haifa, Haifa, Israel
muradtuk@gmail.com, dannyf.post@gmail.com
[2] MIT CSAIL, Cambridge, USA
{baykal,rus}@mit.edu

Abstract. We present an efficient coreset construction algorithm for large-scale Support Vector Machine (SVM) training in Big Data and streaming applications. A coreset is a small, representative subset of the original data points such that a models trained on the coreset are provably competitive with those trained on the original data set. Since the size of the coreset is generally much smaller than the original set, our preprocess-then-train scheme has potential to lead to significant speedups when training SVM models. We prove lower and upper bounds on the size of the coreset required to obtain small data summaries for the SVM problem. As a corollary, we show that our algorithm can be used to extend the applicability of any off-the-shelf SVM solver to streaming, distributed, and dynamic data settings. We evaluate the performance of our algorithm on real-world and synthetic data sets. Our experimental results reaffirm the favorable theoretical properties of our algorithm and demonstrate its practical effectiveness in accelerating SVM training.

1 Introduction

Popular machine learning algorithms are computationally expensive, or worse yet, intractable to train on massive data sets, where the input data set is so large that it may not be possible to process all the data at one time. A natural approach to achieve scalability when faced with Big Data is to first conduct a preprocessing step to summarize the input data points by a significantly smaller, representative set. Off-the-shelf training algorithms can then be run efficiently on this compressed set of data points. The premise of this two-step learning procedure is that the model trained on the compressed set will be provably competitive with the model trained on the original set – as long as the data summary, i.e., the *coreset*, can be generated efficiently and is sufficiently representative.

Coresets are small weighted subsets of the training points such that models trained on the coreset are approximately as good as the ones trained on the original (massive) data set. Coreset constructions were originally introduced in the

This research was supported in part by the U.S. National Science Foundation (NSF) under Awards 1723943 and 1526815, Office of Naval Research (ONR) Grant N00014-18-1-2830, Microsoft, and JP Morgan Chase.

M. Tukan and C. Baykal—These authors contributed equally to this work.

ⓒ Springer Nature Switzerland AG 2020
J. Chen et al. (Eds.): TAMC 2020, LNCS 12337, pp. 287–299, 2020.
https://doi.org/10.1007/978-3-030-59267-7_25

context of computational geometry [1] and subsequently generalized for applications to other problems, such as logistic regression, neural network compression, and mixture model training [5,6,10,17,19] (see [9] for a survey).

A popular coreset construction technique – and the one that we leverage in this paper – is to use importance sampling with respect to the points' *sensitivities*. The sensitivity of each point is defined to be the worst-case relative impact of each data point on the objective function. Points with high sensitivities have a large impact on the objective value and are sampled with correspondingly high probability, and vice-versa. The main challenge in generating small-sized coresets often lies in evaluating the importance of each point in an accurate and computationally-efficient way.

1.1   Our Contributions

In this paper, we propose an efficient coreset construction algorithm to generate compact representations of large data sets to accelerate SVM training. Our approach hinges on bridging the SVM problem with that of k-means clustering. As a corollary to our theoretical analysis, we obtain theoretical justification for the widely reported empirical success of using k-means clustering as a way to generate data summaries for large-scale SVM training. In contrast to prior approaches, our approach is both (i) provably efficient and (ii) naturally extends to streaming or dynamic data settings. Above all, our approach can be used *to enable the applicability of any off-the-shelf SVM solver* – including gradient-based and/or approximate ones, e.g., Pegasos [25], to streaming and distributed data settings by exploiting the *composibility* and *reducibility* properties of coresets [9].

In particular, this paper contributes the following:

1. A coreset construction algorithm for accelerating SVM training based on an efficient importance sampling scheme.
2. An analysis proving lower bounds on the number of samples required by any coreset construction algorithm to approximate the input data set.
3. Theoretical guarantees on the efficiency and accuracy of our coreset construction algorithm.
4. Evaluations on synthetic and real-world data sets that demonstrate the effectiveness of our algorithm in both streaming and offline settings.

2   Related Work

Training SVMs requires $\mathcal{O}(n^3)$ time and $\mathcal{O}(n^2)$ space in the offline setting where n is the number of training points. Towards the goal of accelerating SVM training in the offline setting, [26,27] introduced the Core Vector Machine (CVM) and Ball Vector Machine (BVM) algorithms, which are based on reformulating the SVM problem as the Minimum Enclosing Ball (MEB) problem and Enclosing Ball (EB) problem, respectively, and by leveraging existing coreset constructions for each; see [4]. However, CVM's accuracy and convergence properties

have been noted to be at times inferior relative to those of existing SVM implementations [20]; moreover, unlike the algorithm presented in this paper, neither
the CVM, nor the BVM algorithm extends naturally to streaming or dynamic
settings where data points are continuously inserted or deleted. Similar geometric
approaches, including extensions of the MEB formulation, those based on convex
hulls and extreme points, among others, were investigated by [2,11,13,15,22,24].
Another class of related work includes the use of canonical optimization algorithms such as the Frank-Wolfe algorithm [7], Gilbert's algorithm [7,8], and a
primal-dual approach combined with Stochastic Gradient Descent (SGD) [14].

SGD-based approaches, such as Pegasos [25], have been a popular tool of
choice in approximately-optimal SVM training. Pegasos is a stochastic sub-
gradient algorithm for obtaining a $(1 + \varepsilon)$-approximate solution to the SVM
problem in $\tilde{\mathcal{O}}(dn\lambda/\varepsilon)$ time for a linear kernel, where λ is the regularization
parameter and d is the dimensionality of the input data points. In contrast
to our method, these approaches and their corresponding theoretical guarantees do not feasibly extend to dynamic data sets and/or streaming settings. In
particular, gradient-based approaches cannot be trivially extended to streaming
settings since the arrival of each input point in the stream results in a change of
the gradient.

There has been prior work in streaming algorithms for SVMs, such as those
of [2,13,23,24]. However, these works generally suffer from poor practical performance in comparison to that of approximately optimal SVM algorithms in the
offline (batch) setting, high difficulty of implementation and application to practical settings, and/or lack of strong theoretical guarantees. Unlike the algorithms
of prior work, our method is simultaneously simple-to-implement, exhibits theoretical guarantees, and naturally extends to streaming and dynamic data settings, where the input data set is so large that it may not be possible to store
or process all the data at one time.

3   Problem Definition

Let $P = \{(x, y) : x \in \mathbb{R}^d \times 1, y \in \{\pm 1\}\}$ denote a set of n input points. Note
that for each point $p = (x, y) \in P$, the last entry $x_{d+1} = 1$ of x accounts for
the bias term embedding into the feature space[1]. To present our results with full
generality, we consider the setting where the input points P may have weights
associated with them. Hence, given P and a weight function $u : P \to \mathbb{R}_{\geq 0}$, we
let $\mathcal{P} = (P, u)$ denote the weighted set with respect to P and u. The canonical
unweighted case can be represented by the weight function that assigns a uniform
weight of 1 to each point, i.e., $u(p) = 1$ for every point $p \in P$. For every $T \subseteq P$,
let $U(T) = \sum_{p \in T} u(p)$. We consider the scenario where n is much larger than
the dimension of the data points, i.e., $n \gg d$.

For a normal to a separating hyperplane $w \in \mathbb{R}^{d+1}$, let $w_{1:d}$ denote vector
which contains the first d entries of w. The last entry of w (w_{d+1}) encodes the

---

[1] We perform this embedding for ease of presentation later on in our analysis.

bias term $b \in \mathbb{R}$. Under this setting, the hinge loss of any point $p = (x, y) \in P$ with respect to a normal to a separating hyperplane, $w \in \mathbb{R}^{d+1}$, is defined as $h(p, w) = [1 - y\langle x, w\rangle]_+$, where $[\cdot]_+ = \max\{0, \cdot\}$. As a prelude to our subsequent analysis of sensitivity-based sampling, we quantify the contribution of each point $p = (x, y) \in P$ to the SVM objective function as

$$f_\lambda(p, w) = \frac{1}{2U(P)} \|w_{1:d}\|_2^2 + \lambda h(p, w), \tag{1}$$

where $\lambda \in [0, 1]$ is the SVM regularization parameter, and $h(p, w) = [1 - y\langle x, w\rangle]_+$ is the hinge loss with respect to the query $w \in \mathbb{R}^{d+1}$ and point $p = (x, y)$. Putting it all together, we formalize the λ-regularized SVM problem as follows.

Definition 1 (λ-regularized SVM Problem). *For a given weighted set of points $\mathcal{P} = (P, u)$ and a regularization parameter $\lambda \in [0, 1]$, the λ-regularized SVM problem with respect to $\mathcal{P}$ is given by*

$$\min_{w \in \mathbb{R}^{d+1}} F_\lambda(\mathcal{P}, w),$$

where

$$F_\lambda(\mathcal{P}, w) = \sum_{p \in \mathcal{P}} u(p) f(p, w). \tag{2}$$

We let w^* denote the optimal solution to the SVM problem with respect to $\mathcal{P}$, i.e., $w^* \in \operatorname{argmin}_{w \in \mathbb{R}^{d+1}} F_\lambda(\mathcal{P}, w)$. A solution $\hat{w} \in \mathbb{R}^{d+1}$ is an ξ-approximation to the SVM problem if $F_\lambda(\mathcal{P}, \hat{w}) \leq F_\lambda(\mathcal{P}, w^*) + \xi$. Next, we formalize the *coreset guarantee* that we will strive for when constructing our data summaries.

Coresets. A coreset is a compact representation of the full data set that provably approximates the SVM cost function (2) for *every* query $w \in \mathbb{R}^{d+1}$ – including that of the optimal solution w^*. We formalize this notion below for the SVM problem with objective function $F_\lambda(\cdot)$ as in (2) below.

Definition 2 (ε-coreset). *Let $\varepsilon \in (0, 1)$ and let $\mathcal{P} = (P, u)$ be the weighted set of training points as before. A weighted subset $\mathcal{S} = (S, v)$, where $S \subset P$ and $v : S \to \mathbb{R}_{\geq 0}$ is an ε-coreset for $\mathcal{P}$ if*

$$\forall w \in \mathbb{R}^{d+1} \quad |F_\lambda(\mathcal{P}, w) - F_\lambda(\mathcal{S}, w)| \leq \varepsilon F_\lambda(\mathcal{P}, w). \tag{3}$$

This strong guarantee implies that the models trained on the coreset $\mathcal{S}$ with *any* off-the-shelf SVM solver will be approximately (and provably) as good as the optimal solution w^* obtained by training on the entire data set $\mathcal{P}$. This also implies that, if the size of the coreset is provably small, e.g., logartihmic in n (see Sect. 5), then an approximately optimal solution can be obtained much more quickly by training on $\mathcal{S}$ rather than $\mathcal{P}$, leading to computational gains in practice for both offline and streaming data settings (see Sect. 6).

The difficulty in constructing coresets lies in constructing them (i) *efficiently*, so that the preprocess-then-train pipeline takes less time than training on the full data set and (ii) *accurately*, so that important data points – i.e., those that are imperative to obtaining accurate models – are not left out of the coreset, and redundant points are eliminated so that the coreset size is small. In the following sections, we introduce and analyze our coreset algorithm for the SVM problem.

4   Method

Our coreset construction scheme is based on the unified framework of [10,17] and is shown in Algorithm 1. The crux of our algorithm lies in generating the importance sampling distribution via efficiently computable upper bounds (proved in Sect. 5) on the importance of each point (Lines 1–10). Sufficiently many points are then sampled from this distribution and each point is given a weight that is inversely proportional to its sample probability (Lines 11–12). The number of points required to generate an ε-coreset with probability at least $1 - \delta$ is a function of the desired accuracy ε, failure probability δ, and complexity of the data set (t from Theorem 1). Under mild assumptions on the problem at hand (see Sect. 5.3), the required sample size is polylogarithmic in n.

Algorithm 1: $\text{CORESET}(P, u, \lambda, \xi, k, m)$

Input : A set of training points $P \subseteq \mathbb{R}^{d+1} \times \{-1, 1\}$ containing n points, weight function $u : P \to \mathbb{R}_{\geq 0}$, a regularization parameter $\lambda \in [0, 1]$, an approximation factor $\xi > 0$, a positive integer k, a sample size m

Output: An weighted set (S, v) which satisfies Theorem 1

1 $\tilde{w} \leftarrow$ An ξ-approximation for the optimal SVM of (P, u);

2 $\widetilde{opt}_\xi \leftarrow F_\lambda(\mathcal{P}, \tilde{w}) - \xi$;

3 **for** $y \in \{-, +\}$ **do**

4      $P_y \leftarrow$ all the points in P that are associated with the label y;

5      $\left(c_y^{(i)}, P_y^{(i)}\right)_{i=1}^{k} \leftarrow \text{K-MEANS++}(\mathcal{P}, k)$;

6   **for** *every* $i \in [k]$ **do**

7          $\alpha_y^{(i)} \leftarrow \dfrac{U(P \backslash P_y^{(i)})}{2\lambda U(P) U(P_y^{(i)})}$;

8   **for** *every* $p = (x, y) \in P_y^{(i)}$ **do**

9              $p_\Delta \leftarrow c_y^{(i)} - yx$;

10             $\gamma(p) \leftarrow$

            $\dfrac{u(p)}{U(P_y^{(i)})} + \lambda u(p) \dfrac{9}{2} \max\left\{ \dfrac{4}{9} \alpha_y^{(i)}, \sqrt{4\left(\alpha_y^{(i)}\right)^2 + \dfrac{2\|p_\Delta\|_2^2}{9\widetilde{opt}_\xi}} - 2\alpha_y^{(i)} \right\}$;

11 $t \leftarrow \sum_{p \in P} \gamma(p)$;

12 $(S, v) \leftarrow m$ weighted samples from $\mathcal{P} = (P, u)$ where each point $p \in P$ is sampled with probability $q(p) = \frac{\gamma(p)}{t}$ and, if sampled, has weight $v(p) = \frac{u(p)}{mq(p)}$;

13 **return** (S, v);

Our algorithm is an importance sampling procedure that first generates a judicious sampling distribution based on the structure of the input points and samples sufficiently many points from the original data set. The resulting weighted set of points $\mathcal{S} = (S, v)$, serves as an unbiased estimator for $F_\lambda(\mathcal{P}, w)$ for any query $w \in \mathbb{R}^{d+1}$, i.e., $\mathbb{E}[F_\lambda(\mathcal{S}, w)] = F_\lambda(\mathcal{P}, w)$. Although sampling points uniformly with appropriate weights can also generate such an unbiased estimator, it turns out that the variance of this estimation is minimized if the points are sampled according to the distribution defined by the ratio between each point's sensitivity and the sum of sensitivities, i.e., $\gamma(p)/t$ on Line 12 [3].

4.1 Computational Complexity

Coresets are intended to provide efficient and provable approximations to the optimal SVM solution. However, the very first line of our algorithm entails computing an (approximately) optimal solution to the SVM problem. This seemingly eerie phenomenon is explained by the merge-and-reduce technique [12] that ensures that our coreset algorithm is only run against small partitions of the original data set [6,12,21]. The merge-and-reduce approach leverages the fact that coresets are composable and reduces the coreset construction problem for a (large) set of n points into the problem of computing coresets for $\frac{n}{2|S|}$ points, where $2|S|$ is the minimum size of input set that can be reduced to half using Algorithm 1 [6]. Assuming that the sufficient conditions for obtaining polylogarithmic size coresets implied by Theorem 1 hold, the overall time required is approximately linear in n.

5 Analysis

In this section, we analyze the sample-efficiency and computational complexity of our algorithm. The outline of this section is as follows: we first formalize the importance (i.e., *sensitivity*) of each point and summarize the necessary conditions for the existence of small coresets. We then present the negative result that, in general, sublinear coresets do not exist for *every* data set (Lemma 1). Despite this, we show that we can obtain accurate approximations for the sensitivity of each point via an approximate k-means clustering (Lemmas 2 and 3), and present non-vacuous, data-dependent bounds on the sample complexity (Theorem 1).

5.1 Preliminaries

We will henceforth state all of our results with respect to the weighted set of training points $\mathcal{P} = (P, u)$, $\lambda \in [0, 1]$, and SVM cost function F_λ (as in Sect. 3). The definition below rigorously quantifies the *relative contribution* of each point.

Definition 3 (Sensitivity [6]). *The sensitivity of each point $p \in P$ is given by*

$$s(p) = \sup_w \frac{u(p)f_\lambda(p, w)}{F_\lambda(\mathcal{P}, w)}. \tag{4}$$

Note that in practice, exact computation of the sensitivity is intractable, so we usually settle for (sharp) upper bounds on the sensitivity $\gamma(p) \geq s(p)$ (e.g., as in Algorithm 1). Sensitivity-based importance sampling then boils down to normalizing the sensitivities by the normalization constant – to obtain an importance sampling distribution – which in this case is the *sum of sensitivities* $t = \sum_{p \in P} s(p)$. It turns out that the required size of the coreset is at least linear in t [6], which implies that one immediate necessary condition for sublinear coresets is $t \in o(n)$.

5.2   Lower Bound for Sensitivity

The next lemma shows that a sublinear-sized coreset cannot be constructed for *every* SVM problem instance. The proof of this result is based on demonstrating a hard point set for which the sum of sensitivities is $\Omega(n\lambda)$, ignoring d factors, which implies that sensitivity-based importance sampling roughly boils down to uniform sampling for this data set. This in turn implies that if the regularization parameter is too large, e.g., $\lambda = \theta(1)$, and if $d \ll n$ (as in Big Data applications) then the required number of samples for property (3) to hold is $\Omega(n)$.

Lemma 1. *For an even integer $d \geq 2$, there exists a set of weighted points $\mathcal{P} = (P, u)$ such that*

$$s(p) \geq \frac{n\lambda + d^2}{n\left(\lambda + d^2\right)} \qquad \forall p \in P \qquad and \qquad \sum_{p \in P} s(p) \geq \frac{n\lambda + d^2}{\left(\lambda + d^2\right)}.$$

We next provide upper bounds on the sensitivity of each data point with respect to the complexity of the input data. Despite the non-existence results established above, our upper bounds shed light into the class of problems for which small-sized coresets are ensured to exist.

5.3   Sensitivity Upper Bound

In this subsection we present sharp, data-dependent upper bounds on the sensitivity of each point. Our approach is based on an approximate solution to the k-means clustering problem and to the SVM problem itself (as in Algorithm 1). To this end, we will henceforth let k be a positive integer, $\xi \in [0, F_\lambda(\mathcal{P}, w^*)]$ be the error of the (coarse) SVM approximation, and let $(c_y^{(i)}, P_y^{(i)})$, $\alpha_y^{(i)}$ and p_Δ for every $y \in \{+, -\}$, $i \in [k]$ and $p \in P$ as in Lines 4–9 of Algorithm 1.

Lemma 2. *Let k be a positive integer, $\xi \in [0, F_\lambda(\mathcal{P}, w^*)]$, and let $\mathcal{P} = (P, u)$ be a weighted set. Then for every $i \in [k]$, $y \in \{+, -\}$ and $p \in P_y^{(i)}$,*

$$s(p) \leq \frac{u(p)}{U(P_y^{(i)})} + \lambda u(p) \frac{9}{2} \max \left\{ \frac{4}{9} \alpha_y^{(i)}, \sqrt{4\left(\alpha_y^{(i)}\right)^2 + \frac{2\|p_\Delta\|_2^2}{9\widetilde{opt}_\xi} - 2\alpha_y^{(i)}} \right\} = \gamma(p).$$

Lemma 3. *In the context of Lemma 2, the sum of sensitivities is bounded by*

$$\sum_{p\in P} s(p) \leq t = 4k + \sum_{i=1}^{k} \frac{3\lambda\, Var_+^{(i)}}{\sqrt{2\widetilde{opt}_\xi}} + \frac{3\lambda\, Var_-^{(i)}}{\sqrt{2\widetilde{opt}_\xi}},$$

where $Var_y^{(i)} = \sum_{p\in P_y^{(i)}} u(p) \|p_\Delta\|_2$ for all $i \in [k]$ and $y \in \{+,-\}$.

Theorem 1. *For any $\varepsilon \in (0,1/2), \delta \in (0,1)$, let m be an integer satisfying*

$$m \in \Omega\left(\frac{t}{\varepsilon^2}\left(d\log t + \log(1/\delta)\right)\right),$$

where t is as in Lemma 3. Invoking CORESET *with the inputs defined in this context yields a ε-coreset $\mathcal{S} = (S, v)$ with probability at least $1 - \delta$ in $\mathcal{O}(nd + T)$ time, where T represents the computational complexity of obtaining an ξ-approximated solution to SVM and applying k-means++ on P_+ and P_-.*

Sufficient Conditions and the Effect of k-Means on Our Sensitivity. Theorem 1 immediately implies that, for reasonable ε and δ, coresets of poly-logarithmic (in n) size can be obtained if $d = \mathcal{O}(\text{polylog}(n))$, which is usually the case in our target Big Data applications, and if $\sum_{i=1}^{k} \frac{3\lambda Var_+^{(i)}}{\sqrt{2\widetilde{opt}_\xi}} + \frac{3\lambda Var_-^{(i)}}{\sqrt{2\widetilde{opt}_\xi}} = \mathcal{O}(\text{polylog}(n))$.

Despite the fact that any k-partitioning of the data can be applied instead of k-means for achieving upper bound on the sensitivities of the points, its important to note that k-means actually acts as a trade-off mechanism between the *raw contribution* and the *actual contribution* (the weight term and the max term from Lemma 2, respectively). Choosing the best k can be done via binary search over the values of k that minimize the sensitivity. We refer the reader to literature on the Silhouette and Elbow methods [16] as ways to pick the optimal k.

6   Results

In this section, we present experimental results that demonstrate and compare the effectiveness of our algorithm on a variety of synthetic and real-world data

Table 1. The number of input points and measurements of the total sensitivity computed empirically for each data set in the offline setting. The sum of sensitivities is significantly less than n for virtually all of the data sets, which, by Theorem 1, ensures the sample-efficiency of our approach on the evaluated scenarios.

Measurements	Dataset					
	HTRU	Credit	Pathol.	Skin	Cod	W1
Number of data-points (n)	17,898	30,000	1,000	245,057	488,565	49,749
Sum of Sensitivities (t)	475.8	1,013.0	77.6	271.5	2,889.2	24,231.6
t/n (Percentage)	2.7%	3.4%	7.7%	0.1%	0.6%	51.3%

sets in offline and streaming data settings [18]. Our empirical evaluations demonstrate the practicality and wide-spread effectiveness of our approach: our algorithm consistently generated more compact and representative data summaries, and yet incurred a negligible increase in computational complexity when compared to uniform sampling.

Evaluation. We considered 6 real-world data sets of varying size and complexity as depicted in Table 1. For each data set of size n, we selected a set of $M = 15$ geometrically-spaced subsample sizes $m_1, \ldots, m_M \subset [\log n, n^{4/5}]$. For each sample size m, we ran each algorithm (Algorithm 1 or uniform sampling) to construct a subset $\mathcal{S} = (S, v)$ of size m. We then trained the SVM model as per usual on this subset to obtain an optimal solution with respect to the coreset $\mathcal{S}$, i.e., $w_{\mathcal{S}}^* = \mathrm{argmin}_w F_\lambda(\mathcal{S}, w)$. We then computed the relative error incurred by the solution computed on the coreset $(w_{\mathcal{S}}^*)$ with respect to the ground-truth optimal solution computed on the entire data set (w^*): $|F_\lambda(P, w_{\mathcal{S}}^*) - F_\lambda(P, w^*)|/F_\lambda(P, w^*)$. The results were averaged across 100 trials.

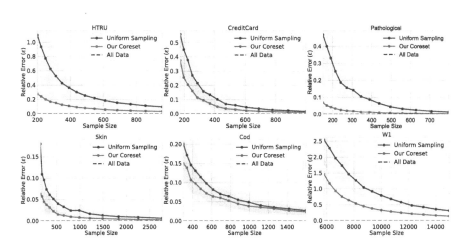

Fig. 1. The relative error of query evaluations with respect uniform and coreset subsamples for the 6 data sets in the offline setting. Shaded region corresponds to values within one standard deviation of the mean.

Figures 1 and 2 depict the results of our comparisons against uniform sampling in the offline setting. In Fig. 1, we see that the coresets generated by our algorithm are much more representative and compact than the ones constructed by uniform sampling: across all data sets and sample sizes, training on our coreset yields significantly better solutions to SVM problem when compared to those generated by training on a uniform sample. For certain data sets, such as HTRU, Pathological, and W1, this relative improvement over uniform sampling is at least an order of magnitude better, especially for small sample sizes. Figure 1 also shows that, as a consequence of a more informed sampling scheme, the variance of each model's performance trained on our coreset is much lower than that of uniform sampling for all data sets.

Figure 2 shows the total computational time required for constructing the sub-sample (i.e., coreset) $\mathcal{S}$ and training the SVM on the subset $\mathcal{S}$ to obtain $w_{\mathcal{S}}^*$. We observe that our approach takes significantly less time than training on the original model when considering non-trivial data sets (i.e., $n \geq 18{,}000$), and underscores the efficiency of our method: we incur a negligible cost in the overall SVM training time due to a more involved coreset construction procedure, but benefit heavily in terms of the accuracy of the models generated (Fig. 1).

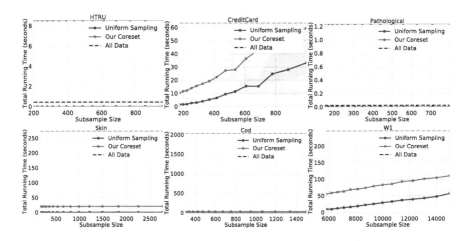

Fig. 2. The *total* computational cost of constructing a coreset and training the SVM model on the coreset, plotted as a function of the size of the coreset.

Next, we evaluate our approach in the streaming setting, where data points arrive one-by-one and the entire data set cannot be kept in memory, for the same 6 data sets. The results of the streaming setting are shown in Fig. 3. Figures 3 portray a similar trend as the one we observed in our offline evaluations: our approach significantly outperforms uniform sampling for all of the evaluated data sets and sample sizes, with negligible computational overhead.

In sum, our empirical evaluations demonstrate the practical efficiency of our algorithm and reaffirm the favorable theoretical guarantees of our approach: the additional computational complexity of constructing the coreset is negligible relative to that of uniform sampling, and the entire preprocess-then-train pipeline is significantly more efficient than training on the original massive data set.

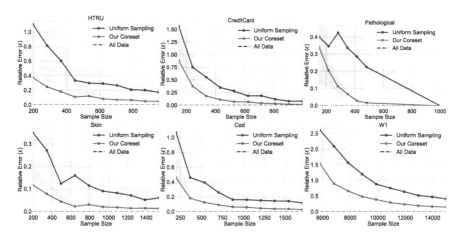

Fig. 3. The relative error of query evaluations with respect uniform and coreset sub-samples for the 6 data sets in the streaming setting. The figure shows that our method tends to fare even better in the streaming setting (cf. Fig. 1).

7   Conclusion

We presented an efficient coreset construction algorithm for generating compact representations of the input data points that are provably competitive with the original data set in training Support Vector Machine models. Unlike prior approaches, our method and its theoretical guarantees naturally extend to streaming settings and scenarios involving dynamic data sets, where points are continuously inserted and deleted. We established instance-dependent bounds on the number of samples required to obtain accurate approximations to the SVM problem as a function of input data complexity and established dataset dependent conditions for the existence of compact representations. Our experimental results on real-world data sets validate our theoretical results and demonstrate the practical efficacy of our approach in speeding up SVM training. We conjecture that our coreset construction can be extended to accelerate SVM training for other classes of kernels and can be applied to a variety of Big Data scenarios.

References

1. Agarwal, P.K., Har-Peled, S., Varadarajan, K.R.: Geometric approximation via coresets. Comb. Comput. Geom. **52**, 1–30 (2005)
2. Agarwal, P.K., Sharathkumar, R.: Streaming algorithms for extent problems in high dimensions. In: Proceedings of the Twenty-First Annual ACM-SIAM Symposium on Discrete Algorithms, pp. 1481–1489. Society for Industrial and Applied Mathematics (2010)
3. Bachem, O., Lucic, M., Krause, A.: Practical coreset constructions for machine learning. arXiv preprint arXiv:1703.06476 (2017)

4. Badoiu, M., Clarkson, K.L.: Smaller core-sets for balls. In: Proceedings of the Fourteenth Annual ACM-SIAM Symposium on Discrete Algorithms, pp. 801–802. Society for Industrial and Applied Mathematics (2003)

5. Baykal, C., Liebenwein, L., Gilitschenski, I., Feldman, D., Rus, D.: Data-dependent coresets for compressing neural networks with applications to generalization bounds. arXiv preprint arXiv:1804.05345 (2018)

6. Braverman, V., Feldman, D., Lang, H.: New frameworks for offline and streaming coreset constructions. arXiv preprint arXiv:1612.00889 (2016)

7. Clarkson, K.L.: Coresets, sparse greedy approximation, and the Frank-Wolfe algorithm. ACM Trans. Algorithms (TALG) 6(4), 63 (2010)

8. Clarkson, K.L., Hazan, E., Woodruff, D.P.: Sublinear optimization for machine learning. J. ACM (JACM) 59(5), 23 (2012)

9. Feldman, D.: Core-Sets: An Updated Survey. Wiley Interdisciplinary Reviews: Data Mining and Knowledge Discovery, p. e1335 (2019)

10. Feldman, D., Langberg, M.: A unified framework for approximating and clustering data. In: Proceedings of the Forty-Third Annual ACM Symposium on Theory of Computing, pp. 569–578. ACM (2011)

11. Gärtner, B., Jaggi, M.: Coresets for polytope distance. In: Proceedings of the Twenty-Fifth Annual Symposium on Computational Geometry, pp. 33–42. ACM (2009)

12. Har-Peled, S., Mazumdar, S.: On coresets for k-means and k-median clustering. In: Proceedings of the Thirty-Sixth Annual ACM Symposium on Theory of Computing, pp. 291–300. ACM (2004)

13. Har-Peled, S., Roth, D., Zimak, D.: Maximum margin coresets for active and noise tolerant learning. In: IJCAI, pp. 836–841 (2007)

14. Hazan, E., Koren, T., Srebro, N.: Beating SGD: learning SVMs in sublinear time. In: Advances in Neural Information Processing Systems, pp. 1233–1241 (2011)

15. Joachims, T.: Training linear SVMs in linear time. In: Proceedings of the 12th ACM SIGKDD International Conference on Knowledge Discovery and Data Mining, pp. 217–226. ACM (2006)

16. Kodinariya, T.M., Makwana, P.R.: Review on determining number of cluster in k-means clustering. Int. J. 1(6), 90–95 (2013)

17. Langberg, M., Schulman, L.J.: Universal ε-approximators for integrals. In: Proceedings of the Twenty-First Annual ACM-SIAM Symposium on Discrete Algorithms, pp. 598–607. SIAM (2010)

18. Lichman, M.: UCI machine learning repository (2013). http://archive.ics.uci.edu/ml

19. Liebenwein, L., Baykal, C., Lang, H., Feldman, D., Rus, D.: Provable filter pruning for efficient neural networks. arXiv preprint arXiv:1911.07412 (2019)

20. Loosli, G., Canu, S.: Comments on the core vector machines: fast SVM training on very large data sets. J. Mach. Learn. Res. 8(Feb), 291–301 (2007)

21. Lucic, M., Faulkner, M., Krause, A., Feldman, D.: Training mixture models at scale via coresets. arXiv preprint arXiv:1703.08110 (2017)

22. Nandan, M., Khargonekar, P.P., Talathi, S.S.: Fast SVM training using approximate extreme points. J. Mach. Learn. Res. 15(1), 59–98 (2014)

23. Nathan, V., Raghvendra, S.: Accurate streaming support vector machines. arXiv preprint arXiv:1412.2485 (2014)

24. Rai, P., Daumé III, H., Venkatasubramanian, S.: Streamed learning: one-pass SVMs. arXiv preprint arXiv:0908.0572 (2009)

25. Shalev-Shwartz, S., Singer, Y., Srebro, N., Cotter, A.: Pegasos: primal estimated sub-gradient solver for SVM. Math. Program. 127(1), 3–30 (2011)

26. Tsang, I.W., Kocsor, A., Kwok, J.T.: Simpler core vector machines with enclosing balls. In: Proceedings of the 24th International Conference on Machine Learning, pp. 911–918. ACM (2007)
27. Tsang, I.W., Kwok, J.T., Cheung, P.M.: Core vector machines: fast SVM training on very large data sets. J. Mach. Learn. Res. **6**(Apr), 363–392 (2005)

Tractabilities for Tree Assembly Problems

Feng Shi[✉], Jie You, Zhen Zhang, and Jingyi Liu

School of Computer Science and Engineering, Central South University,
Changsha 410083, People's Republic of China
fengshi@csu.edu.cn

Abstract. Calculating the "distance" between two given objects with respect to a designated "editing" operation is a hot research area in bioinformatics, where the "distance" is always defined as the minimum number of the "editing" operations required to transform one object into the other one. One of the famous problems in the area is the Minimum Common String Partition problem, which is the simplified Minimum Tree Cut/Paste Distance problem. Within the paper, we consider another simplified version of the Minimum Tree Cut/Paste Distance problem, named Tree Assembly problem, of which the edge-deletion operations are specified. More specifically, the Tree Assembly problem aims to transform a given forest into a given tree by edge-addition operations only. In our investigations, we present a fixed-parameter algorithm with runtime $2^{O(k \log k)} n^{O(1)}$ for the Tree Assembly problem, where k and n are the numbers of trees and nodes in the forest, respectively. Additionally, we give a polynomial-time algorithm for a restricted variant of the problem.

Keywords: Tree assembly problem · Fixed-parameter algorithm · Tree editing distance · Subforest isomorphism

1 Introduction

The Subgraph Isomorphism problem is a classical decision problem in the area of theoretical computer science, where its input comprises two graphs H and G, and the goal is to decide whether H is a subgraph of G. As we know, a plenty of well-known NP-hard problems are restricted versions of the problem, such as the k-clique problem, which decides if there exists a complete graph with at least k vertices in G. The Subgraph Isomorphism problem remains NP-hard on many constrained graph classes [7,8,12], and even some of them are quite specific, e.g., H is a tree and G is a graph of tree-width two, and at most one node in either of them has degree larger than three [13]. Within the paper, we consider a variant of the Subgraph Isomorphism problem, where H is forest and G is a tree, and they have the same number of nodes. Note that if H is further restricted to a tree, then the problem is actually the Tree Isomorphism problem, which can be

This work is supported by the National Natural Science Foundation of China under Grants 61802441, 61672536, 61420106009, 61872450.

ⓒ Springer Nature Switzerland AG 2020
J. Chen et al. (Eds.): TAMC 2020, LNCS 12337, pp. 300–312, 2020.
https://doi.org/10.1007/978-3-030-59267-7_26

solved in linear runtime [1]. However, if H is a forest, then the problem becomes NP-complete even if each tree in H is a path or star [2,10]. Additionally, You et al. [16] showed that when H comprises a collection of paths (or stars), the problem is *fixed-parameter tractable* by the number of trees in H. A problem is *fixed-parameter tractable* (FPT) by a parameter k, if it admits an algorithm with runtime $f(k) \times n^{O(1)}$ for some function f depending on k only [5].

Given a forest F and a tree T, the Tree Assembly (TA) problem studied in the paper asks whether a tree that is isomorphic to T can be obtained by assembling the trees in F together with k edge-addition operations. It is easy to see that k is exactly the number of trees in F minus one; in this sense, the problem is not an optimization problem, but a decision problem. One closely related problem is the Minimum Tree Cut/Paste Distance problem [10], which considers two trees T_1 and T_2, and aims to use the minimum number of edge-deletion and edge-addition operations to transform T_1 into T_2. By a quick inspection on the definition, the TA problem is indeed the processed Minimum Tree Cut/Paste Distance problem, of which all edge-deletion operations are specified. You et al. [15] showed that the Minimum Common String Partition problem [3,4] is also a restricted version of the Minimum Tree Cut/Paste Distance problem, which has been shown to be FPT by the partition size [3]. However, whether the Minimum Tree Cut/Paste Distance problem is FPT, is unknown at so far. Thus we believe that the research on the TA problem can impel the work on the fixed-parameter tractability of the Minimum Tree Cut/Paste Distance problem.

With respect to the Minimum Common String Partition problem, the input consists of two strings (note that each character has the same number of occurrences in the two input strings), and it asks whether we can use at most $2k$ "string-cut" operations (at most k operations on each string) to make them become the same collection of substrings. The problem remains NP-hard, even if one of the two strings has already been cut into substrings, which is called the Exact Block Cover problem [9]: Given a string X and a collection of k substrings, it asks whether the k substrings can form a string that is identical with X. Jiang et al. [9] gave a fixed-parameter algorithm with runtime $O^*(2^k)$ for the problem. Again, the reduction due to You et al. [15] indicates that the TA problem considered in the paper is a generalization of the Exact Block Cover problem. This inspires us that the algorithm proposed by Jiang et al. [9] may solve the TA problem directly. Unfortunately, it does not work. In their algorithm, they reduced the problem into the Traveling Salesman problem [11], however, this fails on the TA problem as the number of possible edge-addition operations between two trees cannot be upper bounded by a constant. Fortunately, a simple reduction rule that can downsize the considered trees is observed. Based on this, we give a fixed-parameter algorithm with runtime $2^{O(k \log k)} n^{O(1)}$ for the TA problem in the paper, indicating that it is FPT by the number of trees in F.

Baumbach et al. [2] showed that if each tree in F is a star, and the types of stars can be bounded by a constant, then the TA problem is polynomial-time solvable. Hence, it is natural to consider whether the TA problem is polynomial-time solvable if the types of trees in F are limited. However, it is not easy to answer: Let F consist of a set of paths and T be a spider tree (only one internal node can have degree larger than 2, and the path from each leaf to the internal node has at least one edge), then the problem is a variant of the Bin-packing problem [14], where the number of different volumes can be bounded by a constant. In the paper, we consider a restricted variant of the TA problem, named Level-2 $(1, k)$-Tree Assembly problem (L2-$(1, k)$-TA problem), where the considered forest F comprises two kinds of trees, called A-tree and B-tree, and F has exactly one A-tree and k B-trees. The problem decides whether the k B-trees can be attached to the A-tree by k edge-addition operations such that the obtained tree is isomorphic to T. This variant has some restrictions, and a polynomial-time algorithm for it is proposed that is surprisingly tricky.

2   Preliminary

A tree T is *rooted* if there is a unique ancestor-descendant relationship defined in the tree, and with respect to which there is exactly one common ancestor for all its nodes, where the common ancestor is called the *root* of the tree. More specifically, a node u is an *ancestor* of a node v (or v is a *descendant* of u) if u is visited by the path from the root of T to v; in this sense, v is an ancestor (or descendant) of itself. All trees considered in the paper are rooted. Denote by $V(T)$ and $E(T)$ the node set and edge set of T respectively, where $|V(T)|$ is the *size* of T, marked as $n(T)$. An edge e of T is denoted by a node pair $[u, v]$, where u and v are endpoints of e, and u is an ancestor of v. For any subset $U \subset V(T)$, $T[U]$ denotes the *induced subforest* of T comprising all nodes of U and all edges with both endpoints in U. In particular, $T[U]$ is a *subtree* of T if $T[U]$ is connected. For a node v in $V(T)$, denote by $d_T(v)$ the number of edges incident to v in T, which is called the *degree* of v in T; by $C_T(v)$ the set of descendants that are adjacent to v (i.e., the *children* of v); and by $T(v)$ the *pendant subtree* rooted at v, which is induced by all descendants of v in T. A tree T_1 is *isomorphic* to another tree T_2, denoted by $T_1 = T_2$, if there is a bijection Φ from $V(T_1)$ to $V(T_2)$ such that $[v, v']$ is an edge of T_1 if and only if $[\Phi(v), \Phi(v')]$ is an edge of T_2. All forests considered in the paper are rooted, where a forest is rooted if all its trees are rooted. Denote by $|F|$ the number of trees in a forest F, and by $n(F)$ the number of nodes in F. For two forests F_1 and F_2, $F_1 \cup F_2$ denotes the forest consisting of all the trees in F_1 and F_2. For a forest F and an edge-set E, $F + E$ denotes the graph obtained by adding the edges of E into F.

Given two nodes u and v of T_1 and T_2, respectively, an *attach operation* on u and v adds a new edge between u and v into the resulting tree. As the resulting tree should be rooted, the additional edge between u and v is required to claim an ancestor-descendant relationship. W.l.o.g., assume that u is an ancestor of v, and then, we say that T_2 is *attached to* T_1. Obviously, v should be the root of T_2; otherwise, there is no common ancestor in the resulting tree. It is necessary to point out that u can be any node in T_1. An *attaching approach* for some forest F is a set of $|F| - 1$ edges, with which one can assemble the trees in F as a tree by $|F| - 1$ attach operations. Let $T' = F + E_+$ be the resulting tree for some attaching approach E_+ of F. For two trees T_1 and T_2 of F, if the path, from the root of T' to that of T_2, visits the root of T_1, then we say that T_2 is *attached below* T_1 with respect to E_+. Precisely, T_2 is attached to T_1 if T_2 is below T_1 and no other root is visited by the path from the root of T_1 to that of T_2 in T'.

The parameterized version of the Tree Assembly problem considered in the paper is formally given as follows.

Tree Assembly problem (TA problem)

INPUT: A forest F and a target tree T^* with $n(F) = n(T^*)$;
PARAMETER: $k = |F|$;
OUTPUT: Return YES if there is an attaching approach E_+ making
$\qquad F + E_+ = T^*$; otherwise, return NO.

An attaching approach E_+ of F is of *level-2*, if there is a tree T in F such that all the other trees are attached to it with respect to E_+, where T is called the *top subtree* with respect to E_+. Next we define a restricted version of the TA problem as follows, where the considered forest F consists of two kinds of trees, denoted by A-tree and B-tree.

Level-2 $(1, k)$-Tree Assembly problem (L2-$(1, k)$-TA problem)

INPUT: A forest F consisting of one A-tree and k B-trees, and a target
$\qquad$ tree T^* with $n(F) = n(T^*)$;
OUTPUT: Return YES if there is a level-2 attaching approach E_+ making
$\qquad F + E_+ = T^*$, where the A-tree is the top subtree with respect
$\qquad$ to E_+; otherwise, return NO.

3 An FPT Algorithm for TA Problem

Given two (rooted) trees with size n, we have the critical lemma given below (for more details, please refer to pp. 84-85 in [1]). Thus given an instance (F, T^*) of the TA problem, if $|F| = 1$, then it is easy to derive whether (F, T^*) is YES.

Lemma 1 [1]. *It takes runtime $O(n)$ to decide whether or not two rooted trees with size n are isomorphic.*

3.1 Simple Observations

In the following discussion, we assume that (F, T^*) is YES and $|F| \geq 2$. There is an attaching approach E_+ making $F + E_+ = T^*$, which we call a *feasible* attaching approach for (F, T^*). Meanwhile, the feasible attaching approach E_+ implies a bijection Φ from $V(F)$ to $V(T^*)$ such that if $[v, v']$ is an edge of F then $[\Phi(v), \Phi(v')]$ is an edge of T^*. Remark that in the remaining text, we also simply say that $\Phi(v)$ is the image of v (or v maps to $\Phi(v)$) with respect to E_+.

Let v be a node of one tree T in F that maps to $v^* \in V(T^*)$ with respect to Φ. Observe that $|C_T(v)| \leq |C_{T^*}(v^*)|$; otherwise, contradicting to the assumption that (F, T^*) is YES. For the children of v and v^*, we have the following two propositions (Fig. 1).

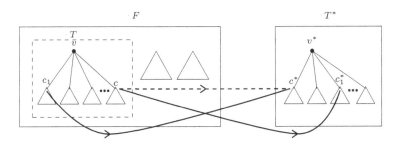

Fig. 1. The mapping relation defined in the proof of Proposition 2. The left solid box represents the forest F, in which the dashed one represents the tree T. The right solid box represents the target tree T^*. The solid thick line from c_1 to c^* represents the bijection from c_1 to c^* with respect to Φ, which applies to the one from c to c_1^*. The dashed thick line from c to c^* represents the bijection from c to c^* with respect to Φ'.

Proposition 1. *Each node in $C_T(v)$ maps to one in $C_{T^*}(v^*)$ with respect to Φ.*

Proposition 2. *If there are two nodes $c \in C_T(v)$ and $c^* \in C_{T^*}(v^*)$ satisfying that $T(c) = T^*(c^*)$, then there is a feasible attaching approach for (F, T^*), with respect to which the nodes in $V(T(c))$ maps to that in $V(T^*(c^*))$.*

With respect to the feasible attaching approach E_+ for the instance (F, T^*), an *auxiliary* (rooted) tree $\mathcal{T}$ can be constructed, where each node $\nu \in V(\mathcal{T})$ corresponds to a unique tree in F, which is marked as B_ν, and $\mathcal{T}$ has an edge $[\nu, \nu']$ if and only if E_+ contains an edge with endpoints in B_ν and $B_{\nu'}$ respectively. The auxiliary tree $\mathcal{T}$ can be regarded as the "skeleton" of T^*, and it specifies the ancestor-descendant relationship among the trees in F with respect to E_+.

Lemma 2. *The auxiliary tree $\mathcal{T}$ for (F, T^*) with respect to E_+ can be obtained in $k^{O(k)}$ runtime.*

Suppose that we have found the auxiliary tree $\mathcal{T}$ for the instance (F, T^*) with respect to E_+. Let γ and r^* be the roots of $\mathcal{T}$ and T^*, respectively, and $C_{\mathcal{T}}(\gamma) = \{\nu_1, \ldots, \nu_h\}$ and $C_{T^*}(r^*) = \{u_1, \ldots, u_p\}$. Let r be the root of B_γ (recall that B_γ is the tree in F that corresponds to the node γ in $\mathcal{T}$), and $C_{B_\gamma}(r) = \{v_1, \ldots, v_q\}$. Observe that r maps to r^* with respect to E_+, and $q \leq p$. Please refer to Fig. 2. By Proposition 2, we have the following reduction rule.

Rule 1. For any v_i of $C_{B_\gamma}(r)$ with $1 \leq i \leq q$, if there is a child u of r^* satisfying that $B_\gamma(v_i) = T^*(u)$, then remove all nodes of $B_\gamma(v_i)$ and $T^*(u)$ from F and T^*, respectively.

Lemma 3. *If (F, T^*) is YES with respect to $\mathcal{T}$, and Rule 1 is not applicable on (F, T^*), then $p \leq k - 1$, where $p = |C_{T^*}(r^*)|$ and $k = |F|$.*

Now a new reduction rule is given for (F, T^*), on which Rule 1 is assumed to be not applicable.

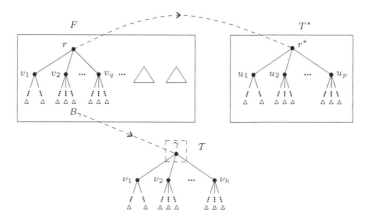

Fig. 2. The top left solid box represents the forest F, and the top right one represents the target tree T^*. The tree located at the bottom is the auxiliary tree $\mathcal{T}$ with respect to a feasible attaching approach E_+ for (F, T^*). Observe that for the root γ of $\mathcal{T}$, the root r of its corresponding tree B_γ in F maps to the root r^* of T^*.

Rule 2. If $q = |C_{B_\gamma}(r)| = 1$ and $p = |C_{T^*}(r^*)| = 1$, then remove the nodes r and r^* from F and T^*, respectively; if $q = |C_{B_\gamma}(r)| = 0$ and $p = |C_{T^*}(r^*)| = 1$, then remove the nodes r, r^*, and γ from F, T^*, and $\mathcal{T}$, respectively.

Note that $q = |C_{B_\gamma}(r)|$ and $p = |C_{T^*}(r^*)|$ cannot be 0 at the same time, as F is assumed to have at least two trees.

Lemma 4. *For the instance I' and auxiliary tree $\mathcal{T}'$ obtained by an application of Rule 2 on (F, T^*), I' is* YES *with respect to $\mathcal{T}'$ if and only if (F, T^*) is* YES *with respect to $\mathcal{T}$.*

3.2 Constructing Sub-instances by Orderings

Now we assume that the values of q and p do not meet the cases given in Rule 2. By Lemma 3, for each child v of r, there is at least one tree in $F \setminus B_\gamma$ that is attached to some node of $B_\gamma(v)$ with respect to the feasible attaching approach E_+ for (F, T^*). In the following, we give the way to find the trees in $F \setminus B_\gamma$ that are attached below $B_\gamma(v)$, and the node in $C_{T^*}(r^*)$ that is the image of v with respect to E_+, based on the orderings of the children of r, r^*, and γ.

We first fix an arbitrary ordering for the children of r, $\{v_1, \ldots, v_q\}$, where $q \leq p \leq h \leq k - 1$. A notion that is critical in the following lemma is introduced here. For an ordering of the children $u_1, \ldots, u_p$ of r^*, it is *adjusted* if the nodes in $\Phi(C_{B_\gamma}(r))$ are always on the left of $C_{T^*}(r^*) \setminus \Phi(C_{B_\gamma}(r))$ (recall that Φ is a bijection from $V(F)$ to $V(T^*)$ specified by E_+).

Lemma 5. *If (F, T^*) is* YES*, then there exists an adjusted ordering $\rho = u_1, \ldots, u_p$ for the children of r^* and an ordering $\sigma = \nu_1, \ldots, \nu_h$ for the children of γ satisfying the following two properties:*

(1). $\Phi(v_i) = u_i$ for all $1 \le i \le q$;
(2). A partition $\mathcal{P} = \{\mathcal{X}_1, \ldots, \mathcal{X}_p\}$ for $v_1, \ldots, v_h$ can be found in polynomial-time such that for each node v' of $\mathcal{T}(v)$, where v is in $\mathcal{X}_i$ (for any $1 \le i \le p$), the images of the nodes in $B_{v'}$ are in $T^*(u_i)$ with respect to Φ.

Proof. For Property (1), as an ordering for the children of r is fixed, we can find an ordering $\rho = u_1, \ldots, u_p$ for the children of r^* such that $\Phi(v_i) = u_i$ for all $1 \le i \le q$ by enumerating all possible orderings of the children of r^*. Note that the ordering is adjusted.

Now we consider Property (2). By Lemma 3, for any child v of γ, the image of B_v is a subtree in some $T^*(u_i)$ $(1 \le i \le p)$. Moreover, by the definition of $\mathcal{T}$, if for the child v of γ, the image of B_v is a subtree in $T^*(u_i)$, then for every node v' in $\mathcal{T}(v)$, the image of $B_{v'}$ is as well a subtree in $T^*(u_i)$. Additionally, there may be more than one child of γ such that their images are in the same $T^*(u_i)$. Hence, there is an ordering $\sigma = v_1, \ldots, v_h$ for the children of γ such that all nodes v_j with the image of B_{v_j} in $T^*(u_i)$ are consecutive in σ; moreover, these consecutive nodes form a group of the partition $\mathcal{P}$, and the ordering of the groups corresponds to that of $u_1, \ldots, u_p$. Now we give the way to find the partition for $v_1, \ldots, v_h$.

By the discussion given above, if (F, T^*) is YES, then there are p disjoint index intervals $I_1 = [1, y_1], I_2 = [y_1 + 1, y_2], \ldots, I_p = [y_{p-1} + 1, h]$ satisfying that for any $1 \le i \le q$ (recall that $q \le p$),

$$n(B_\gamma(v_i)) + \sum_{j \in I_i} \sum_{v \in V(\mathcal{T}(v_j))} n(B_v) = n(T^*(u_i)).$$

If $q = p$, then the proof is done. Otherwise, for any $q + 1 \le i \le p$, $|I_i| = 1$ and

$$\sum_{j \in I_i} \sum_{v \in V(\mathcal{T}(v_j))} n(B_v) = n(T^*(u_i)).$$

Note that for $q + 1 \le i \le p$, B_{v_j} is attached to r, where $I_i = \{j\}$. It is easy to verify that the p index intervals form a partition $\mathcal{P}$ for nodes in σ, and they can be easily found in polynomial-time. $\square$

Assume that we have gotten a partition $\mathcal{P}$ for nodes $v_1, \ldots, v_h$ with respect to two "correct" orderings σ and ρ. Now we are ready to *split* the instance (F, T^*) into p sub-instances as follows.

(1). For each $1 \le i \le q$, let $F_i = \{B_\gamma(v_i)\} \cup \{B_v | v \in V(\mathcal{T}(v')), v' \in \mathcal{X}_i\}$, and $T_i^* = T^*(u_i)$. The auxiliary tree $\mathcal{T}_i$ for (F_i, T_i^*) can be obtained by attaching every $\mathcal{T}(v)$ with $v \in \mathcal{X}_i$ to a new root γ_i, where the tree B_{γ_i} corresponding to γ_i is $B_\gamma(v_i)$.
(2). For each $q + 1 \le i \le p$, let $F_i = \{B_v | v \in V(\mathcal{T}(v')), v' \in \mathcal{X}_i\}$ and $T_i^* = T^*(u_i)$. The auxiliary tree $\mathcal{T}_i$ for (F_i, T_i^*) is $\mathcal{T}(v_j)$, where $\mathcal{X}_i = \{v_j\}$.

By Lemm 5, we have the following lemma.

Lemma 6. *The instance (F, T^*) is YES if and only if for all $1 \le i \le p$, the sub-instance (F_i, T_i^*) is YES with respect to the auxiliary tree $\mathcal{T}_i$.*

3.3   Algorithm Presentation

Now we give an algorithm with an instance (F, T^*) and an auxiliary tree $\mathcal{T}$ for (F, T^*) as input. Please refer to Fig. 3. Remark that the inputed auxiliary tree $\mathcal{T}$ is obtained by enumerating all possible situations for the k trees in the forest F, no matter whether the inputed instance (F, T^*) is YES or NO.

Algorithm Alg-TA$(F, T^*; \mathcal{T})$
Input: A forest F and a target tree T^* with $n(F) = n(T^*)$, and an auxiliary tree $\mathcal{T}$;
Output: Return YES if (F, T^*) is YES with respect to $\mathcal{T}$; otherwise, return NO.

1. **if** $|F| = 1$ **then if** F is isomorphic to T^* **then** return YES;
   else return NO;
2. exhaustively apply Rule 1 to B_γ and T^*;   //γ is the root of $\mathcal{T}$
3. **if** Rule 2 is applicable on $(F, T^*; \mathcal{T})$ **then** apply it and go to Step 1;
4. fix an ordering for the children of the root r of B_γ, $\{v_1, \ldots, v_q\}$;
5. **for** each possible ordering ρ of the children $u_1, \ldots, u_p$ of the root r^* of T^* **do**
6.   **for** each possible ordering σ of the children $v_1, \ldots, v_h$ of γ **do**
7.   **if** a partition for $v_1, \ldots, v_h$ satisfying the conditions of Lemma 5 can be found **then**
8.   create sub-instances $(F_1, T_1^*), \ldots, (F_p, T_p^*)$ and the corresponding auxiliary trees $\mathcal{T}_1, \ldots, \mathcal{T}_p$;
9.   **if** $\bigwedge_{i=1}^{p}$ Alg-TA$(F_i, T_i^*; \mathcal{T}_i)$ **then** return YES;
10. return NO.

Fig. 3. The algorithm for the TA problem with a given auxiliary tree.

Theorem 1. *There exists an algorithm with runtime $k^{O(k)} n^{O(1)}$ that can decide whether or not the instance (F, T^*) of the TA problem is YES, where $k = |F|$ and n is the size of the input.*

Proof. By Lemma 2, if (F, T^*) is YES then we can obtain an auxiliary tree $\mathcal{T}$ in k^{2k} runtime with respect to some feasible attaching approach for (F, T^*). Now we use Fig. 3 for illustration. The correctness of Step 1 is obvious. Proposition 2 and Lemma 4 show the correctness of Steps 2 and 3, respectively. If (F, T^*) is YES, by Lemma 5, there are orderings for $u_1, \ldots, u_p$ and $v_1, \ldots, v_h$ such that one can find a partition for $v_1, \ldots, v_h$ in polynomial-time. The correctness of Step 9 is guaranteed by Lemma 6. If the algorithm cannot return YES for all orderings, then Step 10 has to return NO.

   Now we analyze the runtime of the algorithm. For Step 6, it actually makes an ordering for $\mathcal{T}$: In each iteration, Step 6 finds an ordering for the children of a distinct node in $\mathcal{T}$. Since $\mathcal{T}$ contains k nodes, the total number of loops of Step 6 can be bounded by k^k. In the following, we consider the upper bound on the times of calling the algorithm Alg-TA to get a decision (YES or NO) for (F, T^*) with respect to a fixed ordering for the auxiliary tree $\mathcal{T}$ (i.e., Step 6 can be ignored at the moment), which is denoted by $f(k)$. If $k = 1$, then $|F| = 1$, and the algorithm ends at Step 1, i.e., $f(1) = 1$. If $k = 2$, then $p \leq k - 1 = 1$, implying that Rule 2 is applicable, and $f(2) = 1$. For $k \geq 3$, we have $f(k) = p! \times \sum_{i=1}^{p} f(k_i)$, where k_i is the parameter of the i-th sub-instance created by Step 8, for any $1 \leq i \leq p \leq k - 1$. It is necessary to point out that since Rule 1 cannot be applied, if $p = q$, then each $k_i \geq 2$ for all $1 \leq i \leq p$; otherwise (i.e, $p > q$), $k_i \geq 2$ for all $1 \leq i \leq q$, and $k_i \geq 1$ for all $q + 1 \leq i \leq p$. Moreover, $\sum_{i=1}^{p} k_i = k - 1 + q$. Let $k_{max} = \max\{k_1, \ldots, k_p\}$. If $q = p$, then

$$k_{max} \leq \sum_{i=1}^{q} k_i - 2q + 2 = k - q + 1 = k - p + 1.$$

Otherwise,

$$k_{max} \leq \sum_{i=1}^{q} k_i - (p + q) + 2 = k - p + 1.$$

Therefore, we always have $k_{max} \leq k - p + 1$. Then

$$f(k) \leq p! \times p \times f(k_{max}) \leq p! \times p \times f(k - p + 1).$$

Observe that $p \geq 1$ can take different values at different levels of recursion, and that the value of $f(k)$ keeps the same value if $p = 1$. Thus, w.l.o.g., we assume that the value p_s taken by p at the s-th recursion is always not less than 2 for any $1 \leq s \leq t$, where t denotes the depth of the recursion and $t \leq k$. Then

$$\sum_{s=1}^{t} p_s = k + t, \text{ and } f(k) \leq \prod_{s=1}^{t} p_s! \times p_s \leq \prod_{s=1}^{t} (p_s + 1)! \leq (k + t)! \leq 4^k \times k^{2k}.$$

Since Steps 1–4 and 7–8 can be done in polynomial-time, we have that the algorithm Alg-TA can be done in $k^{5k} \times 4^k \times n^{O(1)}$ time, i.e., the runtime can be bounded by $k^{6k} \times n^{O(1)}$ (if $k \geq 4$). It is worthy to point out that we can enumerate all auxiliary trees with all feasible orderings in runtime k^{2k}. Thus the runtime of the algorithm can be improved to $k^{5k} \times n^{O(1)}$. The proof is done. □

4   A Polynomial-Time Algorithm for L2-$(1, k)$-TA Problem

Consider an instance (F, T^*) of the L2-$(1, k)$-TA problem, where F comprises one A-tree T_A and k B-trees. The following proposition can be easily derived.

Proposition 3. *The instance (F, T^*) of the L2-$(1, k)$-TA problem is YES if and only if T^* has a subtree $T^\dagger$ that is isomorphic to the A-tree and has the same root with T^*, such that removing the nodes in $T^\dagger$ from T^* results in k B-trees.*

In the following, we reduce the problem of finding the subtree $T^\dagger$ defined in Proposition 3 to the *Maximum Common Subtree Isomorphism* (MCSI) problem. Several related notions are given below, which follow the ones given in [6]. Given two trees T_1 and T_2 that have subtrees T_1' and T_2', respectively, if T_1' and T_2' are isomorphic with respect to a bijective function Ψ from $V(T_1')$ to $V(T_2')$, then T_1' (or T_2') is a *common subtree* of T_1 and T_2.

Let f be a weight function defined on $(V(T_1) \times V(T_2)) \cup (E(T_1) \times E(T_2)) \to \mathbb{R}$. Then the *weight* of Ψ is defined as

$$\sum_{v \in V(T_1')} f(v, \Psi(v)) + \sum_{[u,v] \in E(T_1')} f([u,v], [\Psi(u), \Psi(v)]).$$

The MCSI problem on T_1 and T_2 asks for a subtree isomorphism with the *maximum* weight among all subtree isomorphisms between T_1 and T_2.

Now an instance $(T_A, T^*; f)$ of the MCSI problem can be constructed, where the setting approach for the weight function f defined on $(V(T_A) \times V(T^*)) \cup (E(T_A) \times E(T^*))$ is given as follows. Let V_{T_A} be the subset of $V(T_A)$ such that $T_A(v)$ is isomorphic to the B-tree for each node $v \in V_{T_A}$, and E_{T_A} be the set containing the edges between the nodes in V_{T_A} and their parents in T_A. Note that V_{T_A} may be empty. Similarly, let V_{T^*} be the subset of $V(T^*)$ such that $T^*(v)$ is isomorphic to the B-tree for each node $v \in V_{T^*}$, T' be the tree $T^* - \bigcup_{v \in V_{T^*}} T^*(v)$, and E_{T^*} be the set containing the edges between the nodes in V_{T^*} and their parents in T^*. Observe that $|V_{T^*}| \geq k$; otherwise, the instance (F, T^*) of the L2-$(1, k)$-TA problem is obviously NO.

Case I. $|V_{T^*}| = k$.

Clearly each node in V_{T^*} is the image of the root of one B-tree in F if (F, T^*) is YES. That is, for any node $v \in V_{T^*}$, the nodes in $T^*(v)$ are the images of that in one B-tree of F if (F, T^*) is YES. Thus if $T_A = T'$, then (F, T^*) is YES; otherwise, NO. The weight function f is set with the following approach.

Setting Approach I. For any node $v \in V(T_A)$ and $v' \in V(T^*)$, if $v' \in V(T')$, then $f(v, v') = 1$; otherwise, $f(v, v') = -1$. For any edge $e \in E(T_A)$ and $e' \in E(T^*)$, $f(e, e') = 0$.

Lemma 7. *For Case I, (F, T^*) is YES if and only if $(T_A, T^*; f)$ has an optimal solution with weight $n(T_A)$.*

Case II. $|V_{T^*}| > k$.

Remark that $|V_{T^*}| > k$ indicates $|V_{T_A}| > 0$. Let $c = |V_{T^*}| - k$, and n_{B^*} be the size of the B-tree. Thus $n(T_A) = n(T') + c \cdot n_{B^*}$. For this case, the weight function f is set with the following approach.

Setting Approach II. For any node $v \in V(T_A)$ and $v' \in V(T^*)$, if $v' \in V(T')$, then $f(v, v') = n$; otherwise, $f(v, v') = -1$. For any edge $e \in E(T_A)$ and $e' \in E(T^*)$, if $e \in E_{T_A}$ and $e' \in E_{T^*}$, then $f(e, e') = 1$; otherwise, $f(e, e') = 0$.

Proposition 4. *Any optimal solution to $(T_A, T^*; f)$ cannot have a weight greater than $n \cdot n(T') - c \cdot n_{B^*} + c$.*

Lemma 8. *For Case II, (F, T^*) is YES if and only if $(T_A, T^*; f)$ has an optimal solution with weight $n \cdot n(T') - c \cdot n_{B^*} + c$.*

Now we are ready to present the algorithm Alg-L2-TA for the L2-$(1,k)$-TA problem, which is given in Fig. 4.

Theorem 2. *Algorithm Alg-L2-TA can solve the L2-$(1,k)$-TA problem in runtime $O(n^3)$, where n is the size of the input.*

Algorithm Alg-L2-TA(F, T^*)
Input: A forest F comprises one A-tree T_A and k B-trees, and a target tree T^*
       with $n(F) = n(T^*)$;
Output: Return YES if there is a level-2 attaching approach E_+ making $F + E_+ = T^*$,
       where T_A is the top subtree of T^+; otherwise, return NO.

1. if $|F| = 1$ then if F is isomorphic to T^* then return YES else return NO;
2. let $V_{T^*} = \{v | v \in V(T^*), T^*(v) \text{ is isomorphic to the } B\text{-tree}\}$;
3. let T' be the tree obtained by removing $V(T^*(v))$ from T^* for all $v \in V_{T^*}$;
4. if $|V_{T^*}| = k$ then
5.   set the weight function f with Setting Approach I;
6.   call the algorithm given in [6] to solve $(T_A, T^*; f)$;
7.   if the returned solution has weight $n(T_A)$ then return YES else return NO;
  else
8.   set the weight function f with Setting Approach II;
9.   call the algorithm given in [6] to solve $(T_A, T^*; f)$;
10.   if the returned solution has weight $n \cdot n(T') - c \cdot n_{B^*} + c$ then return YES
    else return NO.

Fig. 4. The algorithm for the L2-$(1,k)$-TA problem

Proof. The correctness of Steps 1 is obvious. For Steps 4–10, the discussion given in Lemmata 7 and 8 guarantee their correctness. For the runtime of the algorithm, Steps 6 and 9 take runtime $O(n^3)$ [6], and the other steps take runtime $O(n)$. Together all, we have that the algorithm can be done in time $O(n^3)$. □

5 Conclusion

Within the paper, we investigated the Tree Assembly problem, which is a simplified version of the Minimum Tree Cut/Paste Distance problem (with specified edge-deletions), and a generalized version of the Exact Block Cover problem. By the natural parameter, the number of trees in the considered forest, we gave the first fixed-parameter algorithm with runtime $2^{O(k \log k)} n^{O(1)}$, indicating that the Tree Assembly problem is FPT. Inspired by the work of Baumbach et al. [2], we also presented a variant of the Tree Assembly problem, named Level-2 $(1,k)$-Tree Assembly problem. By transforming it to the Maximum Common Subtree Isomorphism problem, we successfully gave an algorithm with runtime $O(n^3)$ for the Level-2 $(1,k)$-Tree Assembly problem.

We considered a restricted version of the Tree Assembly problem, where F contains exactly c kinds of trees (c is a constant). Although Baumbach et al. [2] showed that the problem is polynomial-time solvable when F is a set of stars, the case becomes quite complicated if we have no restriction on the structures of the trees in F. Hence, studying the hardness of the Level-2 (k_1, k_2)-Tree Assembly problem is a stepping stone to answer the problem mentioned above, where the considered F comprises k_1 A-trees and k_2 B-trees, and the aim is to find a level-2 attaching approach such that all the other trees are attached to a A-tree. Additionally, Future work on the fixed-parameter tractability for the unrooted version of the Tree Assembly problem would be very interesting.

References

1. Aho, A.V., Hopcroft, J.E.: The Design and Analysis of Computer Algorithms. Pearson Education, Chennai (1974)
2. Baumbach, J., Guo, J., Ibragimov, R.: Covering tree with stars. In: Du, D.-Z., Zhang, G. (eds.) COCOON 2013. LNCS, vol. 7936, pp. 373–384. Springer, Heidelberg (2013). https://doi.org/10.1007/978-3-642-38768-5_34
3. Bulteau, L., Komusiewicz, C.: Minimum common string partition parameterized by partition size is fixed-parameter tractable. In: ACM-SIAM Symposium on Discrete Algorithms, pp. 102–121. SIAM (2014)
4. Chrobak, M., Kolman, P., Sgall, J.: The greedy algorithm for the minimum common string partition problem. ACM Trans. Algorithms **1**(2), 350–366 (2005)
5. Cygan, M., et al.: Parameterized Algorithms, vol. 4. Springer, Cham (2015). https://doi.org/10.1007/978-3-319-21275-3
6. Droschinsky, A., Kriege, N.M., Mutzel, P.: Faster algorithms for the maximum common subtree isomorphism problem. In: 41st International Symposium on Mathematical Foundations of Computer Science, MFCS 2016, Kraków, Poland, 22–26 August 2016, pp. 33:1–33:14 (2016)
7. Eppstein, D.: Subgraph isomorphism in planar graphs and related problems. In: Graph Algorithms and Applications I, pp. 283–309. World Scientific (2002)
8. Garey, M.R., Johnson, D.S.: Computers and Intractability: A Guide to the Theory of NP-Completeness, vol. 29. WH Freeman, New York (2002)
9. Jiang, H., Su, B., Xiao, M., Xu, Y., Zhong, F., Zhu, B.: On the exact block cover problem. In: Gu, Q., Hell, P., Yang, B. (eds.) AAIM 2014. LNCS, vol. 8546, pp. 13–22. Springer, Cham (2014). https://doi.org/10.1007/978-3-319-07956-1_2
10. Kirkpatrick, B., Reshef, Y., Finucane, H., Jiang, H., Zhu, B., Karp, R.M.: Comparing pedigree graphs. J. Comput. Biol. **19**(9), 998–1014 (2012)
11. Lawler, E.L., Lenstra, J.K., Kan, A.R., Shmoys, D.B.: The Traveling Salesman Problem: A Guided Tour of Combinatorial Optimization, vol. 3. Wiley, New York (1985)
12. Lingas, A.: Subgraph isomorphism for biconnected outerplanar graphs in cubic time. Theoret. Comput. Sci. **63**(3), 295–302 (1989)
13. Matoušek, J., Thomas, R.: On the complexity of finding iso-and other morphisms for partial k-trees. Discrete Math. **108**(1–3), 343–364 (1992)
14. McCormick, S.T., Smallwood, S.R., Spieksma, F.C.: A polynomial algorithm for multiprocessor scheduling with two job lengths. Math. Oper. Res. **26**(1), 31–49 (2001)

15. You, J., Wang, J., Feng, Q.: Parameterized algorithms for minimum tree cut/paste distance and minimum common integer partition. In: Chen, J., Lu, P. (eds.) FAW 2018. LNCS, vol. 10823, pp. 99–111. Springer, Cham (2018). https://doi.org/10.1007/978-3-319-78455-7_8
16. You, J., Wang, J., Feng, Q., Shi, F.: Kernelization and parameterized algorithms for covering a tree by a set of stars or paths. Theoret. Comput. Sci. **607**, 257–270 (2015)

On Characterization of Petrie Partitionable Plane Graphs

Xin He[1(✉)] and Huaming Zhang[2]

[1] Department of Computer Science and Engineering,
State University of New York at Buffalo, Buffalo, NY 14260, USA
xinhe@buffalo.edu
[2] Department of Computer Science, The University of Alabama in Huntsville,
Huntsville, USA
hzhang@cs.uah.edu

Abstract. Given a plane graph $G = (V, E)$, a *Petrie tour* of G is a tour P of G that alternately turns left and right at each step. A *Petrie tour partition* of G is a collection $\mathcal{P} = \{P_1, \ldots, P_q\}$ of Petrie tours so that each edge of G is in exactly one tour $P_i \in \mathcal{P}$. A Petrie tour is called a *Petrie cycle* if all its vertices are distinct. A *Petrie cycle partition* of G is a collection $\mathcal{C} = \{C_1, \ldots, C_p\}$ of Petrie cycles so that each vertex of G is in exactly one cycle $C_i \in \mathcal{C}$. In this paper, we characterize 3-regular (4-regular, resp.) plane graphs with Petrie cycle (tour, resp.) partitions. Given a 4-regular plane graph $G = (V, E)$, a *3-regularization* of G is a 3-regular plane graph G_3 obtained from G by splitting every vertex $v \in V$ into two degree-3 vertices. G is called *Petrie partitionable* if it has a 3-regularization that has a Petrie cycle partition. In this paper, we present an elegant characterization of Petrie partitionable graphs. The general version of this problem is motivated by a data compression method, *tristrip*, used in computer graphics.

1 Introduction

Throughout this paper, $G = (V, E)$ denotes a connected 3- or 4-regular plane graph. Given a vertex v and an edge $e = (u, v)$, the *left-edge* of e (at v) is the edge $e_1 = (u_1, v)$ that follows e (at v) in clockwise (cw) direction, the *right-edge* of e (at v) is the edge $e_2 = (u_2, v)$ that follows e (at v) in counter-clockwise (ccw) direction. A *walk* of G is a sequence $P = v_0 e_1 v_1 e_2 \ldots e_k v_k$ where $v_i \in V$ are vertices (may be repeated) and $e_j = (v_{j-1}, v_j) \in E$ are distinct edges of G. If $v_0 = v_k$, P is called a *tour*. A walk (tour, resp.) consisting of distinct vertices is called a *path* (*cycle*, resp.) A walk is called a *Petrie walk* if the edge e_{i+1} is alternately the left- and the right-edge of e_i for $1 \leq i < k$. A tour P is called a *Petrie tour* if it is a Petrie walk and the alternating left- and right-edge condition also holds for e_{k-1}, e_k and e_1. *Petrie paths* and *Petrie cycles* are defined similarly.

Consider a 3-regular plane graph $G = (V, E)$ and a Petrie cycle partition $\mathcal{C} = \{C_1, \ldots, C_p\}$ of G. If $\mathcal{C}$ consists of a single cycle C_1, it is called a *Petrie Hamiltonian cycle*. Consider a 4-regular plane graph $G = (V, E)$ and a Petrie

© Springer Nature Switzerland AG 2020
J. Chen et al. (Eds.): TAMC 2020, LNCS 12337, pp. 313–326, 2020.
https://doi.org/10.1007/978-3-030-59267-7_27

tour partition $\mathcal{P} = \{P_1, \ldots, P_q\}$ of G. If $\mathcal{P}$ consists of a single tour P_1, it is called a *Petrie Eulerian tour*. Given a 4-regular plane graph $G = (V, E)$, a *3-regularization* of G is a 3-regular plane graph G_3 obtained from G by splitting every vertex $v \in V$ into two degree-3 vertices. G is called *Petrie partitionable* if it has a 3-regularization G_3 that has a Petrie cycle partition. In this paper, we study the properties of Petrie partitionable graphs. The general version of this problem is motivated by a data compression method, *tristrip*, used in computer graphics. We present an elegant characterization of these graphs.

The paper is organized as follows. Section 2 presents the definitions and the motivation from computer graphics. The properties of 3-regular plane graphs with Petrie Hamiltonian cycle have been studied in [6–8]. In Sect. 3, we generalize these results and give a characterization of 3-regular plane graphs with Petrie cycle partitions. The properties of 4-regular plane graphs with Petrie Eulerian tours have been studied in [10,13]. In Sect. 4, we generalize these results and give a characterization of 4-regular plane graphs with Petrie tour partitions. The results in Sect. 3 and Sect. 4 are relatively easy generalizations of known results in [6–8,10,13]. To the best of our knowledge, they have not been published in literature. Since they are of independent interests and also needed by the development of our main results, we include these results here. Section 5 discusses our main problem and presents a simple characterization of Petrie partitionable graphs (Theorem 6). Section 6 concludes the paper.

2   Preliminaries and Motivations

We use standard terminology [3]. A graph $G = (V, E)$ is called 3-regular (4-regular, resp.), if $deg(v) = 3$ ($deg(v) = 4$, resp.) for all $v \in V$. A *plane* graph G is a graph embedded in the plane without edge crossings (i.e. an embedded planar graph). Let $\mathcal{F}$ denote the set of the faces of G. A plane graph G is called a *triangulation* (*quadrangulation*, resp.), if $deg(F) = 3$ ($deg(F) = 4$, resp.) for all faces $F \in \mathcal{F}$. The *dual graph* $G^* = (V^*, E^*)$ of a plane graph $G = (V, E)$ is defined as follows: Each face F of G corresponds to a vertex v_F in V^*. Each edge e in G corresponds to an edge $e^* = (v_{F_1}, v_{F_2})$ in G^*, called the *dual edge* of e, where F_1 and F_2 are the two faces of G with e on their common boundary.

Motivations: The problem studied in this paper is motivated by a data compression technique used in computer graphics. 3D objects are often represented by *triangular mashes* in computer graphics. For our purpose, this is just a plane triangulation $\tilde{G} = (\tilde{V}, \tilde{E})$. Its vertex set $\tilde{V} = \{1, 2, \ldots, n\}$ are called *points*. An important problem in computer graphics is how to represent $\tilde{G}$ efficiently. A naive method represents each face of $\tilde{G}$ by listing its three boundary points. If $\tilde{G}$ has N faces, this method uses $3N$ points. For large 3D objects, this takes too much space. The *tristrips* representation of $\tilde{G}$ was discussed in [15]. A tristrip is a sequence $\mathcal{T} = F_1 F_2 \ldots F_t$ of faces in $\tilde{G}$, which can be represented by a sequence $S_\mathcal{T} = v_1 v_2 \ldots v_{t+2}$ of points of $\tilde{G}$ in such a way that, for each i ($1 \leq i \leq t$), the three points $v_i v_{i+1} v_{i+2}$ are the boundary points of the face F_i. An example is shown in Fig. 1(a). A tristrip $\mathcal{T} = F_1 \ldots F_t$ is called a *tristrip-cycle*, represented by the point sequence $\mathcal{S}_\mathcal{T} = v_1 v_2 \ldots v_t$, if both $\mathcal{T}$ and $\mathcal{S}_\mathcal{T}$ are regarded as cyclic sequences and every three consecutive points $v_i v_{i+1} v_{i+2}$ $(1 \le i \le t)$ are the boundary points of the face F_i. (Here we define $t + 1 = 1$ and $t + 2 = 2$). An example of tristrip-cycle is shown in Fig. 1(b). Thus, by using a tristrip, t faces in $\mathcal{T}$ are represented by $\mathcal{S}_\mathcal{T}$ of $t + 2$ points (t points for a tristrip-cycle).

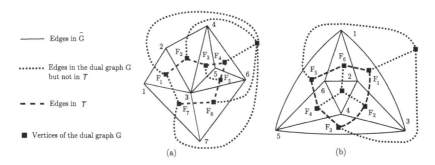

Fig. 1. (a) A tristrip $\mathcal{T} = F_1 F_2 F_3 F_4 F_5 F_6 F_7$ represented by $\mathcal{S}_\mathcal{T} = 123456371$. (b) A tristrip-cycle $\mathcal{T} = F_1 F_2 F_3 F_4 F_5 F_6$ represented by $\mathcal{S}_\mathcal{T} = 123456$.

If all faces of $\tilde{G}$ are included in one tristrip (or tristrip-cycle), we can reduce the space for representing $\tilde{G}$ by a factor of 3 [5]. However, a typical triangular mesh $\tilde{G}$ cannot be included in one tristrip (or tristrip-cycle). It is then a natural question: how to find the fewest disjoint tristrips (or tristrip-cycles) that cover all faces of $\tilde{G}$? This minimization problem is known as *Stripification* problem in computer graphics. It was shown to be NP-hard in [5]. Various heuristic and exact (exponential time) algorithms have been studied in [11,14,15].

The Stripification problem is closely related to the Petrie cycle partition problem as follows. Let $G = (V, E)$ be the dual graph of $\tilde{G}$. Clearly G is 3-regular. For each face F of $\tilde{G}$, let v_F denote the vertex in G corresponding to F. It is easy to see that a sequence of faces $\mathcal{T} = F_1 \ldots F_t$ of $\tilde{G}$ is a tristrip (or tristrip-cycle) if and only if the corresponding sequence $v_{F_1} \ldots v_{F_t}$ is a Petrie path (or Petrie cycle) in G (see Fig. 1(a) and (b)). Hence the problem of finding a minimum tristrip-cycle partition for the faces of $\tilde{G}$ is the same as the problem of finding a minimum Petrie cycle partition for G.

In computer graphics, 3D objects are also represented by *quadrangular meshes* (see [1,2,4]). For our purpose, this is just a plane quadrangulation $\tilde{G} = (\tilde{V}, \tilde{E})$. If we add a chord into each face F of $\tilde{G}$, $\tilde{G}$ becomes a plane triangulation $\tilde{G}_3$ which is called a *triangular extension* of $\tilde{G}$. Since each face F of $\tilde{G}$ has degree 4, there are two ways to add a chord into F. If $\tilde{G}$ has $\tilde{f}$ faces, it has $2^{\tilde{f}}$ triangular extensions. One way to represent $\tilde{G}$ is: first convert it to a plane triangular extension $\tilde{G}_3$ by adding chords into its faces; then represent $\tilde{G}_3$ by using tristrips or tristrip-cycles [5]. The question is: for which of those $2^{\tilde{f}}$ triangular extensions, its faces can be covered by disjoint tristrips/tristrip-cycles?

A special version of this problem is closely related to the Petrie tour partition problem. Consider the dual graph $G = (V, E)$ of $\tilde{G}$. Clearly G is 4-regular. Consider a vertex $v \in V$ corresponding to a face F in $\tilde{G}$, with four incident edges e_1, e_2, e_3, e_4 in cw order. The *split operation* at v splits v into two degree-3 vertices v' and v'' as shown in Fig. 2. There are two ways to split v corresponding to the two ways of adding a chord into F. Let G_3 be a 3-regularization of G obtained by performing split operation at every vertex of G. The edge (v', v'') of G_3 for splitting a vertex $v \in V$ is denoted by $e(v)$ and called a *split edge*.

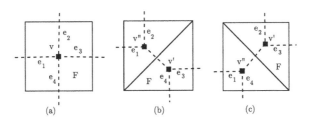

(a)   (b)   (c)

Fig. 2. (a) A vertex v corresponding to a face F in $\tilde{G}$. (b) and (c) Two ways to split v.

Suppose G has a Petrie tour partition $\mathcal{P} = \{P_1, \ldots, P_q\}$. Consider any vertex $v \in V$ with four incident edges $e_1, e_2, e_3, e_4 \in E$. Two tours P_i and P_j in $\mathcal{P}$ visit v (possibly $P_i = P_j$). We split v so that P_i and P_j are still tours after splitting. (See Fig. 2(b) and (c).) Do this at every vertex $v \in V$. Let G_3 be the resulting 3-regularization of G. It is easy to see that $\mathcal{P} = \{P_1, \ldots, P_q\}$ is a Petrie cycle partition of G_3. Thus, if G has a Petrie tour partition, then G has a 3-regularization G_3 with a Petrie cycle partition. In this construction, every edge e in G belongs to a Petrie cycle in the Petrie cycle partition $\mathcal{P}$ of G_3. For its application in computer graphics, this restriction is not necessary. Figure 6(a) shows a 4-regular plane graph G which has no Petrie tour partition (as we will see later). However, it has a 3-regularization G_3 (shown in Fig. 6(b)) which has a Petrie cycle partition with a single Petrie Hamiltonian cycle C. (Some edges of G are in C. Some are not). This motivates:

Definition 1. *A 4-regular plane graphs G is called* Petrie partitionable *if it has a 3-regularization with a Petrie cycle partition.*

The main interest of this paper is to characterize the Petrie partitionable graphs. In computer graphics applications, the problem is to find a 3-regularization of G whose faces can be partitioned into tristrips and/or tristrip-cycles. The NP-hardness result in [5] suggests that this problem might be NP-hard also. The problem considered in this paper is a restricted version of the general problem: partition into Petrie cycles only. In contrast to the more general problem, this restriction leads to a simple characterization. The focus of our study is on the graph-theoretical properties of these problems. The insights obtained here may help for solving the general problem.

3   Characterization of 3-Regular Plane Graphs with Petrie Cycle Partition

In this section, G denotes a connected 3-regular simple (i.e. no self-loops nor parallel edges) plane graph. Suppose G has a Petrie cycle partition $\mathcal{C} = \{C_1, \ldots, C_p\}$. For any vertex $v \in V$, two edges incident to v belong to a cycle $C_i \in \mathcal{C}$ and its third incident edge is not in any $C_j \in \mathcal{C}$. We call the third edge a *bridge edge* with respect to $\mathcal{C}$. A 3-regular plane graph G is called a *multi-3-gon* if all of its faces have degrees divisible by 3. The following Lemma is known:

Lemma 1. *[7, 8] If a 3-regular plane graph has a Petrie Hamiltonian cycle, then it must be a multi-3-gon.*

The following lemma generalizes Lemma 1 (proof omitted).

Lemma 2. *If a 3-regular plane graph G has a Petrie cycle partition, then G must be a multi-3-gon.*

If G has a 3-edge-coloring $\lambda : E \to \{1, 2, 3\}$, the *Heawood valuation* (or simply valuation) associated with λ is a mapping $\lambda^* : V \to \{-1, 1\}$ defined as follows. For any $v \in V$, if the three edges incident to v are colored 1,2,3 in cw order, then $\lambda^*(v) = 1$. Otherwise $\lambda^*(v) = -1$. The following Lemma is well known.

Lemma 3. *[12] A 3-regular plane graph $G = (V, E)$ has a 3-edge-coloring if and only if there exists a mapping $\kappa : V \to \{-1, 1\}$ such that the sum of the values $\kappa(v)$ for all vertices on the boundary of any face F of G is divisible by 3. If κ is such a mapping, then there exists a 3-edge-coloring λ of G such that its associated valuation $\lambda^* = \kappa$.*

The following two theorems characterize graphs with Petrie cycle partitions.

Theorem 1. *Every connected 3-regular multi-3-gon $G = (V, E)$ has exactly three Petrie cycle partitions. (The proof is omitted).*

Theorem 2. *A connected 3-regular plane graph G has a Petrie cycle partition if and only if it is a multi-3-gon. Such G has exactly three Petrie cycle partitions, which can be found in linear time.*

Proof. The proof follows from Lemma 2 and Theorem 1. To implement the algorithm, we first construct a 3-edge-coloring λ of G such that $\lambda^*(v) = 1$ for all $v \in V$. (Pick any vertex v and color its three incident edges 1, 2, 3 in cw order. Then we can propagate the colors to all edges uniquely by the condition $\lambda^*(u) = 1$ for all $u \in V$). Then construct the Petrie cycle partition $\mathcal{C}_{12}$ consisting of the edges of color 1 and 2. Similarly construct $\mathcal{C}_{13}$ and $\mathcal{C}_{23}$. All these steps can be easily done in linear time. $\square$

4   Characterization of 4-Regular Plane Graphs with Petrie Tour Partitions

In this section, we study Petrie tour partitions of a 4-regular plane graph G. The special case of this problem (the Petrie tour partition contains only one tour, i.e. a Petrie Eulerian tour) was studied in [13]. We generalize the results in [13]. Throughout this section, $G = (V, E)$ denotes a 4-regular plane graph without self-loops, but possibly with parallel edges. Let $P = e_1 \ldots e_k$ be a Petrie walk of G. Let $P^* = e_1^* \ldots e_k^*$ be the sequence of the dual edges e_i^* in the dual graph G^*. The following simple observation is crucial to our results.

Observation 1. *[13] If P is a Petrie tour of G, then the sequence P^* of the dual edges is a Petrie tour of the dual graph G^*.*

This observation is illustrated in Fig. 3(a). Based on Observation 1, [13] showed that a 4-regular plane graph with a Petrie Eulerian tour must be bipartite. The following lemma generalizes this result (proof omitted.)

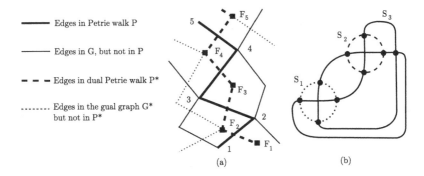

Fig. 3. (a) An example of Observation 1. (b) G and its S-tours $\mathcal{S}(G) = \{S_1, S_2, S_3\}$.

Lemma 4. *If $G = (V, E)$ has a Petrie tour partition, then G must be bipartite.*

Note a 4-regular graph has exactly $2n$ edges, while a bipartite plane graph with no multiple edges has at most $2n - 4$ edges. This explains why we consider multi-graphs in this section.

Consider a tour P of G. Since G is 4-regular, at every vertex of P, we can continue the tour in three ways: go left, straight, or right. A tour S of G consisting of only going-straight steps is called a *straight tour* (or an *S-tour*). Clearly the edge set of G can be uniquely partitioned into S-tours. Denote this partition by $\mathcal{S}(G) = \{S_1, \ldots, S_k\}$. An S-tour may visit a vertex of G twice. An S-tour is called *simple* if it is a cycle in G. Figure 3(b) shows a 4-regular plane graph G. $\mathcal{S}(G)$ contains three S-tours: S_1 and S_2 are simple. S_3 is not. Two S-tours are *independent* if they do not intersect. The following theorem was proved in [9]:

Theorem 3. *[9] Let $G = (V, E)$ be a 4-regular plane graph and $\mathcal{S}(G) = \{S_1 \ldots S_k\}$ be the set of S-tours of G. Then G is bipartite if and only if (i) all S-tours $S_i \in \mathcal{S}(G)$ are simple; and (ii) $\mathcal{S}$ can be partitioned into two subsets $\mathcal{S}_1$ and $\mathcal{S}_2$ such that each $\mathcal{S}_i$ $(i = 1, 2)$ consists of mutually independent S-tours.*

By Lemma 4 and Theorem 3, all 4-regular plane graphs with a Petrie tour partition have a special structure: the set $\mathcal{S}(G)$ is partitioned into two subsets $\mathcal{S}_1$ and $\mathcal{S}_2$; $\mathcal{S}_1$ is a collection of independent simple cycles; $\mathcal{S}_2$ is also a collection of independent simple cycles; and the two sets of cycles are *overlaid* with each other. Such graphs can be complex: Even if $\mathcal{S}_1$ has only one cycle S_1 and $\mathcal{S}_2$ has only one cycle S_2, S_1 and S_2 can cross each other many times in complex ways.

In the following we show that every connected 4-regular bipartite plane graph G has exactly two distinct Petrie tour partitions. Since G is bipartite, we can color its vertices by two colors red and green. Since G is 4-regular, we can color its faces by two colors white and black.

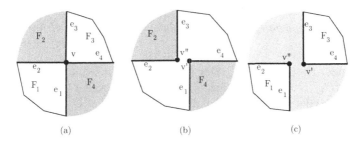

Fig. 4. (a) A vertex v and its incident edges and faces; (b) After the white merge operation at v; (c) After the black merge operation at v.

Definition 2. *Let v be a vertex of G with four incident edges e_i $(1 \leq i \leq 4)$ in cw order and four incident faces F_i $(1 \leq i \leq 4)$ where F_1, F_3 are white and F_2, F_4 are black. Assume e_i, e_{i+1} $(1 \leq i \leq 4)$ are the edges of F_i (see Fig. 4(a).)*

1. *The* white merge *operation at v is (Fig. 4(b)): Replace v by two vertices v' and v''; Make the edges e_1, e_4 incident to v'; and make e_2, e_3 incident to v''.*
2. *The* black merge *operation at v is (Fig. 4(c)): Replace v by two vertices v' and v''; Make the edges e_1, e_2 incident to v''; and make e_3, e_4 incident to v'.*

Note: After the white (black, resp.) merge operation at v, the two white (black, resp.) faces F_1 and F_3 (F_2 and F_4, resp.) become one face.

Definition 3. *1. The* red-white-merge *graph, denoted by G_{rwm} is the graph obtained from G by applying the white merge operation at every red vertex of G and the black merge operation at every green vertex of G. (Fig. 5(b).)*
2. *The* red-black-merge *graph, denoted by G_{rbm} is the graph obtained from G by applying the black merge operation at every red vertex of G and the white merge operation at every green vertex in G. (See Fig. 5(c).)*

Clearly every vertex v in G_{rwm} has degree 2 and the edge set of G_{rwm} one-to-one corresponds to the edge set of G. These properties also hold for G_{rbm}.

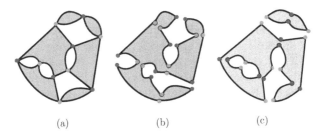

(a)   (b)   (c)

Fig. 5. (a) A 4-regular bipartite plane graph G; (b) G_{rwm}; (c) G_{rbm}. (Color figure online)

Theorem 4. *Every connected 4-regular plane bipartite graph G has exactly two Petrie tour partitions.*

Proof. Consider the graph G_{rbm}. Since every vertex in G_{rbm} has degree 2, G_{rbm} is a set $\mathcal{C} = \{C_1, \ldots, C_q\}$ of disjoint cycles. For each cycle $C_i \in \mathcal{C}$, let P_i be the sequence of the edges of G corresponding to the edges of C_i. Then P_i is a tour of G alternately traveling red and green vertices. Imagine we travel along P_i so that the black faces are on right-side. By the construction of G_{rbm}, P_i always turns left at red vertices and right at green vertices (see Fig. 5(c)). Hence P_i is a Petrie tour of G. Let $\mathcal{P}_{rbm} = \{P_1, \ldots, P_q\}$. Since the edge set of G_{rbm} one-to-one corresponds to the edge set of G, every edge of G belongs to exactly one $P_i \in \mathcal{P}_{rbm}$. So $\mathcal{P}_{rbm}$ is a Petrie tour partition of G. Similarly, the red-white-merge graph G_{rwm} defines another Petrie tour partition $\mathcal{P}_{rwm}$ of G.

Next we show $\mathcal{P}_{rbm}$ and $\mathcal{P}_{rwm}$ are the only Petrie tour partitions of G. Let $\mathcal{Q} = \{Q_1, \ldots, Q_t\}$ be any Petrie tour partition of G. Since G is bipartite, each $Q_i \in \mathcal{Q}$ alternately travels red and green vertices. Consider any tour $Q_i \in \mathcal{Q}$ and three consecutive edges $e_1 = (u, v)$, $e_2 = (v, w)$ and $e_3 = (w, x)$ of Q_i, where u, w are green; v, x are red. Let F_1 be the face with e_1 and e_2 on its boundary. Let F_2 be the face with e_2 and e_3 on its boundary. Depending on if F_1 is a white or black face and if Q_i turns left or right between e_1 and e_2, there are four cases. We only consider the case where Q_i turns left at the red vertex v between e_1 and e_2 and F_1 is a white face. Since Q_i is a Petrie tour, it turns right at the green vertex w between e_2 and e_3. Since F_1 and F_2 share e_2 as common boundary and F_1 is white, F_2 must be black. This corresponds to performing the black merge operation at the red vertex v, and the white merge operation at the green vertex w. Repeating this argument, we see that all $Q_i \in \mathcal{Q}$ are obtained by performing the black merge operation at red vertices and the white merge operation at green vertices of G. Thus $\mathcal{Q}$ is the same as the Petrie tour partition $\mathcal{P}_{rbm}$. (Other cases are similar, resulting either $\mathcal{P}_{rbm}$ or $\mathcal{P}_{rwm}$.) $\qquad\square$

Figure 5(a) shows a 4-regular plane bipartite graph G. Figure 5(b) shows the graph G_{rwm} corresponding to a Petrie tour partition of G with a single Petrie Eulerian tour. Figure 5(c) shows the graph G_{rbm} corresponding to a Petrie tour partition of G with three Petrie tours. The following theorem characterizes the graphs with Petrie tour partitions.

Theorem 5. *A connected 4-regular plane graph G has a Petrie tour partition if and only if it is bipartite. Such G has exactly two Petrie tour partitions, which can be found in linear time.*

Proof. The proof immediately follows from Lemma 4 and Theorem 4. The linear time implementation of the algorithm is straightforward. ☐

5 Characterization of Petrie Partitionable 4-Regular Plane Graphs

In this section, G always denotes a 4-regular plane graph, not necessarily bipartite. $\mathcal{S}(G) = \{S_1, \ldots, S_k\}$ denotes the set of S-tours of G.

Definition 4. *Let G be a 4-regular plane graph, and G_3 be a 3-regularization of G. A Petrie cycle partition $\mathcal{C}_3 = \{C_1, \ldots, C_p\}$ of G_3 is called* full *if every edge of G belongs to a cycle $C_i \in \mathcal{C}_3$. (This implies all split edges of G_3 are bridge edges with respect to $\mathcal{C}_3$).*

A full Petrie cycle partition of G_3 corresponds to a Petrie tour partition of G. So the problem considered in Sect. 4, characterizing G with Petrie tour partitions, is to determine when G has a 3-regularization that has *full* Petrie cycle partitions. In computer graphics, the restriction to full Petrie cycle partitions of G_3 is not necessary. In this section, we study the general problem: characterize the Petrie partitionable graphs. Namely, determine when G has a 3-regularization G_3 that has Petrie cycle partition (full or not). The first observation is such graphs are not necessarily bipartite. This makes the problem more difficult. In this section, we present a simple characterization of such graphs.

Figure 6(a) shows a 4-regular plane graph G with three S-tours S_1, S_2, S_3 where each pair of them intersect. By Lemma 4 and Theorem 3, G has no Petrie tour partitions. Figure 6(b) shows a 3-regularization G_3 of G with a single Petrie Hamiltonian cycle C (denoted by thick lines). Note the bridge edges $(1'', 2''), (2', 6''), (6', 5''), (5', 1')$ in G_3 correspond to the edges of S_3. All edges of S_3 are bridge edges with respect to C.

Lemma 5. *Let G_3 be a 3-regularization of G with a Petrie cycle partition $\mathcal{C}_3 = \{C_1, \ldots, C_p\}$. If any edge $e_1 = (u, v)$ of an S-tour $S_a \in \mathcal{S}(G)$ is a bridge edge with respect to $\mathcal{C}_3$, then all edges of S_a are bridge edges with respect to $\mathcal{C}_3$.*

Proof. Let e_1, e_2, e_3, e_4 be the four edges in G incident to v in cw order where e_1, e_3 are in S_a (Fig. 7(a).) Suppose $e_1 = (u, v)$ is a bridge edge with respect to $\mathcal{C}_3$. In G_3, v is split into two vertices v', v'' connected by a split edge $e(v) = (v', v'')$.

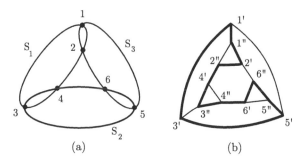

Fig. 6. (a) G; (b) A 3-regularization of G with a Petrie Hamilton cycle.

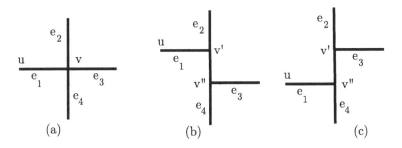

Fig. 7. The proof of Lemma 5.

Case 1: e_1 and e_2 are incident to v' (Fig. 7 (b).) Since e_1 is a bridge edge, e_2 and the split edge (v', v'') must be in a Petrie cycle $C_b \in \mathcal{C}$. Since C_b turns left at v' between e_2 and (v', v''), it must turn right at v''. So e_4 is in S_b and e_3 must be a bridge edge.

Case 2: e_1 and e_4 are incident to v'' (Fig. 7(c).) By similar argument, we can show e_3 must be a bridge edge.

Repeating this argument, we can show all edges in S_a are bridge edges.   □

Given a 3-regularization G_3 and a Petrie cycle partition $\mathcal{C}_3$ (not necessarily full) of G_3, by Lemma 5, the set of S-tours $\mathcal{S}(G)$ can be partitioned into:

- $\mathcal{S}_{bridge} = \{S_i \in \mathcal{S}(G) \mid$ all edges in S_i are bridge edges with respect to $\mathcal{C}_3\}$.
- $\mathcal{S}_{cycles} = \{S_j \in \mathcal{S}(G) \mid$ all edges in S_j belong to some cycles in $\mathcal{C}_3\}$.

Lemma 6. *All S-tours in $\mathcal{S}_{bridge}$ are simple and are mutually independent.*

Proof. If an $S_i \in \mathcal{S}_{bridge}$ visits a vertex v twice, or two $S_i, S_j \in \mathcal{S}_{bridge}$ intersect at v, all four edges incident to v are bridge edges for $\mathcal{C}$. This is impossible.   □

Definition 5. *Let $S_a \in \mathcal{S}(G)$ be an S-tour of G visiting the vertices $v_1, \ldots, v_t$ of G. The graph obtained by expunging S_a, denoted by $G \oslash S_a$, is obtained from G as follows: (a) Delete all edges of S_a from G. (b) In the resulting graph, for each degree 2 vertex v with two incident edges $e' = (u, v)$ and $e'' = (v, w)$, replace e'*

and e'' be a new edge (u, w), (which is called the residual edge *of v_i and denoted by $res(v_i)$.) Repeat this process until the resulting graph is 4-regular again.*

Note all vertices in S_a disappear from $G \oslash S_a$, For example, if G is the graph shown in Fig. 6(a), then $G \oslash S_3$ is the graph with two vertices 3, 4 and four parallel edges connecting them.

$G \oslash S_a$ is clearly a 4-regular plane graph. We can easily obtain the set of S-tours of $G \oslash S_a$ as follows. For any $S_b \in \mathcal{S}(G)$ $(b \neq a)$, define $S_b \oslash S_a$ as:

- If S_b does not pass any vertex $v_i \in S_a$, define $S_b \oslash S_a = S_b$.
- If S_b passes a vertex $v_i \in S_a$, it must pass the two edges $e'_i = (u_i, v_i)$ and $e''_i = (v_i, w_i)$ incident to v_i that are not on S_a. Replace these two edges by the residual edge $res(v_i) = (u_i, w_i)$. Perform this operation for all v_i of S_a visited by S_b. Let $S_b \oslash S_a$ be the resulting tour.

Define $\mathcal{S} \oslash S_a = \{S_b \oslash S_a \mid S_b \in \mathcal{S}(G), b \neq a\}$. Then $\mathcal{S} \oslash S_a$ is the set of S-tours of $G \oslash S_a$. For any subset $\mathcal{S}' \subseteq \mathcal{S}(G)$ of mutually independent S-tours of G, we can expunge the S-tours in $\mathcal{S}'$ from G one by one. The resulting 4-regular plane graph is denoted by $G \oslash \mathcal{S}'$. The following two lemmas are needed by our main theorem.

Lemma 7. *Let G be a 4-regular plane graph and $\mathcal{S}(G) = \{S_1, \ldots, S_k\}$ be the set of its S-tours. Suppose G has a 3-regularization G_3 with a Petrie cycle partition $\mathcal{C}_3 = \{C_1, \ldots, C_p\}$ where all edges of an S-tour $S_a \in \mathcal{S}$ are bridge edges with respect to $\mathcal{C}_3$. Then the graph $G \oslash S_a$ has a 3-regularization $(G \oslash S_a)_3$ with a Petrie cycle partition $\mathcal{C}'_3$ of the same size p. (The proof is omitted.)*

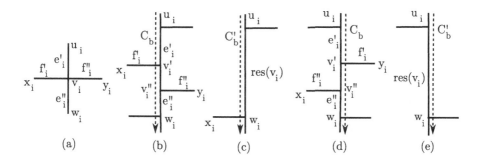

(a)   (b)   (c)   (d)   (e)

Fig. 8. The proof of Lemma 8.

Lemma 8. *Let G be a 4-regular plane graph and $\mathcal{S}(G) = \{S_1, \ldots, S_k\}$ be the set of its S-tours. Let $S_a \in \mathcal{S}(G)$ be an S-tour that visits the vertices $v_1 \ldots v_t$ of G in this order. Suppose that the graph $G \oslash S_a$ has a 3-regularization $(G \oslash S_a)_3$ with a Petrie cycle partition $\mathcal{C}'_3 = \{C'_1, \ldots C'_p\}$ such that every residual edge $res(v_i)$ $(1 \leq i \leq t)$ belongs to some $C'_b \in \mathcal{C}'$. Then G has a 3-regularization G_3 with a Petrie cycle partition $\mathcal{C}_3 = \{C_1, \ldots, C_p\}$ of the same size as $\mathcal{C}'_3$, where all edges of S_a are bridge edges with respect to $\mathcal{C}_3$.*

Proof. We modify each $C'_b \in \mathcal{C}'_3$, so that it becomes a Petrie cycle of G, as follows:
- If C'_b does not pass any residual edge $res(v_i)$, define $C_b = C'_b$.
- If C'_b passes a residual edge $res(v_i) = (u_i, w_i)$, replace $res(v_i)$ by a 3-edge path consisting of $e'_i = (u_i, v'_i), (v'_i, v''_i), e''_i = (v''_i, w_i)$. (For an illustration, see Fig. 8. Figure 8(c) and (f) show the graph before the operation. Figure 8(b) and (d) show the graph after the operation). Perform this operation for all residual edges visited by C'_b. Let C_b be the resulting cycle.

Then $\mathcal{C}_3 = \{C_b \mid C'_b \in \mathcal{C}'_3\}$ is the required Petrie cycle partition of G. □

Now we can state our main result in this section.

Theorem 6. *A 4-regular plane graph G with S-tour set $\mathcal{S}(G)$ is Petrie partitionable if the following hold: (1) All S-tours in $\mathcal{S}(G)$ are simple; and (2) $\mathcal{S}(G)$ can be partitioned into three subsets $\mathcal{S}_1, \mathcal{S}_2, \mathcal{S}_3$ of mutually independent S-tours.*

Proof. Suppose G has a 3-regularization G_3 with a Petrie cycle partition $\mathcal{C}_3$. Let $\mathcal{S}_{bridge}$ be the set of S-tours of G consisting of bridge edges with respect to $\mathcal{C}_3$. By Lemma 6, all S-tours in $\mathcal{S}_{bridge}$ are simple and mutually independent. Let $\mathcal{S}_{cycle} = \mathcal{S}(G) - \mathcal{S}_{bridge}$ and consider the graph $G \oslash \mathcal{S}_{bridge}$. Since the S-tours in $\mathcal{S}_{bridge}$ are mutually independent, we can repeatedly apply Lemma 7 for each $S_a \in \mathcal{S}_{bridge}$. In the end, we get a Petrie tour partition $\mathcal{C}'_3$ of a 3-regularization $(G \oslash \mathcal{S}_{bridge})_3$ of $G \oslash \mathcal{S}_{bridge}$ such that all bridge edges with respect to $\mathcal{C}'_3$ are split edges. In other words, $\mathcal{C}'_3$ corresponds to a Petrie tour partition of $G \oslash \mathcal{S}_{bridge}$. By Theorem 5 and Theorem 3, all S-tours in $\mathcal{S}_{cycle}$ are simple and $\mathcal{S}_{cycle}$ can be partitioned into two subsets $\mathcal{S}_1, \mathcal{S}_2$ of independent S-tours of G. Thus the two conditions stated in the theorem hold.

Suppose the two conditions in theorem hold. Consider the graph $G \oslash \mathcal{S}_3$. It satisfies the conditions in Theorem 3, so $G \oslash \mathcal{S}_3$ is bipartite. By Theorem 4, $G \oslash \mathcal{S}_3$ has a Petrie tour partition. That is: it has a 3-regularization $(G \oslash \mathcal{S}_3)_3$ with a full Petrie cycle partition $\mathcal{C}'_3$. Namely, all bridge edges with respect to $\mathcal{C}'_3$ are the split edges of $(G \oslash \mathcal{S}_3)_3$. We can apply Lemma 8 once to restore one S-tour S_i in $\mathcal{S}_3$. Namely we get a 3-regularization $(G \oslash (\mathcal{S}_3 - S_i))_3$ of $G \oslash (\mathcal{S}_3 - S_i)$ with a Petrie cycle partition $\mathcal{C}'_3$, all of whose bridge edges are the split edges of $(G \oslash \mathcal{S}_3)_3$ and the edges in S_i. Now consider another S-tour S_j in $\mathcal{S}_3$. Since S_i and S_j are independent, we can apply Lemma 8 again to restore S_j. Since the S-tours in $\mathcal{S}_3$ are mutually independent, we can repeat this process and restore all S-tours in $\mathcal{S}_3$ one by one to get a 3-regularization G_3 of G with a Petrie cycle partition $\mathcal{C}_3$ where all edges of each S-tours in $\mathcal{S}_3$ are bridge edges with respect to $\mathcal{C}_3$. This completes the proof. □

6   Conclusion and Open Problems

We studied the properties of 3-regular (resp. 4-regular) plane graphs with Petrie cycle (resp. tour) partitions. We found simple characterizations of these graphs.

We discovered a simple characterization of Petrie partitionable graphs (Theorem 6). Although graph-theoretically elegant, this characterization does not lead to a polynomial time algorithm for recognizing such graphs. Let $H = (V_H, V_H)$ be the graph obtained from G as follows. The node set V_H of H is the set of S-tours of G. Two nodes in V_H are adjacent in H if and only if their corresponding S-tours intersect in G. Testing the condition 2 in Theorem 6 is equivalent to testing if H is 3-vertex-colorable. This, in general, cannot be done in polynomial time. It is interesting to investigate if H possesses special structures which can lead to a polynomial time algorithm for solving this vertex-coloring problem.

The general version of these problems are motivated by applications in computer graphics, which require finding minimum partition of 3-regular plane graphs by Petrie paths and/or Petrie cycles, and finding minimum partition of 4-regular plane graphs by Petrie walks and/or Petrie tours. The general versions of the problem may be NP-hard. It is interesting to see if the insights discovered in this paper can lead to better heuristic algorithms and/or more efficient exact algorithms for solving the general version of these problems.

References

1. Bommes, D., et al.: State of the art in quad meshing. In: Eurographics STARS. http://citeseerx.ist.psu.edu/viewdoc/summary?doi=10.1.1.363.6797 (2012)
2. Bommes, D., Campen, M., Ebke, H.-C., Alliez, P., Kobbelt, L.: Integer-grid maps for reliable quad meshing. ACM Trans. Graph. **32**(4), 98:1–98:12 (2013). Article 98
3. Bondy, J.A., Murty, U.S.R.: Graph Theory with Applications. Macmillan, London (1979)
4. Dong, S., Bremer, P.-T., Garland, M., Pascucci, V., Hart, J.C.: Spectral surface quadrangulation. In: ACM SIGGRAPH 2006, pp. 1057–1066 (2006)
5. Estkowski, R., Mitchell, J.S.B., Xiang, X.: Optimal decomposition of polygonal models into triangle strips. In: Proceedings of the 18th ACM Symposium on Computational Geometry (SoCG 2002), Barcelona, Spain, pp. 254–263, 5–7 June 2002
6. Fouquet, J.L., Jolivet, J.L.: Strong edge-coloring of cubic planar graphs. In: Adrian Bondy, J., Murty, U.S.R. (eds.) Progress in Graph Theory. Proceedings of the Conference on Combinatorics Held at the University of Waterloo, Waterloo, Ontario, pp. 247–264. Academic Press, Cambridge (1982)
7. Ivančo, J., Jendrol̆, S.: On an Eberhard-type problem in cubic polyhedral graphs having Petrie and Hamiltonian cycles. Tatra. Mt. Math. Publ. **18**, 57–62 (1999)
8. Ivančo, J., Jendrol̆, S., Tkšč, M.: Note on Petrie and Hamiltonian cycles in cubic polyhedral graphs. Comment. Math. Univ. Carolin. **35**(2), 413–417 (1994)
9. Jaeger, F., Shank, H.: On the edge-coloring problem for a class of 4-regular maps. J. Graph Theory **5**, 269–275 (1981)
10. Kidwell, M.E., Bruce Richter, R.: Trees and Euler tours in a planar graph and its relatives. Am. Math. Mon. **94**, 618–630 (1987)
11. Porcu, M.B., Scateni, R.: An interactive strpification algorithm based on dual graph operations. In: Eurographics 2003 (2003)
12. Ringel, G.: Färbungsprobleme auf Flächen und Graphen. Berlin (1959)
13. Žitnik, A.: Plane graphs with Eulerian Petrie walks. Discrete Math. **244**, 539–549 (2002)

14. Šíma, J.: Optimal triangle stripifications as minimum energy states in hopfield nets. In: Duch, W., Kacprzyk, J., Oja, E., Zadrożny, S. (eds.) ICANN 2005. LNCS, vol. 3696, pp. 199–204. Springer, Heidelberg (2005). https://doi.org/10.1007/11550822_32
15. Xiang, X., Held, M., Mitchell, J.S.B.: Fast and effective stripification of polygonal surface models. In: 1999 Symposium on Interactive 3D Graphics Layout, pp. 71–78 (1999)

Disjunctive Propositional Logic and Scott Domains

Longchun Wang[1,2] and Qingguo Li[1(✉)]

[1] School of Mathematics, Hunan University, Changsha, China
longchunw@163.com, liqingguoli@aliyun.com
[2] School of Mathematical Sciences, Qufu Normal University, Qufu, China

Abstract. Based on the investigation of the proof system of a disjunctive propositional logic, this paper establishes a purely, syntactic representation of Scott-domains. More precisely, a category of certain proof systems with consequence relations is shown to be equivalent to that of Scott-domains with Scott-continuous functions.

Keywords: Domain theory · Scott-domain · Disjunctive sequent calculus · Categorical equivalence

1 Introduction

Domains introduced by D. Scott have been the objects of interest on which great progress has been made by computer scientists and mathematicians [5,6,10,21, 22]. One aspect of domains is that they can be presented by logical language, and a great deal of possible logical representations for domains have been demonstrated, ranging from Scott's information systems [16] to Amramsky's *domain theory in logic form* [2].

Scott's information system is a set of tokens endowed with a consistency predicate and an entailment relation. This simple structure provides a convenient means of presenting the category of Scott domains with Scott-continuous functions [15]. Recently, many scholars established several kinds of information systems for the representations of various domains [11,12,17,18,20]. Note that although each of these information systems has many features of a logic, only the atomic formulae is taken into account.

In [1], Abramsky devised a complete logical system for Scott-domains which was deliberately suggestive of semantics theory. Scott's information system, as well as Abramsky's domain logic, is made by extracting an appropriate logical language from the category of Scott domains with Scott-continuous functions. Abramsky also provided a logical representation for SFP-domains in a similar way [2]. Following Abramsky's idea, Jung, Kegelmann and Moshier [13] presented a coherent sequent calculus which is a logic corresponding to strong proximity lattices. For a variety of results, see [14,19]. In [3], Chen and Jung built a

This Research Supported by the National Natural Science Foundation of China (11771134).

© Springer Nature Switzerland AG 2020
J. Chen et al. (Eds.): TAMC 2020, LNCS 12337, pp. 327–339, 2020.
https://doi.org/10.1007/978-3-030-59267-7_28

disjunctive propositional logic, which is sound and complete with respect to dD-semilattices. They showed how to use D-semilattices and stable D-semilattices for descriptions of L-dcpos and algebraic L-domains, respectively.

In this paper, we aim to provide a new representation of Scott domains in the framework of the disjunctive propositional logic. Note that our approach differs from the style of Scott's information system and Abramsky's logical form. We neither rely on consistent predicates and atomic formulae to make inference, nor focus on the logical algebras.

As usual logic [7], in Sect. 3, we define several notions in a disjunctive propositional logic, such as tautology, contradiction, satisfiable formula, conjunction, etc. Then we introduce the notions of consistent disjunctive sequent calculi and logical states. We show that the set of logical states of a consistent disjunctive sequent calculus under the set inclusion forms a Scott domain, and each Scott domain can be obtained in this way, up to isomorphism.

On the side of Scott domains, Scott-continuous functions are typically used as morphisms to form a cartesian closed category **SD**. In domain theory, the category **SD** is the most appropriate candidate for the denotational semantics of functional programming languages. In Sect. 4, we introduce the notion of consequence relations, and establish a category equivalent to **SD**.

2   Preliminaries

We first recall some basic definitions and notations of domain theory. Most of them come from [8,9]. For any set X, the symbol $A \sqsubseteq X$ indicates that A is a finite subset of X. A nonempty subset D of a poset P is said to be *directed* if every pair of elements of D has an upper bound in D. A *complete lattice* is a poset in which every subset of it has a supremum. With respect to a poset P, we write $\bigsqcup X$ for the supremum of X. We use $\downarrow X$ to denote the down set $\{d \in P \mid (\exists x \in X) d \leq x\}$, where X is a subset of P. Similarly, we write $\uparrow X$ for the upper set $\{d \in P \mid (\exists x \in X) x \leq d\}$. If X is a singleton $\{x\}$, then we just write $\downarrow x$ or $\uparrow x$. X is a *pairwise inconsistent* subset of a poset P if $\uparrow x \cap \uparrow y = \emptyset$ for all $x \neq y \in X$. The least element of a poset P is denoted by $\bot$.

A *dcpo* P is a poset in which every directed subset D has a supremum $\bigsqcup D$. Let P be a dcpo and $x \in P$. Then x is called a *compact element* of P if for any directed subset D of P the relation $x \leq \bigsqcup D$ always implies the existence of some $d \in D$ with $x \leq d$. We write $K(P)$ for the set of compact elements of P and write $K^*(P)$ for $K(P) - \{\bot\}$.

Definition 1. *(1) A pointed dcpo P is called an algebraic domain if every element x of P is the directed supremum of the compact elements below x.*

(2) An algebraic domain is called a Scott domain if any two elements of it which are bounded above has a supremum.

Definition 2. *Let P and Q be algebraic domains. A function $f : P \to Q$ is Scott-continuous if and only if for all directed subset D of P, $f(\bigsqcup D) = \bigsqcup\{f(x) \mid x \in D\}$.*

In [3, Definition 2.1], a disjunctive propositional logic was introduced by deviating from classical propositional logic in some way. A sequent in a disjunctive propositional logic is an object $\Gamma \vdash \varphi$, where Γ is a finite set of formulae and φ is a single formula. As usual logic, the interpretation of a sequent $\Gamma \vdash \varphi$ is that the conjunction of the formulae in Γ implies φ.

Definition 3. *([3]) Let P be a set, every element of which we call an atomic (disjunctive) formula. Likewise, let $\mathcal{A}_P$ be a set of sequents of the form $p_1, p_2, \ldots, p_n \vdash F$ where the p_i are atomic formulae, and F is the syntactic constant for "false". Each element of $\mathcal{A}_P$ is called an atomic disjointness assumptions, and the pair $(P, \mathcal{A}_P)$ is called a disjunctive basis.*

The class $\mathcal{L}(P)$ of (disjunctive) formulae, and the set $\mathbf{T}(P)$ of valid sequents are generated by mutual transfinite induction by the following rules:

- *Disjunctive formulae*

$$(At) \frac{\phi \in P}{\phi \in \mathcal{L}(P)} \qquad\qquad (Const) \frac{}{T, F \in \mathcal{L}(P)}$$

$$(Conj) \frac{\phi, \psi \in \mathcal{L}(P)}{\phi \wedge \psi \in \mathcal{L}(P)}$$

$$(Disj) \frac{\phi_i \in \mathcal{L}(P)(all\ i \in I) \qquad \phi_i, \phi_j \vdash F(all\ i \neq j \in I)}{\dot{\bigvee}_{i \in I} \phi_i \in \mathcal{L}(P)}$$

- *Valid sequents*

$$(Ax) \frac{(\Gamma \vdash F) \in \mathcal{A}_P}{\Gamma \vdash F} \qquad\qquad (Id) \frac{\phi \in \mathcal{L}(P)}{\phi \vdash \phi}$$

$$(Lwk) \frac{\Gamma \vdash \psi \qquad \phi \in \mathcal{L}(P)}{\Gamma, \phi \vdash \psi} \qquad\qquad (Cut) \frac{\Gamma \vdash \phi \qquad \Delta, \phi \vdash \psi}{\Gamma, \Delta \vdash \psi}$$

$$(LF) \frac{\phi \in \mathcal{L}(P)}{F \vdash \phi} \qquad\qquad (RT) \frac{}{\vdash T}$$

$$(L\wedge) \frac{\Gamma, \phi, \psi \vdash \theta}{\Gamma, \phi \wedge \psi \vdash \theta} \qquad\qquad (R\wedge) \frac{\Gamma \vdash \phi \qquad \Delta \vdash \psi}{\Gamma, \Delta \vdash \phi \wedge \psi}$$

$$(L\dot{\vee}) \frac{\Gamma, \phi_i \vdash \theta(all\ i \in I) \qquad \phi_i, \phi_j \vdash F(all\ i \neq j \in I)}{\Gamma, \dot{\bigvee}_{i \in I} \phi_i \vdash \theta}$$

$$(R\dot{\vee}) \frac{\Gamma \vdash \phi_{i_0}(some\ i_0 \in I) \qquad \phi_i, \phi_j \vdash F(all\ i \neq j \in I)}{\Gamma \vdash \dot{\bigvee}_{i \in I} \phi_i}.$$

Given a concrete disjunctive propositional logic, the set $\mathbf{T}(P)$ can be defined by a relation $\vdash$, that is,

a sequent $\Gamma \vdash \varphi \in \mathbf{T}(P)$ if and only if $(\Gamma, \varphi) \in \vdash$.

For convenience, we call the proof system of the disjunctive propositional logic defined above a disjunctive sequent calculus and denoted it by $(\mathcal{L}(P), \vdash)$. So when we check a pair $(\mathcal{L}(P), \vdash)$ is a disjunctive sequent calculus, we need only to verify that $(\mathcal{L}(P), \vdash)$ satisfies all the rules of disjunctive formulae and valid sequents defined in Definition 3.

Proposition 1. ([3]) Let $(\mathcal{L}(P), \vdash)$ be a disjunctive consequence calculus.

(1) $\Gamma, \varphi, \psi \vdash \phi$ is a valid sequent if and only if $\Gamma, \varphi \wedge \psi \vdash \phi$ is a valid sequent.
(2) $\Gamma \vdash \varphi$ and $\Gamma \vdash \psi$ are valid sequents if and only if $\Gamma \vdash \varphi \wedge \psi$ is a valid sequent.
(3) Assuming $\phi_i, \phi_j \vdash F$ are valid sequents for all $i \neq j \in I$, then $\Gamma, \phi_i \vdash \theta$ are valid sequents if and only if $\Gamma, \dot{\bigvee}_{i \in I} \phi_i \vdash \theta$ is a valid sequent.

3   Logical Representations of Scott Domains

The purpose of this section is to show how the disjunctive propositional logic can be used to represent Scott domains.

Definition 4. Let $(\mathcal{L}(P), \vdash)$ be a disjunctive sequent calculus and $\varphi \in \mathcal{L}(P)$.

(1) φ is called a tautology if $\mathrm{T} \vdash \varphi$ is a valid sequent.
(2) φ is called a contradiction if $\varphi \vdash F$ is a valid sequent.
(3) φ is called a (nontrivial) satisfiable formula if it is neither a tautology nor a contradiction.

The sets of tautologies and contradictions of a disjunctive sequent calculus $(\mathcal{L}(P), \vdash)$ are denoted by $\mathrm{Tau}(P)$ and $\mathrm{Cont}(P)$, respectively.

Definition 5. Let $(\mathcal{L}(P), \vdash)$ be a disjunctive sequent calculus and $\varphi, \psi \in \mathcal{L}(P)$. φ and ψ are said to be logically equivalent, in symbols, $\varphi \approx \psi$, if both $\varphi \vdash \psi$ and $\psi \vdash \varphi$ are valid sequents.

Definition 6. Consider a disjunctive sequent calculus.

(1) A satisfiable formula built up from atomic formulae only by conjunctive connectives is called a conjunction.
(2) A satisfiable formula is said to be a flat formula if it has the form $\dot{\bigvee}_{i \in I} \mu_i$, where μ_i is a conjunction with $\mu_i, \mu_j \vdash F$ is valid for any $i \neq j \in I$.

Proposition 2. ([3]) Every satisfiable formula is logically equivalent to a flat formula.

We write $\mathcal{N}(P)$ for the set of flat formulae.

Definition 7. *Let $(\mathcal{L}(P), \vdash)$ be a disjunctive sequent calculus.*

(1) *A conjunction μ is said to be irreducible if, whenever $\mu \vdash \bigvee_{i \in I} \phi_i$ is valid, where $\phi_i, \phi_j \vdash F$ are valid for all $i \neq j \in I$, then $\mu \vdash \phi_{i_0}$ is valid for some $i_0 \in I$.*

(2) *A flat formula $\bigvee_{i \in I} \mu_i$ is irreducible if each conjunction μ_i is irreducible.*

We denote the set of all irreducible conjunctions by $\mathcal{C}(P)$.

Definition 8. *A disjunctive sequent calculus $(\mathcal{L}(P), \vdash)$ is said to be consistent if, every conjunction in it is irreducible, and for any satisfiable formula ψ there exists an irreducible flat formula $\bigvee_{i \in I} \mu_i$ such that $\psi \vdash \bigvee_{i \in I} \mu_i$ and $\mu_i \vdash \psi$ are valid for all $i \in I$.*

For any $X \subseteq \mathcal{L}(P)$, we make the convention that

$$X[\vdash] = \{\varphi \in \mathcal{L}(P) \mid (\exists \Gamma \sqsubseteq X)\Gamma \vdash \varphi \in \mathbf{T}(P)\}. \tag{3.1}$$

Definition 9. *Let $(\mathcal{L}(P), \vdash)$ be a consistent disjunctive sequent calculus. A nonempty proper subset S of $\mathcal{L}(P)$ is called a logical state of $(\mathcal{L}(P), \vdash)$ if it satisfies the following conditions:*

(S1) $S[\vdash] \subseteq S$.

(S2) *If $\bigvee_{i \in I} \mu_i \in S \cap \mathcal{N}(P)$, then there exists some $i_0 \in I$ such that $\mu_{i_0} \in S$.*

We denote the collection of all the logical states of a consistent disjunctive sequent calculus $(\mathcal{L}(P), \vdash)$ by $|(\mathcal{L}(P), \vdash)|$, which is ordered by set inclusion. The following proposition is clear.

Proposition 3. *Let S be a logical state. Then the following statements hold.*

(1) $S = S[\vdash]$.

(2) *The constant F does not belong to S.*

(3) *If $\varphi, \psi \in S$, then $\varphi \wedge \psi \in S$.*

(4) *If $\varphi \in S$ and $\varphi \wedge \psi \vdash F$ is valid, then $\psi \notin S$.*

Proposition 4. *Let $(\mathcal{L}(P), \vdash)$ be a consistent disjunctive sequent calculus.*

(1) $\mathrm{Tau}(P)$ *is a logical state but $\mathrm{Cont}(P)$ is not.*

(2) *The union of a directed subset of logical states is a logical state.*

Proof. (1) According to definitions of a tautology and $\mathrm{Tau}(P)[\vdash]$, it is easy to see that $\mathrm{Tau}(P)[\vdash] \subseteq \mathrm{Tau}(P)$. And $\mathrm{Tau}(P)$ naturally fulfills condition (S2) since $\mathrm{Tau}(P) \cap \mathcal{N}(P) = \emptyset$. Consequently, $\mathrm{Tau}P$ is a logical state.

$\mathrm{Cont}(P)$ is not a logical state because of part (3) of Proposition 3.

(2) Suppose that $\{S_i \mid i \in I\}$ is a directed set of logical states. Put $S = \bigcup\{S_i \mid i \in I\}$. We show that S is a logical state by checking that S satisfies conditions (S_1) and (S_2).

For condition (S_1), let $\varphi \in S[\vdash]$. Then, by Eq. (3.1), there exists some $\Gamma \sqsubseteq S$ such that $\Gamma \vdash \varphi$. From the fact that $\Gamma \sqsubseteq S$ and the set $\{S_i \mid i \in I\}$ is directed, it follows that $\Gamma \sqsubseteq S_{i_0}$ for some $i_0 \in I$. Since S_{i_0} is a logical state, we have $\varphi \in S_{i_0} \subseteq S$. Therefore, $S[\vdash] \subseteq S$.

For condition (S_2), let $\dot{\bigvee}_{j \in J} \mu_j \in S \cap \mathcal{N}(P)$. Then $\dot{\bigvee}_{j \in J} \mu_j \in S_i \cap \mathcal{N}(P)$ for some $i \in I$. Because S_i is a logical state, there exists some $j_i \in J$ such that $\mu_{j_i} \in S_i \subseteq S$.

Proposition 5. *Let $(\mathcal{L}(P), \vdash)$ be a consistent disjunctive sequent calculus and S a logical state. Then for any subset X of S,*

$$[X]_S = \bigcap \{W \in |(\mathcal{L}(P), \vdash)| \mid X \subseteq W \subseteq S\} \tag{3.2}$$

is also a logical state.

Proof. Assume that $\varphi \in [X]_S[\vdash]$. Then there exists some $\Gamma \sqsubseteq [X]_S$ such that $\Gamma \vdash \varphi$. Thus $\Gamma \sqsubseteq W$ for all $W \in |(\mathcal{L}(P), \vdash)|$ with $X \subseteq W \subseteq S$. Since W is a logical state and $\Gamma \vdash \varphi$, it follows that $\varphi \in W$, and therefore $\varphi \in [\Gamma]_S$. This implies that $[X]_S$ satisfies condition (S1).

For condition (S2), assume that $\dot{\bigvee}_{i \in I} \mu_i \in [X]_S \cap \mathcal{N}(P)$. Then $\dot{\bigvee}_{i \in I} \mu_i \in W$ for all $W \in |(\mathcal{L}(P), \vdash)|$ with $X \subseteq W \subseteq S$. Since W is a logical state, there exists some $i_W \in I$ such that $\mu_{i_W} \in W \subseteq S$. Thus $\{\mu_{i_W} \mid W \in |(\mathcal{L}(P), \vdash)|, X \subseteq W \subseteq S\}$ is a subset of S. Note that $\mu_i, \mu_j \vdash F$ is valid for all $i \neq j \in I$ and S is a logical state, the set $\{\mu_{i_W} \mid W \in |(\mathcal{L}(P), \vdash)|, X \subseteq W \subseteq S\}$ must be a singleton, say $\{\mu_{i_0}\}$. Consequently, there exists some $i_0 \in I$ such that $\mu_{i_0} \in [X]_S$.

Next, we show that each Scott domain can be realized as a poset $|(\mathcal{L}(P), \vdash)|$ of logical states built from a suitable consistent disjunctive sequent calculus $(\mathcal{L}(P), \vdash)$.

Theorem 1. *Let $(\mathcal{L}(P), \vdash)$ be a consistent disjunctive sequent calculus. Then $(|(\mathcal{L}(P), \vdash)|, \subseteq)$ is a Scott domain.*

Proof. By part (2) of Proposition 4, for any directed subset $\{S_i \mid i \in I\}$ of $((\mathcal{L}(P), \vdash)|, \subseteq)$, the union $\bigcup \{S_i \mid i \in I\}$ is a logical state. Then $\bigsqcup \{S_i \mid i \in I\}$ exists in $(|(\mathcal{L}(P), \vdash)|, \subseteq)$, and

$$\bigsqcup \{S_i \mid i \in I\} = \bigcup \{S_i \mid i \in I\}.$$

Part (1) of Proposition 4 has shown that $\mathrm{Tau}(P)$ is a logical state. As a result, $(|(\mathcal{L}(P), \vdash)|, \subseteq)$ forms a dcpo.

We now prove that $(|(\mathcal{L}(P), \vdash)|, \subseteq)$ is a Scott domain, which, by Definition 1, can be divided into three steps.

Step 1: We claim that, for each logical state S and finite subset Γ of S, the logical state $[\Gamma]_S$ defined by Eq. 3.2 is a compact element of the dcpo $(|(\mathcal{L}(P), \vdash)|, \subseteq)$. In fact, let $\{S_i \mid i \in I\}$ be a directed subset in the dcpo $(|(\mathcal{L}(P), \vdash)|, \subseteq)$ with $[\Gamma]_S \subseteq \bigcup \{S_i \mid i \in I\}$. Since $\bigcup \{S_i \mid i \in I\}$ is a logical state, it is evident that $[\Gamma]_S = [\Gamma]_{\bigcup \{S_i \mid i \in I\}}$. Because $\Gamma \sqsubseteq \bigcup \{S_i \mid i \in I\}$, there exists some $i_0 \in I$ such that $\Gamma \subseteq S_{i_0}$. Whence, $[\Gamma]_S = [\Gamma]_{\bigcup \{S_i \mid i \in I\}} \subseteq S_{i_0}$.

Step 2: We show that for any logical state S, the set $\{[\Gamma]_S \mid \Gamma \sqsubseteq S\}$ is directed and S is its union.

Let $\Gamma_1, \Gamma_2 \sqsubseteq S$. Then $[\Gamma_1 \cup \Gamma_2]_S \in \{[\Gamma]_S \mid \Gamma \sqsubseteq S\}$. Obviously, $[\Gamma_1]_S, [\Gamma_2]_S \subseteq [\Gamma_1 \cup \Gamma_2]_S$, which implies the set $\{[\Gamma]_S \mid \Gamma \sqsubseteq S\}$ is directed.

We next show $S = \bigcup\{[\Gamma]_S \mid \Gamma \sqsubseteq S\}$. It is clear that $\bigcup\{[\Gamma]_S \mid \Gamma \sqsubseteq S\} \subseteq S$, since $[\Gamma]_S \subseteq S$ for any $\Gamma \sqsubseteq S$. Conversely, suppose that $\varphi \in S$. Then $\varphi \in [\{\varphi\}]_S \subseteq \{[\Gamma]_S \mid \Gamma \sqsubseteq S\}$. This yields that $S \subseteq \bigcup\{[\Gamma]_S \mid \Gamma \sqsubseteq S\}$.

Step 3: It remains to check that any two logical states which are bounded above have a supremum. Let S_1, S_2 and S_3 be logical states with $S_1, S_2 \subseteq S_3$. We now verify that

$$S = \{\varphi \in \mathcal{L}(P) \mid (\exists \mu_1 \in S_1 \cap \mathcal{C}(P), \exists \mu_2 \in S_2 \cap \mathcal{C}(P))\mu_1 \wedge \mu_2 \vdash \varphi \in \mathbf{T}(P))\} \tag{3.3}$$

is also a logical state and that is the supremum of S_1 and S_2.

Clearly, $S_1, S_2 \subseteq S$ and $S \neq \emptyset$ because of Eq. (3.3). Since $S_1, S_2 \subseteq S_3$, by part (3) of Proposition 3, $\mu_1 \wedge \mu_2 \in S_3$ for any $\mu_1 \in S_1 \cap \mathcal{C}_0(P)$ and $\mu_2 \in S_2 \cap \mathcal{C}_0(P)$. This implies that $S \subseteq S_3$ and $\mu_1 \wedge \mu_2$ is a conjunction, and thus, $\mu_1 \wedge \mu_2$ is irreducible by assumption.

It is obvious that $S[\vdash] \subseteq S$, so that to prove S is a logical state we need only to verify S satisfies condition (S2). For this, let $\bigvee_{i \in I} \mu_i \in S \cap \mathcal{N}(P)$. Then there exist $\nu_1 \in S_1 \cap \mathcal{C}(P)$ and $\nu_2 \in S_2 \cap \mathcal{C}(P)$ such that $\nu_1 \wedge \nu_2 \vdash \bigvee_{i \in I} \mu_i$ is valid. Since $\nu_1 \wedge \nu_2$ is an irreducible conjunction, there exists some $i_0 \in I$ such that $\nu_1 \wedge \nu_2 \vdash \mu_{i_0}$ is valid. As a result, $\mu_{i_0} \in S$.

Assume that S_4 is any other logical state with $S_1, S_2 \subseteq S_4$. According to Eq. (3.3) and the fact that S_4 is a logical state, it is easy to see that $S \subseteq S_4$. As a consequence, S is the supremum of S_1 and S_2 in $|(\mathcal{L}(P), \vdash)|$.

Theorem 1 has shown that each consistent disjunctive sequent calculus generates a Scott domain. We now turn things round and show that every Scott domain $(D, \leq)$ can also associate a consistent disjunctive sequent calculus $(\mathcal{L}(P)_D, \vdash)$.

Given a Scott domain $(D, \leq)$, put

$$\mathcal{U}(D) = \{\uparrow A \mid A \text{ is a pairwise inconsistent subset of } K(D)\} \cup \{\emptyset\}. \tag{3.4}$$

It is clear that the set $\mathcal{U}(D)$ is closed under finite intersections $\cap$ and arbitrary disjoint unions $\dot{\bigcup}$, since $(D, \leq)$ is a Scott domain. Thus we can make the following definition.

Definition 10. *Let $(D, \leq)$ be a Scott domain. Each element of the set $P_D = \{\uparrow x \mid x \in K^*(D)\}$ is said to be an atomic formula in $\mathcal{L}(P_D)$, and the set $\mathcal{L}(P_D)$ is defined by induction as follows:*

(1) Each atomic formula is an element of $\mathcal{L}(P_D)$, and the constant connectives T and F are elements of $\mathcal{L}(P_D)$.

(2) if $\varphi, \psi \in \mathcal{L}(P_D)$, then $\varphi \wedge \psi \in \mathcal{L}(P_D)$,

(3) if $\{\varphi_i \mid i \in I\} \in \mathcal{L}(P)$ with $\widehat{\varphi_i} \cap \widehat{\varphi_j} = \emptyset$ for any distinct pair $(i,j) \in I$, then $\dot{\bigvee}_{i \in I} \varphi_i \in \mathcal{L}(P_D)$, where $\widehat{\varphi}$ is the set replacing the connectives $\mathrm{F}, \mathrm{T}, \wedge$ and $\dot{\bigvee}$ in φ with $\emptyset, D, \cup$ and $\bigcup$, respectively.

Proposition 6. *Let $(D, \leq)$ be a Scott domain. Define a relation $\vdash$ as following*

$$\psi_1 \wedge \psi_2 \wedge \cdots \wedge \psi_n \vdash \varphi \text{ if and only if } \widehat{\psi_1} \cap \widehat{\psi_2} \cap \cdots \cap \widehat{\psi_n} \subseteq \widehat{\varphi},$$

where $\psi_1, \psi_2, \cdots, \psi_n, \varphi \in \mathcal{L}(P_D)$.
Then $(\mathcal{L}(P_D), \vdash)$ is a consistent disjunctive sequent calculus.

Proof. It is easy to see that the relation $\vdash$ satisfies all the rules of disjunctive formulae and valid sequents defined in Definition 3, then the pair $(\mathcal{L}(P_D), \vdash)$ is a disjunctive sequent calculus.

Let $\{x_1, x_2, \cdots, x_n\} \sqsubseteq K^*(D)$. Then the formula $\uparrow x_1 \wedge \uparrow x_2 \wedge \cdots \wedge \uparrow x_n$ of $(\mathcal{L}(P_D), \vdash)$ is irreducible if and only if $\bigsqcup\{x_1, x_2, \cdots, x_n\} \in D$. Note that D is a Scott domain, it follows that each conjunction is irreducible.

For any satisfiable formula ψ of $(\mathcal{L}(P_D), \vdash)$, there is a nonempty pairwise inconsistent subset A of $K^*(D)$ such that $\widehat{\psi} = \bigcup_{a \in A} \uparrow a$. Therefore, $\dot{\bigvee}_{a \in A} \uparrow a$ is a flat formula such that $\psi \vdash \dot{\bigvee}_{a \in A} \uparrow a$ and $\uparrow a \vdash \psi$ are valid for all $a \in A$.

For a logical state of the consistent disjunctive sequent calculi $(\mathcal{L}(P_D), \vdash)$, we have the following characterization.

Proposition 7. *Given a Scott domain $(D, \leq)$, a subset S of $\mathcal{L}(P_D)$ is a logical state of $(\mathcal{L}(P_D), \vdash)$ if and only if there exists some $d \in D$ such that $S = \{\varphi \in \mathcal{L}(P_D) \mid \widehat{\varphi} \in \mathcal{U}(D), d \in \widehat{\varphi}\}$, where $\mathcal{U}(D)$ is defined by Eq. (3.4).*

Proof. Let $d \in D$ and $S = \{\varphi \in \mathcal{L}(P_D) \mid \widehat{\varphi} \in \mathcal{U}(D), d \in \widehat{\varphi}\}$. We prove that S is a logical state of $(\mathcal{L}(P_D), \vdash)$ by showing S satisfies conditions (S1) and (S2). Assume that $\psi \in S[\vdash]$. Then there exists some finite subset Γ of S such that $\Gamma \vdash \psi$. If $\Gamma = \emptyset$, then $\psi \in \mathrm{Tau}(\mathbb{P})$, and thus $d \in \widehat{\psi} = D$. If $\Gamma \neq \emptyset$, then $d \in \widehat{\varphi}$ for all $\varphi \in \Gamma$. This implies that $d \in \widehat{\psi}$ and hence $\psi \in S$. Condition (S1) follows. For condition (S2), assume that $\dot{\bigvee}_{i \in I} \mu_i$ is a flat formula in S. Since $d \in \bigcup_{i \in I} \mu_i$, there exists some $i_0 \in I$ such that $d \in \mu_{i_0}$. Therefore, $\mu_{i_0} \in S$.

For the converse implication, assume that S is a logical state of $(\mathcal{L}(P_D), \vdash)$. Then $S \subseteq \mathcal{L}(P_D)$. We are now ready to look for an element d_S of D such that $S = \{\varphi \in \mathcal{L}(P_D) \mid \widehat{\varphi} \in \mathcal{U}(D), d_S \in \widehat{\varphi}\}$ in three stages.

First, for any $\psi \in S$, noting that $\widehat{\psi} \neq \emptyset$, there exists some pairwise inconsistent subset A of $K(D)$ such that $\widehat{\psi} = \bigcup_{a \in A} \uparrow a$. Then $\dot{\bigvee}_{a \in A} \uparrow a \in S$. By condition (S2), we have $\uparrow a_0 \in S$ for some $a_0 \in A$. This implies that $\bigcap \widehat{S} = \bigcap\{\uparrow a \mid \uparrow a \in S\}$, where $\widehat{S} = \{\widehat{\varphi} \in \mathcal{U}(D) \mid \varphi \in S\}$.

Second, we prove $\{a \mid \uparrow a \in S\}$ is a directed set of D. Let $a_1, a_2 \in \{a \mid \uparrow a \in S\}$. Since $\uparrow a_1, \uparrow a_2 \in S$, by part (3) of Proposition 3, $\uparrow a_1 \wedge \uparrow a_2 \in S$. This implies $\uparrow a_1 \cap \uparrow a_2 \neq \emptyset$. Suppose $\uparrow a_1 \cap \uparrow a_2 = \bigcup_{b \in B} \uparrow b$, where B is a nonempty set of pairwise incompatible elements of $K(D)$. By condition (S2), $\uparrow b \in S$ for some $b \in B$. Therefor, $b \in \{a \mid \uparrow a \in S\}$ and $a_1, a_2 \leq b$.

Finally, put $d_S = \bigsqcup\{a \mid {\uparrow}a \in S\}$. Then $d_S \in D$ and $\bigcap \widehat{S} = \bigcap\{{\uparrow}a \mid {\uparrow}a \in S\} = {\uparrow}d_S$. Thus $d_S \in U$ for any $U \in \widehat{S}$. Therefore, $S \subseteq \{\varphi \in \mathcal{L}(P_D) \mid \widehat{\varphi} \in \mathcal{U}(D), d_S \in \widehat{\varphi}\}$. Conversely, for any $\varphi \in \mathcal{L}(P_D)$ with $d_S \in \widehat{\varphi} \in \mathcal{U}(D)$, there exists some $d \in K(D)$ such that $d_S \in {\uparrow}d \subseteq \widehat{\varphi}$. Since $d_S = \bigsqcup\{a \mid {\uparrow}a \in S\}$ and $\{a \mid {\uparrow}a \in S\}$ is directed, $d \le a_0$ for some $a_0 \in \{a \mid {\uparrow}a \in S\}$. This yields ${\uparrow}a_0 \subseteq \widehat{\varphi}$, and hence ${\uparrow}a_0 \vdash \varphi$ is a valid sequent. By condition (S1), we have $\varphi \in S$.

With the above preparations, we obtain the main result of this section.

Theorem 2. *Each Scott domain $(D, \le)$ is isomorphic to $(|(\mathcal{L}(P_D), \vdash)|, \subseteq)$.*

Proof. By Proposition 7, the following mapping is not only well-defined but also onto,

$$f : (D, \le) \to (|(\mathcal{L}(P_D), \vdash)|, \subseteq), \text{ by } d \mapsto \{\varphi \in \mathcal{L}(P_D) \mid \widehat{\varphi} \in \mathcal{U}(D), d \in \widehat{\varphi}\},$$

where $\mathcal{U}(D)$ is defined by Eq. 3.4.

Further, trivial checks verify that $d_1 \le d_2$ if and only if $\{\varphi \in \mathcal{L}(P_D) \mid \widehat{\varphi} \in \mathcal{U}(D), d_1 \in \widehat{\varphi}\} \subseteq \{\varphi \in \mathcal{L}(P_D) \mid \widehat{\varphi} \in \mathcal{U}(D), d_2 \in \widehat{\varphi}\}$. As a consequence, the mapping f is an order-isomorphism from $(D, \le)$ to $(|(\mathcal{L}(P_D), \vdash)|, \subseteq)$. Whence $(D, \le)$ is isomorphic to $(|(\mathcal{L}(P_D), \vdash)|, \subseteq)$.

4   A Categorical View

The purpose of this section is to extend to the relationship between consistent disjunctive sequent calculi and Scott domains to a categorical equivalence.

Definition 11. *Let $\mathbb{P} = (\mathcal{L}(P), \vdash_P)$ and $\mathbb{Q} = (\mathcal{L}(Q), \vdash_Q)$ be consistent disjunctive sequent calculi. A consequence relation Θ from $\mathbb{P}$ to $\mathbb{Q}$, written as $\Theta : \mathbb{P} \to \mathbb{Q}$, is a binary relation between $\mathcal{C}(P)$ and $\mathcal{L}(Q)$ satisfies the following conditions:*

(C1) *If there is some $\nu \in \mathcal{C}(P)$ such that $\mu \vdash_P \nu$ is valid and $(\nu, \psi) \in \Theta$, then $(\mu, \psi) \in \Theta$.*
(C2) *If $(\mu, \varphi) \in \Theta$ and $\varphi \vdash_Q \psi$ is valid, then $(\mu, \psi) \in \Theta$.*
(C3) *If $(\mu, \psi) \in \Theta$, then there is some $\nu \in \mathcal{C}(Q)$ such that $\nu \vdash_Q \psi$ is valid and $(\mu, \nu) \in \Theta$.*

Consider a consequence relation $\Theta : \mathbb{P} \to \mathbb{Q}$. For any subset X of $\mathcal{L}(P)$, put

$$\Theta[X] = \{\varphi \in \mathcal{L}(Q) \mid (\exists \mu \in X \cap \mathcal{C}(P))(\mu, \varphi) \in \Theta\}. \tag{4.1}$$

Proposition 8.

(1) If $X_1, X_2 \subseteq \mathcal{L}(P)$ with $X_1 \subseteq X_2$, then $\Theta[X_1] \subseteq \Theta[X_2]$.
(2) If S is a logical state of $\mathbb{P}$, then $\Theta[S]$ is a logical state of $\mathbb{Q}$.
(3) If $\mu \in \mathcal{C}(P)$, then $\Theta[\{\mu\}] = \Theta[\{\mu\}[\vdash_P]]$.

Proof. (1) Straightforward from Eq. (4.1).

(2) Assume that $\varphi \in (\Theta[S])[\vdash_Q]$. Then there exist $\mu \in S \cap \mathcal{C}(P)$ and $\psi \in \mathcal{L}(Q)$ such that $(\mu, \psi) \in \Theta$ and $\psi \vdash_Q \varphi$ is valid. By condition (C2), we have $(\mu, \varphi) \in \Theta$. This implies that $\varphi \in \Theta[S]$ and hence $(\Theta[S])[\vdash_Q] \subseteq \Theta[S]$.

Assume that $\bigvee_{i \in I} \mu_i \in \Theta[S] \cap \mathcal{N}(P)$. Then there exists $\mu \in S \cap \mathcal{C}(P)$ such that $(\mu, \bigvee_{i \in I} \mu_i) \in \Theta$. By condition (C3), there exists some $\nu \in \mathcal{C}(Q)$ such that $(\mu, \nu) \in \Theta$ and $\nu \vdash_Q \bigvee_{i \in I} \mu_i$ is valid. Since ν is an irreducible conjunction, $\nu \vdash_Q \mu_{i_0}$ is valid for some $i_0 \in I$. Using condition (C2) again, $(\mu, \mu_{i_0}) \in \Theta$, and thus $\mu_{i_0} \in \Theta[S]$.

(3) According to part (1), it is clear $\Theta[\{\mu\}] \subseteq \Theta[\{\mu\}[\vdash_P]]$. Conversely, let $\varphi \in \Theta[\{\mu\}[\vdash_P]]$. Then there exists some $\nu \in \{\mu\}[\vdash_P] \cap \mathcal{C}(P)$ such that $(\nu, \varphi) \in \Theta$. But $\nu \in \{\mu\}[\vdash_P]$ implies that $\mu \vdash_P \nu$ is valid. By condition (C1), we have $(\mu, \varphi) \in \Theta$, and thus $\varphi \in \Theta[\{\mu\}]$.

Now we show that there is a one-to-one correspondence between consequence relations from $\mathbb{P}$ to $\mathbb{Q}$ and Scott-continuous functions from $|\mathbb{P}|$ to $|\mathbb{Q}|$.

Theorem 3. *Let $\mathbb{P}$ and $\mathbb{Q}$ be consistent disjunctive sequent calculi.*

(1) *For any consequence relation $\Theta : \mathbb{P} \to \mathbb{Q}$, define a function $f_\Theta : |\mathbb{P}| \to |\mathbb{Q}|$ by*

$$f_\Theta(S) = \Theta[S]. \tag{4.2}$$

Then f_Θ is Scott-continuous.

(2) *For any Scott-continuous function $f : |\mathbb{P}| \to |\mathbb{Q}|$, define $\Theta_f \subseteq \mathcal{C}(P) \times \mathcal{L}(Q)$ by*

$$(\mu, \psi) \in \Theta_f \Leftrightarrow \psi \in f(\{\mu\}[\vdash_P]). \tag{4.3}$$

Then Θ_f is a consequence relation from $\mathbb{P}$ to $\mathbb{Q}$.

(3) $\Theta_{f_\Theta} = \Theta$ *and* $f_{\Theta_f} = f$.

Proof. (1) By part (2) of Proposition 8, the function f_Θ is well-defined. And part (1) of Proposition 8 yields that the function f_Θ is monotone. Let $\{S_i \mid i \in I\}$ be a directed subset of logical states. Then $\bigcup_{i \in I} S_i$ is a logical state. Since $\bigcup_{i \in I} f_\Theta(S_i) \subseteq f_\Theta(\bigcup_{i \in I} S_i)$ is clear, to prove the function f_Θ is Scott-continuous, it suffices to show that $f_\Theta(\bigcup_{i \in I} S_i) \subseteq \bigcup_{i \in I} f_\Theta(S_i)$. If $\varphi \in f_\Theta(\bigcup_{i \in I} S_i) = \Theta[\bigcup_{i \in I} S_i]$, then there exists some $\mu \in \bigcup_{i \in I} S_i \cap \mathcal{C}(P)$ such that $(\mu, \varphi) \in \Theta$. From $\mu \in \bigcup_{i \in I} S_i$, it follows that $\mu \in S_{i_0}$ for some $i_0 \in I$. Thus $\varphi \in f_\Theta(S_{i_0})$, and therefore, $f_\Theta(\bigcup_{i \in I} S_i) \subseteq \bigcup_{i \in I} f_\Theta(S_i)$.

(2) It suffices to show that Θ_f satisfies conditions (C1–C3).

For condition (C1), assume that $\nu \in \mathcal{C}(P)$ such that $\mu \vdash_P \nu$ is valid and $(\nu, \psi) \in \Theta_f$. Then $\psi \in f(\{\nu\}[\vdash_P])$. From $\mu \vdash_P \nu$ it follows that $\{\nu\}[\vdash_P] \subseteq \{\mu\}[\vdash_P]$. Since f is monotone, $\psi \in f(\{\mu\}[\vdash_P])$. This means that $(\mu, \psi) \in \Theta_f$.

For condition (C2), assume that $(\mu, \varphi) \in \Theta_f$ and $\varphi \vdash_Q \psi$ is valid. Then $\varphi \in f(\{\mu\}[\vdash_P])$. Since $f(\{\mu\}[\vdash_P])$ is a logical state and $\varphi \vdash_Q \psi$ is valid, $\psi \in f(\{\mu\}[\vdash_P])$. That is $(\mu, \psi) \in \Theta_f$.

For condition (C3), assume that $(\mu, \psi) \in \Theta_f$. Then $\psi \in f(\{\mu\}[\vdash_P])$. Since the disjunctive sequent calculus $\mathbb{Q}$ is consistent, there exists an irreducible flat formula $\bigvee_{i \in I} \mu_i$ such that $\psi \vdash \bigvee_{i \in I} \mu_i$ and $\mu_i \vdash \psi$ are all valid for all $i \in I$. Note that $f(\{\mu\}[\vdash_P])$ is a logical state, it follows that $\bigvee_{i \in I} \mu_i \in f(\{\mu\}[\vdash_P])$. Thus $\mu_{i_0} \in f(\{\mu\}[\vdash_P])$ for some $i_0 \in I$. Let $\mu_{i_0} = \nu$. Then we obtain some $\nu \in \mathcal{C}_0(Q)$ such that $(\mu, \nu) \in \Theta$ and $\nu \vdash_Q \psi$ is valid.

(3) For any $\mu \in \mathcal{C}(P)$ and $\varphi \in \mathcal{L}(Q)$, we have

$$
\begin{aligned}
(\mu, \varphi) \in \Theta_{f_\Theta} &\Leftrightarrow \varphi \in f_\Theta(\{\mu\}[\vdash_P]) \\
&\Leftrightarrow \varphi \in \Theta[\{\mu\}[\vdash_P]] \\
&\Leftrightarrow (\exists \nu \in \mathcal{C}(Q))((\mu, \varphi) \in \Theta, \mu \vdash_P \nu \in \mathbf{T}(P))) \\
&\Leftrightarrow (\mu, \varphi) \in \Theta.
\end{aligned}
$$

This proves that $\Theta_{f_\Theta} = \Theta$.

For any $S \in |\mathbb{P}|$, we have

$$
\begin{aligned}
f_{\Theta_f}(S) = \Theta_f[S] &= \{\varphi \in \mathcal{L}(Q) \mid (\exists \mu \in S \cap \mathcal{C}(P))(\mu, \varphi) \in \Theta_f\} \\
&= \{\varphi \in \mathcal{L}(Q) \mid (\exists \mu \in S \cap \mathcal{C}(P))\varphi \in f(\{\mu\}[\vdash_P])\} \\
&= \bigcup\{f(\{\mu\}[\vdash_P]) \mid \mu \in S \cap \mathcal{C}(P)\} \\
&= f(\bigcup\{\{\mu\}[\vdash_P] \mid \mu \in S \cap \mathcal{C}(P)\} \\
&= f(S).
\end{aligned}
$$

This proves that $f_{\Theta_f} = f$.

Let Θ be a consequence relation from $\mathbb{P}$ to $\mathbb{Q}$, and Θ' a consequence relation from $\mathbb{Q}$ to $\mathbb{R}$. Define $\Theta' \circ \Theta \subseteq \mathcal{C}(P) \times \mathcal{L}(R)$ by

$$(\mu, \varphi) \in \Theta' \circ \Theta \Leftrightarrow (\exists \nu \in \mathcal{C}(Q))((\mu, \nu) \in \Theta, (\nu, \varphi) \in \Theta'), \tag{4.4}$$

and $\mathrm{id}_P \subseteq \mathcal{C}(P) \times \mathcal{L}(P)$ by

$$(\mu, \varphi) \in \mathrm{id}_P \Leftrightarrow \varphi \in \{\mu\}[\vdash_P]. \tag{4.5}$$

Then routine checks verify that $\Theta' \circ \Theta$ is a consequence relation from $\mathbb{P}$ to $\mathbb{R}$ and id_P is a consequence relation from $\mathbb{P}$ to itself.

Using the same argument as checking the associative law of a traditional relation composition, we can carry out the composition $\circ$ defined by Expression (4.4) is associative. Conditions (C1) and (C2) yield that id_P is the identity morphism of $\mathbb{P}$.

So Consistent disjunctive sequent calculi with consequence relations form a category $\mathbf{CDSC}$.

Moreover, $\mathcal{G} : \mathbf{CDSC} \to \mathbf{SD}$ is a functor which maps every consistent disjunctive sequent calculi $\mathbb{P}$ to $(|\mathbb{P}|, \subseteq)$ and consequence relation $\Theta : \mathbb{P} \to \mathbb{Q}$ to f_Θ, where f_Θ is defined by Eq. (4.2).

Theorem 4. CDSC *and* **SD** *are categorically equivalent.*

Proof. According to Theorem 2, it suffices to show that the functor $\mathcal{G}$ is full and faithful.

For any Scott-continuous function $f : |\mathbb{P}| \to |\mathbb{Q}|$, by Theorem 3, the relation Θ_f defined by Eq. (4.3) is a consequence relation from $\mathbb{P}$ to $\mathbb{Q}$ and $\mathcal{G}(\Theta_f) = f_{\Theta_f} = f$. This implies that $\mathcal{G}$ is full.

Let $\Theta_1, \Theta_2 : \mathbb{P} \to \mathbb{Q}$ be two consequence relations with $f_{\Theta_1} = f_{\Theta_2}$, where f_{Θ_1} and f_{Θ_2} are defined by Eq. (4.2). For any $\mu \in \mathcal{C}(\mathbb{P})$, since

$$(\mu, \varphi) \in \Theta_1 \Leftrightarrow \varphi \in \Theta_1[\{\mu\}]$$
$$\Leftrightarrow \varphi \in \Theta_1[\{\mu\}[\vdash_P]]$$
$$\Leftrightarrow \varphi \in f_{\Theta_1}(\{\mu\}[\vdash_P])$$
$$\Leftrightarrow \varphi \in f_{\Theta_2}(\{\mu\}[\vdash_P])$$
$$\Leftrightarrow (\mu, \varphi) \in \Theta_2,$$

it follows that $\Theta_1 = \Theta_2$, and hence $\mathcal{G}$ is faithful.

References

1. Abramsky, S.: Domain theory and the logic of observable properties. Ph.D. thesis, University of London (1987)
2. Abramsky, S.: Domain theory in logical form. Ann. Pure Appl. Logic **51**, 1–77 (1991)
3. Chen, Y., Jung, A.: A logical approach to stable domains. Theoret. Comput. Sci. **368**, 124–148 (2006)
4. Edalat, A., Smthy, M.B.: Information categories. Appl. Categor. Struct. **1**, 197–323 (1993)
5. Erné, M.: Categories of locally hypercompact spaces and quasicontinuous posets. Appl. Categor. Struct. **26**(5), 823–854 (2018). https://doi.org/10.1007/s10485-018-9536-0
6. Ésik, Z.: Residuated park theories. J. Logic Comput. **25**(2), 453–471, 2015
7. Gallier, J.H.: Logic for Computer Science: Foundations of Automatic Theorem Proving. Courier Dover Publications, New York (2015)
8. Gierz, G., Hofmann, K.H., Keimel, K., Lawson, J.D., Mislove, M., Scott, D.S.: Continuous Lattices and Domains. Cambridge University Press, Cambridge (2003)
9. Goubault-Larrecq, J.: Non-Hausdorff Topology and Domain Theory, Volume 22 of New Mathematical Monographs. Cambridge University Press, Cambridge (2013)
10. Ho, W., Goubault-Larrecq, J., Jung, A., Xi, X.: The Ho-Zhao problem. Logical Methods Comput. Sci. **14**(1:7), 1–19 (2018)
11. Hoofman, R.: Continuous information systems. Inf. Comput. **105**, 42–71 (1993)
12. Huang, M., Zhou, X., Li, Q.: Re-visiting axioms of information systems. Inf. Comput. **247**, 130–140 (2015)
13. Jung, A., Kegelmann, M., Moshier, M.A.: Multi lingual sequent calculus and coherent spaces. Fundamenta Informaticae **37**, 369–412 (1999)
14. Jung, A.: Continuous domain theory in logical form. In: Coecke, B., Ong, L., Panangaden, P. (eds.) Computation, Logic, Games, and Quantum Foundations. The Many Facets of Samson Abramsky. LNCS, vol. 7860, pp. 166–177. Springer, Heidelberg (2013). https://doi.org/10.1007/978-3-642-38164-5_12

15. Larsen, K.G., Winskel, G.: Using information systems to solve reoursive domain equations effectively. In: Kahn, G., MacQueen, D.B., Plotkin, G. (eds.) SDT 1984. LNCS, vol. 173, pp. 109–129. Springer, Heidelberg (1984). https://doi.org/10.1007/3-540-13346-1_5

16. Scott, D.S.: Domains for denotational semantics. In: Nielsen, M., Schmidt, E.M. (eds.) ICALP 1982. LNCS, vol. 140, pp. 577–610. Springer, Heidelberg (1982). https://doi.org/10.1007/BFb0012801

17. Spreen, D., Xu, L., Mao, X.: Information systems revisited: the general continuous case. Theoret. Comput. Sci. **405**, 176–187 (2008)

18. Vickers, S.: Entailment systems for stably locally compact locales. Theoret. Comput. Sci. **316**, 259–296 (2004)

19. Wang, L., Li., Q.: A representation of proper BC domains based on conjunctive sequent calculi. Math. Struct. Comput. Sci. 1–13 (2020). https://doi.org/10.1017/S096012951900015X

20. Wu, M., Guo, L., Li, Q.: A representation of L-domains by information system. Theoret. Comput. Sci. **612**, 126–136 (2016)

21. Yao, W.: A categorical isomorphism between injective stratified fuzzy T_0 spaces and fuzzy continuous lattices. IEEE Trans. Fuzzy Syst. **24**(1), 131–139 (2016)

22. Zhang, G.-Q.: Disjunctive systems and L-domains. In: Kuich, W. (ed.) ICALP 1992. LNCS, vol. 623, pp. 284–295. Springer, Heidelberg (1992). https://doi.org/10.1007/3-540-55719-9_81

Dispersing and Grouping Points on Segments in the Plane

Xiaozhou He[1], Wenfeng Lai[2], Binhai Zhu[3]([✉]), and Peng Zou[3]

[1] Business School, Sichuan University, Chengdu, China
xiaozhouhe126@qq.com
[2] College of Computer Science and Technology, Shandong University,
Qingdao, China
2290892069@qq.com
[3] Gianforte School of Computing, Montana State University,
Bozeman, MT 59717, USA
bhz@montana.edu, peng.zou@student.montana.edu

Abstract. Motivated by (continuous) facility location, we consider the problem of dispersing and grouping points on a set of segments (of streets) in the plane. In the former problem, given a set of n disjoint line segments in the plane, we consider the problem of computing a point on each of the n segments such that the minimum Euclidean distance between any two of these points is maximized. We prove that this 2D dispersion problem is NP-hard, in fact, the problem is NP-hard even if all the segments are parallel and are of unit length. This is in contrast to the polynomial solvability of the corresponding 1D problem by Li and Wang (2016), where the intervals are in 1D and are all disjoint. With this result, we also show that the Independent Set problem on Colored Linear Unit Disk Graph (meaning the convex hulls of points with the same color form disjoint line segments) remains NP-hard, and the parameterized version of it is in W[2]. In the latter problem, given a set of n disjoint line segments in the plane, we consider the problem of computing a point on each of the n segments such that the maximum Euclidean distance between any two of these points is minimized. We present a factor-1.1547 approximation algorithm which runs in $O(n \log n)$ time. Our results can be generalized to the Manhattan distance.

Keywords: Dispersion problem · NP-hardness · FPT · Manhattan distance · Geometric optimization

1 Introduction

Dispersion problems belong to the classic facility location problem and have been extensively studied. The goal of such a problem is to build facilities so that they are as far as possible. A typical example is to build a chain of convenience stores such that they should be far from each other to cover more customers. As

© Springer Nature Switzerland AG 2020
J. Chen et al. (Eds.): TAMC 2020, LNCS 12337, pp. 340–351, 2020.
https://doi.org/10.1007/978-3-030-59267-7_29

a matter of fact, a series of research has been done, either over a point set or over a weighted graph [2,5,6,9,16–18].

In [11,12], Li and Wang considered an interesting variation of the problem, where one is given a set of disjoint intervals in 1D and the objective is to put one point on each interval such that the minimum distance between any two of these computed points is maximized. Assuming the intervals are sorted, an optimal linear time greedy algorithm was given (but the analysis is non-trivial). The scenario corresponding to this problem can be considered as constructing resting areas along a highway, where each interval is some section suitable for constructing a resting area.

A natural question arises: what if we are given some disjoint (rectilinear) segments (where each segment is part of a street)? (Here the objective function is the same while the distance could be either Euclidean (L_2) or Manhattan (L_1).) We show that this problem is NP-hard; in fact, NP-hard even when all the segments are parallel (i.e., along one direction) and are of a unit length. It turns out that this is related to the Independent Set (IS) problem on a unit disk graph (UDG) [3,13]; in fact, our NP-hardness proof implies that the problem remains NP-hard even when the unit disk graph is *colored* and *linear* (meaning the convex hulls of points of the same color form disjoint line segments). We suspect that the parameterized version remains to be W[1]-hard, though we are only able to show that it is in W[2] at this point.

The symmetric problem of grouping points on a set of disjoint segments in the plane, i.e., selecting one point on each segment such that the maximum distance between the selected points is minimized, is motivated by constructing commodity distribution centers within a road network. These centers should be close to each other to reduce the distribution or transportation costs. It is not known whether the problem is NP-hard yet, though we are able to show that this problem admits a factor-1.1547 approximation running in $O(n \log n)$ time.

This paper is organized as follows. In Sect. 2, we give the preliminaries. In Sect. 3, we prove that the 2D dispersion problem is NP-hard. In Sect. 4, we consider briefly the independent set problem on colored linear unit disk graphs and prove its W[2] membership. In Sect. 5, we give a simple polynomial time approximation algorithm for the 2D grouping problem. We conclude the paper in Sect. 6.

2 Preliminaries

2.1 Definitions

Given two points $a = (x_a, y_a), b = (x_b, y_b)$ in the plane (2D), the Euclidean or L_2 distance $d(a, b) = d_2(a, b)$ is defined as $d(a, b) = ((x_a - x_b)^2 + (y_a - y_b)^2)^{1/2}$. The Manhattan or L_1 distance $d_1(a, b)$ is defined as $d_1(a, b) = |x_a - x_b| + |y_a - y_b|$. A line segment with endpoints a and b is denoted as $l = (a, b)$.

Finally, a unit disk graph is one where each vertex is a given point of an input set of planar points, two vertices u, v share an edge if two disks of radii R centered at u, v intersect each other. (Note that the standard unit disk graph definitions require that $R = 1/2$, in our definition R could be more general.) It is known that while most NP-hard problems on general graphs remain NP-hard on unit disk graphs [4,7], there are exceptions (e.g., the maximum clique problem is polynomially solvable [7]). Notably, the parameterized version of the independent set problem on unit disk graphs, parameterized by the size of the solution, is known to be W[1]-hard [13]. The more restricted colored version, parameterized by the number of colors, remains to be W[1]-hard [3].

2.2   Problems

The problems studied in this paper are defined as follows:

2D Dispersion Problem: Given a set of disjoint line segments $L = \{L_1, L_2, ..., L_n\} \subset R^2$, the goal is to find n points $V = \{v_1, v_2, ..., v_n\}$ on the n line segments respectively such that the minimum distance among two points in V is maximized, i.e.,

$$\max_{V} \min_{v_i, v_j \in V} d(v_i, v_j).$$

2D Grouping Problem: Given a set of disjoint line segments $L = \{L_1, L_2, ..., L_n\} \subset R^2$, the goal is to find n points $V = \{v_1, v_2, ..., v_n\}$ on the n line segments respectively such that the maximum distance among two points in V is minimized, i.e.,

$$\min_{V} \max_{v_i, v_j \in V} d(v_i, v_j).$$

We show in the next section that the 2D dispersion problem is NP-hard.

3   NP-Hardness for the 2D Dispersion Problem

We reduce the planar 3-SAT problem to the 2D Dispersion problem. The planar 3-SAT problem is a special case of 3-SAT where the input is a conjunction of a set of disjunctive clauses, each with three literals. Moreover, if we create a graph with

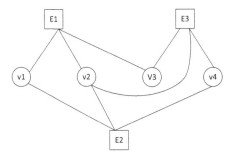

Fig. 1. The planar graph for the planar 3-SAT instance $\phi = (v_1 \lor v_2 \lor \bar{v}_3) \land (\bar{v}_1 \lor v_2 \lor v_4) \land (\bar{v}_2 \lor v_3 \lor \bar{v}_4)$.

edges connecting a literal x and clause c where x appears in c, then the resulting graph is planar. Such a planar graph embedding can be computed in linear time. Given a planar 3-SAT instance $\phi = (v_1 \vee v_2 \vee \bar{v}_3) \wedge (\bar{v}_1 \vee v_2 \vee v_4) \wedge (\bar{v}_2 \vee v_3 \vee \bar{v}_4)$, its corresponding planar graph is shown in Fig. 1.

When we reduce planar 3-SAT to the 2D Dispersion problem, all the segments in U are with the unit length 1 and are horizontal.

Theorem 1. *The 2D Dispersion Problem is NP-hard.*

Proof. First, we describe the variable gadget and clause gadget in detail. Note that the dashed segments in these two gadgets are only for illustration purpose. In fact, they are only used to illustrate the distance between the points. Each variable is corresponding to 4 parallel unit segments with length 1 and the distance between every two closest segments is $\sqrt{3}/3$, as shown in Fig. 2. The length of the dashed segments is $\ell = 2\sqrt{3}/3$. The red points are selected for the 'True' assignment of this variable; and symmetrically, the green points are chosen for the corresponding 'False' assignment.

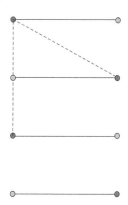

Fig. 2. The variable gadget. (Color figure online)

Next, we depict the clause gadget, which is shown in Fig. 3. The length of the dashed segments is also ℓ. In each clause gadget, if all three green points are selected, then placing one more point on the segment in the middle, the minimum distance between any two of these four points would be smaller than ℓ. To achieve such a distance of at least ℓ, we need to select at least one red point x and another point y on the middle segment—according to which x is selected. (For instance, if x is the top-right red point then y could be the mid-point of the middle segment.)

A complete example of the construction for the clause $(v_1 \vee v_2 \vee \bar{v}_3)$ is shown in Fig. 4. Note that all the segments connecting the variable and clause gadgets are horizontal; moreover, an additional segment is added to ensure the appearance of $\bar{v}_i$ in the corresponding clause gadget.

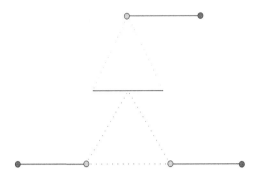

Fig. 3. The clause gadget. (Color figure online)

We claim that the planar 3-SAT instance ϕ has a valid truth assignment if and only in the converted 2D Dispersion instance the minimum distance of any two chosen points is equal to ℓ.

'If' part: If the planar 3-SAT instance ϕ has a truth assignment, in all variable gadgets, either all red points or all green points would be selected. The closest

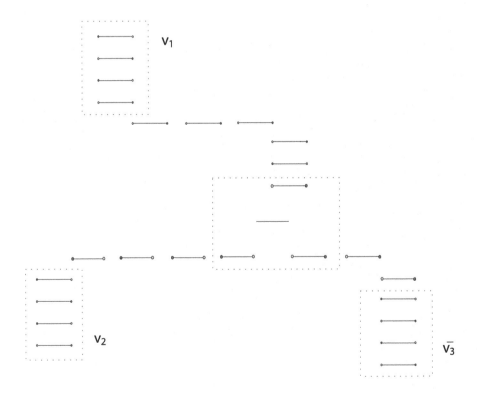

Fig. 4. A complete construction for the clause $(v_1 \vee v_2 \vee \bar{v}_3)$. (Color figure online)

distance of two chosen points in the variable gadgets is equal to ℓ. In the clause gadget, at least one variables needs to be assigned 'True' for the clause. Hence, at least one green points is not selected in the corresponding clause gadget. The minimum distance of any two chosen points in the clause gadget is at least ℓ. Therefore, in the converted instance for 2D Dispersion, the minimum distance of any two selected points is ℓ.

'Only if' part: Suppose that the converted instance for 2D Dispersion has a minimum distance of ℓ between two selected points. Firstly, we note that in all the variable gadgets, if we want to maximize the minimum distance of any two selected points, then either all green points or all red points must be chosen. At this point the minimum distance of any two such selected points is exactly ℓ. In the clause gadget, if the maximum of the minimum distance of any two chosen points is at least ℓ, then at least one red point x needs to be selected. This implies that in the corresponding clause, the corresponding variable v_x is assigned 'True' or the corresponding literal $\bar{v}_x$ is assigned 'False', which means the corresponding clause is evaluated 'True'. Therefore, if in the converted 2D dispersion instance the minimum distance of any two chosen points is at least ℓ, the corresponding planar 3-SAT instance ϕ is satisfied. $\square$

Since the problem is NP-hard even when all the input segments are horizontal and are of unit length, the general version of this problem (where disjoint segments are of various lengths and are with arbitrary orientations) is also NP-hard. We comment that the NP-hardness holds even when the L_1 (or Manhattan) distance is used, this can be done by adjusting the distance between parallel segments accordingly.

4   Hardness for IS on Colored Linear Unit Disk Graphs

The NP-hardness result in the previous section has a direct implication on the Independent Set (IS) problem on Colored Linear Unit Disk Graph, which is a unit disk graph such that the convex hull of the points in the same color form a line segment, no two such segments intersect and the problem is to select one node (disk) in each color such that they form an independent set. We briefly go through the implication in this section.

Before doing that, we note that finding k-multicolored clique or independent set was initially motivated in proving the W[1]-hardness of some graph problems [8]. For geometric intersection graphs (specifically, unit disk graphs) Marx showed the Independent Set problem is W[1]-hard, which implies it is not possible to obtain an FPT (fixed-parameter tractable) algorithm unless FPT=W[1] [13]. Moreover, the W[1]-hardness of the problem implies that it is not possible to obtain an EPTAS (efficient PTAS, namely the running time is $f(\epsilon) \cdot n^c$, where n is the input size). Consequently the PTAS by Hunt et al. (with running time $O(n^{1\epsilon})$) [10] cannot be further improved to have an EPTAS. Bereg et al. considered the k-multicolored independent set problem on unit disk graphs and proved its W[1]-hardness and with that, they proved that the Largest Closest

Pair Color-Spanning Set problem is W[1]-hard [3]. We now briefly sketch our results.

Theorem 2. *The Independent Set problem on Colored Linear Unit Disk Graph is NP-hard, even when all the segments are parallel and are of a unit length.*

Proof. We just use the reduction for Theorem 1. The two changes are: (1) put $1/\epsilon$ points on each segment, (2) the corresponding unit disk graph is formed by drawing an open disk centered at each point of each segment with a radii $\ell/2 = \sqrt{3}/3$; moreover, all the disks centered at the same segment have the same color. Let U be the set of unit-length segments. Calling the resulting unit disk graph G_U, we clearly still have the following statement: the planar 3-SAT instance ϕ is satisfiable if and only if G_U has an independent set of size $|U|$. The details are standard and omitted.   □

An immediate question is the FPT tractability of the parameterized version of the problem, where the parameter k is the number of segments, or the number of colors. Note that the general version, where the points of the same color could be arbitrarily distributed, is W[1]-hard [3]. However, in the current version the unit disk graph is more restricted. Nevertheless we show below that it is in W[2].

Theorem 3. *The Independent Set problem on Colored Linear Unit Disk Graph, parameterized on the number of colors (e.g. segments), is in W[2].*

Proof. Let the number of segments be k and let the set of linear points on segment L_i be ordered as $V_i = \{v_{i,j} | j = 1..P(i)\}$. We need to decide whether a value R exists such that the set S of points selected, one for each segment (color), has the property that the closest pair has distance at least R (or, equivalently, the unit disk graph with radii $R/2$ on all these points has a colorful independent set, i.e., one for each color).

We construct a circuit C as follows: the input are variables corresponding to all the points (for convenience, we still use $v_{i,j}$'s as variables, with a value one is assigned to $v_{i,j}$ meaning that point is selected). For all the points on the same segment L_i, we construct a large OR ($\vee$) gate. Here, 'large' means the input to the OR gate could be greater than 2; and to make the OR gate output a true value, one of these points must be selected.

For two points on two segments $v_{i,i'}$ and $v_{j,j'}$, if their distance is shorter than R then we cannot select both of them. This can be interpreted as $\neg(v_{i,i'} \wedge v_{j,j'})$. In fact, as the distance function from point $v_{i,i'}$ to all (sorted) points on L_j (i.e., V_j) is unimodal. We could construct these nested $\neg$ and $\wedge$ gates in one pass when $v_{i,i'}$ and V_j are fixed.

Finally, we connect all these OR ($\vee$) gates and NOT ($\neg$) gates to a large AND ($\wedge$) gate, which is the output of this circuit. It is easy to see that the IS problem on Colored Linear Unit Disk Graph (with radii $R/2$) has a solution if and only if k variables are selected to have a 'True' output. As from any input gate to the final output we have at most two wefts (large gates), the W[2] membership is hence shown (Fig. 5).   □

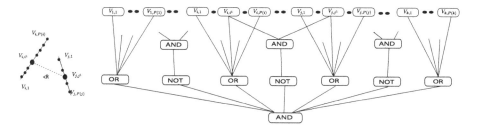

Fig. 5. The circuit for the IS problem on Colored Linear Unit Disk Graph.

We comment that this proof is similar to the one given for the Minimum Diameter Color-Spanning Set problem by Pruente [15]. However, it is tantalizingly open whether this version of the IS problem on unit disk graphs is W[1]-hard. We comment that the unimodality property between $v_{i,i'}$ and V_j might be used either to show its W[1]-hardness or its membership in FPT.

5 Approximation for the 2D Grouping Problem

In this section, we design a factor-1.1547 polynomial-time approximation algorithm for the 2D grouping problem. It turns out that this problem is also related to unit disk graph. Recall that the input to the problem is a set of disjoint line segment $L = \{L_1, L_2, ..., L_n\}$. Assuming that we have an optimal solution $V^* = \{v_1, v_2, ..., v_n\}$ where each v_i is selected from L_i and the maximum distance between v_i, v_j is d^*. Let $C_i(r)$ be a closed disk centered at v_i with radius r. Then, using the so-called *intersection model* for a unit disk graph, if we draw a disk $C_i(d^*/2)$ at each v_i, these disks would have a common intersection.

5.1 The Minimum Intersecting Disk Problem

Based on this property, we define a related Minimum Intersecting Disk problem:

Minimum Intersecting Disk (MID) Problem: Given a set of disjoint line segments $L = \{L_1, L_2, ..., L_n\} \subset R^2$, compute a disk $C(c, r)$ with center c and radius r such that all L_i's intersect $C(c, r)$ and r is minimized (at r^+).

We will present an polynomial-time algorithm to solve the MID problem in this subsection. We first obtain a simple decision procedure $Decide(r)$ which decides whether a disk of radius r exists that intersect all segments in L. Note that a segment completely in $C(c, r)$ is also considered being intersected by $C(c, r)$.

Algorithm 1. Decide(r): decides if there is a disk of radius r intersecting L

1: For each segment L_i, compute the Minkowski sum of L_i with a disk of radius $C(r)$, denoted as $L_i \oplus C(r)$.
2: Compute the common intersection $F(r)$ of all $L_i \oplus C(r)$, i.e., $F(r) = \cap_i (L_i \oplus C(r))$.
3: If $F(r)$ is not empty then return YES and any point in $F(r)$ as a witness for the center of such a disk; otherwise, return NO.

Note that all $L_i \oplus C(r)$ are convex, so their common intersection can be computed using a standard method in computational geometry, e.g., the incremental construction method (i.e., maintain the current common intersection, insert the next one, and update to have the new common intersection). Due to convexity, each update takes $O(\log n)$ time. Hence, $F(r)$ can be computed in $O(n \log n)$ time.

To compute r^+, we notice that all possible values of r^+ can be computed in $O(n^3)$ time: each possible value is decided by either two or three segments, in the former case it is half of the distance between the closest distance between two segments and in the latter case it is the radius of the circle tangent to three segments. We can then obtain the following algorithm.

Algorithm 2. Solution for MID

1: Compute a sorted list $\mathcal{L}$ which contains the radii of all possible circles which are decided by two or three segments.
2: Binary search in $\mathcal{L}$ using $Decide(r)$ to compute the disk $C(c^+, r^+)$ with center c^+ and with the minimum radius r^+, which intersects all segments by construction.
3: Return $C(c^+, r^+)$.

We note that $|\mathcal{L}| = O(n^3)$, hence $\mathcal{L}$ can be constructed and then sorted in $O(n^3 \log n)$ time. Using the binary search, with $D(r)$ as a subroutine, the cost is $O(n \log n) \times O(\log n^3) = O(n \log^2 n)$. The total cost is therefore $O(n^3 \log n) + O(n \log^2 n) = O(n^3 \log n)$. Note that this algorithm, though each to implement, might be too slow when n is large.

We could in fact use the farthest Voronoi diagram for line segments [1,14] to improve this time bound. Simply compute such a Voronoi diagram and the center of MID must lie either on a Voronoi edge or at a Voronoi vertex. In either case, a candidate intersecting disk can be computed in $O(1)$ time. We thus have the following theorem.

Theorem 4. *The Minimum Intersecting Disk problem can be solved in* $O(n \log n)$ *time.*

Proof. The farthest line-segment Voronoi diagram can be computed in $O(n \log n)$ time with n input segments [1,14]. Moreover, such a diagram has a linear size (i.e., $O(n)$ number of edges and vertices). Hence, with a linear search we could identify $O(n)$ candidate intersecting disks and return the smallest one. Therefore, the MID problem can be solved in $O(n \log n)$ time. $\square$

5.2   Approximation Factor Analysis

Our approximation algorithm for the 2D Grouping problem is to compute the minimum intersecting disk $C^+ = C(c^+, r^+)$ for L, and return the maximum distance $d(C^+)$ between the two or three points defining this disk $C(c^+, r^+)$.

Let C^* be the minimum radius disk, with radius $r(C^*)$, enclosing all the selected points of L in an optimal solution for the 2D grouping problem on L. Let $d(C^*)$ be the maximum distance between the two or three points defining C^*. Let Opt be the the maximum distance between the selected points in the optimal 2D grouping problem.

We have the following lemma.

Lemma 1. $r(C^*) \leq \frac{d(C^*)}{\sqrt{3}}$.

Proof. When C^* is defined by two points, apparently we have $r(C^*) = \frac{d(C^*)}{2}$.

When C^* is defined by three points, we have $r(C^*) \leq \frac{d(C^*)}{\alpha}$. This α is minimized when the three defining points for C^* form a regular triangle, where we have $\alpha = \sqrt{3}$.   □

Since Opt is the maximum distance between the selected points on L (which are inside the circle C^*) and $d(C^*)$ is the maximum distance between the two or three points defining C^*, we have $d(C^*) \leq Opt$. Let App be the solution of the approximation solution. We then combine all the arguments together as follows.

$$
\begin{aligned}
App =& d(C^+) \\
\leq& 2 \cdot r^+ \\
\leq& 2 \cdot r(C^*) && \text{(by the optimality of } r^+) \\
\leq& \frac{2}{\sqrt{3}} d(C^*) && \text{(by Lemma 1)} \\
\leq& \frac{2}{\sqrt{3}} Opt && \text{(by the optimality of Opt)} \\
\leq& 1.1547 \cdot Opt.
\end{aligned}
$$

Therefore, we have the following theorem.

Theorem 5. *There is a factor-1.1547 approximation algorithm for the 2D grouping problem which runs in $O(n \log n)$ time.*

We comment that the algorithm would still work if L_1 or Manhattan distance is used, with the approximation factor increased by an additional $\sqrt{2}$ factor.

6   Concluding Remarks

We consider a general version of the 2D dispersing problem, whose 1D counterpart was recently studied by Li and Wang [11,12]. We prove that the 2D dispersion problem is NP-hard, in fact, NP-hard even when all segments are parallel and are of unit length. The proof can be applied to show the maximum independent set problem on colored linear unit disk graph remains to be NP-hard. We also consider the symmetric problem of grouping a set of n points, one on each input segment, such that the maximum distance between any two selected ones is minimized. For the latter problem, we give a factor-1.1547 approximation which runs in $O(n \log n)$ time.

We have several open problems related to this paper:

1. Does the 2D dispersion problem admit a constant factor approximation?
2. Is the Independent Set problem on Colored Linear Unit Disk Graph W[1]-hard?
3. Is the 2D grouping problem polynomial-time solvable?

Acknowledgments. This research is supported by NSF of China under project 61628207. XH is supported by the Fundamental Research Funds for the Central Universities under Project 2012017yjsy219. We thank anonymous reviewers for several useful comments which greatly improve the presentation of this paper.

References

1. Aurenhammer, F., Drysdale, R.L.S., Kraser, H.: Farthest line segment Voronoi diagrams. Infor. Proc. Lett. **100**, 220–225 (2006)
2. Baur, C., Fekete, S.P.: Approximation of geometric dispersion problems. In: Jansen, K., Rolim, J. (eds.) APPROX 1998. LNCS, vol. 1444, pp. 63–75. Springer, Heidelberg (1998). https://doi.org/10.1007/BFb0053964
3. Bereg, S., Ma, F., Wang, W., Zhang, J., Zhu, B.: On some matching problems under the color-spanning model. Theor. Comput. Sci. **786**, 26–31 (2019)
4. Breu, H.: Algorithmic aspects of constrained unit disk graphs. Ph.D. dissertation, Department of Computer Science, UBC, Canada (1996)
5. Cevallos, A., Eisenbrand, F., Zenklusen, R.: Local search for max-sum diversification. In: Proceedings of SODA 2017, pp. 130–142 (2017)
6. Chandra, B., Halldorsson, M.: Approximation algorithms for dispersion problems. J. Algorithms **38**, 438–465 (2001)
7. Clark, B., Colbourn, C., Johnson, D.: Unit disk graphs. Discret. Math. **86**(1–3), 165–177 (1990)
8. Fellows, M., Hermelin, D., Rosamond, F., Vialette, S.: On the parameterized complexity of multiple-interval graph problems. Theor. Comput. Sci. **410**(1), 53–61 (2009)
9. Hassin, R., Rubinstein, S., Tamir, A.: Approximation algorithms for maximum dispersion. Oper. Res. Lett. **21**, 133–137 (1997)
10. Hunt III, H.B., Marathe, M.V., Radhakrishnan, V., Ravi, S.S., Rosenkrantz, D.J., Stearns, R.E.: NC-approximation schemes for NP- and PSPACE-hard problems for geometric graphs. J. Algorithms **26**(2), 238–274 (1998)
11. Li, S., Wang, H.: Dispersing points on intervals. In: Proceedings of ISAAC 2016. LIPIcs, vol. 64, pp. 52:1–52:12 (2016)
12. Li, S., Wang, H.: Dispersing points on intervals. Discret. Appl. Math. **239**, 106–118 (2018)

13. Marx, D.: Efficient approximation schemes for geometric problems? In: Brodal, G.S., Leonardi, S. (eds.) ESA 2005. LNCS, vol. 3669, pp. 448–459. Springer, Heidelberg (2005). https://doi.org/10.1007/11561071_41
14. Papadopoulou, E., Dey, S.K.: On the farthest line-segment Voronoi diagram. Int. J. Comput. Geom. and Appl. **23**(6), 443–460 (2013)
15. Pruente, J.: Minimum diameter color-spanning sets revisited. Discret. Optim. **34**, 100550 (2019)
16. Ravi, R., Rosenkrantz, D., Tayi, G.: Heuristic and special case algorithms for dispersing problems. Oper. Res. **42**, 299–310 (1994)
17. Sydow, M.: Approximation guarantees for max sum and max min facility dispersion with parameterised triangle inequality and applications in result diversification. Mathematica Applicanda **42**, 241–257 (2014)
18. Wang, D.W., Kuo, Y.-S.: A study on two geometric location problems. Inf. Process. Lett. **28**, 281–286 (1988)

Synchronizing Words and Monoid Factorization: A Parameterized Perspective

Jens Bruchertseifer and Henning Fernau$^{(\boxtimes)}$ (iD)

Fachber. 4 – Abteilung Informatikwissenschaften, Universität Trier,
54286 Trier, Germany
{s4jebruc,fernau}@uni-trier.de

Abstract. The concept of a synchronizing word is a very important notion in the theory of finite automata. We consider the associated decision problem to decide if a given DFA possesses a synchronizing word of length at most k, where k is the standard parameter. We show that this problem DFA-SW is equivalent to the problem MONOID FACTORIZATION introduced by Cai, Chen, Downey and Fellows. Apart from the known W[2]-hardness results, we show that these problems belong to A[2], W[P] and WNL. This indicates that DFA-SW is not complete for any of these classes and hence, we suggest a new parameterized complexity class W[Sync] as a proper home for these (and more) problems.

Keywords: Synchronizing word · Deterministic Finite Automaton (DFA) · Parameterized complexity

1 Introduction

Černý's conjecture is arguably the most famous open combinatorial problem concerning deterministic finite automata (DFA), somehow dating back to [7]. Recently, a particular Special Issue was dedicated to this conjecture being around for more than five decades; see [29]. This Special Issue also contains an English translation of Černý's paper [8]. The key notion is that of a synchronizing word. A word x is called *synchronizing* for a DFA A, if there is a state s_f, also called the *synchronizing state* of A, such that if A reads x starting in any state, it will end up in s_f. The Černý conjecture states that every n-state DFA can be synchronized by a word of length $(n-1)^2$ if it can be synchronized at all [9]. Although this bound was proven for several classes of finite-state automata, the general case is still widely open. The currently best upper bound is cubic, and only very little progress has been made; see [19,24,26,27].

The notion of a synchronizing word is not only important from a mathematical perspective, offering a nice combinatorial question, but it is quite important in a number of application areas, simply because synchronization is an important concept for many applied areas: parallel and distributed programming, system and protocol testing, information coding, robotics, etc. Therefore, it is also

© Springer Nature Switzerland AG 2020
J. Chen et al. (Eds.): TAMC 2020, LNCS 12337, pp. 352–364, 2020.
https://doi.org/10.1007/978-3-030-59267-7_30

interesting to compute a shortest synchronizing word. Unfortunately, as it was shown by Ryststov and Eppstein in [14,25], the corresponding decision problem DFA-SW (defined in the following) is NP-complete. Possible applications of this problem are explained in [21]. The problem has also been considered from the viewpoint of approximation [1] and parameterized complexity [4,15,17,23].

DFA-SW
Input: DFA A, $k \in \mathbb{N}$
Question: Is there a synchronizing word w for A with $|w| \leq k$?

We will continue to study this problem from the point of parameterized complexity. The standard parameter for this problem is the length upper bound k, which we assume to be the case without further mentioning in this paper. W.l.o.g., we assume that k is given in unary. It was shown in [4,15,23] that this problem is W[2]-hard, even when restricted to quite particular (and restricted) forms of finite automata. Also, other parameters have been studied, in particular, in [15]. Two decades ago, in [5], Cai, Chen, Downey and Fellows introduced the following algebraic problem.

MONOID FACTORIZATION (see [5])
Input: A finite set Q, a collection $F = \{f_0, f_1, \ldots, f_m\}$ of mappings $f_i : Q \to Q$, $k \in \mathbb{N}$
Question: Is there a selection of at most k mappings $f_{i_1}, \ldots, f_{i_{k'}}$, $k' \leq k$, with $i_j \in \{1, \ldots, m\}$ for $j = 1, \ldots, k'$, such that $f_0 = f_{i_1} \circ f_{i_2} \circ \cdots \circ f_{i_{k'}}$?

Again, k is the standard parameter which we will also consider (exclusively) in this paper. In [5], it was proven that MONOID FACTORIZATION is W[2]-hard. We prove in this paper that both problems are in fact equivalent in a parameterized sense. Furthermore, we exhibit three parameterized complexity classes to which both problems belong to, namely, A[2], W[P] and WNL. This indicates that DFA-SW is not complete for any of these classes and hence, we suggest a new parameterized complexity class W[Sync] as a proper home for these two parameterized problems (and more, as we will show).

Throughout this paper, we assume the reader to be familiar with some concepts from parameterized complexity. In particular, a *parameterized reduction* is a many-one reduction that consumes FPT-time (in our cases, it mostly uses only polynomial time) and translates a parameter value k to a parameter value of $f(k)$ (of the target problem), for some computable function f. A parameterized complexity class can be characterized by one (complete) problem, assuming the class is closed under parameterized reductions. Examples comprise the following classes; for the typical problems, the parameter will be always called k:

W[1] Given a nondeterministic single-tape Turing machine and $k \in \mathbb{N}$, does it accept the empty word within at most k steps?

W[2] Given a nondeterministic multi-tape Turing machine and $k \in \mathbb{N}$, does it accept the empty word within at most k steps?

A[2] Given an alternating single-tape Turing machine whose initial state is existential and that is allowed to switch only once into the set of universal states and $k \in \mathbb{N}$, does it accept the empty word within at most k steps?

WNL Given a nondeterministic single-tape Turing machine and some integer $\ell \geq 0$ in unary and $k \in \mathbb{N}$, does it accept the empty word within at most ℓ steps, visiting at most k tape cells?

W[P] Given a nondeterministic single-tape Turing machine and some integer $\ell \geq 0$ in unary and $k \in \mathbb{N}$, does it accept the empty word within at most ℓ steps, thereby making at most $k \leq \ell$ nondeterministic steps?

More details can be found in textbooks like [12,18]. The *Turing way* to these complexity classes is described also in [10,20]. Further interesting complexity classes (in our discussion) are: FPT, W[3], W[SAT], A[3], para-NP and XP. From the literature, the following relations are known:

- FPT $\subseteq$ W[1] $\subseteq$ W[2] $\subseteq$ W[3] $\subseteq \cdots \subseteq$ W[SAT] $\subseteq$ W[P] $\subseteq$ (para-NP $\cap$ XP);
- FPT $\subseteq$ W[1] $=$ A[1] $\subseteq$ W[2] $\subseteq$ A[2] $\subseteq$ A[3] $\subseteq \cdots \subseteq$ AW[P] $\subseteq$ XP.

Each of the inclusions that we have explicitly written is conjectured but not known to be strict. Also, no non-trivial inter-relations are known between the A- and W-hierarchies, apart from W[t] $\subseteq$ A[t] for each t.

Guillemot defined WNL in [20] in the same way as we described it above. Interesting formal language problems complete for WNL include BOUNDED DFA-INTERSECTION, given k DFAs, with parameter k, plus the length of the word that should be accepted by all k automata, or LONGEST COMMON SUBSEQUENCE, parameterized by the number of given strings. WNL is situated above all levels of the W-hierarchy, because the last two mentioned problems are known to be hard for W[t] for any $t \geq 1$, see [3,30]. This proves the first part of the following theorem that we include also for the ease of reference.

Theorem 1. $\bigcup_{t \geq 1} W[t] \subseteq WNL \subseteq (para\text{-}NP \cap XP)$.

Proof. Clearly, by a standard product automaton construction, BOUNDED DFA-INTERSECTION can be tested in time $\mathcal{O}(n^k)$, where n is the maximum number of states of the input DFAs. Hence, WNL is included in XP.

Recall that membership of BOUNDED DFA-INTERSECTION (parameterized by the number of automata) in WNL follows by guessing an input word letter-by-letter, keeping track of the DFAs by writing their k current states, plus a counter for the number of steps, on the tape of the Turing machine M. We can do so by using as many letters as there are states in the automata, plus q (which is given in unary). Alternatively, when counting the number of bits needed to write down the tape contents using the alphabet $\{0,1\}$, this amounts in $\mathcal{O}(k \log(n))$ many bits, if n upper-bounds the size (number of bits) of (an encoding) of a BOUNDED DFA-INTERSECTION instance. Assuming that M has s many states, then there are obviously no more than $s \cdot 2^{\mathcal{O}(k \log n)}$ many configurations of M. With the help of an additional counter, using $\log(s \cdot 2^{\mathcal{O}(k \log n)}) = \log(s) + \mathcal{O}(k \log n)$ many additional bits, we can ensure that such a nondeterministic Turing machine M' (simulating M) would need no more time than $s \cdot 2^{\mathcal{O}(k \log n)}$ when moving through the configuration graph, avoiding visiting configurations twice. This proves the claimed membership in para-NP. Hence, WNL is included in para-NP.   □

As a final remark concerning this detour to parameterized complexity, observe that WNL is also closely linked to the class $N[f\text{ poly}, f\text{ log}]$ of parameterized problems that can be solved nondeterministically (at the same time) obeying some time bound $f(k) \cdot n^c$ for some constant c and some space bound $f(k) \cdot \log(n)$, where f is some (computable) function, k is the parameter (value) and n gives the instance size, as discussed in [13]. Our reasoning also shows that BOUNDED DFA-INTERSECTION (parameterized by the number of automata) lies in $N[f\text{ poly}, f\text{ log}]$. Hence, WNL can be seen as the closure of $N[f\text{ poly}, f\text{ log}]$ under parameterized reductions. So, although one can argue that $N[f\text{ poly}, f\text{ log}]$ (and also some other classes introduced by Elberfeld, Stockhusen and Tantau) is a better model of parameterized space complexity, WNL fits better into the landscape depicted in Fig. 1, being closed under parameterized reductions by its definition. Elberfeld, Stockhusen and Tantau [13] chose other types of reductions.

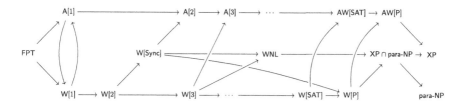

Fig. 1. Visualization of the complexity classes ('A → B' means 'A is contained in B')

2   Finding a Home for DFA-SW

As mentioned above, DFA-SW is known to be W[2]-hard. However, no complexity class was hitherto suggested to which DFA-SW belongs. In this section, we will describe three different memberships.

Theorem 2. DFA-SW *is contained in the classes* WNL *and* W[P].

Proof. Given a DFA A with state set Q and input alphabet Σ, where, w.l.o.g., $Q \cap \Sigma = \emptyset$, together with a bound k on the length of a synchronizing word, a Turing machine M is constructed that works as follows: (1) M writes a word of length at most k over the alphabet Σ on its tape, followed by some letter over the alphabet Q. (2) For each $q \in Q$ (this information can be hard-coded in the finite-state memory of M), M first moves its head to the left end of its tape and then starts reading the tape content from left to right. Each time a symbol $a \in \Sigma$ is read, M updates the current state it stores according to the transition function of A. Finally, M will read a symbol from Q, and it will only continue working if this symbol equals the current state stored in the finite memory of M. Notice that (2) works deterministically. (3) Only if M has completely processed the loop described in (2) (without abort), M will accept. This verifies that the guessed word over Σ is indeed synchronizing, always leading into the state that was also previously guessed. Hence, M will accept the empty word if and only if there is a possibility to guess a synchronizing word of length at most k. It is also clear that the Turing machine makes at most $(|Q|+1)(2k+1)$ many steps, visiting at most $k+1$ tape cells, thereby making at most $k+1$ guesses.   □

We failed when trying to put DFA-SW into W[SAT]. By observing that the switch between phases (1) and (2) of the description of the Turing machine M in the previous proof can be also viewed as switching between existentially and universally quantified states, M can be also re-interpreted to show:

Theorem 3. DFA-SW *is contained in the class* A[2].

3   How to Factor Monoids

We are now going to prove that MONOID FACTORIZATION is FPT-equivalent to DFA-SW.

Lemma 1. *There is a polynomial-time computable parameterized many-one reduction from* MONOID FACTORIZATION *to* DFA-SW.

Proof. Let $F = \{f_0, f_1, \ldots, f_m\}$ be a collection of mappings $f_i : Q \to Q$ and $k \in \mathbb{N}$. Define $\hat{Q} = Q \times Q \cup \{s_0, \ldots, s_k, s_{k+1}, f\}$. Let $\Sigma = \{a_1, \ldots, a_m, \sigma, \tau\}$ and define the transition function $\delta : \hat{Q} \times \Sigma \to \hat{Q}$ as follows.

$$\delta(p, x) = \begin{cases} (q_1, f_i(q_2)) & \text{if } p = (q_1, q_2), x = a_i \text{ for some } 1 \le i \le m \\ (q_1, q_1) & \text{if } p = (q_1, q_2), x = \sigma, \text{ or } x = \tau \text{ and } q_2 \neq f_0(q_1) \\ f & \text{if } p = (q_1, q_2), x = \tau \text{ and } q_2 = f_0(q_1) \\ s_0 & \text{if } p = s_0, x \neq \sigma \\ s_1 & \text{if } p = s_0, x = \sigma \\ s_0 & \text{if } p = s_i, x = \tau, i = 1, \ldots, k \\ s_{i+1} & \text{if } p = s_i, x \neq \tau, i = 1, \ldots, k \\ f & \text{if } p = s_{k+1}, x = \tau \\ s_{k+1} & \text{if } p = s_{k+1}, x \neq \tau \\ f & \text{if } p = f, x \in \Sigma \end{cases}$$

This describes the interesting aspects of the automaton A_F. We claim that (F, k) is a YES-instance of MONOID FACTORIZATION if and only if $(A_F, k+2)$ is a YES-instance of DFA-SW.

Namely, if (F, k) is a YES-instance of MONOID FACTORIZATION, then there exists a selection of at most k mappings $f_{i_1}, \ldots, f_{i_{k'}}$, $k' \le k$, with $i_j \in \{1, \ldots, m\}$ for $j = 1, \ldots, k'$, such that $f_0 = f_{i_1} \circ f_{i_2} \circ \cdots \circ f_{i_{k'}}$. Then, $w = \sigma^{k-k'+1} a_{i_1} \cdot a_{i_2} \cdots a_{i_{k'}} \tau$ synchronizes A_F. Clearly, w begins with $\sigma^{k-k'+1}$. When started in some (q_1, q_2), A_F will be in state (q_1, q_1) after digesting $\sigma^{k-k'+1}$. The word $a_{i_1} \cdot a_{i_2} \cdots a_{i_{k'}}$ will then drive A_F into some state (q_1, q_2'). Now, upon reading τ, A_F could only enter (the only) synchronizing state f if $q_2' = f_0(q_1)$ was true. If A_F starts reading w in any of the states $\{s_0, \ldots, s_k, s_{k+1}, f\}$, it is straightforward to check that A_F will be in state f thereafter.

Conversely, if w is any word of length at most $k+2$ that is synchronizing for A_F, then it must be of length exactly $k+2$, as this is the shortest path length from s_0 down to f, which is a sink state and must be hence the synchronizing state. This also enforces w to start with σ and to end with τ. Also, w cannot contain another occurrence of τ, as this would lead to s_0 again (from any of the states s_i) and hence prevent w from entering f, because the states s_i should be walked through one-by-one, hence counting up to $k+2$. Let us study the longest suffix $v\tau$ of w that satisfies $v \in \{a_1, \ldots, a_m\}^*$. By the structure of w that we analyzed before, we must have $w = u\sigma v\tau$, for some possibly empty word u such that $u\sigma$ starts with σ. In particular, $|v| \leq k$, as $|u| + |v| = k$. Hence, after reading the symbol σ preceding v, A_F will be in one of the states (q, q) or s_i (for some $|u| + 1 \leq i \leq k+1$) or f. Now, digesting v leads us into one of the states s_{k+1} or f or (q, p), with $p = (f_{i_1} \circ f_{i_2} \circ \cdots \circ f_{i_{k'}})(q)$, from which we can enter f only (after reading τ) if $f_0(q) = p$. This shows that, if $u = a_{i_1} \cdot a_{i_2} \cdots a_{i_{k'}}$, then $f_0 = f_{i_1} \circ f_{i_2} \circ \cdots \circ f_{i_{k'}}$. $\qquad\square$

Lemma 2. *There is a polynomial-time computable parameterized reduction that produces, given some DFA A and some integer k as an instance of DFA-SW, an equivalent instance (A', k') of DFA-SW such that A' possesses a sink state (which is then also the unique possible synchronizing state).*

Proof. Consider the DFA $A = (Q, \Sigma, \delta, q_0, F)$. Without loss of generality, assume $\Sigma \cap Q = \emptyset$ and $\sigma \notin \Sigma \cup Q$. Let $\Sigma' = \Sigma \cup Q \cup \{\sigma\}$ be the input alphabet of the DFA A' that we are going to construct. Let $s_0, \ldots, s_k, f \notin Q$ be fresh states. Let $Q' = Q \cup \{s_0, \ldots, s_k, f\}$ be the states of A'. Define the transition function δ' as:

$$\delta'(p, x) = \begin{cases} \delta(p, x) & \text{if } p \in Q, x \in \Sigma \\ p & \text{if } p \in Q, x = \sigma \vee x \in Q \setminus \{p\} \\ f & \text{if } p \in Q, x = p \\ s_0 & \text{if } p = s_i, x \in Q, i = 0, \ldots, k-1 \\ s_{i+1} & \text{if } p = s_i, x \notin Q, i = 0, \ldots, k-1 \\ f & \text{if } p = s_k, x \in Q \\ s_k & \text{if } p = s_k, x \notin Q \\ f & \text{if } p = f, x \in \Sigma' \end{cases}$$

This describes the interesting aspects of the automaton A'. We claim that, letting $k' = k + 1$, then A has a synchronizing word of length at most k if and only if A' has a synchronizing word of length (at most and exactly) k'.

Let $w \in \Sigma^*$ be a synchronizing word, leading A into state $q_f \in Q$, with $|w| \leq k$. Then it is easy to observe that the word $w' = \sigma^{k-|w|} w q_f$ leads A' into the sink state f, wherever A' starts. Hence, w' is a synchronizing word of length k' as claimed. Notice that due to the sequence of states $s_0, \ldots, s_k, f$, there cannot be any shorter synchronizing word in A'.

Conversely, let w' be a synchronizing word of length at most k' for A'. As f is a sink state, it must be the synchronizing state. Since in particular $\delta'^*(s_0, w') = f$, $|w'| = k' = k+1$, and for the same reason, $w' = w''q$ for some $w'' \in (\Sigma \cup \{\sigma\})^k$ and $q \in Q$. Observe that the special letter σ either loops (on $Q \cup \{f\}$) or advances as any other letter from Σ (on $Q' \setminus Q$). Therefore, if w' is synchronizing for A', then so is $\sigma^{k-|w|}wq$, where w is obtained from w'' by deleting all occurrences of σ, i.e., $w \in \Sigma^*$. As σ acts as the identity on Q, and because the final letter q indicates that, upon starting in some state from Q, the automaton must have reached state q (as w' is leading to the sink state f), we can see that w is indeed a synchronizing word for A; moreover, $|w| \leq k$. □

Theorem 4. MONOID FACTORIZATION *is (parameterized and polynomial-time) equivalent to* DFA-SW.

Proof. By Lemma 1, we can reduce MONOID FACTORIZATION to DFA-SW. Conversely, by Lemma 2, we need to consider only instances of DFA-SW that have a sink state. With some background knowledge on transition monoids, it is clear that by interpreting a given DFA $A = (Q, \Sigma, \delta, q_0, F)$ with sink state s_f as a collection F_A of $|\Sigma|$ many mappings $f_a : Q \to Q$, by setting $f_a(q) = \delta(q, a)$, we can solve a DFA synchronization problem given by (A, k) by solving the instance (F, k) of MONOID FACTORIZATION, where $F = \{f_0 = s_f\} \cup F_A$ and the aim is to represent the constant target map $f_0 = s_f$. □

This motivates us to suggest a new parameterized complexity class W[Sync] as the class of parameterized problems that can be reduced to DFA-SW (Fig. 1).

Corollary 1. MONOID FACTORIZATION *is* W[Sync]*-complete.*

4   More Problems Complete for or Contained in W[Sync]

Theorem 5. BOUNDED DFA-INTERSECTION, *parameterized by the length of the commonly accepted string, is complete for* W[Sync].

Previously [30], only W[2]-hardness was known for this parameterized problem.

Proof. By Lemma 2, we need to consider only an instance $A = (Q, \Sigma, \delta, q_0, F)$ of DFA-SW with a sink state s_f. Observe that A has a synchronizing word of length at most k if and only if A has a synchronizing word of length exactly k, because wu is a synchronizing word if w is. Define $A_q = (Q, \Sigma, \delta, q, \{s_f\})$. Observe that $\bigcap_{q \in Q} L(A_q)$ contains some word $w \in \Sigma^k$ if and only if A has a synchronizing word of length exactly k.

Conversely, if $\{A_i \mid 1 \leq i \leq \ell\}$ is a collection of DFAs $A_i = (Q_i, \Sigma, \delta_i, q_{0,i}, F_i)$, then construct an equivalent instance of DFA-SW as follows. First, assume that the state sets Q_i are pairwise disjoint. Then, take two new letters a, b to form

$$\delta(p,x) = \begin{cases} \delta_i(p,x) & \text{if } p \in Q_i, x \in \Sigma \\ q_{0,i} & \text{if } p \in Q_i, x = \sigma \\ f & \text{if } p \in \left(\bigcup_{i=1}^{\ell} F_i\right), x = \tau \\ p & \text{if } p \in \left(\bigcup_{i=1}^{\ell}(Q_i \setminus F_i)\right), x = \tau \\ s_0 & \text{if } p = s_i, x = \tau, i = 0, \dots, k \\ s_{i+1} & \text{if } p = s_i, x \neq \tau, i = 0, \dots, k \\ f & \text{if } p = s_{k+1}, x = \tau \\ s_{k+1} & \text{if } p = s_{k+1}, x \neq \tau \\ f & \text{if } p = f, x \in \Sigma' \end{cases}$$

Fig. 2. Transition function δ of the constructed W[Sync] instance.

$\Sigma' = \Sigma \cup \{\sigma, \tau\}$. Let $Q' = \left(\bigcup_{i=1}^{\ell} Q_i\right) \cup \{s_0, \dots, s_k, s_{k+1}, f\}$ be the state set of the DFA A that we construct. Define the transition function δ as in Fig. 2.

This describes the interesting aspects of the automaton A. We claim that, letting $k' = k + 2$, then $\bigcap_{i=1}^{\ell} L(A_i)$ contains some word $w \in \Sigma^k$ if and only if A has a synchronizing word of length (at most and exactly) k', namely $w' = \sigma w \tau$. More precisely, similar to the construction from Lemma 1, the states s_i force to consider a word from $\{\sigma\}\Sigma^k\{\tau\}$ if there should be a synchronizing word of length k' for A at all. One could move only from the part A_i of A to f when reading τ, which also forces to have been in the set of final states F_i before. Digesting σ as the first letter lets A_i start in the initial state $q_{0,i}$.   □

We now discuss the well-known Longest Common Subsequence problem. The input consists of ℓ strings $x_1, \dots, x_\ell$ over an alphabet Σ, and the task is to find a string $w \in \Sigma^k$ occurring in each of the x_i as a subsequence. As explained in [30], by building an automaton A_i for each x_i that accepts all subsequences of x_i, it is not hard to solve a Longest Common Subsequence instance by a Bounded DFA-Intersection instance, preserving our parameter. Hence:

Proposition 1. Longest Common Subsequence $\in$ W[Sync].

We do not know if Longest Common Subsequence is also hard for W[Sync]. We only know W[2]-hardness from [3], further membership results were unknown hitherto, so the previous proposition remedies this situation a bit.

One could also think of many ways to restrict the inputs of Bounded DFA-Intersection. For instance, observe that the automata constructed in the argument of Proposition 1 are all accepting finite languages. Is there a converse reduction from such a Bounded DFA-Intersection instance to some Longest Common Subsequence instance? Is this leading to another complexity class between W[2] and W[Sync]?

In [4], we discussed the restriction of DFAs to so-called TTSPL graphs for instances of DFA-SW. While we could prove W[2]-hardness also for such restricted instances, it is open if this leads to a problem that is still complete for W[Sync]. As shown by Möhring [22], there are quite close connections between

TTSP(L) graphs and so-called series-parallel partial orders. Without going into any details here, observe that the mappings $Q \to Q$ that can be associated to input letters are monotone with respect to the series-parallel partial order corresponding to the TTSPL automaton graph. Our earlier constructions show:

Corollary 2. DFA-SW, *restricted to DFAs with TTSPL automata graphs, is parameterized and polynomial-time equivalent to* MONOID FACTORIZATION, *restricted to collections of mappings F that are monotone with respect to a given series-parallel partial order on the finite ground set Q.*

This discussion also entails the (open) question concerning the complexity status of MONOID FACTORIZATION, restricted to collections of mappings F that are monotone with respect to a given partial order on the finite ground set Q.

5 Further Comments

The problems that we considered in this paper have quite a rich structure and many variations. We will comment on these variations in this section.

5.1 Variations on MONOID FACTORIZATION

Observe that it is important that the monoid used in MONOID FACTORIZATION is only implicitly given, not by a multiplication table. A variation could be:

Input: A finite set M, a binary operation $\circ$ given in the form of a multiplication table, such that $(M, \circ)$ forms a finite monoid, with neutral element $e \in M$, a target element $t \in M$, a finite subset $B \subseteq M$, $k \in \mathbb{N}$
Question: Is there a selection of at most k elements $b_1, \ldots, b_{k'}$, $k' \leq k$, from B, such that $t = b_1 \circ b_2 \circ \cdots \circ b_{k'}$?

However, an explicit representation of the multiplication table of $(Q^Q, \circ)$ (where Q^Q is the set of all mappings from Q to Q) would already take $\mathcal{O}^*(|Q|^{2|Q|})$ space and hence allow to construct an arc-labeled directed graph with a vertex for each mapping $Q \to Q$ and an arc labeled f_i from f to g if $f \circ f_i = g$, where f_i is from the explicit set of generators $F' = \{f_1, \ldots, f_m\}$. Now, the representability of f_0 with at most k mappings from F' can be solved by looking for a path of length at most k in the directed graph we just described, leading from the identity mapping Δ_Q to f_0. Hence, when the monoid is given in an explicit form, then the factorization problem can be solved in polynomial time. It might be interesting to study other implicitly given monoids with respect to the factorization question. Let us mention one more example. Assume that our implicitly given monoid operation is *set union*. Then, the corresponding factorization problem would take subsets $\{X_0, X_1, \ldots, X_m\}$ of a given finite set S as an input, and the question is to pick at most k sets from $\{X_1, \ldots, X_m\}$, say, $X_{i_1}, \ldots, X_{i_{k'}}$, where $k' \leq k$, such that $X_0 = \bigcup_{j=1}^{k'} X_{i_j}$. Obviously, this corresponds to SET COVER, which hence gives an example of a monoid factorization problem which, when parameterized by k, is complete for W[2]. It might be interesting to investigate further implicitly given monoids from this parameterized perspective. We only mention as a last example from the literature PERMUTATION GROUP FACTORIZATION, which is known to be W[1]-hard but is lacking a precise classification; see [3,12].

5.2   Extension Variants

Following [6,16], we are now defining so-called extension problems, depending on the chosen partial order $\prec$ on Σ^*. Maybe surprisingly, the complexity status of these problems heavily depends on this choice.

EXT DFA-SW-$\prec$

Input: DFA A with input alphabet Σ, $u \in \Sigma^*$

Question: Is there a $w \in \Sigma^*$, $u \prec w$, such that w is minimal for the set of synchronizing words for A with respect to $\prec$?

We are focussing on the length-lexicographical ordering $\leq_{ll}$ and the subsequence ordering $|$ in the following. For further orderings, we refer to [16]. We consider $|u|$ to be the standard parameter.

Theorem 6. EXT DFA-SW-$\leq_{ll}$ *is contained in* co-WNL $\cap$ co-W[P] $\cap$ co-A[2], *but hard for* co-W[Sync].

Proof. For membership in co-WNL$\cap$co-W[P]$\cap$co-A[2], we can modify the proofs of Theorems 2 or 3, building a nondeterministic Turing machine M as follows, given A and u. As before, the machine can first guess a possible word $w \leq_{ll} u$ and verify if it is synchronizing. If such a word is found, then (A, u) is a NO-instance. The reduction itself checks if A is synchronizable at all; then we also have that if M does not find a synchronizing word $w \leq_{ll} u$, then (A, u) is a YES-instance, because as A is synchronizable, there must be a synchronizing word v, and according to the previous tests, $u \leq_{ll} v$ must hold.

For the hardness claim, consider a DFA A on input alphabet Σ, together with k, as an instance of DFA-SW. We can first check in polynomial time if A is synchronizable at all. If A is not synchronizable, then (A, k) (clearly) is a NO-instance of DFA-SW, so our reduction will produce some fixed NO-instance of EXT DFA-SW-$\leq_{ll}$. Hence, we now assume that A is synchronizable. Let $c \notin \Sigma$ be a fresh letter. Consider an arbitrary ordering $<$ on Σ, extended by $c < x$ for all $x \in \Sigma$ towards an ordering on $\hat{\Sigma} = \Sigma \cup \{c\}$. We are going to define the DFA $\hat{A}$ as an extension of A, working on the same state set Q. Let c simply act as the identity on Q. Hence, no word from c^* is synchronizing for $\hat{A}$. As A is synchronizable, $\hat{A}$ is also synchronizable. Consider $\hat{A}$ together with $u = c^{k+1}$ as an instance of EXT DFA-SW-$\leq_{ll}$. If $\hat{A}$ has a synchronizing word $w \in \Sigma^{\leq k}$, then clearly u is not extendible, as $|w| < |u|$. Otherwise, as $\hat{A}$ is synchronizable, $\hat{A}$ must have some synchronizing word w with $|w| \geq |u|$, and any synchronizing word of $\hat{A}$ is of length at least $|u|$. As u is the smallest of all words in $\hat{\Sigma}^*$ of length at least $|u|$, any synchronizing word will hence extend u. Hence, if $\hat{A}$ has no synchronizing word of length at most k, then u is extendible. $\square$

Theorem 27 in [16] converted an instance of EXT HITTING SET into an instance of EXT DFA-SW-|. With [2], this proves that EXT DFA-SW-|, is W[3]-hard, lifting it beyond another rarely mentioned complexity class; see [11]. This construction can be also adapted for TTSPL automata graphs.

5.3   Minimum Synchronizable Sub-automata

> DFA-MSS (referring to a minimum synchronizable sub-automaton)
>
> Input: DFA A with input alphabet Σ, $k \in \mathbb{N}$
>
> Question: Is there a sub-alphabet $\hat{\Sigma} \subseteq \Sigma$, $|\hat{\Sigma}| \leq k$, such that the restriction of A to $\hat{\Sigma}$ is synchonizable, i.e., is there a synchronizing word over $\hat{\Sigma}$?

In [28], Türker and Yenegün asked to extract a synchronizable sub-automaton that is as small as possible, obtained by deleting letters from its specification. They formalized this idea as a weighted minimization problem. Here, it is sufficient to consider the unweighted variant (defined in the box). Their NP-hardness proof can be re-interpreted as a result on parameterized complexity.

Corollary 3. DFA-MSS *is W[2]-hard.*

Without proof, we mention the following membership result. However, membership in A[2] is open, nor to we know about WNL-hardness. It might be also the case that DFA-MSS is W[Sync]-hard.

Theorem 7. DFA-MSS *is contained in* $\mathsf{WNL} \cap \mathsf{W[P]}$.

A Short Summary. We looked at various W[2]-hard problems where a proper classification is still missing. In particular, problems rooting in Formal Languages offer interesting sample problems. Many questions are still open about W[Sync].

References

1. Berlinkov, M.V.: Approximating the minimum length of synchronizing words is hard. Theory Comput. Syst. **54**(2), 211–223 (2014)
2. Bläsius, T., Friedrich, T., Lischeid, J., Meeks, K., Schirneck, M.: Efficiently enumerating hitting sets of hypergraphs arising in data profiling. In: Algorithm Engineering and Experiments (ALENEX), pp. 130–143. SIAM (2019)
3. Bodlaender, H., Downey, R.G., Fellows, M.R., Wareham, H.T.: The parameterized complexity of sequence alignment and consensus. Theor. Comput. Sci. **147**, 31–54 (1995)
4. Bruchertseifer, J., Fernau, H.: Synchronizing series-parallel automata with loops. In: Freund, R., Holzer, M., Sempere, J.M. (eds.) Eleventh Workshop on Non-Classical Models of Automata and Applications, NCMA, pp. 63–78. Österreichische Computer Gesellschaft (2019)
5. Cai, L., Chen, J., Downey, R., Fellows, M.: On the parameterized complexity of short computation and factorization. Arch. Math. Logic **36**, 321–337 (1997)

6. Casel, K., Fernau, H., Ghadikolaei, M.K., Monnot, J., Sikora, F.: On the complexity of solution extension of optimization problems. Technical report arXiv:1810.04553 [cs.CC], Cornell University, arXiv (2018)

7. Černý, J.: Poznámka k homogénnym experimentom s konečnými automatmi. Matematicko-fyzikálny časopis **14**(3), 208–216 (1964)

8. Černý, J.: A note on homogeneous experiments with finite automata. J. Automata Lang. Comb. **24**(2–4), 123–132 (2019)

9. Černý, J., Pirická, A., Rosenauerová, B.: On directable automata. Kybernetika **7**(4), 289–298 (1971)

10. Cesati, M.: The Turing way to parameterized complexity. J. Comput. Syst. Sci. **67**, 654–685 (2003)

11. Chen, J., Zhang, F.: On product covering in 3-tier supply chain models: Natural complete problems for W[3] and W[4]. Theor. Comput. Sci. **363**(3), 278–288 (2006)

12. Downey, R.G., Fellows, M.R.: Fundamentals of Parameterized Complexity. Texts in Computer Science. Springer, Heidelberg (2013). https://doi.org/10.1007/978-1-4471-5559-1

13. Elberfeld, M., Stockhusen, C., Tantau, T.: On the space and circuit complexity of parameterized problems: classes and completeness. Algorithmica **71**(3), 661–701 (2015)

14. Eppstein, D.: Reset sequences for monotonic automata. SIAM J. Comput. **19**(3), 500–510 (1990)

15. Fernau, H., Heggernes, P., Villanger, Y.: A multi-parameter analysis of hard problems on deterministic finite automata. J. Comput. Syst. Sci. **81**(4), 747–765 (2015)

16. Fernau, H., Hoffmann, S.: Extensions to minimal synchronizing words. J. Automata Lang. Comb. **24**, 287–307 (2019)

17. Fernau, H., Krebs, A.: Problems on finite automata and the exponential time hypothesis. Algorithms **10**(24), 1–25 (2017)

18. Flum, J., Grohe, M.: Parameterized Complexity Theory. Springer, Heidelberg (2006). https://doi.org/10.1007/3-540-29953-X

19. Frankl, P.: An extremal problem for two families of sets. Eur. J. Comb. **3**(2), 125–127 (1982)

20. Guillemot, S.: Parameterized complexity and approximability of the longest compatible sequence problem. Discret. Optim. **8**(1), 50–60 (2011)

21. Kisielewicz, A., Kowalski, J., Szykuła, M.: Computing the shortest reset words of synchronizing automata. J. Comb. Optim. **29**(1), 88–124 (2013). https://doi.org/10.1007/s10878-013-9682-0

22. Möhring, R.H.: Computationally tractable classes of ordered sets. In: Rival, I. (ed.) Algorithms and Order: Proceedings of the NATO Advanced Study Institute. NATO Science Series C, vol. 255, pp. 105–194. Springer, Heidelberg (1989). https://doi.org/10.1007/978-94-009-2639-4_4

23. Andres Montoya, J., Nolasco, C.: On the synchronization of planar automata. In: Klein, S.T., Martín-Vide, C., Shapira, D. (eds.) LATA 2018. LNCS, vol. 10792, pp. 93–104. Springer, Cham (2018). https://doi.org/10.1007/978-3-319-77313-1_7

24. Pin, J.E.: On two combinatorial problems arising from automata theory. Ann. Discret. Math. **17**, 535–548 (1983)

25. Rystsov, I.K.: On minimizing the length of synchronizing words for finite automata. In: Theory of Designing of Computing Systems, pp. 75–82. Institute of Cybernetics of Ukrainian Acad. Sci. (1980). (in Russian)

26. Shitov, Y.: An improvement to a recent upper bound for synchronizing words of finite automata. J. Automata Lang. Combin. **24**(2–4), 367–373 (2019)

27. Szykuła, M.: Improving the upper bound on the length of the shortest reset word. In: Niedermeier, R., Vallée, B. (eds.) 35th Symposium on Theoretical Aspects of Computer Science, STACS. LIPIcs, vol. 96, pp. 56:1–56:13. Schloss Dagstuhl - Leibniz-Zentrum für Informatik (2018)

28. Türker, U.C., Yenigün, H.: Complexities of some problems related to synchronizing, non-synchronizing and monotonic automata. Int. J. Found. Comput. Sci. **26**(1), 99–122 (2015)

29. Volkov, M.V.: Preface: special issue on the Černý conjecture. J. Automata Lang. Comb. **24**(2–4), 119–121 (2019)

30. Todd Wareham, H.: The parameterized complexity of intersection and composition operations on sets of finite-state automata. In: Yu, S., Păun, A. (eds.) CIAA 2000. LNCS, vol. 2088, pp. 302–310. Springer, Heidelberg (2001). https://doi.org/10.1007/3-540-44674-5_26

Hidden Community Detection on Two-Layer Stochastic Models: A Theoretical Perspective

Jialu Bao[1], Kun He[2(✉)], Xiaodong Xin[2], Bart Selman[3], and John E. Hopcroft[3]

[1] Department of Computer Science, University of Wisconsin-Madison,
Madison, WI 50706, USA
[2] School of Computer Science and Technology, Huazhong University of Science
and Technology, Wuhan 430074, China
`brooklet60@hust.edu.cn`
[3] Department of Computer Science, Cornell University, Ithaca, NY 14853, USA

Abstract. Hidden community is a new graph-theoretical concept recently proposed by [3], in which the authors also propose a meta-approach called HICODE (Hidden Community Detection) for detecting hidden communities. HICODE is demonstrated through experiments that it is able to uncover previously overshadowed weak layers and uncover both weak and strong layers at a higher accuracy. However, the authors provide no theoretical guarantee for the performance. In this work, we focus on theoretical analysis of HICODE on synthetic two-layer networks, where layers are independent to each other and each layer is generated by stochastic block model. We bridge their gap through two-layer stochastic block model networks in the following aspects: 1) we show that partitions that locally optimize modularity correspond to grounded layers, indicating modularity-optimizing algorithms can detect strong layers; 2) we prove that when reducing found layers, HICODE increases absolute modularities of all unreduced layers, showing its layer reduction step makes weak layers more detectable. Our work builds a solid theoretical base for HICODE, demonstrating that it is promising in uncovering both weak and strong layers of communities in two-layer networks.

Keywords: Hidden community · Multi-layer stochastic block model · Modularity optimization · Social network

1 Introduction

Community detection problem has occurred in a wide range of domains, from social network analysis to biological protein-protein interactions, and numerous algorithms have been proposed, based on the assumption that nodes in the same community are more likely to connect with each other. While many real-world

---

J. Bao—Portion of the work was done while at Cornell University.

© Springer Nature Switzerland AG 2020
J. Chen et al. (Eds.): TAMC 2020, LNCS 12337, pp. 365–376, 2020.
https://doi.org/10.1007/978-3-030-59267-7_31

social networks satisfy the assumption, their communities can overlap in interesting ways: communities based on schools can overlap as students attend different schools; connections of crime activities often hide behind innocuous social connections; proteins serving multiple functions can belong to multiple function communities. In any of these networks, communities can have more structures than random overlappings. For example, communities based on schools may be divided into primary school, middle school, high school, college and graduate school layers, where each layer are approximately disjoint. This observation inspires us to model real world networks as having multiple layers.

To simulate real-world networks, researchers also build generative models such as single-layer stochastic block model $G(n, n_1, p, q)$ $(p > q)$. It can be seen as Erdős-Rényi model with communities—$G(n, n_1, p, 1)$ has n nodes that belongs to n_1 disjoint blocks/communities (we use them interchangeably in the following), and any node pair internal to a community has probability p to form an edge, while any node pair across two communities have q probability to form an edge. We propose a multi-layer stochastic block model $G(n, n_1, p_1, ..., n_L, p_L)$, where each layer l consists of n_l disjoint communities, and communities in different layers are independent to each other. Each layer l is associated with one edge probability p_l, determining the probability that a node pair internal to a community in that layer forms an edge. In this ideal abstraction, we assume that each node belongs to exactly one community in each layer, and an edge is generated only through that process, i.e. all edges outgoing communities of one layer are generated as internal edges in some other layers. Note that our model is different to the multi-layer stochastic blockmodel proposed by Paul et al. [6], where they have different types of edges, and each type of edges forms one layer of the network.

He et al. [3,4] first introduce the concept of hidden communities, remarked as a new graph-theoretical concept [7]. He et al. propose the Hidden Community Detection (HICODE) algorithm for networks containing both strong and hidden layers of communities, where each layer consists of a set of disjoint or slightly overlapping communities. A hidden community is a community most of whose nodes also belong to other stronger communities as measured by metrics like modularity [2]. They showed through experiments that HICODE uncovers grounded communities with higher accuracy and finds hidden communities in the weak layers. However, they did not provide any theoretical support.

In this work, we provide solid theoretical analysis that demonstrates the effectiveness of HICODE on two-layer stochastic models. One important step in HICODE algorithm is to reduce the strength of one partition when the partition is found to approximate one layer of communities in the network. Since communities in different layers unavoidably overlap, both internal edges and outgoing edges of remaining layers have a chance to be reduced while reducing one layer. It was unclear how the modularity of remaining layer would change. Through rigorous analysis of three layer weakening methods they suggested, we prove that using any one of *RemoveEdge*, *ReduceEdge* and *ReduceWeight* on one layer increases the modularity of the grounded partition in the unreduced layer. Thus, we provide evidence that HICODE's layer reduction step makes weak layers more detectable.

In addition, through simulation, we show that on two-layer stochastic block model networks, partitions with locally maximal modularity roughly correspond to planted partitions given by grounded layers. As a result, modularity optimizing community detection algorithms such as Louvain [1] can approximate layers fairly accurately in a two-layer stochastic block model, even when layers are almost equally strong and non-trivially overlapped. This indicates the previous proof's assumption that one layer of communities is reduced exactly is reasonable. We also illustrate how the modularity of randomly sampled partitions change as HICODE iterates, and our plots show that not only absolute modularity but also relative modularity of unreduced layers increases as HICODE reduces one found layer.

2   Preliminary

In this section, we first introduce metrics that measure community partition quality. Then, we summarize important components in HICODE, the iterative meta-approach we are going to analyze, and in particular, how it reduce layers of detected communities during the iterations. Also, we define the multi-layer stochastic block model formally, and the rationale why it is a reasonable abstraction of generative processes of real world networks.

2.1   Modularity Metric

In determining plausible underlying communities in a network, we rely on metrics measuring quality of community partitions. Usually, nodes sharing common communities are more likely to develop connections with each other, so in single-layer networks, we expect that most edges are internal to one grounded community, instead of outgoing edges whose two endpoints belong to two communities. It thus gives rise to metrics measuring the similarity between an arbitrary partition and the grounded partition based on the fraction of internal edges to outgoing edges. One widely-used metric of this kind is "modularity" [2]. We define the modularity of one community in multi-layer networks as follows:

Definition 1 (Modularity of a community). *Given a graph $G = (V, E)$ with a total of e edges and multiple layers of communities, where each layer of communities partitions all nodes in the graph, for a community i in layer l, let e_{ll}^i denote i's internal edges, and e_{lout}^i denote the number of edges that have exactly one endpoint in community i. Let d_l^i be the total degree of nodes in community i ($d_l^i = 2e_{ll}^i + e_{lout}^i$). Then the modularity of community i in layer l is $Q_l^i = \frac{e_{ll}^i}{e} - \left(\frac{d_l^i}{2e}\right)^2$.*

Roughly, the higher fraction of internal edges a community has among all edges, the higher its modularity in graph, indicating that members in that community are more closely connected.

When optimizing modularity, the algorithm concerns the modularity of a partition instead of one community. The modularity of a partition is defined as follows, which is consistent with the original definition of Girvan et al. [2]:

Definition 2 (Modularity of a partition/layer). *Given a network $G = (V, E)$ with multiple layers of communities, for any layer l, say l partitions all the nodes into disjoint communities $\{1, \ldots, N\}$, then the layer modularity is $Q_l = \sum_{i=1}^{N} Q_l^i$.*

Whether in single-layer network or multi-layer ones, the ground truth community partition is expected to have high modularity when compared to other possible partitions.

2.2 HIdden COmmunity DEtection (HICODE) Algorithm

Informally, given a state-of-the-art community detection algorithm $\mathcal{A}$ for single layer networks, HICODE($\mathcal{A}$) finds all layers in multi-layer networks through careful alternations of detecting the strongest layer in the remaining graph using $\mathcal{A}$ and reducing found layers on the network. Given a network $G = (V, E)$, He *et al.* [3] proposed three slightly different methods for reducing layers in HICODE:

1. **RemoveEdge:** Given one layer l that partitions G, *RemoveEdge* removes all internal edges of layer l from G.
2. **ReduceEdge:** Given one layer l that partitions G, *ReduceEdge* approximates the background density q of edges contributed by all other layers, and then removes $1 - q$ fraction of internal edges of layer l from network G. We will detail the computation of q after introducing multi-layer stochastic block model.
3. **ReduceWeight:** This is the counterpart of ReduceEdge on weighted graphs. Given one layer l that partitions network G, *ReduceWeight* approximates the background density q of edges contributed by all other layers, and then reduces the weight of all internal edges to a q fraction of its original values.

For detailed description of HICODE, see [3].

2.3 Multi-layer Stochastic Block Model

Before defining the general multi-layer Stochastic Block Model (SBM), consider the case where there is exactly two layers.

Definition 3 (Two-layer Stochastic Block Model). *A synthetic network $G(n, n_1, p_1, n_2, p_2)$ generated by two-layer stochastic block model has n nodes, where $n, n_1, n_2 \in N^+, n_1, n_2 \geq 3$. For $l = 1$ or 2, layer l of G consists of n_l planted communities of size $s_l = \frac{n}{n_l}$ with internal edge probability $p_l \in (0, 1]$. Communities in different layers are grouped independently, so they are expected to intersect with each other by $r = \frac{n}{n_1 n_2}$ nodes.*

Each community of layer l is expected to have $p_l \cdot \frac{1}{2} s_1^l$ internal edges[1]. The model represents an ideal scenario when there is no noise and all outgoing edges

---

[1] For simplicity, we allow self-loops.

of one layer are the result of them being internal edges of some other layers. We will detail the expected number of outgoing edges and the size of the intersection block of layers in Lemma 1 in the next section.

For example, in $G(200, 4, 5, p_1, p_2)$, layer 1 contains four communities $C_1^1 = \{1, 2, ..., 50\}$, $C_1^2 = \{51, 52, ..., 100\}$, $C_1^3 = \{101, 102, ..., 150\}$, $C_1^4 = \{151, 152, ..., 200\}$, and layer 2 contains five communities $C_2^1 = \{1, 6, ..., 196\}$, $C_2^2 = \{2, 7, ..., 197\}$, $C_2^3 = \{3, 8, ..., 198\}$, $C_2^4 = \{4, 9, ..., 199\}$, $C_2^5 = \{5, 10, ..., 200\}$. Each community is modeled as an Erdős-Rényi graph. Each C_1^i in layer 1 is expected to have $0.5 \cdot 50^2 p_1$ internal edges, and each C_2^i in layer 2 are expected to have $0.5 \cdot 40^2 p_1$ internal edges.

Each community in layer 1 overlaps with each community in layer 2. Each overlap consists of 20% of the nodes of layer 1 community and 25% of the nodes of layer 2 community. Figure 1 (a) and (b) show the adjacency matrix when nodes are ordered by $[1, ..., n]$ for layer 1, and $[1, 6, ..., 196, 2, 7, ..., 197, 5, 10, ..., 200]$ for layer 2, respectively (Here we set $p_1 = 0.12, p_2 = 0.10$). Fig. 1 (c) and (d) show an enlarged block for each layer. Edges in layer 1 are plotted in red, edges in layer 2 are plotted in blue and the intersected edges are plotted in green.

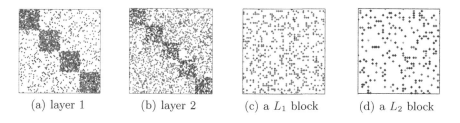

(a) layer 1   (b) layer 2   (c) a L_1 block   (d) a L_2 block

Fig. 1. The stochastic blocks in two layers. (Color figure online)

More generally, we can define a multi-layer stochastic block model.

Definition 4 (Multi-layer Stochastic Block Model). *A multi-layer stochastic block model $G(n, n_1, p_1, ..., n_L, p_L)$ generates a network with L layers, and each layer l has n_l communities of size $\frac{n}{n_l}$ with internal edge probability p_l. All layers are independent with each other.*

2.4 Background Edge Probability for Multi-layer SBM

Given a layer, observed edge probability within its grounded communities would be higher than its grounded edge generating probability, because other layers could also generate edge internal to this layer. When we are interested in the grounded edge generating probability of a layer, we can consider edges generated by all other layers as background noise. Since layers are independent to each other, these background noise edges are uniformly distributed among communities of layer l, so we can expect background noise edge probability the same on node pairs either internal to or across layer l communities. Thus, the observed edge probability $\widehat{p}$ of communities in a layer l equals $p + \widehat{q} - p \cdot \widehat{q}$, where p is the grounded edge generating probability of layer l, $\widehat{q}$ is the observed edge probability across layer 1 communities. Thus, we can estimate the actual edge probability by $p = \frac{\widehat{p} - \widehat{q}}{1 - \widehat{q}}$.

3   Theoretical Analysis on Two-Layer SBM

In this section, we show that on networks generated by two-layer stochastic block model, weakening one layer would not decrease the quality of communities in any other layer even when they considerably overlap with each other. We will prove on two-layer stochastic block models that absolute modularity of unreduced layer must increase after performing RemoveEdge, ReduceEdge, or ReduceWeight. For simplicity, we make the assumption that the base algorithm can uncover a layer exactly – every time it finds a layer to reduce, it does not make mistakes on community membership. This is a strong assumption, but later on we will justify why our result still holds if the base algorithm only approximates layers and why the base algorithm can almost always find some approximate layers.

For each community in layer l, let s_l denote the size of each community in layer l, and m_l denote the number of node pairs in the community. Since we allow self-loops, $m_l = \frac{1}{2}s_l^2$. Also, with the assumption that all communities in one layer are equal sized, their expected numbers of internal (or outgoing) edges are the same. Thus, we can use e_{ll}, e_{lout} to respectively denote the expected number of internal, outgoing edges for each community i in layer l. Then, let $d_l = 2e_{ll} + e_{lout}$ denote the expected total degree of any community in layer l.

Lemma 1. *In the synthetic two-layer block model network $G(n, n_1, n_2, p_1, p_2)$, for a given community i in layer 1, the expected number of its internal edges as well as outgoing edges, and layer 1's modularity are as follows:*

$$e_{11} = \left(1 - \frac{1}{n_2}\right) m_1 p_1 + \frac{1}{n_2} m_1 p_{12}, \tag{1}$$

$$e_{1out} = \frac{p_2}{n_2} s_1(n - s_1), \tag{2}$$

$$Q_1 = 1 - \frac{1}{n_1} - \frac{e_{1out}}{d_1}, \tag{3}$$

where $p_{12} = p_1 + p_2 - p_1 \cdot p_2$. Symmetrically, given a community i in layer 2, the expected number of its internal edges as well as outgoing edges, and layer 2's modularity are as follows:

$$e_{22} = \left(1 - \frac{1}{n_1}\right) m_2 p_2 + \frac{1}{n_1} m_2 p_{12}, \tag{4}$$

$$e_{2out} = \frac{p_1}{n_1} s_2(n - s_2), \tag{5}$$

$$Q_2 = 1 - \frac{1}{n_2} - \frac{e_{2out}}{d_2}. \tag{6}$$

We omitted the proof because of the space limit.

Lemma 2. *For layer l in a two-layer stochastic block model, if the layer weakening method (e.g. RemoveEdge, ReduceEdge, ReduceWeight) reduces a bigger percentage of outgoing edges than internal edges, i.e. the expected number of internal and outgoing edges after weakening, e'_{ll}, e'_{lout}, satisfy $\frac{e'_{lout}}{e_{lout}} < \frac{e'_{ll}}{e_{ll}}$, then the modularity of layer l increases after the weakening method.*

We omitted the proof because of the space limit.

For a synthetic stochastic block model network G with set of layers $\mathcal{L}$, let S_l be the set of edges whose underlying node pairs are only internal to layer $l \subseteq \mathcal{L}$, let $S_{l_1 l_2}$ be the set of edges internal to both layers $l_1, l_2 \subseteq \mathcal{L}$. Concretely, in the two-layer stochastic block model, $\mathcal{L} = \{1, 2\}$. S_1 is the set of edges only internal to layer 1, S_2 is the set of edges only internal to layer 2, and S_{12} is the set of edges internal to both layer 1 and layer 2.

Lemma 3. *In a two-layer stochastic blockmodel network $G(n, n_1, n_2, p_1, p_2)$, before any weakening procedure.*

$$e_{11} = \frac{|S_{12}| + |S_1|}{n_1}, \qquad\qquad e_{1out} = \frac{2}{n_1}|S_2|,$$

$$e_{22} = \frac{|S_{12}| + |S_2|}{n_2}, \qquad\qquad e_{2out} = \frac{2}{n_2}|S_1|.$$

We omitted the proof due to space limit.

Using the above three lemmas, we can prove the following theorems.

Theorem 1. *For a two-layer stochastic blockmodel network $G(n, n_1, n_2, p_1, p_2)$, the modularity of a layer increases if we apply RemoveEdge on communities in the other layer.*

Proof. If we remove all internal edges of communities in layer 1, both $|S_{12}|$ and $|S_1|$ become 0, then the remaining internal edges of layer 2 is $e'_{22} = \frac{1}{n_2}(|S_{12}| + |S_2|) = \frac{|S_2|}{n_2} > 0$. There is no outgoing edge of layer 2, so $e'_{2out} = 0$. Thus, $\frac{e'_{2out}}{e_{2out}} = 0 < \frac{e'_{22}}{e_{22}}$, and applying Lemma 2, we have that the modularity of layer 2 after RemoveEdge on layer 1 $Q'_2 > Q_2$.

Similarly, the modularity of layer 1 after RemoveEdge on layer 2, Q'_1, is greater than Q_1.

RemoveEdge not only guarantees to increase the absolute modularity of layer 2 but also guarantees that layer 2 would have higher modularity than any possible partition of n nodes into n_2 communities in the reduced network.

Theorem 2. *For a two-layer stochastic blockmodel network $G(n, n_1, n_2, p_1, p_2)$, If no layer 2 community contains more than half of the total edges inside it after applying RemoveEdge on layer 1, then layer 2 has the highest modularity among all possible partitions of n nodes into n_2 communities.*

Proof. After applying RemoveEdge on layer 1, there are no outgoing edges of any community in layer 2. It means that for any community i, $e_{2out}^i = 0$ and $d_2^i = 2e_{22}^i$. Thus, the modularity of layer 2 is:

$$Q_2 = \sum_{i \in \text{layer 2}} Q_2^i = \sum_{i \in \text{layer 2}} \left[\frac{e_{22}^i}{e} - \left(\frac{d_2^i}{2e} \right)^2 \right]$$

$$= \sum_{i \in \text{layer 2}} \left[\frac{4e \cdot e_{22}^i - (2e_{22}^i)^2}{4e^2} \right] = n_2 \left(\frac{e \cdot e_{22} - (e_{22})^2}{e^2} \right).$$

For any one partition, we can transform layer 2 partition to it by moving a series of nodes across communities. Every time we move one node from one community i to another community j, both e_{2out}^i, e_{2out}^j will increase by 1, e_{22}^i will decrease by 2 while e_{22}^j remains the same. Let $e_{2out}^{\prime i}, e_{22}^{\prime i}$ denote corresponding values after all movements. The following always holds no matter how many times we move:

$$2 \sum_{i \in \text{layer 2}} (e_{22}^i - e_{22}^{\prime i}) = \sum_{i \in \text{layer 2}} e_{2out}^{\prime i}$$

Now Q_2', the modularity of the new partition after moving, is:

$$Q_2' = \sum_{i \in \text{layer 2}} \frac{e_{22}^{\prime i}}{e} - \left(\frac{d_2^{\prime i}}{2e} \right)^2$$

$$= \sum_{i \in \text{layer 2}} \frac{4e \cdot e_{22}^{\prime i}}{4e^2} - \sum_{i \in \text{layer 2}} \frac{(2e_{22}^{\prime i} + e_{2out}^{\prime i})^2}{4e^2}.$$

Let $e_{22}^i - e_{22}^{\prime i} = \Delta_i$. Because of $(a+b)^2 \geq a^2 + b^2$ for any $a, b \geq 0$, we have:

$$Q_2' \leq \sum_{i \in \text{layer 2}} \frac{4e \cdot e_{22}^{\prime i}}{4e^2} - \sum_{i \in \text{layer 2}} \frac{(2e_{22}^{\prime i})^2 + (e_{2out}^{\prime i})^2}{4e^2}$$

$$= \frac{4e \cdot \sum e_{22}^{\prime i} - \sum 4(e_{22}^{\prime i})^2 - \sum (e_{2out}^{\prime i})^2}{4e^2}$$

$$= \frac{4e \cdot \sum (e_{22}^i - \Delta_i) - \sum 4(e_{22}^i - \Delta_i)^2 - \sum (e_{2out}^{\prime i})^2}{4e^2}$$

$$= Q_2 + \frac{8 \sum \Delta_i e_{22}^i - 4e \cdot \sum \Delta_i - \sum (e_{2out}^{\prime i})^2 - 4 \sum \Delta_i^2}{4e^2}$$

Let T abbreviate $8 \sum \Delta_i e_{22}^i - 4e \cdot \sum \Delta_i - \sum (e_{2out}^{\prime i})^2 - 4 \sum \Delta_i^2$, then $Q_2' = Q_2 + \frac{T}{4e^2}$. When no layer 2 community contains more than half of the total edges after applying RemoveEdge on layer 1, i.e., $e_{22}^i \leq \frac{e}{2}$,

$$T = 8 \sum \Delta_i e_{22}^i - 4e \cdot \sum \Delta_i - \sum (e_{2out}^{\prime i})^2 - 4 \sum \Delta_i^2$$

$$\leq 4e \cdot \sum \Delta_i - 4e \cdot \sum \Delta_i - \sum (e_{2out}^{\prime i})^2 - 4 \sum \Delta_i^2 \leq 0.$$

Finally, we have $Q_2' \leq Q_2 + \frac{T}{4e^2} \leq Q_2$. Hence, layer 2 has the highest modularity among all possible partitions of n nodes into n_2 communities. In this way, RemoveEdge makes the unreduced layer easier for the base algorithm to detect.

Theorem 3. *For a two-layer stochastic blockmodel network $G(n, n_1, n_2, p_1, p_2)$, the modularity of a layer increases if we apply ReduceEdge on all communities in the other layer.*

Proof. In ReduceEdge of layer 1, we keep edges in the given community with probability $q_1' = \frac{1-\hat{p}}{1-\hat{q}}$, where $\hat{p}$ is the observed edge probability within the detected community and $\hat{q}$ is the observed background noise.

ReduceEdge on layer 1 would only keep q_1' fraction of edges in S_{12} and S_1, so after ReduceEdge,

$$e_{22}' = \frac{1}{n_2}(|S_2| + |S_{12}| \cdot q_1') > \frac{1}{n_2}(|S_2| + |S_{12}|) \cdot q_1' = e_{22} \cdot q_1',$$

$$e_{2out}' = \frac{2}{n_1}|S_1| \cdot q_1' = e_{2out} \cdot q_1'.$$

Thus, $\frac{e_{2out}'}{e_{2out}} < \frac{e_{22}'}{e_{22}}$, and Lemma 2 indicates that $Q_2 < Q_2'$. Similarly, for the modularity of layer 1 after ReduceEdge on layer 1, $Q_1' > Q_1$.

Theorem 4. *For a synthetic two-layer block model network $G(n, n_1, n_2, p_1, p_2)$, the modularity of a layer increases if we apply ReduceWeight on all communities in the other layer.*

Proof. According to [3], ReduceWeight on layer 1 multiplies the weight of edges in layer 1 community by $q_1' = 1 - \frac{1-\hat{p}}{1-\hat{q}}$ percent. In weighted network, the weight sum of internal edges of a community i in layer 2 is $e_{22} = \frac{1}{2}\sum_{u,v \in i} w_{uv} \cdot A_{uv}$ where w_{uv} is the weight of edge (u,v). By construction, ReduceWeight on layer 1 reduces weight of all edges in S_{12} or S_1, but does not change weight of edges in S_2. Thus,

$$e_{22}'^i = \frac{1}{2} \sum_{u,v \in i,\, (u,v) \in S_{12}} w_{uv} \cdot A_{uv} \cdot q_1' + \frac{1}{2} \sum_{u,v \in i,\, (u,v) \in S_2} w_{uv} \cdot A_{uv}$$

$$> \left(\frac{1}{2} \sum_{u,v \in i,\, (u,v) \in S_{12}} w_{uv} \cdot A_{uv} + \frac{1}{2} \sum_{u,v \in i,\, (u,v) \in S_2} w_{uv} \cdot A_{uv} \right) \cdot q_1'$$

$$= e_{22}^i \cdot q_1'$$

$$e_{2out}^i = \frac{1}{2} \sum_{u \in i, v \notin i} w_{uv} A_{uv}$$

$$e_{2out}'^i = \frac{1}{2} \sum_{u \in i, v \notin i} w_{uv} A_{uv} \cdot q_1' = e_{2out}^i \cdot q_1'$$

Thus, $\frac{e_{2out}'}{e_{2out}} < \frac{e_{22}'}{e_{22}}$, and combined with Lemma 2, this proves that $Q_2' > Q_2$, the modularity increases after ReduceWeight.

Similarly, the modularity of layer 1 after RemoveEdge on layer 1, $Q_1' > Q_1$.

The analysis shows that weakening one layer with any one of the methods (RemoveEdge, ReduceEdge, ReduceWeight) increases the modularity of the other layer. These results follow naturally from Lemma 2, which is in some way a stronger claim that the modularity of the remaining layer increases as long as a larger percentage of outgoing edges is reduced than internal edges.

4   Simulation of Relative Modularity

To show whether reducing layers makes other layers more detectable when running HICODE, we simulate how grounded layers' relative modularity changes as the weakening method iterates on two-layer stochastic block models, and compare the grounded layers' modularity value with other partitions' modularity values. The number of possible partitions of n nodes is exponential, so it would be computationally unrealistic just to enumerate them, let alone calculate modularity for all of them. So we employ sampling of partitions. We calculate modularity for all sampled partitions and plot them on a 2-dimensional plane based on their similarities with the grounded layer 1 and layer 2, and show the modularity values through the colormap with nearest interpolation.

4.1   Sampling Method

We sample 2000 partitions similar to layer 1 (or 2) by starting from layer 1 (or 2), and then exchange a pair of nodes or change the membership of one node for $k = 1, ..., 500$ times. We also include 1200 partitions that mixed layer 1 and layer 2 by having k randomly selected nodes getting assigned to their communities in layer 1 and the rest $200 - k$ nodes getting assigned to their communities in layer 2. As planted communities in different layers are independent, this sampling method gives a wide range of partitions while being relatively fast. To measure the similarity between two partitions, we adapt normalized mutual information (NMI) [5] for overlapping communities (The definition of NMI is in Appendix C.). Partitions of nodes are inherently high-dimensional. To place them on 2-dimensional plane for the plotting purpose, we use its NMI similarity with layer 1 as the x-coordinate, and NMI similarity with layer 2 as the y-coordinate.

At each iteration, We use the modularity optimization based fast community detection algorithm [1] as the base algorithm to uncover a single layer of communities.

4.2   Simulation on ReduceEdge

Figure 2 presents the simulated results on a two-layer block model $G(600, 15, 12, 0.1, 0.12)$ using ReduceEdge as the weakening method. In this network, layer 2 is the dominant layer (communities are bigger and denser) and layer 1 is the hidden layer. The modularity of layer 2 is 0.546, while the modularity of layer 1 is 0.398. We plot the modularity of the estimated layer and other sampled partitions at different iterations of HICODE. On each subfigure, the dark red

Total node number = 600, layer 1 size = 50, layer 2 size = 40, p1 = 0.1, p2 = 0.12

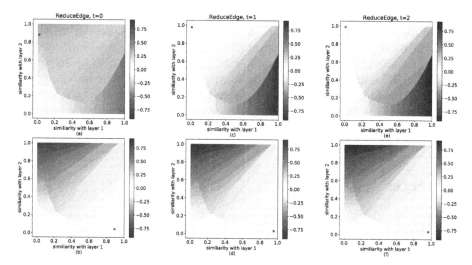

Fig. 2. Simulation results of ReduceEdge on $G(600, 15, 12, 0.1, 0.12)$. (Color figure online)

cross sign denotes where the estimated layer projects on the 2-dimensional plane. Simulations using RemoveEdge and ReduceWeight yield similar results.

1. Initially, two grounded layers here have similar modularity values, contributing to the two local peaks of modularity, one at the right-bottom and the other at the left-top.
2. (a): At iteration $t = 0$:, the base algorithm finds an approximate layer 2, whose NMI similarity with layer 2 is about 0.90.
3. (b): After reducing that partition, the modularity local peak at the left-top sinks and the modularity peak at right-bottom rises, and the base algorithm finds an approximate layer 1 whose NMI similarity with layer 1 is about 0.89. ReduceEdge then reduces this approximated layer 1 and makes it easier to approximate layer 2.
4. (c) and (d): At $t = 1$, the base algorithm finds an approximate layer 2 having 0.97 NMI similarity with layer 2, which is a significant improvement. As that more accurate approximation of layer 2 is reduced, the base algorithm is able to find a better approximation of layer 1 too. In our run, it finds an approximation that has 0.96 NMI similarity with layer 1.
5. (e) and (f): As HICODE iterates, at $t = 2$, the base algorithm is able to uncover an approximate layer 2 with 0.98 NMI similarity, and an approximate layer 1 with 0.97 NMI similarity.

5   Conclusion

In this work, we provide a theoretical perspective on the hidden community detection meta-approach HICODE, on multi-layer stochastic block models. We prove that in synthetic two-layer stochastic blockmodel networks, the modularity of a layer will increase, after we apply a weakening method (RemoveEdge, ReduceEdge, or ReduceWeight) on all communities in the other layer, which boosts the detection of the current layer when the other layer is weakened. A simulation of relative modularity during iterations is also provided to illustrate on how HICODE weakening method works during the iterations. Our work builds a solid theoretical base for HICODE, demonstrating that it is promising in uncovering both hidden and dominant layers of communities in two-layer stochastic block model networks. In future work, we will generalize the theoretical analysis to synthetic networks with more than two stochastic block model layers.

References

1. Blondel, V.D., Guillaume, J.L., Lambiotte, R., Lefebvre, E.: Fast unfolding of communities in large networks. J. Stat. Mech. Theory Exp. **2008**(10), P10008 (2008)
2. Girvan, M., Newman, M.E.: Community structure in social and biological networks. Proc. Natl. Acad. Sci. **99**(12), 3807–3870 (2015)
3. He, K., Li, Y., Soundarajan, S., Hopcroft, J.E.: Hidden community detection in social networks. Inf. Sci. **425**, 92–106 (2018)
4. He, K., Soundarajan, S., Cao, X., Hopcroft, J.E., Huang, M.: Revealing multiple layers of hidden community structure in networks. CoRR abs/1501.05700 (2015)
5. McDaid, A.F., Greene, D., Hurley, N.: Normalized mutual information to evaluate overlapping community finding algorithms. arXiv preprint arXiv:1110.2515 (2011)
6. Paul, S., Chen, Y.: Consistent community detection in multi-relational data through restricted multi-layer stochastic blockmodel. Electron. J. Stat. **10**(2), 3807–3870 (2016)
7. Teng, S.H., et al.: Scalable algorithms for data and network analysis. Found. Trends Theor. Comput. Sci. **12**(1–2), 1–274 (2016)

A Primal-Dual Algorithm for Euclidean k-Means Problem with Penalties

Chunying Ren[1], Dachuan Xu[1], Donglei Du[2], and Min Li[3(✉)]

[1] Department of Operations Research and Information Engineering, Beijing University of Technology, Beijing 100124, People's Republic of China
[2] Faculty of Management, University of New Brunswick, Fredericton, NB E3B 5A3, Canada
[3] School of Mathematics and Statistics, Shandong Normal University, Jinan 250014, People's Republic of China
liminemily@sdnu.edu.cn

Abstract. In the classical k-means problem, we are given a data set $\mathcal{D} \subseteq \mathbb{R}^{\ell}$ and an integer k. The object is to select a set $S \subseteq \mathcal{R}^{\ell}$ of size at most k such that each point in $\mathcal{D}$ is connected to the closet cluster in S with minimum total squared distances. However, in some real-life applications, it is more desirable and beneficial to pay a small penalty for not connecting some outliers in $\mathcal{D}$ that are too far away from most points. As a result, we are motivated to study the k-means problem with penalties, for which we propose a $(6.357+\varepsilon)$-approximation algorithm via the primal-dual technique, improving the previous best approximation ratio of $19.849 + \epsilon$ in [7] also by using the primal-dual technique.

Keywords: Approximation algorithm · k-means problem · Penalties · Primal-dual

1 Introduction

The classic k-means problem has been extensively studied in operations research and computer science [4,8,10,11,13–15]. Given an integer k and a data set $\mathcal{D} \subseteq \mathcal{R}^{\ell}$ of n points, the k-means problem selects a subset S of at most k center points in $\mathcal{R}^{\ell}$ to minimize $\sum_{j \in \mathcal{D}} c(j, S)$, where $c(j, S)$ is the squared Euclidean distances from j to the nearest point in S.

In general the problem is NP-hard [2,5]. Several approximation algorithms exist in the literature based on different techniques. One popular algorithm due to Lloyd-Forgy [12] is called the "k-means" algorithm. This algorithm performs very well in practice but its theoretical guarantee is very poor. In order to improve the performance guarantee, Arthur and Vassilvitskii [3] proposed a modified "k-means" algorithm, called k-means++, by randomly selecting the first initial k centers with a specific probability and show that this new algorithm has approximation ratio $O(\ln k)$. The first constant polynomial time approximation algorithm for the k-means problem is given in [10] with an approximation

© Springer Nature Switzerland AG 2020
J. Chen et al. (Eds.): TAMC 2020, LNCS 12337, pp. 377–389, 2020.
https://doi.org/10.1007/978-3-030-59267-7_32

ratio $9 + \varepsilon$ based on local search technique. Currently, the best approximation factor for the k-means problem is $6.357 + \varepsilon$ [1].

The main focus of this work is to consider the k-means problem with penalties. Formally, we are given an integer k, and a data set $\mathcal{D} \subseteq \mathcal{R}^\ell$ of n points where each point $j \in \mathcal{D}$ is associated with a penalty cost p_j. The objective is to select a set $S \subseteq \mathcal{R}^\ell$ of size k and a subset $\mathcal{D}_p \subseteq \mathcal{D}$ such that the total squared distance of connected points and the total penalty cost of non-connected point, $\sum_{j \in \mathcal{D} \setminus \mathcal{D}_p} c(j, S) + \sum_{j \in \mathcal{D}_p} p_j$, is minimized. The first constant approximation algorithm for this problem has an approximation ratio $25 + \varepsilon$ via local search technique [16]. The current best approximation ratio $19.849 + \epsilon$ is based on primal-dual technique [7].

Our main contribution in this article is to give a quasi-polynomial time algorithm with the best $(6.357+\varepsilon)$-approximation ratio so far by adopting the primal-dual method, improving the previous best result.

The outline of this article is as follows. Section 2 describes preliminaries. Section 3 presents how to obtain an integer feasible solution to the k-means problem with penalties through several algorithms, and analyzes the approximation ratio. Finally, the concluding remarks are given in Sect. 4.

2   Preliminaries

Given a k-means problem with penalties instance, we apply the standard discrete techniques [6] to obtain a corresponding discrete k-means problem with penalties by sacrificing a small approximate ratio loss; That is, instead of arbitrary choosing k points in $\mathcal{R}^\ell$, we select k points in the discrete set $\mathcal{F} \subseteq \mathcal{R}^\ell$. From now on, we simply refer to the discrete k-means problem with penalties as the k-means problem with penalties.

Definition 1 (k-means problem with penalties). *Given an instance $(\mathcal{D}, \mathcal{F}, p, d, k)$ of the problem, where $\mathcal{D}$ stands for the set of clients, $\mathcal{F}$ stands for the set of facilities, p stands for the set of penalty costs, each of which is assume to satisfy $p_j \geq 1, \forall j \in \mathcal{D}$, d is a metric distance, and k is an integer, the object is to select a set S of k facilities in $\mathcal{F}$ and a set $\mathcal{D}_p$ in $\mathcal{D}$, so as to minimize the following objective function:*

$$\sum_{j \in \mathcal{D} \setminus \mathcal{D}_p} c(j, S) + \sum_{j \in \mathcal{D}_p} p_j,$$

where $c(j, S) = d^2(j, S)$.

Next, we introduce the linear programming (LP) formulation for the k-means problem with penalties, along with the corresponding LP relaxation, Lagrangian relaxation, and its dual program.

The standard linear programming formulation of the k-means problem with penalties has three sets of indicator variables:

– $y_i = 1$, if facility $i \in \mathcal{F}$ is opened; and $y_i = 0$, otherwise.

- $r_j = 1$, if client $j \in \mathcal{D}$ is not connected to a facility; and $r_j = 0$, otherwise.
- $x_{ij} = 1$, if client $j \in \mathcal{D}$ is connected to facility $i \in \mathcal{F}$; and $x_{ij} = 0$, otherwise.

We relax each indicator variable to the positive real space to obtain the standard LP relaxation as follows:

$$\text{OPT}_k := \min \sum_{i \in \mathcal{F}} \sum_{j \in \mathcal{D}} x_{ij} c(i,j) + \sum_{j \in \mathcal{D}} p_j \cdot r_j$$

$$\text{s.t} \quad \sum_{i \in \mathcal{F}} x_{ij} + r_j \geq 1, \quad \forall j \in \mathcal{D},$$

$$x_{ij} \leq y_i, \quad \forall j \in \mathcal{D}, \forall i \in \mathcal{F},$$

$$\sum_{i \in \mathcal{F}} y_i \leq k,$$

$$x_{ij}, y_i, r_j \geq 0, \quad \forall j \in \mathcal{D}, \forall i \in \mathcal{F}.$$

The first constraint indicates that each client should be either connected to at least one facility or penalized. The second constraint says that clients can only be connected to opened facilities. The third constraint enforces that at most k facilities are opened.

We now establish a connection between the facility location problem with penalties and the k-means problem with penalties by employing Lagrangian relaxation (e.g., [9]), where a Lagrange multiplier λ is introduced to the third constraint which can be included in the objective function.

$$\text{LP-P}(\lambda) := \min \sum_{i \in \mathcal{F}} \sum_{j \in \mathcal{D}} x_{ij} c(i,j) + \sum_{j \in \mathcal{D}} p_j \cdot r_j + \lambda \left(\sum_{i \in \mathcal{F}} y_i - k \right)$$

$$\text{s. t.} \quad \sum_{i \in \mathcal{F}} x_{ij} + r_j \geq 1, \quad \forall j \in \mathcal{D},$$

$$x_{ij} \leq y_i, \quad \forall j \in \mathcal{D}, \forall i \in \mathcal{F},$$

$$x_{ij}, y_i, r_j \geq 0, \quad \forall j \in \mathcal{D}, \forall i \in \mathcal{F}.$$

$$\text{DUAL-P}(\lambda) := \max \sum_{j \in \mathcal{D}} \alpha_j - \lambda \cdot k$$

$$\text{s. t.} \quad \sum_{j \in \mathcal{D}} \beta_{ij} \leq \lambda, \quad \forall i \in \mathcal{F},$$

$$\alpha_j - c(i,j) \leq \beta_{ij}, \quad \forall j \in \mathcal{D}, \forall i \in \mathcal{F},$$

$$\alpha_j \leq p_j, \quad \forall j \in \mathcal{D},$$

$$\alpha_j, \beta_{ij} \geq 0, \quad \forall j \in \mathcal{D}, \forall i \in \mathcal{F}.$$

Evidently, for any $\lambda \geq 0$, the optimal primal value LP-P(λ) is at most the optimal OPT$_k$.

Note that if the term $\lambda \cdot k$ in the objective function of LP-P(λ) and DUAL-P(λ) is ignored, LP-P(λ) and DUAL-P(λ) can be regarded as the standard linear program and its dual of the penalized facility location problem where the open cost of each facility is equal to λ and the connection cost is equal to $c(\cdot,\cdot) = d^2(\cdot,\cdot)$. If the number of facilities opened does not exceed k by selecting the appropriate λ, then we have a feasible solution to the original problem.

Based on this idea we give a quasi-polynomial time $(6.357 + \varepsilon)$-approximation algorithm. This algorithm employs the JV algorithm [9] as the basic algorithm and contains two sub-routines: first enumerate λ to obtain a sequence facility solution S and then process the sequence S so that the number of open facilities is exactly k.

3 A Quasi-Polynomial Time Approximation Algorithm

Before describing the main algorithm, we first present a new algorithm, denoted as JV-P(δ), which is a modified primal-dual algorithm based on the JV(δ) algorithm in [9] to solve the facility location problem with penalties where the open cost of each facility is equal to λ.

3.1 JV-P(δ) Algorithm

This algorithm contains two phases: the dual-growth phase (Algorithm 1) and the pruning phase (Algorithm 2).

In the dual-growth phase, we construct a feasible dual solution (α, β) for DUAL-P(λ).

Algorithm 1. JV-P(δ): The dual-growth phase

---

1. Initially, let $\alpha := 0$ and set $\beta_{ij} := [\alpha_j - c(i,j)]^+ = \max\{\alpha_j - c(i,j), 0\}$. Let $A := D$ denote the set of active clients and let θ represent time. Starting from zero, $\forall j \in A$, α_j and θ rise at the same time per unit speed. When $\alpha_j = p_j$, we say that j is frozen, and α_j is not increased. If one of the following events occurs, j is removed from A.

   Event 1. A dual constraint $\sum_{j \in D} [\alpha_j - c(j,i)]^+ = \lambda$ for a facility $i \in \mathcal{F}$. In this case, we say that facility i is tight or temporarily opened. We update A by removing $j \in A$ if $\alpha_j \geq c(i,j)$. We say that facility i is the witness of these removed clients, denoted as $w(j) = i$.

   Event 2. An active client $j \in A$ gets a tight edge, i.e., $\alpha_j - c(j,i) = 0$, to some already tight facility i. In this case, we remove j from A and let i be its witness.

2. When $A = \emptyset$ or j is frozen $\forall j \in A$, the dual-growth phase stops.

The dual-growth phase may have opened more facilities than necessary. The pruning phase will select a subset of these facilities to open. We introduce some notations as follows.

1. $\mathcal{F}_y$: the set of facilities temporarily opened during the dual-growth phase.

2. $N(j) := \{i \in \mathcal{F} : \alpha_j - c(i, j) > 0\}, \forall j \in \mathcal{D}$.
3. $N(i) := \{j \in \mathcal{D} : \alpha_j - c(i, j) > 0\}, \forall i \in \mathcal{F}$.
4. For a temporarily opened facility i, let $t_i := \max_{j \in N(i)} \alpha_j$, and $t_i := 0$ if $N(i) = \emptyset$. So, we have $t_i \geq \alpha_j, \forall j \in N(i)$. For a client j and its witness $w(j)$, $\alpha_j \geq t_{w(j)}$.

We construct the client-facility graph G and the conflict graph H.

– $G = (\mathcal{D} \cup \mathcal{F}_y, E)$: there is an edge between facility i and client j if $i \in N(j)$.

– $H = (\mathcal{F}_y, E)$: there is an edge between facility i and i' if $j \in N(i) \cap N(i')$ and $c(i, i') \leq \delta \min(t_i, t_{i'})$.

Intuitively, the size of δ affects the number of edges in H, and thus determining the size of the maximal independent set in H. Therefore, by properly adjusting δ, we can obtain a better approximation ratio of the JV-P(δ) algorithm.

Algorithm 2. JV-P(δ): The pruning phase

Find a maximal independent set IS of H and open these facilities. Define the penalized set of clients $\mathcal{D}_p$,

$$\mathcal{D}_p := \{j \in \mathcal{D} : j \notin N(i), \forall i \in \text{IS} \text{ and } \alpha_j = p_j\}.$$

The remaining clients are connected to the closest facility in IS.

Lemma 1. *Let d be an Euclidean metric on $\mathcal{D} \cup \mathcal{F}$ and define $c(j, i) := d(j, i)^2$ for every $i \in \mathcal{F}$ and $j \in \mathcal{D}$. Then, for any $\lambda \geq 0$, Algorithm JV-P(δ) constructs a solution α to DUAL-P(λ) and outputs a set IS of opened facilities and $\mathcal{D}_p$ such that*

$$\sum_{j \in \mathcal{D} \setminus \mathcal{D}_p} d(j, \text{IS})^2 + \sum_{j \in \mathcal{D}_p} p_j \leq \rho \cdot \left(\sum_{j \in \mathcal{D}} \alpha_j - \lambda |\text{IS}| \right), \tag{1}$$

where $\delta \geq 2$ is a parameter which minimizes

$$\rho(\delta) = \max \left\{ (1 + \sqrt{\delta})^2, \frac{1}{\delta/2 - 1} \right\}.$$

It can be verified that $\delta \approx 2.315$ and $\rho \approx 6.357$.

Proof. According to Algorithm 1, α is a feasible solution to DUAL-P(λ). Now we just need to prove

$$
\frac{c(j,\mathrm{IS})}{\rho} \le \alpha_j - \sum_{j \in N(j) \cap \mathrm{IS}} (\alpha_j - c(j,i))
$$

$$
= \alpha_j - \sum_{i \in \mathrm{IS}} [\alpha_j - c(j,i)]^+, \quad \forall j \in \mathcal{D} \setminus \mathcal{D}_p, \tag{2}
$$

$$
\frac{p_j}{\rho} \le \alpha_j - \sum_{i \in \mathrm{IS}} [\alpha_j - c(j,i)]^+, \quad \forall j \in \mathcal{D}_p. \tag{3}
$$

Then summing up over all clients and by $[\alpha_j - c(j,i)]^+ \ge 0$ for any $i \in \mathrm{IS}, j \in \mathcal{D}_p$, we have

$$
\frac{1}{\rho} \left(\sum_{j \in \mathcal{D} \setminus \mathcal{D}_p} d(j,\mathrm{IS})^2 + \sum_{j \in \mathcal{D}_p} p_j \right) \le \sum_{j \in \mathcal{D}} \alpha_j - \sum_{j \in \mathcal{D} \setminus \mathcal{D}_p} \sum_{i \in \mathrm{IS}} [\alpha_j - c(j,i)]^+.
$$

Moreover, Note that any facility in IS is opened and no client in $\mathcal{D}_p$ contributes positively to any facility in IS. We obtain that $\sum_{j \in \mathcal{D} \setminus \mathcal{D}_p} [\alpha_j - c(j,i)]^+ = \lambda$, for any $i \in \mathrm{IS}$. Therefore, (1) follows.

The inequality (2) follows form Theorem 3.3 in [1]. Here we briefly describe the process.

According to the JV-P(δ) algorithm, we have $|N(j) \cap \mathrm{IS}| = \{0, 1, > 1\}$, $j \in \mathcal{D} \setminus \mathcal{D}_p$. Let $s_j := |N(j) \cap \mathrm{IS}|$, $S_j := N(j) \cap \mathrm{IS}$, $\mathcal{D}_0 := \{j \in \mathcal{D} \setminus \mathcal{D}_p | s_j = 0\}$, $\mathcal{D}_1 := \{j \in \mathcal{D} \setminus \mathcal{D}_p | s_j = 1\}$ and $\mathcal{D}_{>1} := \{j \in \mathcal{D} \setminus \mathcal{D}_p | s_j > 1\}$.

Case 1. $s_j = 0$. $\forall j \in \mathcal{D}_0$, there exists a $w(j) \in \mathcal{F}$ such that $\alpha_j \ge t_{w(j)}$, and $\alpha_j \ge c(j, w(j))$. Furthermore, if $w(j) \in \mathrm{IS}$, then $d(j,\mathrm{IS}) \le d(j, w(j))$; otherwise, there exists an $i \in \mathcal{F}$ such that $c(w(j), i) \le \delta \min\{t_{w(j)}, t_i\} \le \delta \alpha_j$, in which case

$$
d(j,\mathrm{IS}) \le d(j,i) \le d(w(j),j) + d(w(j),i) \le \sqrt{\alpha_j} + \sqrt{\delta \alpha_j} = (1 + \sqrt{\delta})\sqrt{\alpha_j}.
$$

Moreover, $c(j,\mathrm{IS}) \le (1 + \sqrt{\delta})^2 \alpha_j$ and $\sum_{i \in \mathrm{IS}} [\alpha_j - c(j,\mathrm{IS})]^+ = 0$ together imply that

$$
c(j,\mathrm{IS}) \le (1 + \sqrt{\delta})^2 \left\{ \alpha_j - \sum_{i \in \mathrm{IS}} [\alpha_j - c(j,\mathrm{IS})]^+ \right\}.
$$

When $\rho \ge (1 + \sqrt{\delta})^2$, we have

$$
\frac{c(j,\mathrm{IS})}{\rho} \le \alpha_j - \sum_{i \in \mathrm{IS}} [\alpha_j - c(j,\mathrm{IS})]^+.
$$

Case 2. $s_j = 1$. $\forall j \in \mathcal{D}_1$, there is only one facility $i^* \in \mathrm{IS}$ such that $[\alpha_j - c(j,i^*)]^+ > 0$. So, we have

$$
c(j,\mathrm{IS}) \le c(j,i^*) = \alpha_j - (\alpha_j - c(j,i^*)) = \alpha_j - \sum_{i \in \mathrm{IS}} [(\alpha_j - c(j,i)]^+.
$$

Case 3. $s_j > 1$. In an ℓ-dimensional metric space, the following holds: $\forall j \in \mathcal{D}_{>1}$,

$$\sum_{i \in S_j} c(j, i) \geq \sum_{i \in S_j} c(i, \mu) = \frac{\sum_{i \in S_j} \sum_{i' \in S_j} c(i, i')}{2s_j}$$

$$\geq \frac{\sum_{i \in S_j} \sum_{i' \in S_j, i \neq i'} c(i, i')}{2s_j} \geq \frac{\sum_{i \in S_j} \sum_{i' \in S_j, i \neq i'} \delta \alpha_j}{2s_j}$$

$$= \frac{s_j - 1}{2} \delta \cdot \alpha_j.$$

In the above, $\mu = \frac{1}{s_j} \sum_{i \in S_j} i$ is the centroid point of the facility in S_j, and the fourth inequality follows from $c(i, i') > \delta \min\{t_i, t_{i'}\} \geq \delta \alpha_j, \forall i \in S_j$. Hence,

$$\sum_{i \in S_j} (\alpha_j - c(j, i)) \leq \left(s_j - \frac{s_j - 1}{2} \delta \right) \alpha_j \leq \left(s_j \left(1 - \frac{\delta}{2} \right) + \frac{\delta}{2} \right) \alpha_j \leq \left(2 - \frac{\delta}{2} \right) \alpha_j.$$

For $s_j > 1$, $s_j(1 - \frac{\delta}{2}) + \frac{\delta}{2}$ is a non-increasing function of s_j because $\delta \geq 2$. We also know that $c(j, \mathrm{IS}) \leq c(j, i) \leq \alpha_j, \forall i \in S_j$.
So,

$$\left(1 - \left(2 - \frac{\delta}{2} \right) \right) c(j, \mathrm{IS}) = \left(\frac{\delta}{2} - 1 \right) c(j, \mathrm{IS}) \leq \alpha_j - \sum_{i \in S_j} (\alpha_j - c(j, i)).$$

When

$$\rho \geq \frac{1}{\frac{\delta}{2} - 1},$$

we have

$$\frac{c(j, \mathrm{IS})}{\rho} \leq \alpha_j - \sum_{i \in \mathrm{IS}} [\alpha_j - c(j, \mathrm{IS})]^+.$$

Combining all the above cases, let

$$\rho(\delta) := \max \left\{ (1 + \sqrt{\delta})^2, \frac{1}{\frac{\delta}{2} - 1} \right\},$$

where $\delta \geq 2$. Setting $\delta \approx 2.315$, we have $\rho \approx 6.357$, and hence (2) follows.
      For (3), we have $\alpha_j = p_j$, and $\alpha_j \leq c_{ij}, \forall j \in \mathcal{D}_p$ and $\forall i \in \mathrm{IS}$. So $\sum_{i \in \mathrm{IS}} [\alpha_j - c(j, \mathrm{IS})]^+ = 0$. Also $\rho > 1$. Hence (3) is correct. $\square$
      In the JV-P(δ) algorithm, $|\mathrm{IS}| \leq k$ may not be guaranteed unless some information about λ is given. When λ is small, the number of facilities in IS tends to $|F|$. When λ is large, the number of facilities in IS approaches 1. Therefore, we hope to find a suitable λ to make $|\mathrm{IS}| \leq k$. Similar to that in [1], we give a quasi-polynomial algorithm for the k-means problem with penalties by enumerating λ and finding a k-cardinality solution, such that the obtained solution satisfies the constraint that at most k facilities are opened.

3.2   Enumerating λ

We enumerate all the possibilities of $\lambda = 0, 1 \cdot \varepsilon_z, \cdots, L \cdot \varepsilon_z$, where ε_z is a small step size and L is large. By similar analysis in [1], we choose

$$n \gg 1/\epsilon, \varepsilon_z = n^{-3-30\log_{1+\epsilon} n}, L = 4n^7 \cdot \varepsilon_z^{-1}.$$

At the same time, it can be guaranteed that when $\lambda = 0$, the solution obtained by the algorithm is IS $= F$, and when $\lambda = L \cdot \varepsilon_z$, $|\text{IS}| \leq 1$.

We also use the notion of buckets that partition the α-values of the clients.

Definition 2. *For any value $x \in \mathbb{R}$, let*

$$B(x) := \begin{cases} 0, & \text{if } x < 1. \\ 1 + \lfloor \log_{1+\varepsilon}(x) \rfloor, & \text{if } x \geq 1. \end{cases}$$

Here $B(x)$ is the index of the bucket containing x and is also a piecewise function.

Let $n = |\mathcal{D}|$. In order to simplify the algorithm, we preprocess the given instance such that $1 \leq d(j,i)^2 \leq n^6$, $\forall j \in \mathcal{D}$ and $\forall i \in \mathcal{F}$ by only sacrificing the approximate ratio by a constant $(1 + 100/n^2)$. This proof can be found in Appendix C of [1].

The enumeration of λ is given in Algorithm 3. Note that the enumeration of λ iterates L times, and the running time per iteration is $O(n)$, where $n = |\mathcal{D}|$. So the total running time is $n^{O(\epsilon^{-1}\log n)}$.

Algorithm 3. Enumerating λ

---

1. Initially, $\lambda = 0$ and set $\alpha_j^{out} = \min\{\min_{i \in \mathcal{F}} d(j,i)^2, p_j\}$, and IS $= \mathcal{F}$.
2. $\alpha = \alpha^{out}$, $\lambda = \lambda + \varepsilon_z$, $\mathcal{D}_p = \emptyset$, $A = \emptyset$. θ represents time. Starting from zero, when $\theta = \alpha_j, \forall j \in \mathcal{D}$, let $A = A \cup j$. α_j and θ rise at the same time per unit speed. When $\alpha_j = p_j$, we say that j is frozen and α_j is not increased. If one of the following events occurs, j is removed from A.
   Event 1. $\alpha_j = p_j$ and $p_j - c(j,i) \leq 0$ $\forall i \in \mathcal{F}$. Set $\mathcal{D}_p = \mathcal{D}_p \cup j$.
   Event 2. j has a tight edge to a tight facility i with $B(\alpha_j) \geq B(t_i)$. Let i be the witness of j, denoted as $w(j) = i$.
3. While Step 2 is happening, we decrease every client j with $B(\alpha_j) > B(\theta)$ by $|A|$ times the growth rate of θ.
4. When all clients have been removed from A, output $\alpha^{out} = \alpha$ and $\mathcal{D}_p$.
5. When $\lambda = L \cdot \varepsilon_z$, the algorithm stops; Otherwise go back step 2.

Through the algorithm, when j has a tight edge to a facility i, or $\alpha_j = p_j$, then α_j stops increasing. So α^λ is a feasible solution for DUAL-P(λ). In every solution $\alpha = \alpha^\lambda$ ($\lambda = 0, 1 \cdot \varepsilon_z, \cdots, L \cdot \varepsilon_z$), let $\mathcal{D}_p$ be the set of penalized clients. Every client $j \in \mathcal{D} \backslash \mathcal{D}_p$ has a tight edge to a tight facility $w(j) \in \mathcal{F}$ with $B(\alpha_j) \geq B(t_{w(j)})$.

Lemma 2. *For each client $j \in \mathcal{D}\backslash\mathcal{D}_p$, we have*

$$\left|\alpha_j^\lambda - \alpha_j^{\lambda+\varepsilon_z}\right| \leq \frac{1}{n^2}.$$

Proof. When the constraints $\alpha_j = p_j$ is added, Lemma 4.5 in [1] implies the desired result.

3.3 Finding a k-cardinality Solution

To get a feasible solution to the k-means problem with penalties, we need to find a suitable λ so that the number of corresponding IS is k, which will be achieved in Algorithm 4.

Note that in Algorithm 4, the total number of iterations is at most L, and the running time per iteration is at most $O(m)$, where $m = |\mathcal{F}|$. So the total running time after adding algorithm 3 is also $n^{O(\varepsilon^{-1}\log n)}$.

For each solution α^λ, we construct G^λ, H^λ and IS^λ, following the same method as in Sect. 3.1.

Algorithm 4. Finding a k-cardinality solution

---

1. Initially, $\lambda = 0$.
2. Input $G^\lambda = (\mathcal{D} \cup \mathcal{F}_y^\lambda, E^\lambda)$, $G^{\lambda+\varepsilon_z} = (\mathcal{D} \cup \mathcal{F}_y^{\lambda+\varepsilon_z}, E^{\lambda+\varepsilon_z})$, H^λ, IS^λ.
3. Construct a client-facility graph $G^{(\lambda,1)}$ with bipartition $\mathcal{D}$ and $\mathcal{F}_y^\lambda \cup \mathcal{F}_y^{\lambda+\varepsilon_z}$ that has an edge from client j to facility $i \in \mathcal{F}_y^\lambda$ if (j,i) is present in G^λ and to $i \in \mathcal{F}_y^{\lambda+\varepsilon_z}$ if (j,i) is present in $G^{\lambda+\varepsilon_z}$. The opening time t_i of facility i is now naturally set to t_i^λ if $i \in \mathcal{F}_y^\lambda$ and to $t_i^{\lambda+\varepsilon_z}$ if $i \in \mathcal{F}_y^{\lambda+\varepsilon_z}$.
4. Generate the conflict graph $H^{(\lambda,1)}$ from $G^{(\lambda,1)}$ and t, and output a maximal independent set $\text{IS}^{(\lambda,1)}$ of $H^{(\lambda,1)}$ by greedily extending IS^λ.
5. Let $p = 2$.
    while $p \neq |F_y^\lambda| + 2$ do
        Removing a facility $i \in \mathcal{F}_y^\lambda$ from $G^{(\lambda,1)}$, we construct and output a new conflict graph $H^{(\lambda,p)}$ from $G^{(\lambda,p)}$ and its maximal independent set $\text{IS}^{(\lambda,p)}$ by greedily extending $\text{IS}^{(\lambda,p-1)}\backslash\{i\}$.
        if $|\text{IS}^{(\lambda,p)}| = k$, then
            Algorithm 4 stop, and output $\text{IS}^{(\lambda,p)}$.
        else
            $p = p + 1$.
        end if
    end while
6. At the end of the procedure (after $|F_y^\lambda|$ many steps), we have $G^{\lambda+\varepsilon_z} = G^{(\lambda,|F_y^\lambda|+1)}$. Set $H^{\lambda+\varepsilon_z} = H^{(\lambda,|F_y^\lambda|+1)}$, $\text{IS}^{\lambda+\varepsilon_z} = \text{IS}^{(\lambda,|F_y^\lambda|+1)}$ and $\lambda = \lambda + \varepsilon_z$. Go to Step 2.

3.4 Analysis

By Algorithm 4, there must exist some λ such that the number of IS is equal to k. Because at most one point is removed at one time in Algorithm 4, without loss of generality, we assume $|\mathrm{IS}^{(\lambda, l)}| = k$, corresponding to $G^{(\lambda, l)}$, and $H^{(\lambda, l)}$, for $2 \le l \le |F_y^\lambda|$. Let

$$\mathcal{D}_p := \left\{ j \in \mathcal{D} : j \notin N(i), \forall i \in \mathrm{IS} \text{ and } \alpha_j^{\lambda + \varepsilon_z} = p_j \right\},$$

denote the set of penalized clients, where $j \notin N(i)$ for each $i \in \mathrm{IS}$ is interpreted as $\alpha_j^\lambda \le c(i, j)$ if $i \in H^\lambda$, or $\alpha_j^{\lambda + \varepsilon_z} \le c(i, j)$ if $i \in H^{\lambda + \varepsilon_z}$. The other clients are connected to their respective closest facilities in the IS.

Let

$$\alpha_j := \begin{cases} \min \left\{ \alpha_j^\lambda, \alpha_j^{\lambda + \varepsilon_z} \right\}, & j \in \mathcal{D} \backslash \mathcal{D}_p, \\ p_j, & j \in \mathcal{D}_p. \end{cases}$$

Note that $\alpha \le \alpha^{\lambda + \varepsilon_z}$ is a feasible solution of DUAL-P($\lambda + \varepsilon_z$). Since α^λ and $\alpha^{\lambda + \varepsilon_z}$ are close, so $\alpha_j \ge \alpha^\lambda - 1/n^2$, $\alpha_j \ge \alpha^{\lambda + \varepsilon_z} - 1/n^2$ for all j.

For each client j, we define a set of facilities $S_j \subseteq \mathrm{IS}$ to which j contributes as follows: for all $i \in \mathrm{IS}$, we have $i \in S_j$ if $\alpha_j > d(j, i)^2$.

Now we choose $\mathcal{D}_0 = \{ j \in \mathcal{D} \backslash \mathcal{D}_p : S_j = \emptyset \}$, $\mathcal{D}_{>0} = \mathcal{D} \backslash \mathcal{D}_0 \backslash \mathcal{D}_p$, $\beta_{ij} = [\alpha_j - d(j, i)^2]^+$ and similarly $\beta_{ij}^\lambda = [\alpha_j^\lambda - d(j, i)^2]^+$ and $\beta_{ij}^{\lambda + \varepsilon_z} = [\alpha_j^{\lambda + \varepsilon_z} - d(j, i)^2]^+$.

Lemma 3. *For any* $j \in \mathcal{D}_{>0}$, *we have* $d(j, \mathrm{IS})^2 \le \rho \cdot (\alpha_j - \sum_{i \in S_j} \beta_{ij})$.

The proof follows from Lemma 1, and we omit the details here.

Lemma 4. *For every* $j \in \mathcal{D}_0$, *we have* $d(j, \mathrm{IS})^2 \le (1 + 5\varepsilon) \rho \cdot \alpha_j$.

Proof. For every $j \in \mathcal{D}_0$, there exists a facility $w(j) \in \mathcal{F}_y^{\lambda + \varepsilon_z}$ such that

$$B(\alpha_j) \ge B(t_{w(j)}) \Rightarrow (1 + \varepsilon) \alpha_j^{\lambda + \varepsilon_z} \ge t_{w(j)}^{\lambda + \varepsilon_z},$$

$$\alpha_j^{\lambda + \varepsilon_z} \ge c(j, w(j)).$$

Since $\alpha_j = \min \left\{ \alpha_j^\lambda, \alpha_j^{\lambda + \varepsilon_z} \right\}$, α^λ and $\alpha^{\lambda + \varepsilon_z}$ are close and $\alpha_j \ge 1$. From Lemma 2, we have

$$\alpha_j^{\lambda + \varepsilon_z} \le \alpha_j + \frac{1}{n^2} \le \left(1 + \frac{1}{n^2} \right) \alpha_j.$$

If $w(j) \in \mathrm{IS}$, then $d(j, \mathrm{IS}) \le d(j, w(j))$; otherwise, there exists an $i \in \mathrm{IS}$ such that $c(w(j), i) \le \delta \cdot t_{w(j)}^{\lambda + \varepsilon_z}$, and hence

$$d(j, \mathrm{IS}) \le d(j, i) \le d(j, w(j)) + d(w(j), i) \le \sqrt{\alpha_j^{\lambda + \varepsilon_z}} + \sqrt{\delta(1 + \varepsilon) \alpha_j^{\lambda + \varepsilon_z}}$$

$$\le (1 + \sqrt{\delta} + \varepsilon) \sqrt{\alpha_j^{\lambda + \varepsilon_z}} \le (1 + \sqrt{\delta} + \varepsilon) \sqrt{1 + \frac{1}{n^2}} \sqrt{\alpha_j}.$$

$$\le (1 + 2\varepsilon)(1 + \sqrt{\delta}) \sqrt{\alpha_j}.$$

Therefore

$$\frac{c(j, \mathrm{IS})}{(1 + \sqrt{\delta})^2} \leq (1 + 2\varepsilon)^2 \alpha_j \leq (1 + 5\varepsilon)\alpha_j.$$

Recall that $\rho \geq (1 + \sqrt{\delta})^2$, implying the desired result. □

The following result is an immediate consequence of Algorithms 1–4.

Lemma 5. *For every $j \in \mathcal{D}_p$ and $i \in \mathrm{IS}$, we have $p_j = \alpha_j$ and $\beta_{ij} = 0$.*

Lemma 6. *For any $i \in \mathrm{IS}$, we have $\sum_{j \in \mathcal{D} \backslash \mathcal{D}_p} \beta_{ij} \geq \lambda - \frac{1}{n}$.*

Proof. Since $\alpha_j \geq \max\{\alpha_j^{\lambda+\varepsilon_z}, \alpha_j^{\lambda}\} - \frac{1}{n^2}$ for every client $j \in \mathcal{D} \backslash \mathcal{D}_p$, we have

$$\sum_{j \in \mathcal{D} \backslash \mathcal{D}_p} \beta_{ij} \geq \sum_{j \in \mathcal{D} \backslash \mathcal{D}_p} \left(\max\left\{\beta_j^{\lambda+\varepsilon_z}, \beta_j^{\lambda}\right\} - \frac{1}{n^2} \right) \geq \lambda - \frac{1}{n}.$$

The second inequality follows because, for any $i \in \mathrm{IS}$, if $i \in \mathcal{F}_y^{\lambda+\varepsilon_z}$, then $\sum_{j \in \mathcal{D} \backslash \mathcal{D}_p} \beta_{ij}^{\lambda+\varepsilon_z} = \lambda + \varepsilon_z$; otherwise, if $i \in \mathcal{F}_y^{\lambda}$, then $\sum_{j \in \mathcal{D} \backslash \mathcal{D}_p} \beta_{ij}^{\lambda+\varepsilon_z} = \lambda$. □

Theorem 1. *Algorithm 4 outputs IS satisfying*

$$\sum_{j \in \mathcal{D} \backslash \mathcal{D}_p} d(j, \mathrm{IS})^2 + \sum_{j \in \mathcal{D}_p} p_j \leq (\rho + O(\varepsilon)) \cdot \mathrm{OPT}_k,$$

where $\rho \approx 6.357$.

Proof. From Lemmas 3–5, we have

$$\sum_{j \in \mathcal{D} \backslash \mathcal{D}_p} d(j, \mathrm{IS})^2 \leq (1 + 5\varepsilon)\rho \sum_{j \in \mathcal{D} \backslash \mathcal{D}_p} \left(\alpha_j - \sum_{i \in S_j} \beta_{ij} \right),$$

$$\sum_{j \in \mathcal{D}_p} p_j \leq (1 + 5\varepsilon)\rho \sum_{j \in \mathcal{D}_p} \alpha_j.$$

Therefore,

$$\sum_{j \in \mathcal{D} \backslash \mathcal{D}_p} d(j, \mathrm{IS})^2 + \sum_{j \in \mathcal{D}_p} p_j \leq (1 + 5\varepsilon)\rho \left(\sum_{j \in \mathcal{D}} \alpha_j - \sum_{j \in \mathcal{D} \backslash \mathcal{D}_p} \sum_{i \in S_j} \beta_{ij} \right).$$

α is a feasible solution of DUAL-P($\lambda + \varepsilon_z$), by Lemma 6, and $\sum_{i \in \mathrm{IS}} \beta_{ij} = \sum_{i \in S_j} \beta_{ij}$, we have

$$\sum_{j \in \mathcal{D} \setminus \mathcal{D}_p} d(j, \mathrm{IS})^2 + \sum_{j \in \mathcal{D}_p} p_j$$

$$\leq (1 + 5\varepsilon)\rho \left(\sum_{j \in \mathcal{D}} \alpha_j - \sum_{j \in \mathcal{D} \setminus \mathcal{D}_p} \sum_{i \in S_j} \beta_{ij} \right) \leq (1 + 5\varepsilon)\rho \left(\sum_{j \in \mathcal{D}} \alpha_j - |\mathrm{IS}|(\lambda - \frac{1}{n}) \right)$$

$$= (1 + 5\varepsilon)\rho \left(\sum_{j \in \mathcal{D}} \alpha_j - k \cdot (\lambda + \varepsilon_z) + \frac{k}{n} + k \cdot \varepsilon_z \right)$$

$$= (1 + 5\varepsilon)\rho \left(\sum_{j \in \mathcal{D}} \alpha_j - k \cdot (\lambda + \varepsilon_z) + \frac{k}{n} + k \cdot n^{-3 - 30 \log_{1+\varepsilon} n} \right)$$

$$\leq (1 + 5\varepsilon)\rho(\mathrm{OPT}_k + 1 + \frac{1}{n^3}) \leq (1 + 5\varepsilon)\rho(\mathrm{OPT}_k + 2)$$

$$\leq (1 + 5\varepsilon)\rho(1 + 2\varepsilon)\mathrm{OPT}_k$$

$$= (\rho + O(\varepsilon))\mathrm{OPT}_k.$$

The penultimate inequality follows from

$$1 \leq \varepsilon \cdot n \leq \varepsilon \cdot \sum_{j \in \mathcal{D}} \min \left\{ \min_{i \in \mathcal{F}} c(j, i), p_j \right\} \leq \varepsilon \cdot \mathrm{OPT}_k. \qquad \square$$

4   Discussion

In this article, we study the k-means problem with penalties. In the Euclidean space, a quasi-polynomial time $(6.357 + \varepsilon)$-approximation algorithm is given via the primal-dual technique. We note that the same algorithm can be used to obtain an approximate algorithm of $2.633 + \varepsilon$ for the penalized k-median problem in the Euclidean space. We conjecture that our algorithm can be improved to achieve a polynomial time complexity with the same approximation ratio.

Acknowledgements. The first two authors are supported by Natural Science Foundation of China (No. 11871081). The third author is supported by the Natural Sciences and Engineering Research Council of Canada (NSERC) grant 06446, and Natural Science Foundation of China (Nos. 11771386, 11728104). The fourth author is supported by Higher Educational Science and Technology Program of Shandong Province (No. J17KA171) and Natural Science Foundation of Shandong Province (No. ZR2019MA032) of China.

References

1. Ahmadian, S., Norouzi-Fard, A., Svensson, O., Ward, J.: Better guarantees for k-means and Euclidean k-median by primal-dual algorithms. In: Proceedings of FOCS, pp. 61–72 (2017)

2. Aloise, D., Deshpande, A., Hansen, P., Popat, P.: NP-hardness of Euclidean sum-of-squares clustering. Mach. Learn. **75**, 245–248 (2009)
3. Arthur, D., Vassilvitskii, S.: k-means++: the advantages of careful seeding. In: Proceedings of SODA, pp. 1027–1035 (2007)
4. Cohen-Addad, V., Klein, P.N., Mathieu, C.: Local search yields approximation schemes for k-means and k-median in Euclidean and minor-free metrics. SIAM J. Comput. **48**, 644–667 (2019)
5. Drineas, P., Frieze, A., Kannan, R., Vempala, S., Vinay, V.: Clustering large graphs via the singular value decomposition. Mach. Learn. **56**, 9–33 (2004)
6. Feldman, D., Monemizadeh, M., Sohler, C.: A PTAS for k-means clustering based on weak coresets. In: Proceedings of SoCG, pp. 11–18 (2007)
7. Feng, Q., Zhang, Z., Shi, F., Wang, J.: An improved approximation algorithm for the k-means problem with penalties. In: Chen, Y., Deng, X., Lu, M. (eds.) FAW 2019. LNCS, vol. 11458, pp. 170–181. Springer, Cham (2019). https://doi.org/10.1007/978-3-030-18126-0_15
8. Friggstad, Z., Rezapour, M., Salavatipour, M.R.: Local search yields a PTAS for k-means in doubling metrics. SIAM J. Comput. **48**, 452–480 (2019)
9. Jain, K., Vazirani, V.V.: Approximation algorithms for metric facility location and k-median problems using the primal-dual schema and Lagrangian relaxation. J. ACM **48**, 274–296 (2001)
10. Kanungo, T., Mount, D., Netanyahu, N., Piatko, C., Silverma, R.: A local search approximation algorithm for k-means clustering. Comput. Geom. **28**, 89–112 (2004)
11. Li, M., Xu, D., Yue, J., Zhang, D., Zhang, P.: The seeding algorithm for k-means problem with penalties. J. Comb. Optim. **39**, 15–32 (2020)
12. Lloyd, S.: Least squares quantization in PCM. IEEE Trans. Inf. Theory **28**, 129–137 (1982)
13. Makarychev, K., Makarychev, Y., Razenshteyn, I.: Performance of Johnson-Lindenstrauss transform for k-means and k-medians clustering. In: Proceedings of STOC, pp. 1027–1038 (2019)
14. Vazirani, V.V.: Approximation Algorithms. Springer, Heidelberg (2003)
15. Williamson, D.P., Shmoys, D.B.: The Design of Approximation Algorithms. Cambridge University Press, Cambridge (2011)
16. Zhang, D., Hao, C., Wu, C., Xu, D., Zhang, Z.: Local search approximation algorithms for the k-means problem with penalties. J. Comb. Optim. **37**, 439–453 (2019)

The Complexity of the Partition Coloring Problem

Zhenyu Guo$^{(\boxtimes)}$, Mingyu Xiao, and Yi Zhou

University of Electronic Science and Technology of China, Chengdu, China
Harry.Guo@outlook.com, myxiao@gmail.com, zhou.yi@uestc.edu.cn

Abstract. Given a simple undirected graph $G = (V, E)$ and a partition of the vertex set V into p parts, the PARTITION COLORING PROBLEM asks if we can select one vertex from each part of the partition such that the chromatic number of the subgraph induced on the p selected vertices is bounded by k. PCP is a generalized problem of the classical VERTEX COLORING PROBLEM and has applications in many areas, such as scheduling and encoding, etc. In this paper, we show the complexity status of the PARTITION COLORING PROBLEM with three parameters: the number of colors, the number of parts of the partition, and the maximum size of each part of the partition. Furthermore, we give a new exact algorithm for this problem.

Keywords: Graph coloring · Partition coloring · NP-completeness

1 Introduction

Given a simple undirected graph $G = (V, E)$, the vertex coloring is to assign each vertex a color such that no two adjacent vertices have the same color. In the VERTEX COLORING PROBLEM (VCP), a graph G together with an integer k is given, and the goal is to decide whether G can be colored by using at most k colors [17]. VCP is an important problem in both graph theory and practice [10–12].

In this paper, we study a generalized version of VCP, the PARTITION COLORING PROBLEM (PCP), which is also called the SELECTIVE GRAPH COLORING PROBLEM in some references [4]. In PCP, we are given a graph $G = (V, E)$ with a partition $\mathcal{V}$ of the vertex set and an integer k, where $\mathcal{V} = \{V_1, V_2, \cdots, V_p\}$, $V_i \cap V_j = \emptyset$ for all $1 \leq i, j \leq p$, and $\bigcup_{1 \leq i \leq p} V_i = V$. The problem asks whether there is an induced subgraph containing exactly one vertex from each part V_i of the partition $\mathcal{V}$ that is colorable by using k colors. Note that when each part V_i ($1 \leq i \leq p$) is a singleton, i.e., $|V_i| = 1$, PCP is equal to VCP. So VCP is a special case of PCP. Indeed, PCP, together with other extended coloring problems, such as the COLORING SUM PROBLEM [19,21], the EDGE COLORING PROBLEM [14], the MIXED GRAPH COLORING PROBLEM [13], the SPLIT COLORING PROBLEM [7] and so on, have been intensively studied from the view of computational complexity in the last decades [3,16,22,24].

© Springer Nature Switzerland AG 2020
J. Chen et al. (Eds.): TAMC 2020, LNCS 12337, pp. 390–401, 2020.
https://doi.org/10.1007/978-3-030-59267-7_33

1.1   Existing Literature

The literature dealing with PCP is rich and diverse. In terms of applications, PCP was firstly introduced in [20] to solve the wavelength routine and assignment problem, which is to assign a limited number of bandwidths on fiber networks. PCP also finds applications in a wide range in dichotomy-based constraint encoding, antenna positioning, and frequency assignment and scheduling [4].

PCP is NP-Complete in general since the well-known NP-Complete problem VCP is a special case of PCP. In pursuit of fast solution methods for PCP, heuristic searches without guarantee of the optimality represent one of the most popular approaches [9,20,23]. Exact algorithms based on integer linear programming were also investigated for solving problems of small scale [8,9,15]. In terms of computational complexity, the NP-Hardness of this problem on special graph classes, such as paths, circles, bipartite graphs, threshold graphs and split graphs were studied [4,5].

1.2   Our Contributions

In this paper, we further study the computational complexity of PCP. We always use k to denote the number of colors, p to denote the number of parts in the partition $\mathcal{V}$, and q to denote the upper bound of the size of all parts in the partition $\mathcal{V}$. We give some boundaries between P and NPC for this problem with different constant settings on the three parameters. We also consider the parameterized complexity of PCP: we show that PCP parameterized by p is W[1]-hard and PCP parameterized by both p and q is FPT. The main complexity results are summarized in Table 1 and Table 2. In addition, we give a fast exact algorithm for PCP, which is based on subset convolution and runs in $O((\frac{n+p}{p})^p n \log k)$ time. Note that when $p = n$, PCP becomes VCP and the running time bound becomes $O(2^n n \log k)$, the best-known running time bound for VCP.

Due to the limited space, the proof of some lemmas and theorems are omitted, which are marked with $\star$.

2   Preliminaries

Let $G = (V, E)$ stand for a simple and undirected graph with $n = |V|$ vertices and $m = |E|$ edges. For a vertex subset $X \subseteq V$, we use $G[X]$ to denote the subgraph induced by X. For a vertex $v \in V$, the set of vertices adjacent to v is called the set of *neighbors* of v and denoted by $N(v)$. A graph is called a *clique* if there is an edge between any pair of vertices in the graph and a graph is called an *independent set* if there is no edge between any pair of vertices.

Table 1. Complexity results with different constants q and k

q	k		
	$k = 1$	$k = 2$	$k \geq 3$
$q = 1$	P (Theorem 2)	P (Theorem 2)	NPC (Theorem 2)
$q = 2$	P (Theorem 3)	NPC (Corollary 1)	
$q \geq 3$	NPC (Theorem 5)		

Table 2. Complexity results with parameters p and q

p is a constant	P (Theorem 6)
Parameterized by p	W[1]-hard (Theorem 7)
Parameterized by p and q	FPT (Theorem 8)

For a nonnegative integer k, a k-coloring in a graph $G = (V, E)$ is a function $c : V \rightarrow \{1, 2, \cdots, k\}$ such that for any edge vu it holds that $c(v) \neq c(u)$. A graph is k-*colorable* if it allows a k-coloring. The VERTEX COLORING PROBLEM (VCP) is to determine whether a given graph is k-colorable. The smallest integer k to make G k-colorable is called the *chromatic number* of G and denoted by $\chi(G)$.

Given an integer p, a p-*partition* of the vertex set V of G is denoted by $\mathcal{V} = \{V_1, V_2, \cdots, V_p\}$, where $V_i \cap V_j = \emptyset$ for any pair of different i and j in $\{1, 2, \ldots, p\}$ and $\bigcup_{1 \leq i \leq p} V_i = V$. Each subset of a p-partition is also called a *part*. The maximum size of the parts in a p-partition $\mathcal{V} = \{V_1, V_2, \cdots, V_p\}$ is denoted by q, i.e., $q = \max_{i=1}^{p} |V_i|$. A *selection* of a p-partition $\mathcal{V}$ is a subset of vertex $S \subseteq V$ such that $|S \cap V_i| = 1$ for any $i \in \{1, 2, \ldots, p\}$. The PARTITION COLORING PROBLEM is formally defined as follows.

The Partition Coloring Problem (PCP)
Input: a graph $G = (V, E)$, a p-partition $\mathcal{V} = \{V_1, V_2, \cdots, V_p\}$ of V, and an integer k;
Question: Is there a selection S of $\mathcal{V}$ such that the chromatic number of the induced graph $G[S]$ is at most k, i.e., $\chi(G[S]) \leq k$?

We also introduce two known hard problems here, which will be used to prove the hardness results of our problems.

A *Conjunctive Normal Formula* (CNF) ϕ is a conjunction of m given *clauses* $C_1, C_2, \cdots, C_m$ on n boolean variables, where each clause C_i is a disjunction of literals or a single literal and a literal is either a variable or the negation of a variable. A literal x_i and its negation $\overline{x}_i$ are called a pair of *contrary literals*. A *truth assignment* to ϕ is an assignment of the n variables such that every clause in ϕ is true. The k-SATISFIABILITY PROBLEM is defined as follows:

The k-Satisfiability Problem(k-SAT)
Input: A CNF ϕ of m given *clauses* $C_1, C_2, \cdots, C_m$ on n boolean variables $x_1, x_2, \cdots, x_n$, where each clause contains at most k literals.
Question: Is there a truth assignment to ϕ?

The k-SAT is NP-Complete for each fixed integer $k \geq 3$ [17], but polynomially solvable for $k = 1$ or 2 [18]. These results will be used in the proofs of our Theorems 3, 4 and 5.

The INDEPENDENT SET PROBLEM is another famous problem, which is defined as follows:

The Independent Set Problem
Input: a graph $G = (V, E)$, an integer k;
Question: Is there an independent set of size at least k in G?

The INDEPENDENT SET PROBLEM is polynomially solvable when k is a constant and NP-Complete when k is part of the input [17]. Downey and Fellows [6] further showed that the INDEPENDENT SET PROBLEM is W[1]-hard when taking k as the parameter. This W[1]-hardness result implies that the INDEPENDENT SET PROBLEM will not allow an algorithm with running time $f(k)poly(n)$ for any computable function $f(k)$ and polynomial function on the input size $poly(n)$ under the assumption $FPT \neq W[1]$. For more background about parameterized complexity, readers are referred to the monograph [6]. The hardness results of the INDEPENDENT SET PROBLEM will be used to prove the hardness of PCP, say Theorem 7.

3   Complexity of PCP

In general, PCP is known to be NP-hard since it contains the well known NP-hard problem, the VERTEX COLORING PROBLEM as a special case, where each part contains exactly one vertex. In this paper, we will consider the complexity of PCP with respect to the following three parameters:

- the number of colors, k;
- the number of parts in the partition, p;
- the maximum cardinality among all parts in the partition, q.

We will show that PCP is polynomially solvable only when some of the three parameters are small constants. First of all, it is trivially to see that PCP is in NP for any setting of the three parameters. Given an assignment of colors to a subset of vertices, we can easily check whether it is a selection and a feasible k-coloring in polynomial time.

Theorem 1. *PCP is in NP.*

3.1   Parameters q and k

We now discuss the NP-Hardness of PCP with different constant values of q and k. As mentioned above, when $q = 1$, the problem is equal to VCP, which is NP-Complete for each constant $k \geq 3$ and polynomially solvable for each $k \leq 2$ [17]. Therefore, we have the following conclusion.

Theorem 2. *When $q = 1$, PCP is polynomially solvable for each constant $1 \leq k \leq 2$ and NP-Complete for each constant $k \geq 3$.*

Next, we consider the cases where $q \geq 2$.

Theorem 3. *When $q = 2$ and $k = 1$, PCP is polynomially solvable.*

Proof. We show that the case that $k = 1$ and $q = 2$ can be polynomially reduced to the polynomially solvable problem 2-SAT [18].

For an instance of PCP with $q = 2$ and $k = 1$, a graph $G = (V, E)$ and a p-partition $\mathcal{V}$ of V where each part has at most 2 vertices, we construct an instance ϕ of 2-SAT on p variables.

For each part V_j in $\mathcal{V}$, we associate it with a variable x_j. Then we have p variables in ϕ. Furthermore, we associate each vertex in G with a literal (either a variable x or its negative $\overline{x}$): for each part of size 2, say $V_j = \{u_{j1}, u_{j2}\}$, we associate vertex u_{j1} with literal $\ell_{j1} = x_j$ and associate vertex u_{j2} with literal $\ell_{j2} = \overline{x}_j$; for each part of size 1, say $V_j = \{u_{j1}\}$, we associate vertex u_{j1} with literal $\ell_{j1} = x_j$. Next, we construct clauses. We will have p *vertex clauses* and $|E|$ *edge clauses*. For each part V_j of size 1, we construct a vertex clause ℓ_{j1} containing exactly one literal; for each part V_j of size 2, we construct a vertex clause $\ell_{j1} \vee \ell_{j2}$ of size 2. We can see that the second kind of vertex clause will always be true since $\ell_{j2} = \overline{\ell_{j1}}$. However, we keep them for the purpose of the presentation. For each edge $(u, v) \in E$, we construct an edge clause $\overline{\ell_u} \vee \overline{\ell_v}$ of size 2, where u is associated with the literal ℓ_u and v is associated with the literal ℓ_v. Thus, we have $|E|$ edge clauses.

We prove that there is a selection of $\mathcal{V}$ which is 1-colorable if and only if ϕ is satisfied.

The "$\Rightarrow$" part: Assume there is a selection S of $\mathcal{V}$ which is 1-colorable. Then S will form an independent set. For each vertex v in S, we let its associated literal ℓ_v be 1. For any variable left without assigning a value, we simply let it be 1. We claim that this is a truth assignment to ϕ. Since each vertex is associated with a different literal, we know the above assignment of letting the literals associated with vertices in S is feasible. First of all, we know each edge clause is satisfied since each part contains at least one vertex in S and then each edge clause contains at least one literal with value 1. For an edge clause $\overline{\ell_u} \vee \overline{\ell_v}$ corresponding to the edge (u, v), if it is not satisfied, then $\ell_u = \ell_v = 1$ and thus both of u and v are in S, which is a contradiction to the fact that S is an independent set. So all edge clauses are satisfied and ϕ is satisfied.

The "$\Leftarrow$" part: Assume that ϕ is satisfied. For a truth assignment A of ϕ, we select a vertex v into the selection S if and only if its associated literal is assigned 1. Then the set S is a 1-colorable selection. The reason is as follows. Each vertex clause can have at most one literal of value 1 in A. So each part has one vertex being selected into S. For any two vertices $u, v \in S$, if there is an edge between them, then there is an edge clause $\overline{\ell_u} \vee \overline{\ell_v}$. Since $\overline{\ell_u} \vee \overline{\ell_v}$ should be 1 in A, we know that at least one of ℓ_u and ℓ_v is 0 and then at least one of u and v is not in S, a contradiction. So there is no edge between any two vertices in S and such S is an independent set.   □

Theorem 4. *When $q = 2$ and $k = 2$, PCP is NP-Complete.*

Proof. Theorem 1 shows that the problem is in NP. For the NP-Hardness, we give a reduction from the known NP-Complete problem 3-SAT to PCP with $q = 2$ and $k = 2$.

Let ϕ be a 3-SAT formula of m clauses $C_1, C_2, \cdots, C_m$ on n boolean variables $x_1, x_2, \cdots, x_n$, where we can assume that $|C_t| = 3$ holds for each clause C_t. We construct an instance of PCP. The graph $G = (V, E)$ contains $|V| = 9m + 2$ vertices. In the p-partition $\mathcal{V}$, each part has at most $q = 2$ vertices and $p = 6m+2$. We will show that G has a selection S of $\mathcal{V}$ such that the chromatic number of $G[S]$ is at most $k = 2$ if and only if ϕ is satisfiable.

The graph G is constructed in the following way.

First, we introduce two vertices denoted by g and r. Then, for each clause $C_t = (a_1^t \vee a_2^t \vee a_3^t)$ ($t \in \{1, \cdots, m\}$) in ϕ, we introduce 9 vertices that are divided into three layers of three vertices, called the *literal layer*, the *middle layer* and the *conflict layer*. The three vertices in the literal layer are denoted by l_1^t, l_2^t, l_3^t, the three vertices in the middle layer are denoted by m_1^t, m_2^t, m_3^t, and the three vertices in the conflict layer are denoted by c_1^t, c_2^t, c_3^t. For $i \in 1, 2, 3$ the four vertices l_i^t, m_i^t and c_i^t are associated with the literal a_i^t. In total, the graph has $9m + 2$ vertices.

For edges in the graph G, we first add an edge between g and r. Then, for each clause C_t, we introduce 9 edges as follows.

- Connect l_i^t to m_i^t for each each $i \in \{1, 2, 3\}$ (3 edges);
- Connect each vertex m_i^t ($i \in \{1, 2, 3\}$) in the middle layer to g (3 edges);
- Connect each pair of vertices in the conflict layer to form a triangle (3 edges).

Last, for each pair of contrary literals $a_i^{t_1}$ and $a_j^{t_2}$ ($i, j \in \{1, 2, 3\}, t_1, t_2 \in \{1, \cdots, m\}$) in ϕ, add an edge between the two vertices associated with $a_i^{t_1}$ and $a_j^{t_2}$ in the literal layers.

In terms of the p-partition $\mathcal{V}$, we will have $p = 6m + 2$ parts, each of which contains at most $q = 2$ vertices. Vertices g and r form two separated parts containing one vertex, $\{g\}$ and $\{r\}$. For each clause C_t, the 9 vertices associated with it will be divided into 6 parts: $\{l_i^t\}$ and $\{m_i^t, c_i^t\}$ for $i = 1, 2, 3$.

An illustration of the construction is shown in Fig. 1. We now show that CNF ϕ is satisfiable if and only if there is a selection S of $\mathcal{V}$ such that $G[S]$ is 2-colorable.

The "⟸" part: Assume that S is a selection of $\mathcal{V}$ such that $G[S]$ is 2-colorable and let $c : S \rightarrow \{green, red\}$ be a 2-coloring of $G[S]$. Since there is

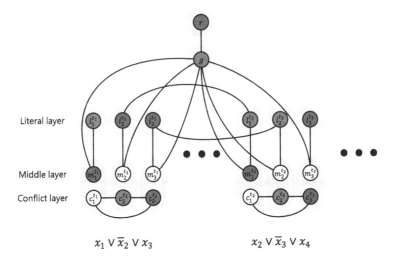

Fig. 1. An example of the construction with two clauses $C_{t_1} = x_1 \vee \overline{x}_2 \vee x_3$ and $C_{t_2} = x_2 \vee \overline{x}_3 \vee x_4$, where x_1 and x_2 have value true, and x_3 and x_4 have value false. In the figure, a grey shadow represents a part of the partition. (Color figure online)

an edge between g and r, we know that $c(g) \neq c(r)$. With loss of generality, we assume that $c(g) = green$ and $c(r) = red$.

Each vertex in the literal layers must be in the selection S since it is in a part of size 1. We claim that

Property 1. For the three vertices l_1^t, l_2^t and l_3^t ($t \in \{1, \cdots, m\}$) in a literal layer associated to the clause C_t, at least one of them is assigned $green$ in the 2-coloring c.

Assume to the contrary that $c(l_1^t) = c(l_2^t) = c(l_3^t) = red$. For this case, none of m_1^t, m_2^t m_3^t can be assigned to either red or $green$ and then none of them is in S. Such all of c_1^t, c_2^t and c_3^t are in S. Note that c_1^t, c_2^t and c_3^t form a triangle. It is impossible to color them by using only 2 colors, a contraction. So Property 1 holds.

For each clause C_t ($t \in \{1, \ldots, m\}$), we select an arbitrary vertex l_i^t in the literal layer with $green$ color in the 2-coloring c and assign the corresponding literal a_i^t value 1. After doing this, if there are still variables without assigned the value, arbitrarily assign 1 or 0 to it. We claim the above assignment of the variables is a truth assignment to ϕ. Note that there is an edge between any pair of contrary literals in the graph G. So it is impossible that two contrary literals are assigned $green$ in c. Therefore, the above assignment is a feasible assignment. Furthermore, by Property 1, we know that each clause will have at least one literal assigned value 1. Therefore, it is a truth assignment to ϕ.

The "$\Rightarrow$" **part:** Assume there is a truth assignment of ϕ. We show that there is a selection S of $\mathcal{V}$ and a 2-coloring c of $G[S]$. First of all, vertices g and r are selected into S. We let $c(g) = green$ and $c(r) = red$. Second, all vertices in the literal layers are selected into S, and a vertex l_i^t in the literal layers is assigned to color *green* (resp., *red*) in c if the corresponding literal a_i^t has value 1 (resp., 0). Third, for the parts containing a vertex in the middle layer and a vertex in the conflict layer, we select the vertex m_i^t in the middle layer into S and assign color *red* to it if the corresponding vertex l_i^t in the literal layer is assigned to color *green*; we select the vertex c_i^t in the conflict layer into S if the corresponding vertex l_i^t in the literal layer is assigned to color *red*. For the color of c_i^t, if no neighbor of c_i^t in $G[S]$ has been assigned a color, we assign color *green* to it, and otherwise, we assign color *red* to it. We argue that the above coloring is a feasible 2-coloring of $G[S]$. After the second step, no adjacent vertices are assigned the same color because there are only edges between pairs of contrary literals. In the third step, we can assign a *red* color to vertices m_i^t in the middle layer because they are only adjacent to vertices T and l_i^t, both of which are assigned *green*. For the vertices c_i^t in the conflict layer, each of them is adjacent to at most one vertex in $G[S]$ because at least one vertex in the literal layer of each clause C_t is assigned *green* by Property 1 and then at most two vertices in the conflict layer of C_t can be selected into S. So our way to color vertices in the conflict layer is correct. □

Readers are refer to Fig. 1 for an illustration of the reduction of the case $q = 2$ and $k = 2$, where ϕ contains at least two clauses $C_{t_1} = x_1 \vee \overline{x}_2 \vee x_3$ and $C_{t_2} = x_2 \vee \overline{x}_3 \vee x_4$. When x_1 and x_2 have value true, x_3 and x_4 have value false, then we assign $l_1^{t_1}$, $l_1^{t_2}$, $l_2^{t_2}$, $c_2^{t_1}$ and $c_2^{t_2}$ *green* and we assign $l_2^{t_2}$, $l_3^{t_1}$, $l_3^{t_2}$, $m_1^{t_1}$, $m_1^{t_2}$, $c_3^{t_1}$ and $c_3^{t_2}$ *red*.

Now let us extend the below theorem to the cases $q \geq 2$ and $k \geq 2$.

Corollary 1. ⋆ *For each constant $q \geq 2$ and for each constant $k \geq 2$, PCP is NP-Complete.*

Theorem 5. ⋆ *When $k = 1$, PCP is NP-Complete for each constant $q \geq 3$.*

We can prove this theorem by reducing from q-SAT to PCP with $k = 1$, where $q \geq 3$. The detailed proof is omitted here.

3.2   Parameter p

Next, we consider the parameter p. It is easy to see that PCP is polynomially solvable when p is a constant. A simple brute-force algorithm runs in polynomial time: by enumerating all vertices in each part to search the selection S we will get at most $\prod_{i=1}^{p} |V_i|$ candidates for S in $\prod_{i=1}^{p} |V_i| \leq n^p$ time; for each candidate we can check whether it is k-colorable in $O(p^k)$ time, where $k \leq p$. When p is a constant, the algorithm runs in polynomial time.

Theorem 6. *When p is a constant, PCP is polynomially solvable.*

Since PCP is NP-hard when p is part of the input and polynomially solvable when p is a constant, it is reasonable to consider whether PCP is fixed-parameter tractable by taking parameter p. We have the following negative result.

Theorem 7. ⋆ *Taking p as the parameter, PCP is W[1]-hard even for each fixed* $k \geq 1$.

This theorem can be proved by reducing from the INDEPENDENT SET PROBLEM. The detailed proof is omitted here.

On the other hand, it is easy to see that PCP is FPT when both of p and q are taking as the parameters.

Theorem 8. *Taking p and q as the parameters, PCP is fixed-parameter tractable.*

Proof. In fact, a simple brute-force algorithm is FPT. If $k > p$, the problem is trivial. Next, we assume that $k \leq p$. We enumerate all possible selections, the number of which is at most q^p. For each candidate selection, there are at most $k^p \leq p^p$ different ways to color them. To check whether a color is feasible can be done in linear time. So the algorithm runs in $O((pq)^p(|V| + |E|))$ time.   □

4   An Exact Algorithm for PCP

In this section, we consider fast exact algorithms for PCP. It is known that VCP can be solved in $O^*(2^n)^1$ time by using subset convolution [1], while traditional dynamic programming algorithms can only lead to running time of $O^*(3^n)$. By using the $O^*(2^n)$-time algorithm for VCP, we can get a simple $O^*((\frac{2n}{p})^p)$-time algorithm for PCP: we enumerate all candidates of the selection and check whether they are k-colorable. The number of candidates of the selection is $\prod_{i=1}^{p} |V_i|$, which is at most $(\frac{n}{p})^p$ by the AM-GM inequality [2]. Each candidate is a part of p vertices and we use the $O^*(2^p)$-time algorithm to check whether it is k-colorable. So in total, the algorithm runs in $O^*((\frac{2n}{p})^p)$ time. Next, we use the subset convolution technique to improve the running time bounded to $O((\frac{n+p}{p})^p n \log k)$. Note that when $p = n$, the problem becomes VCP and the running time bound reaches the best known bound $O^*(2^n)$ for VCP.

Definition 1. *Given a p-partition* $\mathcal{V} = \{V_1, \cdots, V_p\}$, *a vertex subset* $S \subseteq V$ *is called a* semi-selection *if it holds that* $|S \cap V_i| \leq 1$ *for all* $1 \leq i \leq p$.

It is easy to see that a semi-selection S with size $|S| = p$ is a selection. We use $\mathcal{S}$ to denote the set of all semi-selections corresponding to a p-partition $\mathcal{V}$. We have that

$$|\mathcal{S}| \leq (\frac{n+p}{p})^p. \tag{1}$$

Each semi-selection S has at most one vertex in each part V_i. There are $|V_i| + 1$ possibilities for $S \cap V_i$. Such the number of semi-selections is $|\mathcal{S}| = \prod_{i=1}^{p}(|V_i|+1)$. Since $\sum_{i=1}^{p}(|V_i| + 1) = n + p$, by the AM-GM inequality, we have $|\mathcal{S}| \leq (\frac{n+p}{p})^p$.

The set $\mathcal{S}$ is a hereditary family, that is to say, for any semi-selection $S \in \mathcal{S}$, all the subsets of S are also semi-selections. We will use the following lemma to design our algorithm.

---

[1] The notation O^* is a modified big-O notation that suppresses all polynomially bounded factors.

Theorem 9. *Let $S \in \mathcal{S}$ be a semi-selection. Then $G[S]$ is k-colorable if there is a subset $T \subseteq S$ such that $G[T]$ is 1-colorable and $G[S \setminus T]$ is $(k-1)$-colorable.*

Proof. Since $\mathcal{S}$ is a hereditary family, we know that each subset of S is also a semi-selection. For a k-coloring of $G[S]$, the set of vertices with the same color is a satisfied subset T.   □

Before introducing our algorithm, we give the definition of the subset convolution first.

Definition 2. *Let $\mathcal{S}$ be a hereditary family on a set contains n elements and g, h be two integer functions on $\mathcal{S}$, i.e., $g, h : \mathcal{S} \to \mathbb{Z}$. The subset convolution of g and h, denoted by $g * h$, is a function assigning to any $S \in \mathcal{S}$ an integer*

$$(g * h)(S) = \sum_{T \subseteq S} g(T) \cdot h(S \setminus T).$$

The subset convolution can be computed in time $O(n|\mathcal{S}|)$ [1].

We show how to use the subset convolution to solve PCP. Let f be an indicator function on semi-selections $f : \mathcal{S} \to \{0, 1\}$. For any semi-selection S, $f(S) = 1$ if $G[S]$ is 1-colorable, and $f(S) = 0$ otherwise. Define $f^{*k} : \mathcal{S} \to \mathbb{Z}$ as follows:

$$f^{*k} = \underbrace{f * f * \cdots * f}_{k \text{ times}}.$$

We have the following theorem

Theorem 10. ⋆ *For any semi-selection $S \in \mathcal{S}$, the graph $G[S]$ is k-colorable if and only if $f^{*k}(S) > 0$.*

By this theorem, to check whether there is a k-colorable selection, we only need to check whether there is a semi-selection S such that $|S| = n$ and $f^{*k}(S) > 0$. The detailed steps of the algorithm are given below.

Algorithm 1. An exact algorithm for PCP.

Require: A simple undirected graph $G = (V, E)$, a p-partition $\mathcal{V}$ and an integer k.
Ensure: 'yes' or 'no' to indicate wether there exits a selection S such that $G[S]$ is k-colorable.
 1: Enumerate all semi-selections and store them in $\mathcal{S}$;
 2: Check all the semi-selections S whether they are 1-colorable, and let $f(S) = 1$ if S is 1-colorable and $f(S) = 0$ otherwise;
 3: Calculate the subset convolution f^{*k};
 4: If there is a semi-selection S such that $|S| = p$ and $f^{*k}(S) > 0$, stop and return 'yes';
 5: return 'no'.

Theorem 11. ⋆ *Algorithm 1 solves PCP in $O((\frac{n+p}{p})^p n \log k)$ time.*

We have proved that PCP with parameter p is W[1]-hard. It is unlikely to remove n from the exponential part of the running time. Furthermore, when $p = n$, the problem becomes VCP and the running time bound $O^*(2^n)$, which is the best-known result for VCP.

5   Concluding Remarks

In this paper, we have analyzed the computational complexity of the PARTITION COLORING PROBLEM. By reducing from or to the INDEPENDENT SET PROBLEM, the VERTEX COLORING PROBLEM and the k-SAT PROBLEM, we show different NP-hardness results of the PARTITION COLORING PROBLEM with respect to different constant settings of three parameters: the number of colors, the number of parts of the partition, and the maximum size of each part of the partition. We also design polynomial-time algorithms for the remaining cases. It would be interesting to look into the complexity status of the problem in subgraph classes with different settings on the three parameters and find more effective fixed-parameter tractable algorithms for PARTITION COLORING PROBLEM parameterized by both p and q.

References

1. Björklund, A., Husfeldt, T., Kaski, P., Koivisto, M.: Fourier meets möbius: fast subset convolution. In: Proceedings of the Thirty-Ninth Annual ACM symposium on Theory of computing, pp. 67–74. ACM (2007)
2. Cauchy, A.L.B.: Cours d'analyse de l'École Royale Polytechnique. Debure (1821)
3. Damaschke, P.: Parameterized mixed graph coloring. J. Comb. Optim. **38**(2), 362–374 (2019)
4. Demange, M., Ekim, T., Ries, B., Tanasescu, C.: On some applications of the selective graph coloring problem. Eur. J. Oper. Res. **240**(2), 307–314 (2015)
5. Demange, M., Monnot, J., Pop, P., Ries, B.: On the complexity of the selective graph coloring problem in some special classes of graphs. Theoret. Comput. Sci. **540**, 89–102 (2014)
6. Downey, R.G., Fellows, M.R.: Fixed-parameter tractability and completeness II: on completeness for W[1]. Theoret. Comput. Sci. **141**(1–2), 109–131 (1995)
7. Ekim, T., de Werra, D.: On split-coloring problems. J. Comb. Optim. **10**(3), 211–225 (2005). https://doi.org/10.1007/s10878-005-4103-7
8. Frota, Y., Maculan, N., Noronha, T.F., Ribeiro, C.C.: A branch-and-cut algorithm for partition coloring. Networks **55**(3), 194–204 (2010)
9. Furini, F., Malaguti, E., Santini, A.: An exact algorithm for the partition coloring problem. Comput. Oper. Res. **92**, 170–181 (2018)
10. Galinier, P., Hao, J.K.: Hybrid evolutionary algorithms for graph coloring. J. Comb. Optim. **3**(4), 379–397 (1999). https://doi.org/10.1023/A:1009823419804
11. Gamst, A.: Some lower bounds for a class of frequency assignment problems. IEEE Trans. Veh. Technol. **35**(1), 8–14 (1986)
12. Glass, C.A., Prügel-Bennett, A.: Genetic algorithm for graph coloring: exploration of Galinier and Hao's algorithm. J. Comb. Optim. **7**(3), 229–236 (2003). https://doi.org/10.1023/A:1027312403532

13. Hansen, P., Kuplinsky, J., de Werra, D.: Mixed graph colorings. Math. Methods Oper. Res. **45**(1), 145–160 (1997). https://doi.org/10.1007/BF01194253
14. Holyer, I.: The NP-completeness of edge-coloring. SIAM J. Comput. **10**(4), 718–720 (1981)
15. Hoshino, E.A., Frota, Y.A., De Souza, C.C.: A branch-and-price approach for the partition coloring problem. Oper. Res. Lett. **39**(2), 132–137 (2011)
16. Jin, Y., Hamiez, J.-P., Hao, J.-K.: Algorithms for the minimum sum coloring problem: a review. Artif. Intell. Rev. **47**(3), 367–394 (2016). https://doi.org/10.1007/s10462-016-9485-7
17. Karp, R.M.: Reducibility among combinatorial problems. In: Miller, R.E., Thatcher, J.W., Bohlinger, J.D. (eds) Complexity of Computer Computations. The IBM Research Symposia Series, pp. 85–103 (1972). Springer, Boston. https://doi.org/10.1007/978-1-4684-2001-2_9
18. Krom, M.R.: The decision problem for a class of first-order formulas in which all disjunctions are binary. Math. Logic Q. **13**(1–2), 15–20 (1967)
19. Kubicka, E., Schwenk, A.J.: An introduction to chromatic sums. In: Proceedings of the 17th Conference on ACM Annual Computer Science Conference, pp. 39–45. ACM (1989)
20. Li, G., Simha, R.: The partition coloring problem and its application to wavelength routing and assignment. In: Proceedings of the First Workshop on Optical Networks, p. 1. Citeseer (2000)
21. Lin, W., Xiao, M., Zhou, Y., Guo, Z.: Computing lower bounds for minimum sum coloring and optimum cost chromatic partition. Comput. Oper. Res. **109**, 263–272 (2019)
22. Lucarelli, G., Milis, I., Paschos, V.T.: On the max-weight edge coloring problem. J. Comb. Optim. **20**(4), 429–442 (2010). https://doi.org/10.1007/s10878-009-9223-z
23. Pop, P.C., Hu, B., Raidl, G.R.: A memetic algorithm with two distinct solution representations for the partition graph coloring problem. In: Moreno-Diaz, R., Pichler, F., Quesada-Arencibia, A. (eds.) Computer Aided Systems Theory - EUROCAST 2013. EUROCAST 2013. Lecture Notes in Computer Science, vol. 8111, pp. 219–226. Springer, Heidelberg (2013). https://doi.org/10.1007/978-3-642-53856-8_28
24. Zhou, X., Nishizeki, T.: Algorithm for the cost edge-coloring of trees. J. Comb. Optim. **8**(1), 97–108 (2004). https://doi.org/10.1023/B:JOCO.0000021940.40066.0c

FPT Algorithms for Generalized Feedback Vertex Set Problems

Bin Sheng[✉][iD]

Collaborative Innovation Center of Novel Software Technology and Industrialization,
College of Computer Science and Technology,
Nanjing University of Aeronautics and Astronautics,
Nanjing 211106, Jiangsu, People's Republic of China
shengbinhello@nuaa.edu.cn

Abstract. An r-pseudoforest is a graph in which each component can be made into a forest by deleting at most r edges, and a d-quasi-forest is a graph in which each component can be made into a forest by deleting at most d vertices.

In this paper, we study the parameterized tractability of deleting minimum number of vertices to obtain r-pseudoforest and d-quasi-forest, generalizing the well-studied feedback vertex set problem. We first provide improved FPT algorithm and kernelization results for the r-pseudoforest deletion problem, and then we show that the d-quasi-forest deletion problem is also FPT.

Keywords: FPT · Kernelization · Generalized Feedback Vertex Set.

1 Preliminary

The Feedback Vertex Set problem, which asks to delete a minimum number of vertices from a given graph to make it acyclic, is one of the 21 NP-hard problems proved by Karp [11]. It has important applications in bio-computing, artificial intelligence, and so on. The problem has attracted a lot of attention from the parameterized complexity community due to its importance. Both its undirected and directed versions have been well studied [3–5,8].

Feedback vertex set problem is fixed-parameter tractable when parameterized with the solution size, the number of vertices to delete. For the undirected feedback vertex set problem, the state of the art algorithm runs in time $O^*(3.460^k)$ in the deterministic setting [10] and $O^*(3^k)$ in the randomized setting [12], here the O^* notation hides polynomial factors in n.

Several classes of nearly acyclic graphs have been defined in the literature. A graph F is an r-pseudoforest if we can delete at most r edges from each component in F to get a forest. A pseudoforest is a 1-pseudoforest. A graph F is an almost r-forest if we can delete r edges from F to get a forest.

National Natural Science Foundation of China (No. 61802178).

© Springer Nature Switzerland AG 2020
J. Chen et al. (Eds.): TAMC 2020, LNCS 12337, pp. 402–413, 2020.
https://doi.org/10.1007/978-3-030-59267-7_34

As a generalization of feedback vertex set problem, Philip et al. [14] introduced the problem of deleting vertices to get a nearly acyclic graph. Several results have been obtained in this line of research. In [14], the authors gave a $O(c_r^k n^{O(1)})$ algorithm for r-pseudoforest deletion, which asks to delete at most k vertices to get an r-pseudoforest. The c_r here depends on r doubly exponentially. They also gave a $7.56^k n^{O(1)}$ time algorithm for the problem of pseudoforest deletion. Bodlaender et al. [1] gave an improved algorithm for pseudoforest deletion running in time $O(3^k n k^{O(1)})$.

Rai and Saurabh [15] gave a $O^*(5.0024^{(k+r)})$ algorithm for the Almost Forest Deletion problem, which asks to delete minimum vertices to get an almost r-forest. Lin et al. [13] gave an improved algorithm for this problem that runs in time $O^*(5^k 4^r)$.

A d-quasi-forest is a graph in which each connected component admits a feedback vertex set of size at most d. Hols and Kratsch [9] raised this notion and showed that the Vertex Cover problem admits a polynomial kernel when parameterized with distance to d-quasi-forest. They did not show how to obtain such a modulator to d-quasi-forest.

In this paper, we first give an algorithm for r-pseudoforest deletion that runs in time $(1 + (2r + 3)^{r+2})^{k+1} n^{O(1)}$, improving the algorithm in [14]. We also provide an improved kernelization result for r-pseudoforest deletion. We then give an FPT algorithm to obtain a minimum modulator to d-quasi-forest. To the author's knowledge, this is the first nontrivial FPT result for d-quasi-forest deletion.

2   Notations and Terminology

Here we give a brief list of the graph theory concepts used in this paper; for other notations and terminology, we refer readers to [2].

For a graph $G = (V(G), E(G))$, $V(G)$ and $E(G)$ are its vertex set and edge set respectively. A non-empty graph G is *connected* if there is a path between any pair of vertices. Otherwise, we call it *disconnected*.

The *multiplicity* of an edge is the number of its appearances in the multigraph. An edge uv is a *loop* if $u = v$. The *degree* of a vertex is the number of its appearances as end-vertex of some edge. We use $\delta(G)$ to denote the minimum degree of vertices in G. A *forest* is a graph in which there is no cycle. A *tree* is a connected forest. A graph $H = (V(H), E(H))$ is a *subgraph* of a graph $G = (V(G), E(G))$, if $V(H) \subseteq V(G)$ and $E(H) \subseteq E(G)$. A subgraph H of G is an *induced subgraph* of G if for any $u, v \in V(G)$, edge $uv \in E(H)$ if and only if $uv \in E(G)$. We denote H by $G[V(H)]$ if H is induced by the vertex set $V(H)$. We use $E(X, Y)$ to denote the set of edges between vertex sets X and Y. And $N_X(u)$ denotes the neighborhood of u in vertex set $X \subseteq V(G)$.

For a positive integer r, an *r-pseudoforest* is a graph in which each connected component can be made into a tree by deleting at most r edges. A *pseudoforest* is 1-pseudoforest. A *d-quasi-forest* is a graph in which each component admits a feedback vertex set of size at most d.

3   Branching Algorithm for r-pseudoforest Deletion

Definition 1. *Given a graph $G = (V, E)$, a subset $S \subseteq V(G)$ is an r-pseudoforest deletion set of G if $G \setminus S$ is an r-pseudoforest.*

Here is a formal definition of the parameterized r-pseudoforest deletion problem.

r-pseudoforest Deletion

> *Instance:* Graph G, integers k and r.
>
> *Parameter:* k and r.
>
> *Output:* Decide if there exists an r-pseudoforest deletion set X of G with $|X| \leq k$?

For a connected component C in a graph G, we call the quantity $|E(C)| - |V(C)| + 1$ the *excess* of C and denote it by $ex(C)$. Note that $ex(C) \geq 0$ for any connected component. Let $\mathcal{C}$ be the set of components in G. We define the excess of G, denoted by $ex(G)$, as $ex(G) = max_{C \in \mathcal{C}} ex(C)$, i.e., the maximum excess among all components in G. By definition, G is an r-pseudoforest if and only if $ex(G) \leq r$. For vertex subset $S \subseteq V(G)$, let $cc(S) = cc(G[S])$ be the number of components in $G[S]$.

We give the following observations about r-pseudoforest.

Observation 1: If G' is a subgraph of an r-pseudoforest, then G' is also an r-pseudoforest.

Observation 2: If G is an r-pseudoforest, then each component C in G has at most $|V(C)| - 1 + r$ edges.

By Observation 2, checking whether a given graph is an r-pseudoforest can be done in polynomial time.

Now we show how to solve the r-pseudoforest deletion via the approach of iterative compression. As a standard step, we introduce the following disjoint version of r-pseudoforest deletion.

Disjoint r-pseudoforest Deletion

> Input: Graph G with an r-pseudoforest deletion set S, $|S| \leq k + 1$, integers k and r.
>
> Parameter: k and r.
>
> Output: Decide if there exists an r-pseudoforest deletion set X of G with $|X| \leq k$ and $X \cap S = \emptyset$?

To solve the Disjoint r-pseudoforest Deletion, we apply the following reduction rules.

Reduction Rule 1: Let (G, S, k, r) be an instance of Disjoint r-pseudoforest Deletion, if there exists a vertex $v \in V(G) \setminus S$ such that $d_G(v) = 1$, then return $(G - v, S, k, r)$.

We append the safety proof of Reduction Rule 1 in the appendix.

Reduction Rule 2: If there exists $v \in V(G) \setminus S$ such that $G[S \cup v]$ is not an r-pseudoforest, then return $(G - v, S, k - 1, r)$.

Reduction Rule 2 is safe since every r-pseudoforest deletion set disjoint from S must contain v.

Reduction Rule 3: If there exists a vertex $u \in V(G) \setminus S$ of degree two, such that at least one neighbor of u is in $V(G) \setminus S$, then delete u and put a new edge between its two neighbors (even if they are adjacent). If both incident edges of u are to the same vertex, delete u and put a new loop on the adjacent vertex (even if it has a loop already).

The following lemma establishes the correctness of Rule 3, whose proof can be found in the appendix.

Lemma 1. *Reduction Rule 3 is safe.*

Reduction Rule 4: If $k < 0$, then return no.

Reduction Rules 1–4 can be applied in polynomial time. Given an instance, (G, S, k, r) of Disjoint r-pseudoforest Deletion, apply Rules 1–4 whenever possible.

Now we show how to solve the problem when Reduction Rules 1–4 cannot be applied.

Define measure $\phi(I) = k + cc(S) + \Sigma_{C \in \mathcal{C}(G[S])}(r - ex(C))$. Note that initially $\phi(I) \leq k + cc(S) + cc(S)r \leq 2k + (k+1)r + 1 < (k+1)(r+2)$ since $|S| \leq k+1$.

To get a depth bounded search tree, we prove that $\phi(I)$ decreases after each application of the following branching rules.

BR-1. Branching on a vertex $v \notin S$ with $d_S(v) \geq 2$.

In one branch, we put v into the solution and call the algorithm on $(G - \{v\}, S, k - 1, r)$. Note that in this branch, $cc(S)$, $ex(C)$ (for each $C \in \mathcal{C}(G[S])$) remain the same while k decreases by 1. Hence $\phi(I)$ drops by 1.

In the other branch, we put v into S and call the algorithm on $(G, S \cup \{v\}, k, r)$. Let $S' = S \cup \{v\}$. There are the following two possible cases regarding the distribution of $N_S(v)$.

Case 1: $N_S(v)$ belongs to more than one component in $G[S]$, thus $cc(S') \leq cc(S) - 1$. Let $C_1, C_2, \ldots, C_t (t \geq 2)$ be the set of components in $G[S]$ that are adjacent to v.

To compute the difference between excess sums in S and S', denote

$$\sigma_S = \Sigma_{i \in [t]}(r - ex(C_i))$$
$$= rt - \Sigma_{i \in [t]}(ex(C_i))$$
$$= rt - \Sigma_{i \in [t]}(|E(C_i)| - |V(C_i)| + 1)$$
$$= rt - \Sigma_{i \in [t]}|E(C_i)| + \Sigma_{i \in [t]}|V(C_i)| - t,$$

$$\sigma_{S'} = r - ex(G[\cup_{i \in [t]}V(C_i) \cup \{v\}])$$
$$= r - (|E(\cup_{i \in [t]}V(C_i) \cup \{v\})| - |V(\cup_{i \in [t]}V(C_i) \cup \{v\})| + 1)$$
$$= r - (\Sigma_{i \in [t]}|E(C_i)| + d_S(v) - \Sigma_{i \in [t]}|V(C_i)|)$$

Since $\sigma_S - \sigma_{S'} = r(t - 1) + d_S(v) - t \geq r(t-1) \geq r$, $\phi(I)$ drops by at least $1 + \sigma_S - \sigma_{S'} \geq 1 + r(t-1) \geq 1 + r$.

Case 2: All the neighbors of v belong to one component, denoted by C^*. Then $cc(S') = cc(S)$, and $ex(G[V(C^*) \cup \{v\}]) - ex(C^*) \geq 1$ as $d_S(v) \geq 2$. Hence $\phi(I)$ decreases by at least 1.

Therefore, in BR-1, the measure $\phi(I)$ drops by 1 in one case, and at least $1 + r$ or 1 in the other, while remaining non-negative. In the worst case, it gives us a branching vector $(1, 1)$.

After exhaustive applications of Rule 3 and BR-1, every vertex in $G - S$ has degree at least 3. Moreover, if there exists a vertex $u \notin S$ such that $d_{G-S}(u) \leq 1$, then $d_S(u) \geq 2$. And so if $d_S(u) \leq 1$ holds for each $u \in G - S$, then $d_{G-S}(u) \geq 2$, that is $\delta(G - S) \geq 2$. Thus each $u \in G - S$ must be in a cycle.

If G is not an r-pseudoforest, there must be edges between $G - S$ and S, since both $G[S]$ and $G - S$ are r-pseudoforest. In the following, we branch on vertices in $G - S$ adjacent to S.

BR-2. Branching on a vertex $v \notin S$ adjacent to S.

First, consider the case when there is a component C in $G - S$ such that there is only one edge uv between C and S, where $u \in C$ and $v \in S$. Note that if the solution should intersect $V(C)$, then it suffices to contain u. Thus we branch on whether to put u into the solution. In one branch, we put u into the solution, then k decreases and $\phi(I)$ decreases by one. In the other branch, we put C into S. Since each vertex in $G - S$ is of degree at least 3, thus C is not a tree, and so the value of $r - ex(S)$ decreases, thus $\phi(I)$ also decreases by at least one.

Now assume each component in $G - S$ has at least two edges to S. Look at one shortest path P in $G - S$, such that both endvertices of P are adjacent to some vertex in S (we allow P to be an isolated vertex). Note that such a shortest path can be found in polynomial time. We prove that $|V(P)| \leq 2r + 2$. As any component C in $G - S$ is an r-pseudoforest, $|E(C)| - |V(C)| + 1 \leq r$. Note that each vertex in P has degree at least 3 after exhaustive applications of Reduction Rule 3. And observe that no internal vertex of P has an edge to S, otherwise we find a path shorter than P, a contradiction. Let C_0 be the component in $G - S$ containing P, we know $|E(C_0)| \geq 3/2|V(P)| - 2$, and $ex(C_0) \geq ex(G[V(P)]) \geq 3/2|V(P)| - 2 - |V(P)| + 1$. As $ex(C_0) \leq r$, so $|V(P)| \leq 2r + 2$.

We branch on whether to delete any vertex on the path P. Suppose $P = v_1, v_2 \ldots, v_t$. We consider $t + 1$ branches. In branch i, where $i \in [t]$, we delete vertex v_i on path P, and call the algorithm on $(G - \{v_i\}, S, k - 1, r)$. In branch $t+1$, we don't delete any vertex on P. Note that for branch i, where $i \leq t$, $cc(S)$, $ex(C)$ (for any $C \in \mathcal{C}(G[S])$) remain the same while k decreases by at least 1. Thus $\phi(I)$ drops by at least 1.

If $G[S + V(P)]$ is not an r-pseudoforest, then we ignore branch $t + 1$. Otherwise, for branch $t + 1$, we get a new instance (G, S', k, r), where $S' = S \cup V(P)$. If edges between $V(P)$ and S are to the same component C in $G[S]$, then $ex(C \cup V(P)) = ex(C) + 1$, thus in this branch, $\phi(I)$ decreases by 1. Otherwise, the edges between $V(P)$ and S are to different components in $G[S]$. In this case, $cc(S)$ decreases by 1 and $\sigma_S - \sigma_{S'} \geq 0$, so $\phi(I)$ drops by at least 1. This gives us a $(t + 1)$-tuple branching vector $(1, 1, \ldots, 1)$ in which $t \leq 2r + 2$.

According to the branching vectors in BR-1 and BR-2, the algorithm runs in time $O^*((2r+3)^{(k+1)(r+2)})$.

The following lemma states that a fast parameterized algorithm for the disjoint version problem gives a fast algorithm for the original problem.

Lemma 2. *[7] If there is an algorithm solving Disjiont r-pseudoforest Deletion in time $f(k)n^{O(1)}$, then there is an algorithm solving r-pseudoforest Deletion in time $\sum_{i=0}^{i=k} \binom{k+1}{i} f(k-i)n^{O(1)}$.*

So we get an algorithm for r-pseudoforest deletion with a running time $\sum_{i=0}^{i=k} \binom{k+1}{i}(2r+3)^{(k-i+1)(r+2)} \leq (1+(2r+3)^{r+2})^{k+1}$, which improves over the result in [14]. Our result answers the question raised in [1] on whether there is an algorithm for r-pseudoforest deletion running in time $O^*(c_r^k)$.

Theorem 1. *There exists an algorithm for r-pseudoforest deletion with running time $(1+(2r+3)^{r+2})^{k+1}n^{O(1)}$.*

4   Kernelization of r-pseudoforest Deletion

In this section, we give an improved kernel for the r-pseudoforest deletion problem. By exhaustively applying Reduction Rules 1 and 3, we get an instance with minimum degree at least 3.

Lemma 3. *If a graph G has minimum degree at least 3, maximum degree at most d, and an r-pseudoforest deletion set of size at most k, then it has at most $(2dr-d+1)k$ vertices and at most $3kdr-kd$ edges.*

Proof. Let X be an r-pseudoforest deletion set of G of size at most k. Let $F = G - X$. It follows that each component in F can be made into a forest by deleting at most r edges. Suppose there are c components in F, we know that $c \leq kd$, since deleting any vertex of degree t produces at most t components. For each component C_i with $i \in [c]$, we know there are at most $|V(C_i)|-1+r$ edges. Thus, $|E(F)| = \Sigma_{i\in[c]}|E(C_i)| \leq \Sigma_{i\in[c]}(|V(C_i)|-1+r) = |V(F)|+c(r-1)$. By counting the number of edges incident with $V(F)$, we know that

$$3|V(F)| \leq 2(|E(F)|) + |E(X,V(F))| \leq 2(|V(F)|+c(r-1)) + kd.$$

It follows that $|V(F)| \leq 2c(r-1)+kd$. So $|V(G)| \leq |X|+|V(F)| \leq 2c(r-1)+k(d+1) \leq (2dr-d+1)k$. And $|E(G)| \leq |E(F)|+|E(X,V(F))|+|E(G[X])| \leq |V(F)|+c(r-1)+kd < 3c(r-1)+2kd = 3kdr-kd$.   □

We need to make use of the following result.

Theorem 2. *([14]) Given an instance (G,k) of r-pseudoforest Deletion, in polynomial time, we can get an equivalent instance (G',k') such that $k' \leq k$, $|V(G')| \leq |V(G)|$ and the maximum degree of G' is at most $(k+r)(3r+8)$.*

Theorem 3. *The r-pseudoforest deletion problem admits a kernel with $max\{O(k^2r^2), O(kr^3)\}$ vertices and $max\{O(k^2r^2), O(kr^3)\}$ edges.*

Proof. According to Theorem 2, we know that r-pseudoforest deletion admits a kernel with maximum degree at most $d = (k+r)(3r+8)$. Thus by Lemma 3, we can obtain a kernel for r-pseudoforest deletion which has at most $(2r-1)kd+k = \max\{O(k^2r^2), O(kr^3)\}$ vertices, and at most $(3r-1)kd = (3r-1)k(k+r)(3r+8) = \max\{O(k^2r^2), O(kr^3)\}$ edges. $\square$

The kernel in Theorem 3 improves over the $O(ck^2)$ kernel in [14], in which the constant c depends on r exponentially.

5 d-quasi-forest Deletion

Definition 2. *Given a graph $G = (V, E)$, a subset $S \subseteq V(G)$ is a d-**quasi-forest deletion set** of G if $G - S$ is a d-quasi-forest.*

Let us recall the definition of parameterized d-quasi-forest deletion first.

d-quasi-forest deletion
   Instance: An undirected graph G, integers d and k.
   Parameter: d and k.
   Output: Decide if there exists a d-quasi-forest deletion set $X \subseteq V(G)$ with $|X| \leq k$.

Lemma 4. *A yes instance of d-quasi-forest deletion has treewidth at most $k + d + 1$.*

We append the proof of Lemma 4 in the appendix.
   We first point out that the problem is FPT according to Courcelle's theorem by expressing it with Monadic Second Logic. The basic idea of the expression is as follows: $\exists v_1, v_2, \ldots, v_k \in V(G)$ such that $\forall X \subseteq V(G) - \{v_1, v_2, \ldots, v_k\}$, $Conn(X) \to FVS(X) \leq d$. The definition of $FVS(X) \leq d$ is as follows: $\exists y_1, y_2, \ldots, y_d \in X$, such that $\neg ExistsCycle(X - \{y_1, y_2, \ldots, y_d\})$. The definition of $ExistsCycle(X)$ is as follows: $\exists E \subseteq E(G)$, such that $Conn(E)$ and $\forall e \in E$, $\exists u, v \in X$, such that $Inc(u, e)$ and $Inc(v, e)$ and $Deg(u, E) = Deg(v, E) = 2$.

Theorem 4. *(Courcelle [6]) Given a graph G and a formula φ in Monadic Second Logic describing a property of interest, and parameterizing by the combination of $tw(G)$ and the size of the formula φ, it can be determined in time $f(tw(G), |\varphi|)n^{O(1)}$ whether G has the property of interest.*

Theorem 5. *The d-quasi-forest deletion problem is FPT parameterized with k and d.*

Unfortunately, the algorithm implied by Courcelle's Theorem may be several layers exponential. Aiming at fully exploiting the problem structure and design a faster algorithm, we solve the d-quasi-forest deletion problem by the iterative compression approach.

d-quasi-forest deletion compression

Instance: An undirected graph G, an integer k and a d-quasi-forest deletion set $Z \subseteq V(G)$ with $|Z| \leq k + 1$.

Parameter: d and k.

Output: Decide if there exists a d-quasi-forest deletion set $X \in V(G)$ with $|X| \leq k$.

By guessing the intersection of Z and X, we reduce the d-quasi-forest deletion compression problem into the following disjoint version.

Disjoint d-quasi-forest deletion

Instance: An undirected graph G, an integer k and a d-quasi-forest deletion set $Z \subseteq V(G)$ with $|Z| \leq k + 1$.

Parameter: d and k.

Output: Decide if there exists a d-quasi-forest deletion set $X \in V(G)$ with $|X| \leq k, X \cap Z = \emptyset$.

Denote $F = G - Z$, then F is a d-quasi-forest, that is, each connected component in F admits feedback vertex set of size at most d. Note that according to the algorithm in [10], we can check whether a given graph is a d-quasi-forest in time $O^*(3.460^d)$.

Reduction Rule 1: If there is a vertex u of degree at most one, then delete u and return a new instance $(G - u, k)$.

Reduction Rule 2: If there is a vertex u of degree exactly two in G, then delete u and add an edge between the neighbors of u.

After exhaustive applications of Reduction Rules 1–2, the resulting instance has minimum degree at least 3.

Reduction Rule 3: Observe that $G[Z]$ is a d-quasi-forest, otherwise, it is a no-instance. If there is any vertex $u \in V(G) \setminus Z$ such that $G[u \cup V(Z)]$ is not a d-quasi-forest, then delete u and decrease k by one.

Lemma 5. *If (G, Z, k) is a yes instance of Disjoint d-quasi-forest deletion, then for each vertex $u \in Z$, $G - Z$ contains at most $k + d$ components, that contain at least one cycle, adjacent with u.*

Proof. If there are more than $k + d$ components in $G - Z$, which are adjacent to u, containing a cycle, then for any vertex set $X \subseteq V(G - Z)$, such that $|X| \leq k$, the component containing u in $G - X$ is not a d-quasi-forest. Thus (G, Z, k) is a no instance. It follows that for each vertex $u \in Z$, the number of components (which is not a tree) in $G - Z$ that are adjacent to u is at most $k + d$. □

Note that each component in $G - Z$ admits a feedback vertex set of size at most d. By branching on a minimum feedback vertex set of each component in $G - Z$, i.e., either put the vertex into the solution or put it into Z, we obtain new instances in which every component in $G - Z$ is a tree. Moreover, after all the branchings, the size of Z is upper bounded by $k + 1 + (k + 1)(k + d)d = (k + 1)(kd + d^2 + 1)$.

From now on, we assume each component in $G - Z$ is a tree. The following lemma bounds the number of trees in $G - Z$ that has a large neighborhood in Z.

Lemma 6. *If (G, Z, k) is a yes instance of Disjoint d-quasi-forest deletion, then there are at most $2|Z| + d + k + 1$ trees in $G - Z$ that has at least $d + 2$ neighbors in Z.*

Proof. Consider the measure $\mu(I) = cc(Z) + d - fvs'(Z) + \omega'(Z)$, in which $cc(Z)$ is the number of components in $G[Z]$. $\omega'(Z)$ is the maximum number of components in $G[Z - D]$ where D is a minimum feedback vertex set of $G[Z]$. And $fvs'(Z) = \max_{C \in G[Z]} fvs(C)$, where C is a component in $G[Z]$. Note that $\mu(I) \geq 0$ if $G[Z]$ is a d-quasi-forest.

By putting a tree T that has at least $d + 2$ neighbors in Z into Z, $\mu(I)$ decreases, because either $cc(Z)$ decreases, or $fvs'(Z)$ increases, or $\omega'(Z)$ decreases. Indeed, if $cc(Z)$ and $fvs'(Z)$ do not change, then any feedback vertex set D of $G[Z]$ contains no vertex in C. Since D contains at most d vertices, C connects at least two components in $G[Z - D]$, and so $\omega'(Z)$ decreases.

Since originally $cc(Z) + d - fvs(Z) + \omega'(Z) \leq 2|Z| + d$, $G - Z$ of any yes instance contains at most $2|Z| + d + k$ trees that have at least $d + 2$ neighbors in Z. Otherwise, for any vertex subset $X \subseteq V(G) \setminus Z$ with $|X| \leq k$, there are more than $2|Z| + d + 1$ trees in $G - Z - X$ that each has at least $d + 2$ neighbors in Z. If we put all such trees into Z, then $\mu(I)$ becomes negative and so $G - X$ is not a d-quasi-forest. $\qquad\square$

The next lemma shows that we can do some preprocessing to partition these trees(with a large neighborhood in Z) into smaller subtrees.

Lemma 7. *Let (G, Z, k) be a yes instance of Disjoint d-quasi-forest deletion. For each tree T in $G - Z$ that has at least $d + 2$ neighbors in Z, there is a partition of T into less than $2(2|Z| + d + k + 1)$ subtrees, each has at most $d + 1$ neighbors in Z.*

We append the proof of Lemma 7 in the appendix.

Let (G, Z_0, k) be a yes instance of Disjoint d-quasi-forest deletion. For each tree in $G - Z_0$, by branching on the boundaries of its at most $2(2|Z_0| + d + k + 1)$ subtrees(either put the vertex into the solution or put it into Z), we reduce the instance into at most $2^{2|Z_0| + d + k + 1}$ new instances, in which each tree has at most $d + 1$ neighbors in Z. When we have done this for all trees in $G - Z_0$, Z still has bounded size, as for each such tree, we put less than $2(2|Z_0| + d + k + 1)$ vertices into Z. Note that we have $|Z_0| \leq (k + 1)(kd + d^2 + 1)$.

Now we obtain instances in which $G - Z$ consists of only trees each has at most $d + 1$ neighbors in Z, where $|Z| < |Z_0| + 4(2|Z_0| + d + k + 1)^2 = \max\{O(k^4 d^2), O(k^2 d^4)\}$. To design an FPT algorithm, we further reduce the number of trees in $G - Z$.

Definition 3. *Two trees T_1, T_2 in $G - Z$ have same **neighborhood type** in Z if $N_Z(T_1) = N_Z(T_2)$ and for any vertex $u \in N_Z(T_1)$, $|E(u, T_1)| = 1$ if and only if $|E(u, T_2)|$.*

Reduction Rule 4: For each neighborhood type σ, reduce the number of trees in $G - Z$ that have neighborhood type σ in Z to $k + d + 2$.

Lemma 8. *Reduction Rule 4 is safe.*

We append the proof of Lemma 8 in the appendix.

It takes polynomial time to decide the neighborhood type of each tree in $G-Z$, thus Reduction Rule 4 can be applied in polynomial time. By Definition 3, for each $M \subseteq Z$, there are $2^{|M|}$ different neighborhood types with neighborhood M, depending on whether the number of edges between each vertex in M and the trees is exactly one. Since every tree in $G-Z$ has at most $d+1$ neighbors in Z, there are at most $\sum_{1 \le i \le d+1} \binom{|Z|}{i} 2^i$ different neighborhood types. After exhaustive applications of Reduction Rule 4, there are at most $k+d+2$ trees of each neighborhood type.

Now we just need to solve instances in which $|Z|$ is upper bounded and $G-Z$ contains at most k trees, each has at most $d+1$ neighbors in Z.

Since the number of trees in $G-Z$ is upper bounded, we may guess which(at most k) trees intersects the solution. For each guess, check whether putting all the other trees into Z violate the requirement of d-quasi-forest. This checking can be done in time $O^*(3.460^k)$.

Theorem 6. *(Gallai) Given a simple graph G, a set $R \subseteq V(G)$, and an integer s, one can in polynomial time either*

1. *find a family of $s+1$ pairwise vertex-disjoint R-paths, or*
2. *conclude that no such family exists and, moreover, find a set B of at most $2s$ vertices, such that in $G \backslash B$ no connected component contains more than one vertex of R.*

Definition 4. *A vertex $u \in Z$ is **forced**, if u is in every feedback vertex set of size at most d of the component containing u in $G-X$, for every solution X.*

We introduce the notion of force vertex to help bound the size of each tree in $G-Z$, with the observation that there is no need to keep too many neighbors of forced vertices in trees in $G-Z$.

Let G' be constructed from G by deleting u and add $d(u)$ leaves, each adjacent to one neighbor of u. Set $s=k+d$, and let R be the set of added leaves. If G' contains $k+d+1$ pairwise vertex-disjoint R-paths, which can be checked via Gallai's theorem, then u is a forced vertex.

The following lemma shows that we can further bound the size of each tree in $G-Z$.

Lemma 9. *The problem reduces to a bounded number of instances in which $G-Z$ consists of at most k trees and each has at most $O(k^2 d^2)$ neighbors in Z. Moreover, each tree in $G-Z$ contains at most $k\binom{p}{2} + k|N_2| + (k+1)(k\binom{p}{2} + k|N_2|) = O(k^{10}d^8)$ vertices, where $p = max\{O(k^4 d^2), O(k^2 d^4)\}$ and $|N_2| \le k(d+1)$.*

Proof. Since $G-Z$ contains at most k trees, we know that $N_Z(G-Z)$ contains at most $k(d+1)$ vertices. For each vertex $u \in N$, we decide whether it is a forced vertex via Gallai's Theorem. Thus we obtain a partition of $N_Z(G-Z) = N_1 \cup N_2$, where N_1 contains all the forced vertices and $N_2 = N \setminus N_1$. Via Gallai's Theorem, for each vertex $u \in N_2$, we can find a set B_u with at most $2(k + d)$ vertices, such that each component in $G - Z - B_u$ contains at most one neighbor of u. Branching on whether to put each vertex in B_u into the solution, we partition each tree T in $G - Z$ into subtrees, which contains at most one neighbor of each vertex in $B_u \cup N_2$.

We apply reduction rule 4 again to reduce the number of trees. In the reduced instance, $G - Z$ contains at most k trees, each containing at most one neighbor of each vertex in $N_2 \cup (\cup_{u \in N_2} B_u)$. And then we further reduce the instance by guessing how the trees are going to intersecting with the solution.

Now we show how to bound the size of each tree in $G - Z$.

We guess how the components are going to be separated in $G - X$, where X is a minimum d-quasi-forest deletion set of (G, Z, k). There is a bounded number of such guesses. And then, we check whether the guessing can be realized. Regard all vertices in the same component from N_1 as one vertex. For two vertices $u, v \in N_1$ that are guessed to be in two different components, there are at most k vertices in $G - Z$ that are neighbors of both u and v. Otherwise, deleting X cannot separate u and v. Thus the total number of vertices that have neighbors in two guessed components is upper bounded. Suppose we guessed there to be p components, then at most $\binom{p}{2}$ pairs of vertices in the components. It follows that there are at most $k\binom{p}{2}$ vertices in the trees that have neighbors in different guessed components.

And we also know that in each tree, the number of vertices that are adjacent to $N_2 \cup (\cup_{u \in N_2} B_u)$ is upper bounded, since each tree contains at most one neighbor of each vertex in $N_2 \cup (\cup_{u \in N_2} B_u)$.

Thus we only need to bound the number of vertices that are not adjacent to $N_2 \cup (\cup_{u \in N_2} B_u)$ and only have neighbors in just one component of $G[Z]$. We bound this by arguing that there is no need to keep too many such vertices in the same tree, since such vertices only connecting vertices that are guessed to be in the same components, moreover, their neighbors in Z are forced to be in the feedback vertex sets, thus the number of such vertices does not affect the set of feasible solutions. In each tree, for each path connecting neighbors of $N_2 \cup (\cup_{u \in N_2} B_u)$ and vertices with neighbors in more than one component, we just keep at most one vertex that is adjacent to one component in Z (note that there are at most $k + 1$ components).

In the reduced instance, each tree in $G - Z$ contains at most $k\binom{p}{2} + k|N_2| + (k+1)(k\binom{p}{2} + k|N_2|)$ vertices. Moreover, there are at most k trees in $G - Z$. And so we can solve the reduced instance in FPT time by branching on the vertices in $G - Z$, which has a bounded number of vertices. □

Theorem 7. *The d-quasi-forest deletion problem can be solved in time $O^*(c_1^{(k^2 d^2)^{(c_2 k^2 d^2)}})$, where c_1 and c_2 are some constants.*

We append the proof of Theorem 7 in the appendix.

6   Conclusion

In this paper, we provide FPT results for two generalized versions of feedback vertex set problem. It would be interesting to know whether the problem of d-quasi-forest deletion admits a polynomial kernel.

References

1. Bodlaender, H.L., Ono, H., Otachi, Y.: A faster parameterized algorithm for pseudoforest deletion. Discrete Appl. Math. **236**, 42–56 (2018)
2. Bollobás, B.: Modern Graph Theory. GTM, vol. 184. Springer, New York (1998). https://doi.org/10.1007/978-1-4612-0619-4
3. Cao, Y., Chen, J., Liu, Y.: On feedback vertex set: new measure and new structures. Algorithmica **73**(1), 63–86 (2015). https://doi.org/10.1007/s00453-014-9904-6
4. Chen, J., Fomin, F.V., Liu, Y., Songjian, L., Villanger, Y.: Improved algorithms for feedback vertex set problems. J. Comput. Syst. Sci. **74**(7), 1188–1198 (2008)
5. Chitnis, R.H., Cygan, M., Hajiaghayi, M.T., Marx, D.: Directed subset feedback vertex set is fixed-parameter tractable. ACM Trans. Algorithms **11**(4), 28:1–28:28 (2015)
6. Courcelle, B.: The monadic second-order logic of graphs. I. Recognizable sets of finite graphs. Inf. Comput. **85**(1), 12–75 (1990)
7. Cygan, M., et al.: Parameterized Algorithms. Springer, Cham (2015). https://doi.org/10.1007/978-3-319-21275-3
8. Cygan, M., Pilipczuk, M., Pilipczuk, M., Wojtaszczyk, J.O.: Subset feedback vertex set is fixed-parameter tractable. SIAM J. Discrete Math. **27**(1), 290–309 (2013)
9. Hols, E.-M.C., Kratsch, S.: Smaller parameters for vertex cover kernelization. In: 12th International Symposium on Parameterized and Exact Computation, IPEC 2017, 6–8 September 2017, Vienna, Austria, pp. 20:1–20:12 (2017)
10. Iwata, Y., Kobayashi, Y.: Improved analysis of highest-degree branching for feedback vertex set. In: 14th International Symposium on Parameterized and Exact Computation, IPEC 2019, 11–13 September 2019, Munich, Germany, pp. 22:1–22:11 (2019)
11. Karp, R.M.: Reducibility among combinatorial problems. In: Miller, R.E., Thatcher, J.W., Bohlinger J.D. (eds.) Complexity of Computer Computations. The IBM Research Symposia Series, pp. 85–103. Springer, Boston. https://doi.org/10.1007/978-1-4684-2001-2_9
12. Li, J., Nederlof, J.: Detecting feedback vertex sets of size k in $O^*(2.7^k)$ time. CoRR, abs/1906.12298 (2019)
13. Lin, M., Feng, Q., Wang, J., Chen, J., Bin, F., Li, W.: An improved FPT algorithm for almost forest deletion problem. Inf. Process. Lett. **136**, 30–36 (2018)
14. Philip, G., Rai, A., Saurabh, S.: Generalized pseudoforest deletion: algorithms and uniform kernel. SIAM J. Discrete Math. **32**(2), 882–901 (2018)
15. Rai, A., Saurabh, S.: Bivariate complexity analysis of almost forest deletion. Theor. Comput. Sci. **708**, 18–33 (2018)

Fixed-Order Book Thickness with Respect to the Vertex-Cover Number: New Observations and Further Analysis

Yunlong Liu, Jie Chen, and Jingui Huang[(⊠)]

School of Information Science and Engineering, Hunan Normal University,
Changsha 410081, People's Republic of China
{ylliu,jie,hjg}@hunnu.edu.cn

Abstract. The FIXED-ORDER BOOK THICKNESS problem asks, given a graph $G = (V, E)$ and a linear order $\prec$ of V, whether there is a page assignment σ such that $\langle \prec, \sigma \rangle$ is a k-page book embedding of G. Recently, Bhore et al. (GD 2019) provided a parameterized algorithm with respect to the vertex-cover number of G (denoted by τ). In this paper, we first re-analyze Bhore et al.'s algorithm. By introducing a novel analysis approach, we prove a bound of $2^{O(\tau^2 log \tau)} \cdot |V|$ improving on Bhore et al.'s bound of $2^{O(\tau^3)} \cdot |V|$ for its running time. By employing this analysis approach, we also show that the general FIXED-ORDER BOOK THICKNESS problem, in which at most b crossings over all pages are allowed, admits an algorithm running in time $2^{O((\tau^2 + b\tau) log \tau (b+1))} \cdot |V|$.

1 Introduction

The book thickness for graphs, as an important geometric invariant, has been studied extensively [1–3]. It is directly related with the notion of k-page book embedding of graphs. For an integer $k \geq 1$, a *k-page book embedding* of a graph G is to place the vertices linearly on a spine (a line segment) and the edges on k pages (k half planes sharing the spine) so that each edge is embedded in one of the pages without generating edge-crossings [3]. The minimum k such that G admits a k-page book embedding is the *book thickness* of G, denoted by bt(G).

Given a graph $G = (V, E)$ and a positive integer k, the problem BOOK THICK-NESS is to determine whether bt$(G) \leq k$. When the linear order $\prec$ of V (namely, the ordering of vertices in V along the spine) is predetermined and fixed, BOOK THICKNESS is specially called FIXED-ORDER BOOK THICKNESS. Correspondingly, the book thickness of G is called *fixed-order book thickness* and denoted by fo-bt$(G, \prec)$ [4]. The problem FIXED-ORDER BOOK THICKNESS is equivalent to determining whether a given circle graph can be properly vertex-colored by at most k colors [5]. It has been studied in many contexts, such as complexity theory [6,7], graph coloring [8–10], and sorting with parallel stacks [11].

This research was supported in part by the National Natural Science Foundation of China under Grant No. 61572190.

© Springer Nature Switzerland AG 2020
J. Chen et al. (Eds.): TAMC 2020, LNCS 12337, pp. 414–425, 2020.
https://doi.org/10.1007/978-3-030-59267-7_35

The problem FIXED-ORDER BOOK THICKNESS is NP-complete in general [6,7]. Very recently, Bhore et al. [4] provided some parameterized algorithms for this problem. In particular, they considered the vertex-cover number of graph, denoted by τ, as the parameter, and presented an algorithm running in time $2^{O(\tau^3)} \cdot |V|$.

One general version of FIXED-ORDER BOOK THICKNESS, concerns the setting where we allow edges on the same page to cross, with a given budget of at most b crossings over all pages. This problem was posed in [4] and one related problem has been studied by Bannister and Eppstein [12] with the number of pages k restricted to be either 1 or 2.

In this paper, we further study parameterized algorithms for FIXED-ORDER BOOK THICKNESS (abbreviated by FOBT) with respect to the vertex-cover number. We re-analyze the algorithm given by Bhore et al. in [4]. Based on some new observations on this problem, we introduce a noval approach to bound the size of *the record set* in that algorithm. Our result is that the size of the record set can be reduced from $2^{O(\tau^3)}$ to $2^{O(\tau^2 log\tau)}$, which indicates that the problem FOBT admits an algorithm running in time $2^{O(\tau^2 log\tau)} \cdot |V|$.

We also use our approach to investigate the general FIXED-ORDER BOOK THICKNESS (abbreviated by FOBT-CROSS). We first develop an algorithm for FOBT-CROSS by extending the techniques used in solving the problem FOBT, whose feasibility was mentioned by Bhore et al. [4]. Then, we focus on employing our approach to analyze its running time and show that this algorithm can be done in time $2^{O((\tau^2 + b\tau)log\tau(b+1))} \cdot |V|$. Because of the space limit, several proofs are omitted and will be given in the complete version of the paper.

2   Preliminaries

We consider only undirected graphs. For a graph $G = (V, E)$, let $n = |V|$ and $V(G)$ be the vertex set of G. For two vertices u and v, let uv denote the edge between u and v. For $r \in \mathbb{N}$, we use $[1, r]$ to denote the set $\{1, \dots, r\}$.

A *vertex cover* C of a graph $G = (V, E)$ is a subset $C \subseteq V$ such that each edge in E has at least one end-vertex in C. A vertex $v \in V$ is a *cover vertex* if $v \in C$. The *vertex cover number* of G, denoted by $\tau(G)$, is the size of a minimum vertex cover of G. In the rest of this paper, we will use C to denote a minimum vertex cover of size τ, and let $U = V \setminus C$.

Given a graph $G = (V, E)$ and a minimum vertex cover C, we use E_C to denote the set of all edges whose both endpoints lie in C. For each $i \in [1, n - \tau]$, we also use $E_i = \{u_j c \in E \mid j < i, c \in C\}$ to denote the set of all edges with one endpoint outside of C that lies to the left of u_i.

Given an n-vertex graph $G = (V, E)$ with a linear order $\prec$ of V such that $v_1 \prec v_2 \prec \dots \prec v_n$, we assume that $V(G) = \{v_1, v_2, \dots, v_n\}$ is indexed such that $i < j$ if and only if $v_i \prec v_j$. We use $X = \{x \in [1, n - \tau] \mid \exists c \in C : u_x$ is the immediate successor of c in $\prec\}$ to denote the set of indices of vertices in U which occur immediately after a cover vertex; and assume that the integers in X, denoted as $x_1, x_2, \dots, x_z$, are listed in ascending order.

For ease of presentation, we define some special planer graphs. For an integer $k \geq 0$, a graph G is a *k-restricted plane graph with spine L* if G satisfies the following properties: (1) all vertices lie in a horizontal line L with a fixed-order; (2) all edges lie in the half-plane above L; and (3) G contains at most k crossings. A k-restricted plane graph G with spine L is a *k-crossings plane graph with spine L* if each edge of G takes part in generating edge-crossings. Obviously, a k-restricted plane graph G with spine L can be decomposed into a maximal 0-restricted plane graph G_1 with spine L and a k-crossings plane graph G_2 with spine L.

3   Improved Bounds on the Running Time of Algorithm for FOBT

In this section, we re-analyze the parameterized algorithm for FOBT given by Bhore et al. in [4], and prove an improved upper bounds on its running time.

We first restate some notations introduced in [4]. For a given graph $G = (V, E)$ and a minimum vertex cover C in G, let S be the family of all possible non-crossing page assignments of edges in E_C and let $s \in S$. A page assignment $\alpha : E_i \rightarrow [1, k]$ is called a *valid partial page assignment* if $\alpha \cup s$ maps edges to pages in a non-crossing fashion. In a valid partial page assignment $\alpha : E_i \rightarrow [1, k]$, a vertex $c \in C$ is (α, s)-*visible* to u_t (for $t \in [1, n - \tau]$) on page p if it is possible to draw an edge from u_t to c on page p without crossing any other edge mapped to page p by $\alpha \cup s$. For an index $a \in [1, n - \tau]$, a *visibility matrix* $M_i(a, \alpha, s)$ is a $k \times \tau$ matrix, where the entry (p, r) of $M_i(a, \alpha, s)$ is 1 if c_r is (α, s)-visible to u_a on page p and 0 otherwise. It is assumed that $k < \tau$ since the problem FOBT with respect to τ is trivial if $\tau \leq k$.

For a vertex $u_i \in U$, the record set was defined as follows: $\mathcal{R}_i(s) = \{(M_i(i, \alpha, s), M_i(x_1, \alpha, s), M_i(x_2, \alpha, s), \ldots, M_i(x_z, \alpha, s)) \mid \exists$ valid partial page assignment $\alpha : E_i \rightarrow [1, k]\}$. A mapping Λ_i^s from $\mathcal{R}_i(s)$ to a valid partial page assignments of E_i maps $(M_0, \ldots, M_z) \in \mathcal{R}_i(s)$ to some α such that $(M_0, \ldots, M_z) = (M_i(i, \alpha, s), M_i(x_1, \alpha, s), M_i(x_2, \alpha, s), \ldots, M_i(x_z, \alpha, s))$.

The basic idea in the algorithm given by Bhore et al. in [4] is to dynamically process the vertices in U in a left-to-right fashion. For each vertex $u_i \in U$, this algorithm computes a record set $\mathcal{R}_i(s)$ containing at most $2^{\tau^3 + \tau^2}$ records. Correspondingly, all valid partial page assignments of $E_i \cup E_C$ are divided into at most $2^{\tau^3 + \tau^2}$ groups such that all assignments in the same group are "interchangeable" and lead to the same visibility matrices stored in one record in $\mathcal{R}_i(s)$.

The function $2^{\tau^3 + \tau^2}$ was directly deduced from the $\tau^3 + \tau^2$ binary bits [4]. Based on some new observations, we introduce a novel approach to estimate the size of $\mathcal{R}_i(s)$ such that $|\mathcal{R}_i(s)|$ can be upper-bounded by another function.

We first sketch the main idea of our approach. Let $\alpha \cup s$ be a valid page assignment of the edges in $E_i \cup E_C$. Observe that, on each page, the *invisibility* from a vertex $c \in C$ to another vertex $u \in U$ is essentially determined by only one of edges that enclose c (or u). Correspondingly, the visibility matrices in each record in $\mathcal{R}_i(s)$ can be essentially determined by only a part of edges in

$E_i \cup E_C$. Let $V' = C \cup \{u_1, u_{x_1}, u_{x_2}, \ldots, u_{x_z}\}$. Furthermore, we can obtain a simplified assignment $\alpha' \cup s$ by shifting some edge on each page such that the corresponding visibility matrices can be determined by edges whose endpoints are all in the set V'. Note that $|V'| \leq 2\tau + 1$. This means that there are at most $(2\tau + 1)\tau$ edges on each page in $\alpha' \cup s$. We also observe that any page of a k-page book embedding is a planar subgraph, which indicates that the total number of edges in each page is at most $3 \times (2\tau + 1) - 6$ edges. Based on these observations, we can obtain a new function $f(\tau)$ bounding the number of all possible combinations of edges on each page, which results in a bound on the number of records in $\mathcal{R}_i(s)$.

In the following, we formally define the operation on edge-shifting. Let $\alpha \cup s$ be a valid partial page assignment of edges in $E_i \cup E_C$, let u and y be two vertices in U, and let $E(p, y)$ for $p \in [1, k]$ be the set of edges on page p that incident to the vertex y. Suppose that $E(p, y) = \{yc_y^1, yc_y^2, \ldots, yc_y^t\}$, where $1 \leq t \leq \tau$ and $c_y^1 \prec c_y^2 \prec \ldots \prec c_y^t$. The operation $\mathbf{SHIFT}(u, y)$ on shifting edges in $E(p, y)$ from y to u is defined as follows.

For $q = 1$ **to** t **do**: (1) delete the edge yc_y^q; (2) if there exists no edge between u and c_y^q, then add one edge uc_y^q between u and c_y^q.

Let $V'' = \{u_1, u_{x_1}, u_{x_2}, \ldots, u_{x_z}\}$. Assume that the vertices in C are ordered as $c_1 \prec c_2 \prec \ldots \prec c_\tau$. On page p, we shift some edge in E_i such that only some vertex in V'' are adjacent to the vertices in C. Specifically, we distinguish three cases based on the position of u_i in $\prec$.

Case 1: $u_i \prec c_1$. Assume that there are h ($h \geq 2$) vertices lying on the left of u_i, denoted as $u_0^1 \prec u_0^2 \prec \ldots \prec u_0^h \prec u_i$. For each $r \in [2, h]$, if $E(p, u_0^r) \neq \emptyset$ then execute the operation $\mathbf{SHIFT}(u_0^1, u_0^r)$.

Case 2: $c_j \prec u_i \prec c_{j+1}$ (for $j \in [1, \tau - 1]$). Assume that there are h ($h \geq 2$) vertices between c_j and c_{j+1}, denoted as $c_j \prec u_d^1 \prec u_d^2 \prec \ldots \prec u_d^h \prec c_{j+1}$. For each $r \in [2, h]$, if $E(p, u_d^r) \neq \emptyset$ then execute the operation $\mathbf{SHIFT}(u_d^1, u_d^r)$. Additionally, if $j \geq 2$, then execute the procedure as follows. For each $d \in [1, j - 1]$, assume that there are h_d ($h_d \geq 2$) vertices between c_d and c_{d+1}, denoted as $c_d \prec u_d^1 \prec u_d^2 \prec \ldots \prec u_d^{h_d} \prec c_{d+1}$. For each $r \in [2, h_d]$, if $E(p, u_d^r) \neq \emptyset$ then execute the operation $\mathbf{SHIFT}(u_d^1, u_d^r)$.

Case 3: $c_\tau \prec u_i$. Assume that there are h ($h \geq 2$) vertices between c_τ and u_i, denoted as $c_\tau \prec u_\tau^1 \prec u_\tau^2 \prec \ldots \prec u_\tau^h \prec u_i$. For each $r \in [2, h]$, if $E(p, u_\tau^r) \neq \emptyset$ then execute the operation $\mathbf{SHIFT}(u_\tau^1, u_\tau^r)$.

For the assignment $\alpha \cup s$ of edges in $E_i \cup E_C$, we deal with each page by employing the previous shifting process. After shifting all possible edge on each page, we obtain a simplified assignment $\alpha' \cup s$ such that all edges on each page of $\alpha' \cup s$ are incident to at most $2\tau + 1$ vertices. Figure 1 shows one example of edge-shifting.

Next, we show that the visibility matrices for $\alpha' \cup s$ are equals to those for $\alpha \cup s$, respectively.

Lemma 1. *For each* $x \in \{i\} \cup X$, $M_i(x, \alpha, s) = M_i(x, \alpha', s)$.

Proof. Let $c_j \in C$ for $j \in [1, \tau]$ and $u_x \in U$ for $i \leq x$. By the definition of the visibility matrix, it is sufficient to draw a comparison between $\alpha \cup s$ and $\alpha' \cup s$ on

Fig. 1. The illustration on edge-shifting from an original 2-page assignment of $E_5 \cup E_C$ (a) to a simplified 2-page assignment (b).

the visibility from c_j to u_x. Moreover, since each edge in E_C remains unchanged during edge-shifting, we do not consider the edge in E_C separating c_j from u_x.

($\Rightarrow$) Assume that (p, j) is 0 in $M_i(x, \alpha, s)$ ($p \in [1, k]$). Then there must be an edge $c_w u_z$ in E_i separating c_j from u_x on page p in $\alpha \cup s$. In the following, we show that the vertex c_j is still separated from u_x by an edge on page p in $\alpha' \cup s$, indicating that (p, j) is 0 in $M_i(x, \alpha', s)$. We distinguish two cases based on the order of c_w, u_z, c_j, u_x in $\prec$. Case (1): c_w lies between c_j and u_x. Since $c_w u_z \in E_i$ and $i \leq x$, only two subcases are valid, that is, $u_z \prec c_j \prec c_w \prec u_x$ or $u_z \prec u_x \prec c_w \prec c_j$. Evidently, the vertex u_z always lies on the left of u_x. After edge-shifting, the edge $u_z c_w$ either remains unchanged or is replaced by another edge $u_r c_w$ with $u_r \prec u_z$ and $r \in X \cup \{1\}$. Hence, the vertices c_j is still separated from u_x by at least one edge. Case (2): u_z lies between c_j and u_x. Similarly, only two subcases are valid, that is, $c_w \prec c_j \prec u_z \prec u_x$ or $c_j \prec u_z \prec u_x \prec c_w$. Correspondingly, the vertex c_w lies either on the left of c_j or on the right of u_x. After edge-shifting, the edge $u_z c_w$ either remains unchanged or is replaced by another edge $u_r c_w$ with $r \in X \cup \{1\}$ and $c_j \prec u_r$. Hence, the vertex c_j is still separated from u_x by at least one edge.

($\Leftarrow$) Assume that the entry (p, j) is 0 in $M_i(x, \alpha', s)$. Then there must be an edge $c_w u_r$ (for some $r \in X \cup \{1\}$) separating c_j from u_x on page p in $\alpha' \cup s$. We argue that c_j must be separated from u_x by an edge in E_i on page p in $\alpha \cup s$, indicating that (p, j) is 0 in $M_i(x, \alpha, s)$. We distinguish two cases based on the order of c_w, c_j, u_r, u_x in $\prec$. Case (1): c_w lies between c_j and u_x. By the assumption that $i \leq x$ and the fact that $c_w u_r$ is either in E_i or shifted from an edge in E_i, we only consider two valid subcases, that is, $u_r \prec c_j \prec c_w \prec u_x$ or $u_r \prec u_x \prec c_w \prec c_j$. If $c_w u_r \in E_i$, we are done. If $c_w u_r$ is shifted from an edge $c_w u_z$ in E_i, it follows that $u_r \prec u_z \prec c_j$ or $u_r \prec u_z \prec u_x$, respectively. Hence, the vertex c_j is still separated from u_x by $c_w u_z$ in E_i on page p in $\alpha \cup s$. Case (2): u_r lies between c_j and u_x. Similarly, only two subcases need to be considered, that is, $c_w \prec c_j \prec u_r \prec u_x$ or $c_j \prec u_r \prec u_x \prec c_w$. If $c_w u_r \in E_i$, we are done. If $c_w u_r$ is shifted from an edge $c_w u_z$ in E_i, it follows that $c_j \prec u_r \prec u_z \prec u_x$ or $c_j \prec u_r \prec u_z \prec u_x$, respectively. Hence, the vertex c_j is still separated from u_x by $c_w u_z$ in E_i. $\qquad \square$

Based on Lemma 1, we can re-estimate the size of the record set $\mathcal{R}_i(s)$.

Lemma 2. *The size of $\mathcal{R}_i(s)$ can be bounded by $2^{O(\tau^2 \log \tau)}$.*

Proof. Let L be a straight line joining $2\tau + 1$ vertices with a fixed order and let $\mathcal{Q} = \{P \mid P$ is a 0-restricted plane graph with spine $L\}$. We first estimate the size of $\mathcal{Q}$ according to the number of different combinations of its edges. As we know, for a graph with $2\tau + 1$ vertices, there are at most $(2\tau + 1) \times \tau$ edges. Additionally, there are at most $3 \times (2\tau + 1) - 6 = 6\tau - 3$ edges in P since it is a planar graph with $2\tau + 1$ vertices. Hence, $|\mathcal{Q}| = \Sigma_{i=1}^{6\tau-3} \binom{(2\tau+1)\times\tau}{i} \leq (6\tau - 3)(2\tau^2 + \tau)^{6\tau-3}$.

Let $\mathcal{D} = \mathcal{Q}_1 \times \mathcal{Q}_2 \times \cdots \times \mathcal{Q}_k$ ($\mathcal{Q}_r = \mathcal{Q}$, $r \in [1, k]$). Then, $|\mathcal{D}| = (6\tau - 3)^k (2\tau^2 + \tau)^{(6\tau-3)k} = 2^{k\log(6\tau-3)+((6\tau-3)k)(\log(2\tau^2+\tau))} = 2^{O(k\tau\log\tau)}$.

Let $\mathcal{P}_i(s) = \{\alpha \cup s \mid \alpha = \Lambda_i^s(\rho)$ and $\rho \in \mathcal{R}_i(s)\}$, and let $\alpha_1 \cup s$ and $\alpha_2 \cup s$ be two distinct assignments in $\mathcal{P}_i(s)$. By the definition of $\mathcal{R}_i(s)$, there exists at least one $r \in \{i\} \cup X$ such that $M_i(r, \alpha_1, s) \neq M_i(r, \alpha_2, s)$. In the following, we show that there exists an injective function g from $\mathcal{P}_i(s)$ to $\mathcal{D}$. (1) Let $(\alpha \cup s) \in \mathcal{P}_i(s)$. The assignment $\alpha \cup s$ is a k-page book embedding which includes k half-planes. After shifting some edge in $\alpha \cup s$, each half-plane is translated into a 0-restricted plane graph with $2\tau + 1$ vertices. Hence, there exists a unique tuple $(P_1, P_2, \ldots, P_k)$ in $\mathcal{D}$ such that $g(\alpha \cup s) = (P_1, P_2, \ldots, P_k)$. (2) for any two distinct assignments $\alpha_1 \cup s$ and $\alpha_2 \cup s$ in $\mathcal{P}_i(s)$, it holds that $g(\alpha_1 \cup s) \neq g(\alpha_2 \cup s)$. Otherwise, for each $x \in \{i\} \cup X$, $M_i(x, g(\alpha_1), s) = M_i(x, g(\alpha_2), s)$. By Lemma 1, for each $x \in \{i\} \cup X$, $M_i(x, \alpha_1, s) = M_i(x, \alpha_2, s)$, contradicting the fact that $\alpha_1 \cup s$ and $\alpha_2 \cup s$ are two distinct assignments in $\mathcal{P}_i(s)$.

As a consequence, the size of $\mathcal{R}_i(s)$ is no larger than that of $\mathcal{D}$. Note that $k < \tau$. Therefore, the size of $\mathcal{R}_i(s)$ can be bounded by $2^{O(\tau^2\log\tau)}$.

$\square$

Based on Lemma 2 and the fact $|S| < \tau^{\tau^2}$, we draw the following conclusion.

Theorem 1. *There is an algorithm which takes as input a graph $G = (V, E)$ with a vertex order $\prec$, runs in time $2^{O(\tau^2\log\tau)} \cdot |V|$ where τ is the vertex cover number of G, and computes a page assignment σ such that $(\prec, \sigma)$ is a (fo-bt$(G, \prec)$)-page book embedding of G.*

4   On Parameterized Algorithm for the General Problem

The general FIXED-ORDER BOOK THICKNESS, i.e., FOBT-CROSS, parameterized by the vertex-cover number τ and the number of crossings b is formally defined as follows.

Input: a tuple $(G = (V, E), \prec)$, a non-negative integer b.
Parameters: τ, b;
Output: a k-page book drawing $(\prec, \sigma)$ of G such that the number of crossings over all pages in $(\prec, \sigma)$ is no more than b, or no such k-page drawing exists.

For the problem FOBT-CROSS, Bhore et al. [4] mentioned the techniques in their algorithm for FOBT can be extended to it, but they did not elaborate further. In order to facilitate the analysis, we first briefly present a specific algorithm for FOBT-CROSS by extending the techniques in [4]. Then, we pay more attention to employing our approach to analyze its running time.

4.1   Design of the Parameterized Algorithm

We expand some basic notions defined in [4], such as the valid assignment, the visibility matrix, and the record set (restated in Sect. 3).

Given a graph $G = (V, E)$ and a minimum vertex cover C in G, we call s (resp. α) : E_C (resp. E_i) $\rightarrow [1, k]$ a *valid page assignment* if s (resp. $\alpha \cup s$) maps edges to k pages such that the number of crossings over all pages is no more than b. Considering the allowed crossing may be on any one page, we introduce two data tables, i.e., *the crossing number vector* and *the crossing number matrix*, to store the information about the number of related crossings, respectively.

The cross number vector is used to record the number of crossings in each page. More precisely, a crossing number vector $N_i(\alpha, s)$ is a $k \times 1$ matrix, where the entry $N_i(\alpha, s)[p]$ stores the number of crossings on page p in $\alpha \cup s$ ($p \in [1, k]$).

The cross number matrix is used to record the number of crossings produced by some "potential" edges and some assigned edges. More precisely, a $k \times \tau$ crossing number matrix $M_i(a, \alpha, s)$ is one data table in which the entry (p, r) of $M_i(a, \alpha, s)$ records the number of edges in $E_i \cup E_C$ separating c_r from u_a on page p in $\alpha \cup s$. When the number of these edges exceeds $b + 1$, it only record the number $b + 1$. Figure 2 shows the crossing number vector and the crossing number matrix for a 2-page assignment of $E_3 \cup E_C$, respectively.

$$N_3(\alpha, s) = \begin{pmatrix} 2 \\ 1 \end{pmatrix}$$

$$M_3(3, \alpha, s) = \begin{pmatrix} 3 & 2 & 0 & 3 & 3 \\ 1 & 1 & 0 & 0 & 1 \end{pmatrix}$$

Fig. 2. A partial 2-page assignment of a graph G (left), the corresponding crossing number vector (upper right), and the corresponding crossing number matrix (lower right).

By the definition of crossing number matrix, there exists one corresponding crossing number matrix for each vertex in U. However, for some vertices, their corresponding matrices are actually the same one.

Lemma 3. *Let α be a valid partial page assignment of E_i, $x_j \in X$, and x_l be the immediate successor of x_j in X. If $u_i \prec u_{x_j} \prec u_h \prec u_{x_l}$ in $\prec$ and $u_h \notin C$, then $M_i(h, \alpha, s) = M_i(x_j, \alpha, s)$.*

Based on Lemma 3, we can use $\tau + 1$ matrices to capture the complete information about all of the "potential" crossings on each page. Moreover, the information in the crossing number vector plays an important role in distinguishing different page assignments. Hence, we define one *expanded record set* as follows: $\mathcal{R}'_i(s) = \{(N_i(\alpha, s), M_i(i, \alpha, s), M_i(x_1, \alpha, s), M_i(x_2, \alpha, s), \ldots, M_i(x_z, \alpha, s)) \mid \exists$ valid partial page assignment $\alpha : E_i \rightarrow [1, k]\}$. For ease of presentation, the tuple $(N_i(\alpha, s), M_i(i, \alpha, s), M_i(x_1, \alpha, s), M_i(x_2, \alpha, s), \ldots, M_i(x_z, \alpha, s))$ is also called a *matrix queue* for $\alpha \cup s$, and is denoted by $\mathcal{M}_i(\alpha, s)$ in the rest of this paper. Along with $\mathcal{R}'_i(s)$, we also store a mapping Λ'^s_i from $\mathcal{R}'_i(s)$ to a valid partial page assignments of E_i which maps $(N, M_0, \ldots, M_z) \in \mathcal{R}'_i(s)$ to some α such that $(N, M_0, \ldots, M_z) = \mathcal{M}_i(\alpha, s)$.

Checking whether a page assignment is a valid assignment is the main step for FOBT-CROSS. We discard the invalid partial assignment in two cases: (1) if the number of crossings on any page exceeds the allowed value b; (2) if the number of crossings over all page exceeds the allowed value b. Let γ be a valid page assignment of E_{i-1}, and let β be a page assignment of the edges incident to the vertex u_{i-1}. Figure 3 describes the procedure of checking whether $\gamma \cup \beta \cup s$ is a valid partial page assignment of $E_i \cup E_C$, in which $n_{\beta[j]}$ denotes the number of new crossings generated from edges mapped to page j in $\gamma \cup s$ by β.

Procedure Checking(γ, s, β)

1. $j=1$; $t = 0$; con=1;
2. **while** $(j \leq k)$ and (con=1) **do**
2.1   **if** $N_{i-1}(\gamma, s)[j] + n_{\beta[j]} \leq b$ **then** $\{t = t + N_{i-1}(\gamma, s)[j] + n_{\beta[j]};$
                                                    $j = j + 1 \; ; \}$
2.2   **else** con=0;
3. **if** $(t \leq b)$ and (con=1) **then** output a valid partial assignment $\gamma \cup \beta \cup s$;
                     else discard (γ, s, β).

Fig. 3. The main steps in Checking procedure

Let α_1 and α_2 be two valid page assignments of E_{i-1}, let β be a page assignment of the edges incident to the vertex u_{i-1}, and let $\alpha_1 \cup \beta \cup s$ and $\alpha_2 \cup \beta \cup s$ be the corresponding outputs of the Checking procedure.

Lemma 4. *If* $\mathcal{M}_{i-1}(\alpha_1, s) = \mathcal{M}_{i-1}(\alpha_2, s)$, *then* $\mathcal{M}_i(\alpha_1 \cup \beta, s) = \mathcal{M}_i(\alpha_2 \cup \beta, s)$.

Proof. First, we show that for each $p \in [1, k]$, $N_i(\alpha_1 \cup \beta, s)[p] = N_i(\alpha_2 \cup \beta, s)[p]$. Let n_1 (resp. n_2) be the number of crossings generated by adding some edge assigned by β to page p in $\alpha_1 \cup s$ (resp. $\alpha_2 \cup s$). By $M_{i-1}(x, \alpha_1, s) = M_{i-1}(x, \alpha_2, s)$, where $x \in \{i\} \cup X$, it holds that $n_1 = n_2$. By $N_{i-1}(\alpha_1, s)[p] = N_{i-1}(\alpha_2, s)[p]$, it follows that $N_{i-1}(\alpha_1, s)[p] + n_1 = N_{i-1}(\alpha_2, s)[p] + n_2$, i.e., $N_i(\alpha_1 \cup \beta, s)[p] = N_i(\alpha_2 \cup \beta, s)[p]$. Note that in the case $N_{i-1}(\alpha_1, s)[p] + n_1 > b$ (or $N_{i-1}(\alpha_2, s)[p] + n_2 > b$), the assignment β will be discarded by the step 3 in the Checking procedure.

Second, we show that for each $x \in \{i\} \cup X$, $M_i(x, \alpha_1 \cup \beta, s) = M_i(x, \alpha_2 \cup \beta, s)$. Assume that one entry (p, j) in $M_i(h, \alpha_1 \cup \beta, s)$ increased by r due to adding some edge by β, where $h \in \{i\} \cup X$. Then there must be r' ($r' \geq r$) added edges separating c_j from u_h on page p in $\alpha_1 \cup \beta \cup s$, respectively. In the assignment $\alpha_2 \cup \beta \cup s$, each of these edges added by β also separates c_j from u_h on page p. Hence, the corresponding entry (p, j) in $M_i(a, \alpha_2 \cup \beta, s)$ is also increased by r. $\qquad\square$

Based on the framework of dynamic programming in [4] and the Checking procedure above, we can obtain an algorithm for solving FOBT-CROSS, denoted by ALGF. Specifically, the main steps in ALGF can be sketched as follows.

The basic strategy in ALGF is to dynamically generate some page assignment containing at most b crossings in a left-to-right fashion. Assume the record set $\mathcal{R}'_{i-1}(s)$ has been computed. Each page assignment β of edges incident to vertex u_{i-1} and each matrix queue $\rho \in \mathcal{R}'_{i-1}(s)$ are branched. For each such β and $\gamma = \Lambda'^s_{i-1}(\rho)$, the procedure Checking$(\gamma, s, \beta)$ is called. If $\gamma \cup \beta \cup s$ contains at most b crossings, then this procedure outputs a valid partial assignment $\gamma \cup \beta \cup s$. Moreover, the matrix queue $\mathcal{M}_i(\gamma \cup \beta, s)$ is computed and stored, and the mapping Λ'^s_i is set to map this matrix queue to $\gamma \cup \beta$. Otherwise, the tuple (γ, s, β) is discarded.

Based on Lemma 4, we obtain the following conclusion on the algorithm ALGF.

Theorem 2. *If $(G, \prec)$ contains at least one valid assignment, then the algorithm ALGF$((G, \prec), \tau, b)$ returns a valid page assignment.*

4.2   Analysis on the Running Time

We adapt the approach used for FOBT to analyze its running time. First of all, we expand the notion of edge-shifting. Let $\alpha \cup s$ be a valid partial page assignment of edges in $E_i \cup E_C$, let u and y be two vertices in U, and let $E(p, y)$ be the set of edges on page p that incident to the vertex y. Assume that $E(p, y) = \{yc^1_y, yc^2_y, \ldots, yc^t_y\}$. The operation **SHIFT$'$**$(u, y)$ on shifting edges in $E(p, y)$ from y to u is re-defined as follows.

For $q = 1$ **to** t **do**: (1) delete the edge yc^q_y; (2) if the number of edges between u and c^q_y is less than $b + 1$ then add one edge uc^q_y between u and c^q_y.

For the assignment $\alpha \cup s$, we execute the extended edge-shifting along the same lines in the description for FOBT (see Sect. 3). After shifting all possible edges on each page, we obtain a simplified assignment $\alpha' \cup s$ such that the edges on each page are incident only with at most $2\tau + 1$ vertices. Figure 4 shows one example on the extended edge-shifting. Note that the operation **SHIFT$'$** may yield some multiple edges between two vertices on some page.

Fig. 4. The illustration on extended edge-shifting from an original 2-page assignment of $E_6 \cup E_C$ (a) to a simplified 2-page assignment, in which $b = 1$ (b).

Let $i \in [1, n - \tau]$, $M_i(x, \alpha, s)$ be the crossing number matrix for the original assignment $\alpha \cup s$, and $M_i'(x, \alpha', s)$ be the crossing number matrix for the simplified assignment $\alpha' \cup s$.

Lemma 5. *For each $x \in \{i\} \cup X$, $M_i(x, \alpha, s) = M_i(x, \alpha', s)$.*

Proof. Let $c_j \in C$ for $j \in [1, \tau]$ and $u_x \in U$ for $i \leq x$. By the definition of the crossing number matrix, it is sufficient to draw a comparison between $\alpha \cup s$ and $\alpha' \cup s$ on the number of edges separating c_j from u_x. Moreover, since each edge in E_C remains unchanged during the extended edge-shifting, we do not consider the edge in E_C separating c_j from u_x.

($\Rightarrow$) Assume that the entry (p, j) is r in $M_i(x, \alpha, s)$ ($p \in [1, k], r \in [0, b + 1]$). Then there must be h ($h \geq 0$) edges separating c_j from u_x on page p in $\alpha \cup s$. Our aim is to show that (p, j) is also r in $M_i(x, \alpha', s)$. We distinguish two cases based on the value of h. Case (1): $h \leq b + 1$ (correspondingly, $r = h$). Without loss of generality, let e be one of the h edges in E_i. By the proof of Lemma 1, there exists a corresponding edge e' separating c_j from u_x on page p in $\alpha' \cup s$. Moreover, by the assumption that $h \leq b + 1$ and the rule that each repeated edge generating from edge-shifting is replaced by one multiple-edge, there is a one-to-one relationship between e and e'. Hence, there are h edges separating c_j from u_x on page p in $\alpha' \cup s$, indicating that (p, j) is r in $M_i(x, \alpha', s)$. Case (2): $h > b + 1$ (correspondingly, $r = b + 1$). Then there must be at least $b + 2$ edges that separating c_j from u_x on page p in $\alpha \cup s$. After shifting some edge, the value of h either remained unchanged or decreased. For the latter, h is at least $b + 1$ because the number of edges that incident to u_q ($u_q \in U$, $q \in X \cup \{1\}$) is set to $b + 1$. Thus, on page p in $\alpha' \cup s$, the number of edges separating c_j from u_x is at least $b + 1$, indicating that (p, j) is r in $M_i(x, \alpha', s)$.

($\Leftarrow$) Assume that the entry (p, j) is r in $M_i(x, \alpha', s)$ ($r \in [0, b + 1]$). Then there must be h ($h \geq 0$) edges separating c_j from u_x on page p in $\alpha' \cup s$. Our aim is to show that (p, j) is r in $M_i(x, \alpha, s)$. We distinguish two cases based on the value of h. Case (1): $h \leq b + 1$ (correspondingly, $r = h$). Without loss of generality, let e be one of the h edges. Along the same lines in the previous paragraph, it holds that there exists one edge $e' \in E_i$ separating c_j from u_x on page p in $\alpha \cup s$. Moreover, there exists a one-to-one relationship between e and e'. So, (p, j) is r in $M_i(x, \alpha, s)$. Case (2): $h > b + 1$ (correspondingly, $r = b + 1$). Then there must be at least $b + 2$ edges separating c_j from u_x on page p in $\alpha' \cup s$.

By the rule of the extended edge-shifting, on page p in $\alpha \cup s$, the number of edges separating c_j from u_x either remained unchanged or increased. Hence, the entry (p, j) is filled with $b + 1$, indicating that (p, j) is r in $M_i(x, \alpha, s)$.

□

Based on Lemma 5, we obtain a bound on the size of the record set $\mathcal{R}'_i(s)$.

Lemma 6. *The size of* $\mathcal{R}'_i(s)$ *can be bounded by* $2^{O((\tau^2 + 2b\tau)\log\tau(b+1))}$.

Proof. Let L be a straight line joining $2\tau + 1$ vertices with a fixed order, let $\mathcal{P}_1 = \{P_1 \mid P_1$ is a maximal 0-restricted plane graph with spine $L.\}$, and let $\mathcal{P}_2 = \{P_2 \mid P_2$ is a b-crossings plane graph with spine $L.\}$.

We first estimate the size of $\mathcal{P}_1$ and $\mathcal{P}_2$, respectively. Since each element in $\mathcal{P}_1$ is a maximal planar graph, $|\mathcal{P}_1| = \Sigma_{i=1}^{6\tau-3}\binom{(2\tau+1)\tau}{i}(b+1)^i \leq (6\tau - 3)((2\tau + 1)\tau)^{6\tau-3}(b+1)^{6\tau-3}$, in which the factor $(b+1)^i$ indicates that the multiplicities of each edge is at most $b+1$. Since each element in $\mathcal{P}_2$ contains at most b crossings (i.e., at most $2b$ edges), $|\mathcal{P}_2| = \Sigma_{i=1}^{2b}\binom{(2\tau+1)\tau}{i}(b+1)^i \leq 2b((2\tau+1)\tau)^{2b}(b+1)^{2b}$. Let $\mathcal{Q} = \mathcal{P}_1 \times \mathcal{P}_2$. Then, $|\mathcal{Q}| \leq (6\tau - 3)((2\tau + 1)\tau)^{6\tau-3}(b+1)^{6\tau-3} \times 2b((2\tau+1)\tau)^{2b}(b+1)^{2b} = 2b(6\tau - 3)((2\tau + 1)(b+1)\tau)^{6\tau+2b-3}$.

Let N be a $k \times 1$ matrix, and for each $i \in [1, k]$, assume that $N[i] \in [0, b]$. Let $\mathcal{D}' = N \times \mathcal{Q}_1 \times \mathcal{Q}_2 \times \cdots \times \mathcal{Q}_k$ ($\mathcal{Q}_r = \mathcal{Q}, r \in [1, k]$). Then, $|\mathcal{D}'| = (b+1)^k(2b)^k(6\tau - 3)^k((2\tau + 1)(b+1)\tau)^{(6\tau+2b-3)k} = 2^{O((\tau+b)k\log\tau(b+1))}$.

Let $\mathcal{P}'_i(s) = \{\alpha \cup s \mid \alpha = \Lambda'^s_i(\rho)$ and $\rho \in \mathcal{R}'_i(s)\}$. Based on Lemma 5, we can show that there exists an injective function g from $\mathcal{P}'_i(s)$ to $\mathcal{D}'$ along the same lines in the proof of Lemma 2.

As a consequence, the size of $\mathcal{R}'_i(s)$ is no larger than that of $\mathcal{D}'$. Therefore, the size of $\mathcal{R}'_i(s)$ can be bounded by $2^{O((\tau+b)k\log\tau(b+1))}$.

□

Based on Lemma 6 and the fact $|S| < \tau^{\tau^2}$, we obtain the flowing conclusion.

Theorem 3. *The algorithm ALGF for the problem FOBT-CROSS runs in time* $2^{O((\tau+b)k\log\tau(b+1))+\tau^2\log\tau} \cdot |V|$.

Let fo-bt$(G, \prec, b)$ be the minimum k such that $(G, \prec, b)$ is a YES instance of the problem FOBT-CROSS. Since fo-bt$(G, \prec, b) \leq$ fo-bt$(G, \prec)$ and fo-bt$(G, \prec) < \tau$, it follows that fo-bt$(G, \prec, b) < \tau$. Now, we arrive at our main result.

Theorem 4. *There is an algorithm which takes as input a graph* $G = (V, E)$ *with a vertex order* $\prec$, *and an integer* b, *runs in time* $2^{O((\tau^2+b\tau)\log\tau(b+1))} \cdot |V|$, *and computes a page assignment* σ *such that* $(\prec, \sigma)$ *is a* (fo-bt$(G, \prec, b)$)-*page book drawing of* G.

5   Conclusions

We studied parameterized algorithms for the problem FIXED-ORDER BOOK THICKNESS with respect to the vertex-cover number. By introducing a novel analysis approach, we proved an improved running time bound for the parameterized algorithm given by Bhore et al. in [4]. By extending this analysis approach, we also show that the general FIXED-ORDER BOOK THICKNESS problem admits a parameterized algorithm running in time $2^{O((\tau^2+b\tau)log\tau(b+1))} \cdot |V|$.

The main strategy in our analysis approach is to construct an auxiliary graph by shifting some edge on the original graph such that the number of combinations giving rise to the complexity of running time can be estimated conveniently and more accurately. We believe this strategy has potential to derive improved running time bound on some algorithms for other problems.

Acknowledgements. The authors thank the anonymous referees for their constructive suggestions which have resulted in improvement on the presentations.

References

1. Bilski, T.: Optimum embedding of complete graphs in books. Discrete Math. **182**(1–3), 21–28 (1998)
2. Ganley, J.L., Heath, L.S.: The page number of k-trees is $O(k)$. Discrete Appl. Math. **109**(3), 215–221 (2001)
3. Hong, S.-H., Nagamochi, H.: Simpler algorithms for testing two-page book embedding of partitioned graphs. Theoret. Comput. Sci. **725**, 79–98 (2018)
4. Bhore, S., Ganian, R., Montecchiani, F., Nöllenburg, M.: Parameterized algorithms for book embedding problems. In: Archambault, D., Tóth, C.D. (eds.) GD 2019. LNCS, vol. 11904, pp. 365–378. Springer, Cham (2019). https://doi.org/10.1007/978-3-030-35802-0_28
5. Dujmović, V., Wood, D.R.: On linear layouts of graphs. Discrete Math. Theoret. Comput. Sci. **6**, 339–358 (2004)
6. Garey, M.R., Johnson, D.S., Miller, G.L., Papadimitriou, C.H.: The complexity of coloring circular arcs and chords. SIAM J. Alg. Discr. Meth. **1**(2), 216–227 (1980)
7. Unger, W.: The complexity of colouring circle graphs. In: Finkel, A., Jantzen, M. (eds.) STACS 1992. LNCS, vol. 577, pp. 389–400. Springer, Heidelberg (1992). https://doi.org/10.1007/3-540-55210-3_199
8. Gyárfás, A.: On the chromatic number of multiple interval graphs and overlap graphs. Discrete Math. **55**(2), 161–166 (1985)
9. Ageev, A.A.: A triangle-free circle graph with chromatic number 5. Discrete Math. **152**(1–3), 295–298 (1996)
10. Kostochka, A., Kratochvíl, J.: Covering and coloring polygon-circle graphs. Discrete Math. **163**(1–3), 299–305 (1997)
11. Chung, F., Leighton, F., Rosenberg, A.: Embedding graphs in book: a layout problem with applications to VLSI design. SIAM J. Alg. Discr. Meth. **8**(1), 33–58 (1987)
12. Bannister, M.J., Eppstein, D.: Crossing minimization for 1-page and 2-page drawings of graphs with bounded treewidth. J. Graph Algorithms Appl. **22**(4), 577–606 (2018)

Acyclic Edge Coloring Conjecture Is True on Planar Graphs Without Intersecting Triangles

Qiaojun Shu[1,3], Yong Chen[1], Shuguang Han[2,3], Guohui Lin[3(✉)], Eiji Miyano[4], and An Zhang[1,3]

[1] Department of Mathematics, Hangzhou Dianzi University, Hangzhou, China
{qjshu,chenyong,anzhang}@hdu.edu.cn
[2] Department of Mathematics, Zhejiang Sci-Tech University, Hangzhou, China
dawn1024@163.com
[3] Department of Computing Science, University of Alberta, Edmonton, Canada
{qiaojun,shuguang,guohui,az4g}@ualberta.ca
[4] Department of Artificial Intelligence, Kyushu Institute of Technology, Iizuka, Japan
miyano@ces.kyutech.ac.jp

Abstract. An acyclic edge coloring of a graph G is a proper edge coloring such that no bichromatic cycles are produced. The acyclic edge coloring conjecture by Fiamčik (1978) and Alon, Sudakov and Zaks (2001) states that every simple graph with maximum degree Δ is acyclically edge $(\Delta + 2)$-colorable. Despite many milestones, the conjecture is still unknown true or not even for planar graphs. In this paper, we first show by discharging methods that every planar graph without intersecting triangles must have one of the six specified groups of local structures; then by induction on the number of edges we confirm affirmatively the conjecture on planar graphs without intersecting triangles.

Keywords: Acyclic edge coloring · Planar graph · Intersecting triangles · Discharging · Induction

1 Introduction

Only simple and connected graphs are considered in this paper. Let G be such a graph with vertex set $V(G)$ and edge set $E(G)$. For an integer $k \geq 2$, a *(proper) edge k-coloring* is a mapping $c : E(G) \rightarrow \{1, 2, \ldots, k\}$ such that any two adjacent edges receive different colors. (We drop proper in the sequel.) G is *edge k-colorable* if G has an edge k-coloring. The *chromatic index* $\chi'(G)$ of G is the smallest integer k such that G is edge k-colorable. An edge k-coloring c of G is called *acyclic* if there are no bichromatic cycles in G, i.e., the subgraph of G

The research was supported by ZJNSF LY20F030007 and LQ15A010010, NSFC 11601111, 11971139, 11771114 and 11571252, CSC 201508330054 and 201908330090, NSERC, KAKENHI JP17K00016, and JST CREST JPMJR1402.

© Springer Nature Switzerland AG 2020
J. Chen et al. (Eds.): TAMC 2020, LNCS 12337, pp. 426–438, 2020.
https://doi.org/10.1007/978-3-030-59267-7_36

induced by any two colors is a forest. The *acyclic chromatic index* of G, denoted by $a'(G)$, is the smallest integer k such that G is acyclically edge k-colorable. Let $\Delta(G)$ (Δ for short) denote the maximum degree of the graph G. One sees that $\Delta \leq \chi'(G) \leq a'(G)$. Note that $\chi'(G) \leq \Delta + 1$ by Vizing's theorem [16] and that $a'(K_4) = \Delta + 2$. Fiamčik [7] and Alon, Sudakov, and Zaks [2] independently made the following *acyclic edge coloring conjecture* (AECC):

Conjecture 1. (AECC) For any graph G, $a'(G) \leq \Delta + 2$.

On an arbitrary graph G, the following milestones have been achieved: Alon, McDiarmid and Reed [1] proved that $a'(G) \leq 64\Delta$ by a probabilistic argument. The upper bound was improved to 16Δ [10], to $\lceil 9.62(\Delta - 1) \rceil$ [11], to $4\Delta - 4$ [6], and most recently in 2017 to $\lceil 3.74(\Delta - 1) \rceil + 1$ by Giotis et al. [8] using the Lovász local lemma. On the other hand, the AECC has been confirmed true for graphs with $\Delta \in \{3, 4\}$ [3, 4, 15, 19].

When G is planar, i.e., G can be drawn in the two-dimensional plane so that its edges intersect only at their ending vertices, Basavaraju et al. [5] showed that $a'(G) \leq \Delta + 12$; and the upper bound was improved to $\Delta + 7$ by Wang et al. [21] and to $\Delta + 6$ by Wang and Zhang [18]. The AECC has been confirmed true for graphs without i-cycles for each $i \in \{3, 4, 5, 6\}$ in [13, 14, 20, 22], respectively.

A *triangle* is synonymous with a 3-cycle. We say that two triangles are *adjacent* if they share a common edge, and are *intersecting* if they share at least a common vertex. Recall that the truth of the AECC for planar graphs without triangles has been verified in [14]. When a planar graph G contains triangles but no intersecting triangles, Hou, Roussel, and Wu [9] proved the upper bound $a'(G) \leq \Delta + 5$, and Wang and Zhang [17] improved it to $a'(G) \leq \Delta + 3$. This paper focus on planar graphs without intersecting triangles too, and we completely resolve the AECC by showing the following main theorem.

Theorem 1. *The AECC is true for planar graphs without intersecting triangles.*

The rest of the paper is organized as follows. In Sect. 2, we characterize six groups of local structures (also called *configurations*), and by discharging methods we prove that any planar graph without intersecting triangles must contain one of these local structures. Incorporating a known property of edge colorings and bichromatic cycles (Lemma 9), in Sect. 3 we prove by induction on the number of edges that the graph admits an acyclic edge ($\Delta + 2$)-coloring. Due to space limit, most part of the induction is not included here but provided in [12].

2   The Six Groups of Local Structures

Given a graph G, let $d(v)$ denote the degree of the vertex v in G. A vertex of degree k (at least k, at most k, respectively) is called a *k-vertex* (k^+-*vertex*, k^--*vertex*, respectively). Let $n_k(v)$ ($n_{k^+}(v)$, $n_{k^-}(v)$, respectively) denote the number of k-vertices (k^+-vertices, k^--vertices, respectively) adjacent to v in G.

Theorem 2. *Let G be a 2-connected planar graph with $\Delta \geq 5$ and without intersecting triangles. Then G contains one of the following local structures (or configurations, used interchangeably) (A_1)–(A_6), as shown in Fig. 1:*

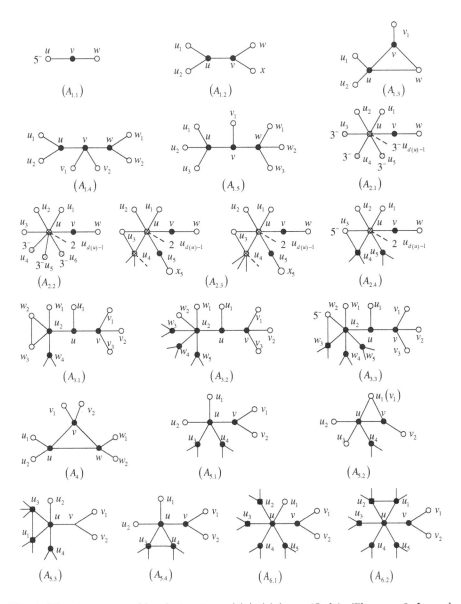

Fig. 1. The six groups of local structures (A_1)–(A_6) specified in Theorem 2. In each local structure, solid vertices have all their neighbors (and thus their degrees) determined while the non-filled vertices could be adjacent to other vertices outside of the local structure. All vertices are labelled starting with a character, and the degree of a non-filled vertex, if known, starts with a digit; each dashed edge indicates existence of zero, one, or more edges.

(A_1) A path uvw such that one of the following holds:

    $(A_{1.1})$ $d(v) = 2$ and $d(u) \leq 5$;

    $(A_{1.2})$ $d(u) = d(v) = 3$;

    $(A_{1.3})$ $d(v) = 3$, $d(u) = 4$, and $uw \in E(G)$;

    $(A_{1.4})$ $d(v) = 4$ and $d(u) = d(w) = 3$;

    $(A_{1.5})$ $d(v) = 3$ and $d(u) = d(w) = 4$.

(A_2) A vertex u with $n_2(u) \geq 1$. Let $u_1, u_2, \ldots, u_{d(u)-1}, v$ be the sorted neighbors of u such that $d(v) = 2$ and $d(u_i) - n_2(u_i) \geq d(u_{i+1}) - n_2(u_{i+1})$ for every $i = 1, 2, \ldots, d(u) - 2$. Then at least one of the following cases holds:

    $(A_{2.1})$ $n_2(u) + n_3(u) \geq d(u) - 2$;

    $(A_{2.2})$ $n_2(u) + n_3(u) = d(u) - 3$, and $n_3(u) \leq 3$;

    $(A_{2.3})$ $n_2(u) = d(u) - 4$, $u_3 u_4 \in E(G)$, and $n_2(u_4) \in \{d(u_4) - 4, d(u_4) - 5\}$;

    $(A_{2.4})$ $d(u_3) \leq 5$, $d(u_4) = 4$, $d(u_5) = 3$, and $u_3 u_4 \in E(G)$.

(A_3) A path $u_2 uv$ with $d(u) = 3$, $d(v) = 4$. Let $w_1, w_2, \ldots, w_{d(u_2)-1}$ be the neighbors of u_2 other than u. Then at least one of the following cases holds:

    $(A_{3.1})$ $d(u_2) = 5$, $d(w_4) = 3$, and $w_2 w_3 \in E(G)$;

    $(A_{3.2})$ $d(u_2) = 6$, $d(w_3) = d(w_4) = d(w_5) = 3$;

    $(A_{3.3})$ $d(u_2) = 6$, $d(w_2) \leq 5$, $d(w_3) = 4$, $d(w_4) = d(w_5) = 3$, and $w_2 w_3 \in E(G)$.

(A_4) A 3-cycle $uvwu$ with $d(u) = d(v) = d(w) = 4$.

(A_5) A 5-vertex u is adjacent to u_1, u_2, u_3, u_4 and a 3-vertex v. Then at least one of the following cases holds:

    $(A_{5.1})$ $d(u_3) = d(u_4) = 3$;

    $(A_{5.2})$ $d(u_4) = 3$ and $u_1 v \in E(G)$;

    $(A_{5.3})$ $d(u_1) = 4$, $d(u_3) = 5$, $d(u_4) = 3$, and $u_1 u_3 \in E(G)$;

    $(A_{5.4})$ $d(u_3) = d(u_4) = 4$, and $u_3 u_4 \in E(G)$.

(A_6) A 6-vertex u is adjacent to u_1, u_2 and four 3-vertices v, u_3, u_4, u_5. Then at least one of the following cases holds:

    $(A_{6.1})$ $d(u_2) = 3$;

    $(A_{6.2})$ $d(u_1) = d(u_2) = 4$ and $u_1 u_2 \in E(G)$.

The rest of the section is devoted to the proof of Theorem 2, by contradiction. That is, we assume to the contrary that G contains none of the local structures (A_1)–(A_6), and derive contradictions. We employ several known local structural properties for planar graphs from [13], which are summarized in Lemma 2.

2.1 Definitions and Notations

Since G is 2-connected, $d(v) \geq 2$ for any $v \in V(G)$. Let G' be the graph obtained by deleting all the 2-vertices of G; let H be a connected component of G'. Clearly, every vertex $v \in V(H)$ has its (called *original*) degree $d(v) \geq 3$ in G, and H is also a planar graph without intersecting triangles. In what follows, we assume that H is embedded in the two dimensional plane such that its edges intersect only at their ending vertices, and we refer H to as a *plane* graph.

For/In the plane graph H, we use the following notations:

- $N_H(v) = \{u \mid uv \in E(H)\}$ and $d_H(v) = |N_H(v)|$—the degree of the vertex $v \in V(H)$;
- similarly define what a k-vertex, a k^+-vertex, and a k^--vertex are;
- $n'_k(v)$ $(n'_{k^+}(v), n'_{k^-}(v),$ respectively)—the number of k-vertices (k^+-vertices, k^--vertices, respectively) adjacent to v;
- $F(H)$—the face set of H;
- $V(f)$—the set of vertices on (the boundary of) the face f;
- a vertex v and a face f are *incident* if $v \in V(f)$;
- $n_k(f)$ $(n_{k^+}(f), n_{k^-}(f),$ respectively)—the number of k-vertices (k^+-vertices, k^--vertices, respectively) in $V(f)$;
- $\delta(f)$—the minimum degree of the vertices in $V(f)$;
- $d(f)$—the degree of the face f, which is the number of edges on the face f with cut edges counted twice, and similarly define what a k-face, a k^+-face, and a k^--face are;
- $F(v) = \{f \in F(H) \mid v \in V(f)\}$;
- $m_k(v)$ $(m_{k^+}(v), m_{k^-}(v),$ respectively)—the number of k-faces (k^+-faces, k^--faces, respectively) in $F(v)$;
- for a vertex $v \in V(H)$ with $d_H(v) = k$, let $u_1, u_2, \ldots, u_k$ be all the neighbors of v in clockwise order, and denote the face containing $u_1 v u_2$ $(u_2 v u_3, \ldots, u_{k-1} v u_k, u_k v u_1,$ respectively) as f_1 $(f_2, \ldots, f_{k-1}, f_k,$ respectively); note that some of them could refer to the same face, i.e., the number of distinct faces could be less than k.

2.2   Structural Properties

Note from the construction of H that $d_H(u) = d(u) - n_2(u)$ and thus $d_H(u) = d(u)$ if $n_2(u) = 0$. The inexistence of $(A_{1.1})$ in G states that no 5^--vertex is adjacent to a 2-vertex in G; thus $3 \le d(u) \le 5$ implies $d_H(u) = d(u)$. The inexistences of $(A_{2.1})$ and $(A_{2.2})$ in G together imply that every 6^+-vertex u is adjacent to at most $(d(u) - 4)$ 2-vertices in G; thus if $d(u) \ge 6$ and $n_2(u) \ge 1$, then $d_H(u) \ge 4$. We summarize these into the following lemma.

Lemma 1. $\delta(H) \ge 3$; if the degree of a vertex u is $3 \le d(u) \le 5$, then $d_H(u) = d(u)$; if $d(u) \ge 6$ and $n_2(u) \ge 1$, then $d_H(u) \ge 4$. $\qquad\square$

Assuming the inexistences of some of (A_1)–(A_6), the three items in the next lemma are proven for planar graphs without 5-cycles in [13], but in fact hold for all planar graphs.

Lemma 2. [13]

(1) $n'_3(u) = n_3(u)$ for any vertex $u \in V(H)$.
(2) If a vertex u has $d_H(u) = 4$ and $n'_3(u) \ge 1$, then $d(u) = 4$.
(3) Let $f = [uvw]$ be a 3-face with $\delta(f) = 3$, then $n_{5^+}(f) = 2$. $\qquad\square$

Lemma 3. *Let $f = [uvw]$ be a 3-face with $\delta(f) = 4$.*

(1) *If $d(v) > d_H(v) = 4$, then $\min\{d_H(u), d_H(w)\} \geq 6$;*
(2) *if $d_H(v) = 4$ and $4 \leq d_H(u) \leq 5$, then $d(v) = 4$;*
(3) *$n_{5+}(f) \geq 1$.*

Proof. (1) and (2) hold due to the inexistence of $(A_{2.3})$.
     For (3), if $n_{5+}(f) = 0$, then $d(u) = d(v) = d(w) = 4$ by (2), implying the existence of (A_4) in G, a contradiction.   □

Lemma 4. *Let u be a vertex with $d_H(u) = 5$. Then $n_3(u) \leq 2$, and if $n_3(u) = 2$ then $d(u) = 5$.*

Proof. If $n_3(u) \geq 3$ then $n_2(u) + n_3(u) \geq d(u) - 5 + 3 \geq d(u) - 2$; since $(A_{2.1})$ does not exist, we have $n_2(u) = 0$ and subsequently $d(u) = 5$, which contradicts to the inexistence of $(A_{5.1})$. We thus prove that $n_3(u) \leq 2$.
     From $n_3(u) = 2$, we have $n_2(u) + n_3(u) = d(u) - 5 + 2 = d(u) - 3$; since $(A_{2.2})$ does not exist, we have $n_2(u) = 0$ and subsequently $d(u) = 5$.   □

Lemma 5. *Let u be a vertex with $d_H(u) = 6$. Then $n_3(u) \leq 4$, and*

(1) *if $n_3(u) \geq 3$, then $d(u) = 6$;*
(2) *if $n_3(u) = 4$, then $n'_{5+}(v) \geq 2$ for any 3-vertex $v \in N_H(u)$.*

Proof. Similar to the proof of Lemma 4, since G contains no $(A_{6.1})$, we have $n_3(u) \leq 4$, and if additionally $n_3(u) \geq 3$ then $d(u) = 6$, that is, (1) holds.
     If $n_3(u) = 4$, then by (1) $d(u) = 6$. Since G contains no $(A_{3.2})$, $n'_{5+}(v) \geq 2$ for any 3-vertex $v \in N_H(u)$, that is, (2) holds.   □

2.3   Discharging to Show Contradictions

To derive a contradiction, we make use of the discharging methods, which are very similar to an amortized analysis. First, by Euler's formula $|V(H)| - |E(H)| + |F(H)| = 2$ and the relation $\sum_{v \in V(H)} d_H(v) = \sum_{f \in F(H)} d(f) = 2|E(H)|$, we have the following equality:

$$\sum_{u \in V(H)} (2d_H(u) - 6) + \sum_{f \in F(H)} (d(f) - 6) = -12. \tag{1}$$

Next, we define an initial weight function $\omega(\cdot)$ on $V(H)$ by setting $\omega(u) = 2d_H(u) - 6$ for each vertex $u \in V(H)$ and setting $\omega(f) = d(f) - 6$ for each face $f \in F(H)$. It follows from Eq. (1) that the total weight is equal to -12. In what follows, we will define a set of discharging rules (R1)–(R4) to move portions of weights from vertices to faces (the function $\tau(u \to f)$). At the end of the discharging, a new weight function $\omega'(\cdot)$ is achieved and we are able to show that $\omega'(x) \geq 0$ for every $x \in V(H) \cup F(H)$ (Lemmas 6–7). This contradicts the negative total weight.
     For a k-vertex $u \in V(H)$ where $k \geq 4$ and $f \in F(u)$, let $\tau(u \to f)$ denote the amount of weight transferred from u to f.

(R1) When $k = 4$, $\tau(u \to f_i) = \frac{1}{2}$ for each $i = 1, 2, 3, 4$.

(R2) When $k \geq 5$ and $f = [vuw]$ is a 3-face,

    (R2.1) if $\delta(f) = 3$, then $\tau(u \to f) = \frac{3}{2}$;

    (R2.2) assume $\delta(f) = 4$ with $d_H(v) = 4$,

        (R2.2.1) if $k \geq 7$, then $\tau(u \to f) = 2$;

        (R2.2.2) if $k = 6$, then

$$\tau(u \to f) = \begin{cases} \frac{3}{2}, & \text{if } n_3(u) \geq 4; \\ \frac{5}{4}, & \text{if } n_3(u) = 3 \text{ and } n_{6+}(f) = 2; \\ \frac{5}{4}, & \text{if } n_3(u) = 3, d_H(w) = 5 \text{ and } n_3(w) \leq 1; \\ 2, & \text{otherwise}; \end{cases}$$

        (R2.2.3) if $k = 5$, then $\tau(u \to f) = \begin{cases} 2, & \text{if } n_3(u) = 0; \\ \frac{5}{4}, & \text{if } n_3(u) = 1; \\ 1, & \text{if } n_3(u) \geq 2; \end{cases}$

    (R2.3) if $\delta(f) \geq 5$, then $\tau(u \to f) = 1$.

(R3) When $k \geq 5$ and $f = [vuyx]$ is a 4-face,

$$\tau(u \to f) = \begin{cases} 1, & \text{if } d(v) = 3 \text{ and, either } d(y) = 3 \text{ or } d(x) = 4; \\ \frac{3}{4}, & \text{if } d(v) = 3, d_H(y) \geq 4 \text{ and } d_H(x) \geq 5; \\ \frac{1}{2}, & \text{if } d_H(v) \geq 4 \text{ and } d_H(y) \geq 4. \end{cases}$$

(R4) When $k \geq 5$ and $f = [\cdot xuy\cdot]$ is a 5-face,

$$\tau(u \to f) = \begin{cases} \frac{1}{2}, & \text{if } d(x) = d(y) = 3; \\ \frac{1}{4}, & \text{if } \max\{d_H(x), d_H(y)\} \geq 4. \end{cases}$$

It remains to validate that $\omega'(x) \geq 0$ for every $x \in V(H) \cup F(H)$.

Lemma 6. *For every face $f \in F(H)$, $\omega'(f) \geq 0$.*

Proof. We distinguish the following four cases for $d(f)$.

Case 1. $d(f) \geq 6$. Since a face never transfers out any weight, in this case $\omega'(f) \geq \omega(f) = d(f) - 6 \geq 0$.

Case 2. $d(f) = 3$ with $f = [uvw]$. In this case, $\omega(f) = d(f) - 6 = -3$.

If $\delta(f) = 3$, then $n_{5+}(f) = 2$ by Lemma 2(3); by (R2.1) each 5^+-vertex in $V(f)$ gives $\frac{3}{2}$ to f and thus $\omega'(f) \geq -3 + 2 \times \frac{3}{2} = 0$.

If $\delta(f) \geq 5$, then by (R2.3) $\tau(u \to f) = 1$ for each $u \in V(f)$, and thus $\omega'(f) \geq -3 + 3 \times 1 = 0$.

If $\delta(f) = 4$ with $d_H(v) = 4$, then $n_{4+}(f) = 3$, $n_{5+}(f) \geq 1$ by Lemma 3, and $n_3(x) \leq d_H(x) - 2$ for each $x \in \{u, w\}$. By (R1) $\tau(v \to f) = \frac{1}{2}$. When $\min\{\tau(u \to f), \tau(w \to f)\} \geq \frac{5}{4}$, $\omega'(f) \geq -3 + 2 \times \frac{5}{4} + \frac{1}{2} = 0$; when $\max\{\tau(u \to f), \tau(w \to f)\} = 2$, $\omega'(f) \geq -3 + 2 + 2 \times \frac{1}{2} = 0$. In the other cases:

    Case 2.1. $d_H(u) = 4$ and, either $d_H(w) = 6$ with $n_3(w) = 4$ or $d_H(w) = 5$ with $n_3(w) \geq 1$. In this subcase, $d(u) = d(v) = 4$ by Lemma 3 and thus one of $(A_{2.4})$, $(A_{5.4})$ and $(A_{6.2})$ exists, a contradiction.

    Case 2.2. $d_H(u) = d_H(w) = 5$, with $n_3(u) \geq 2$ and $n_3(w) \geq 1$. In this subcase, $n_3(u) = 2$ and $d(u) = 5$ by Lemma 4, and $d(v) = 4$ by Lemma 3. Since G contains no $(A_{2.4})$, we have $d(w) = 5$ and thus $(A_{5.3})$ exists, a contradiction.

Case 3. $d(f) = 4$ with $f = [uvxy]$. In this case, $\omega(f) = d(f) - 6 = -2$.

If $\delta(f) \geq 4$, then by (R1) and (R3) $\omega'(f) \geq -1 + 4 \times \frac{1}{2} = 0$.

Consider next $\delta(f) = 3$. Since G contains neither $(A_{1.2})$ nor $(A_{1.4})$, $n_3(f) \leq 2$. If $n_3(f) = 2$, then $n_{5+}(f) = 2$ and by (R3) $\omega'(f) = -2 + 2 \times 1 = 0$.

If $n_3(f) = 1$, then $d(v) = 3$ and, since G contains no $(A_{1.5})$, we have $\max\{d_H(x), d_H(u)\} \geq 5$. Assuming w.l.o.g. $d_H(u) \geq 5$, by (R1) and (R3), if $d(x) = 4$, then we have $\tau(y \to f) \geq \frac{1}{2}$, $\tau(u \to f) = 1$, and $\tau(x \to f) = \frac{1}{2}$, leading to $\omega'(f) \geq -2 + 1 + 2 \times \frac{1}{2} = 0$; or if $d_H(x) \geq 5$, then we have $\tau(u \to f) = \tau(x \to f) = \frac{3}{4}$, leading to $\omega'(f) \geq -2 + 2 \times \frac{3}{4} + \frac{1}{2} = 0$.

Case 4. $d(f) = 5$ with $f = [uvwxy]$. In this case, $\omega(f) = d(f) - 6 = -1$.

By (R1) and (R4), each 4^+-vertex in $V(f)$ gives at least $\frac{1}{4}$ to f. If $n_{4+}(f) \geq 4$, then $\omega'(f) \geq -1 + 4 \times \frac{1}{4} = 0$.

Otherwise, since G contains neither $(A_{1.2})$ nor $(A_{1.4})$ and by Lemma 2(1), we may assume w.l.o.g. $d(u) = d(w) = 3$, $n_{4+}(f) = 3$, and $d_H(v) \geq 5$. Then by (R1) and (R4), we have $\tau(v \to f) \geq \frac{1}{2}$, $\tau(x \to f) \geq \frac{1}{4}$, and $\tau(y \to f) \geq \frac{1}{4}$, leading to $\omega'(f) \geq -1 + 1 \times \frac{1}{2} + 2 \times \frac{1}{4} = 0$. $\qquad \square$

Lemma 7. *For every vertex* $u \in V(H)$, $\omega'(u) \geq 0$.

Proof. Recall from Lemma 1 that $d_H(u) \geq 3$. If $d_H(u) = 3$, then $\omega'(u) = \omega(u) = 2d_H(u) - 6 = 0$, since u never transfers its weight out by our rules.

If $d_H(u) = 4$, then $\omega'(u) = \omega(u) - 4 \times \frac{1}{2} = 2d_H(u) - 6 - 2 = 0$ by (R1).

If $d_H(u) \geq 7$, then using the heaviest weights in (R2)–(R4) we have $\omega'(u) \geq 2d_H(u) - 6 - 2 - 1 \times (d_H(u) - 1) = d_H(u) - 7 \geq 0$.

Below we distinguish two cases where $d_H(u) = 5$ and 6, respectively. Note that $m_3(u) \leq 1$ since G contains no intersecting triangles.

Case 1. $d_H(u) = 5$. In this case, $\omega(u) = 2d_H(u) - 6 = 4$.

By Lemma 4, $n_3(u) \leq 2$, and if $n_3(u) = 2$ then $d(u) = 5$. When $m_3(u) = 0$, if the number of faces with $\tau(u \to f) \leq \frac{1}{2}$ is at least 2, then $\omega'(u) \geq 4 - 2 \times \frac{1}{2} - 3 \times 1 = 0$; otherwise, there is at most one face $f \in F(u)$ such that $\tau(u \to f) \leq \frac{1}{2}$, and thus $n_3(u) = 2$. One sees for the same reason that these two 3-vertices should not be on the same face, and we assume they are u_2 and u_4, i.e., $d(u_2) = d(u_4) = 3$. For each of u_2 and u_4, at least one of its neighbors besides u is a 5^+-vertex since G contains no $(A_{1.5})$. It follows from (R3) and (R4) that $\tau(u \to f_1) + \tau(u \to f_2) \leq 1 + \frac{3}{4}$, $\tau(u \to f_3) + \tau(u \to f_4) \leq 1 + \frac{3}{4}$, $\tau(u \to f_5) \leq \frac{1}{2}$, and thus $\omega'(u) \geq 4 - 2 \times (1 + \frac{3}{4}) - \frac{1}{2} = 0$.

When $m_3(u) = 1$ and assume w.l.o.g. $d(f_1) = 3$ with $f_1 = [u_1 u u_2]$, we let $s_3(u)$ be the number of 3-vertices in $N(u) \setminus \{u_1, u_2\}$ and we have $s_3(u) \leq n_3(u) \leq 2$ by Lemma 4. Using (R2)–(R4), we discuss the following three subcases for three possible values of $s_3(u)$.

Case 1.1. $s_3(u) = 0$. If $\delta(f_1) = 3$ with $d(u_1) = 3$, then $\tau(u \to f_1) = \frac{3}{2}$, $\tau(u \to f_i) \leq \frac{1}{2}$, $i \in \{2, 3, 4\}$, $\tau(u \to f_5) \leq 1$, and $\omega'(u) \geq 4 - \frac{3}{2} - 3 \times \frac{1}{2} - 1 = 0$. If $\delta(f_1) \geq 4$, then $\tau(u \to f_1) \leq 2$, $\tau(u \to f_i) \leq \frac{1}{2}$, $i \in \{2, 3, 4, 5\}$, and $\omega'(u) \geq 4 - 2 - 4 \times \frac{1}{2} = 0$.

Case 1.2. $s_3(u) = 1$. By Lemma 4, if $n_3(u) = 2$ then $d(u) = 5$ and one of u_1 and u_2 is a 3-vertex, which contradicts the inexistence of $(A_{5.2})$. Therefore,

$n_3(u) = 1$, which implies $\delta(f_1) \geq 4$ and further by (R2.2.3) and (R2.3) $\tau(u \to f_1) \leq \frac{5}{4}$. By symmetry, if $d(u_3) = 3$, then $\tau(u \to f_2) + \tau(u \to f_3) \leq 1 + \frac{3}{4}$, $\tau(u \to f_i) \leq \frac{1}{2}$, $i \in \{4,5\}$, leading to $\omega'(u) \geq 4 - \frac{5}{4} - (1 + \frac{3}{4}) - 2 \times \frac{1}{2} = 0$; if $d(u_4) = 3$, then $\tau(u \to f_3) + \tau(u \to f_4) \leq 1 + \frac{3}{4}$, $\tau(u \to f_i) \leq \frac{1}{2}$, $i \in \{1,5\}$, leading to $\omega'(u) \geq 4 - \frac{5}{4} - (1 + \frac{3}{4}) - 2 \times \frac{1}{2} = 0$.

Case 1.3. $s_3(u) = 2$. We have $d(u) = 5$, $n_3(u) = 2$ and $\delta(f_1) \geq 4$, and further by (R2.2.3) and (R2.3) $\tau(u \to f_1) \leq 1$. Since G contains no $(A_{3.1})$, all neighbors of the two 3-vertices in $N(u)$ are 5^+-vertices. By symmetry, if $d(u_3) = d(u_4) = 3$, then $\tau(u \to f_i) \leq \frac{3}{4}$, $i \in \{2,4\}$, $\tau(u \to f_3) \leq 1$, $\tau(u \to f_5) \leq \frac{1}{2}$, leading to $\omega'(u) \geq 4 - 2 \times 1 - 2 \times \frac{3}{4} - \frac{1}{2} = 0$; if $d(u_3) = d(u_5) = 3$, then $\tau(u \to f_j) \leq \frac{3}{4}$, $j \in \{2,3,4,5\}$, leading to $\omega'(u) \geq 4 - 1 - 4 \times \frac{3}{4} = 0$.

Case 2. $d_H(u) = 6$. In this case, $\omega(u) = 2d_H(u) - 6 = 6$. When $m_3(u) = 0$, $\omega'(u) \geq 6 - 1 \times 6 = 0$; when $m_3(u) = 1$, we discuss similarly as in Case 1 by using Lemmas 2–5. We leave the details in [12].

This finishes the proof that for every vertex u, $\omega'(u) \geq 0$.   □

Lemmas 6 and 7 together contradict the negative total weight of -12 stated in Eq. (1), and thus prove Theorem 2.

3   Acyclic Edge Coloring

In this section, we show how to derive an acyclic edge coloring, by an induction on $|E(G)|$. The following lemma gives the starting point.

Lemma 8. ([3,4,15,19]) *If $\Delta(G) \in \{3,4\}$, then $a'(G) \leq \Delta(G) + 2$.*

Assume that c is a partial acyclic edge k-coloring of the graph G using the color set $C = \{1, 2, \ldots, k\}$. For a vertex $v \in V(G)$, let $C(v)$ denote the set of colors assigned the edges incident at v under c. If the edges of a path $P = ux \ldots v$ are alternatively colored with colors i and j, we call it an $(i,j)_{(u,v)}$-path. Furthermore, if $uv \in E(G)$ is also colored by i or j, we call $ux \ldots vu$ an $(i,j)_{(u,v)}$-cycle. For simplicity, we use $\{e_1, e_2, \ldots, e_m\} \to a$ to state that all the edges e_1, e_2, $\ldots$, e_m are (re-)colored with a, $e_i \to S$ ($S \neq \emptyset$) to state that e_i is (re-)colored with a color in S, simply $e_1 \to a$ to state that e_1 is colored with a, and $(e_1, e_2, \ldots, e_m) \to (a_1, a_2, \ldots, a_m)$ to state that each e_j is (re-)colored with a_j respectively. We also use $(e_1, e_2, \ldots, e_m)_c = (a_1, a_2, \ldots, a_m)$ to denote that $c(e_j) = a_j$ respectively.

Lemma 9. [15] *Suppose G has an acyclic edge coloring c, and $P = uv_1v_2\text{-}\ldots\text{-}v_kv_{k+1}$ is a maximal $(a,b)_{(u,v_{k+1})}$-path with $c(uv_1) = a$ and $b \notin C(u)$. Then there is no $(a,b)_{(u,w)}$-path for any vertex $w \notin V(P)$.*

Proof of Theorem 1. The flow of the proof is depicted in Fig. 2. For the base cases, when $|E(G)| \leq \Delta + 2$, each edge is colored distinctly; when $\Delta \leq 4$, we assume an acyclic edge $(\Delta + 2)$-coloring by Lemma 8.

In the sequel we assume that $\Delta \geq 5$ and $|E(G)| \geq \Delta + 3$. Recall that a connected graph has a block-cut tree representation in which each block is a 2-connected component and two blocks overlap at most one cut-vertex; we assume w.l.o.g. that G is 2-connected, since if necessary we may swap any two colors for the edges in a block. Theorem 2 tells that G contains at least one of the configurations (A_1)–(A_6). One sees that each of (A_1)–(A_6) contains an edge uv such that $d(u) + d(v) \leq 8$ or $d(v) \in \{2,3\}$. We pick such an edge uv and let $H = G - uv$ (that is, $H = (V, E \setminus \{uv\})$). The graph H is also planar without intersecting triangles and $\Delta \geq \Delta(H) \geq \Delta - 1 \geq 4$. By the inductive assumption, H has an acyclic edge $(\Delta+2)$-coloring c using the color set $C = \{1, 2, \ldots, \Delta+2\}$. Since $d_H(u) + d_H(v) \leq 6$, or $d_H(u) \leq \Delta - 1$ and $d_H(v) \in \{1,2\}$, we conclude from $|C| = \Delta + 2 > \max\{6, \Delta + 1\}$ that $C \setminus (C(u) \cup C(v)) \neq \emptyset$.

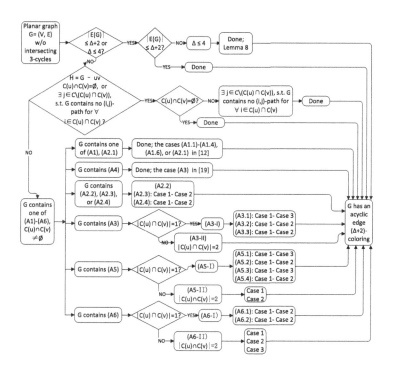

Fig. 2. The flow of the proof of Theorem 1.

If $C(u) \cap C(v) = \emptyset$, then let $uv \to C \setminus (C(u) \cup C(v))$, which gives an acyclic edge $(\Delta + 2)$-coloring for G. Likewise, if there exists $j \in C \setminus (C(u) \cup C(v))$ such that H contains no $(i, j)_{(u,v)}$-path for any $i \in C(u) \cap C(v)$, then let $uv \to j$. In the remaining case, for each $i \in C(u) \cap C(v)$, let $B_i = \{j \mid$ there is an $(i, j)_{(u_s, v_t)}$-path$\}$, where $c(uu_s) = c(vv_t) = i$. One clearly sees that $B_i \subseteq C(u_s) \cap C(v_t)$. In the sequel we continue the proof with the following assumption $(*_1)$:

$(*_1)$ $C(u) \cap C(v) \neq \emptyset$; for any $j \in C \setminus (C(u) \cup C(v))$, H contains an $(i,j)_{(u,v)}$-path for some $i \in C(u) \cap C(v)$; and $C \setminus (C(u) \cup C(v)) \subseteq \bigcup_{i \in C(u) \cap C(v)} B_i$.

It is not hard to see that if G contains (A_1) or $(A_{2.1})$, then the proof for the cases $(A_{1.1})$–$(A_{1.4})$, $(A_{1.6})$ and $(A_{2.1})$ in [13] can be adopted; if G contains (A_4), then the proof for the case (A_3) in [20] can be adopted. In [12] we provide the complete proof for the case where G contains one of the configurations $(A_{2.2})$–$(A_{2.4})$, (A_3), and (A_5), and show that an acyclic $(\Delta + 2)$-edge coloring can be achieved. The rest of the section deals with the case where G contains the configuration (A_6).

Assume $c(uu_i) = i$ for $1 \leq i \leq 5$ and $C_6^* = C \setminus \{1,2,3,4,5\}$. Note that $|C_6^*| \geq 3$, $N(u) \cap N(v) = \emptyset$, and $(C \setminus (C(u) \cup C(v))) \subseteq \bigcup_{i \in C(u) \cap C(v)} B_i$. By Assumption $(*_1)$, $1 \leq |C(u) \cap C(v)| \leq 2$. We thus distinguish two cases for the two possible values of $|C(u) \cap C(v)|$. We demonstrate one case where $|C(u) \cap C(v)| = 2$ (denoted as $(A_6\text{-II})$), i.e., $(vv_1, vv_2)_c = (a,b)$ where $a, b \in C(u)$, while leave the other case (denoted as $(A_6\text{-I})$) to [12].

Let $S_i = \{c(uu_j) \mid d(u_j) = i\}$ for $i \in \{3,4\}$.

If there exists $i \in C_6^* \setminus C(v_1)$, then let $vv_1 \to i$ and reduce the proof to Case $(A_6\text{-I})$. Otherwise, $C_6^* \subseteq C(v_1) \cap C(v_2)$. If for any $l \in C(u) \setminus \{a,b\}$, G contains no $(k,l)_{(u_a,u_l)}$-path for some $k \in C_6^* \setminus C(u_a)$, then let $uu_a \to k$ and reduce the proof to Case $(A_6\text{-I})$. Otherwise, we conclude that:

$(*_{6.1})$ For each $i \in \{a,b\}$, $C_6^* \subseteq C(u_i)$ or for any $k \in C_6^* \setminus C(u_i)$, G contains a $(k,l)_{(u_i,u_l)}$-path for some $l \in C(u) \setminus \{a,b\}$.

Conclusion $(*_{6.1})$ implies that $C(v) \setminus S_3 \neq \emptyset$ and we consider the following three cases. Note that $C_6^* \subseteq C(v_2) \cap C(v_1)$.

Case $(A_6\text{-II-1})$. $(vv_1, vv_2)_c = (1,2)$ for $(A_{6.1})$. If G contains no $(i,2)_{(v_1,v_2)}$-path for some $i \in \{3,4,5\} \setminus C(v_1)$, then let $vv_1 \to i$ and we are done by Conclusion $(*_{6.1})$; otherwise, G contains an $(i,2)_{(v_1,v_2)}$-path for every $i \in \{3,4,5\} \setminus C(v_1)$. It follows that $C(v_1) = C_6^* \cup \{1,2,3\}$ and $C(v_2) = C_6^* \cup \{2,4,5\}$. By Conclusion $(*_{6.1})$, $C(u_2) \cap \{3,4,5\} \neq \emptyset$. Note that $(C_6^* \setminus C(u_2)) \subseteq B_1$. Further we have $(C_6^* \setminus C(u_2)) \subseteq B_i$ if letting $(vv_1, vv_2) \to (j,1)$ for $j \in \{4,5\}$. It implies that $\Delta = 6$, $C(u_2) = \{2,3,6\}$ and $C(u_4) = \{4,7,8\}$, $C(u_5) = \{5,7,8\}$. We let $(uu_4, uv) \to (6,4)$.

Case $(A_6\text{-II-2})$. $(vv_1, vv_2)_c = (1,3)$ for $(A_{6.2})$. By Conclusion $(*_{6.1})$, $C(u_3) \cap \{2,4,5\} \neq \emptyset$ and assume that $7,8 \in B_1 \setminus C(u_3)$. If G contains no $(i,3)_{(v_1,v_2)}$-path for some $i \in \{4,5\} \setminus C(v_1)$, let $vv_1 \to i$ and we are done by Conclusion $(*_{6.1})$; otherwise, $\{4,5\} \subseteq C(v_1)$ or G contains an $(i,3)_{(v_1,v_2)}$-path for every $i \in \{4,5\} \setminus C(v_1)$.

Case 2.1. $C(v_1) = C_6^* \cup \{1,4,5\}$. If there exists an $i \in \{4,5\} \setminus C(v_2)$, letting $(vv_1, vv_2) \to (3,i)$ and we are done by Conclusion $(*_{6.1})$; otherwise, $C(v_2) = C_6^* \cup \{3,4,5\}$. We let $(vv_1, vv_2, uv) \to (3,1,7)$.

Case 2.2. $3 \in C(v_1)$ and G contains an $(i,3)_{(v_1,v_2)}$-path for every $i \in \{4,5\} \setminus C(v_1)$. It follows that $\{4,5\} \setminus C(v_1) \neq \emptyset$ and assume that $5 \notin C(v_1)$. If $1 \notin C(v_2)$, first let $(vv_1, vv_2) \to (5,1)$ and then, $uv \to k$ if there is a $k \in \{7,8\} \setminus B_5$, or else

$(uu_5, uv) \rightarrow (6, 7)$. If $1 \in C(v_2)$, then it follows that $C(v_1) = C_6^* \cup \{1, 3, 4\}$ and $C(v_2) = C_6^* \cup \{3, 1, 5\}$. Letting $vv_1 \rightarrow 2$ and we are done by similar discussion since $2 \notin C(v_2)$.

Case (A_6-II-3). $(vv_1, vv_2)_c = (1, 2)$ for $(A_{6.2})$. If $1 \notin C(v_2)$, then we let $vv_2 \rightarrow \{3, 4, 5\} \setminus C(v_2)$; otherwise, we have $1 \in C(v_2)$, $2 \in C(v_1)$, and there exists $j_1 \in \{3, 4, 5\} \setminus (C(v_1) \cup C(v_2))$. We then let $vv_2 \rightarrow j_1$ to reduce to the above Case (A_6-II-2). $\qquad\square$

References

1. Alon, N., Mcdiarmid, C., Reed, B.: Acyclic coloring of graphs. Random Struct. Algorithms **2**, 277–288 (1991)
2. Alon, N., Sudakov, B., Zaks, A.: Acyclic edge colorings of graphs. J. Graph Theory **37**, 157–167 (2001)
3. Andersen, L.D., Máčajová, E., Mazák, J.: Optimal acyclic edge-coloring of cubic graphs. J. Graph Theory **71**, 353–364 (2012)
4. Basavaraju, M., Chandran, L.S.: Acyclic edge coloring of graphs with maximum degree 4. J. Graph Theory **61**, 192–209 (2009)
5. Basavaraju, M., Chandran, L.S., Cohen, N., Havet, F., Müller, T.: Acyclic edge-coloring of planar graphs. SIAM J. Discrete Math. **25**, 463–478 (2011)
6. Esperet, L., Parreau, A.: Acyclic edge-coloring using entropy compression. Eur. J. Comb. **34**, 1019–1027 (2013)
7. Fiamcik, J.: The acyclic chromatic class of a graph. Math. Slovaca **28**, 139–145 (1978)
8. Giotis, I., Kirousis, L., Psaromiligkos, K.I., Thilikos, D.M.: Acyclic edge coloring through the Lovász local lemma. Theoret. Comput. Sci. **665**, 40–50 (2017)
9. Hou, J., Roussel, N., Wu, J.: Acyclic chromatic index of planar graphs with triangles. Inf. Process. Lett. **111**, 836–840 (2011)
10. Molloy, M., Reed, B.: Further algorithmic aspects of the local lemma (1998)
11. Ndreca, S., Procacci, A., Scoppola, B.: Improved bounds on coloring of graphs. Eur. J. Comb. **33**, 592–609 (2012)
12. Shu, Q., Chen, Y., Han, S., Lin, G., Miyano, E., Zhang, A.: Acyclic edge coloring conjecture is true on planar graphs without intersecting triangles. arXiv (2020)
13. Shu, Q., Wang, W., Wang, Y.: Acyclic edge coloring of planar graphs without 5-cycles. Discrete Appl. Math. **160**, 1211–1223 (2012)
14. Shu, Q., Wang, W., Wang, Y.: Acyclic chromatic indices of planar graphs with girth at least 4. J. Graph Theory **73**, 386–399 (2013)
15. Shu, Q., Wang, Y., Ma, Y., Wang, W.: Acyclic edge coloring of 4-regular graphs without 3-cycles. Bull. Malays. Math. Sci. Soc. **42**, 285–296 (2019). https://doi.org/10.1007/s40840-017-0484-x
16. Vizing, V.G.: On an estimate of the chromatic class of a p-graph. Discret Analiz **3**, 25–30 (1964)
17. Wang, T., Zhang, Y.: Acyclic edge coloring of graphs. Discrete Appl. Math. **167**, 290–303 (2014)
18. Wang, T., Zhang, Y.: Further result on acyclic chromatic index of planar graphs. Discrete Appl. Math. **201**, 228–247 (2016)
19. Wang, W., Ma, Y., Shu, Q., Wang, Y.: Acyclic edge coloring of 4-regular graphs (II). Bull. Malays. Math. Sci. Soc. **42**, 2047–2054 (2019). https://doi.org/10.1007/s40840-017-0592-7

20. Wang, W., Shu, Q., Wang, Y.: Acyclic edge coloring of planar graphs without 4-cycles. J. Comb. Optim. **25**, 562–586 (2013)
21. Wang, W., Shu, Q., Wang, Y.: A new upper bound on the acyclic chromatic indices of planar graphs. Eur. J. Comb. **34**, 338–354 (2013)
22. Wang, Y., Shu, Q., Wu, J.-L., Zhang, W.: Acyclic edge coloring of planar graphs without a 3-cycle adjacent to a 6-cycle. J. Comb. Optim. **28**(3), 692–715 (2014). https://doi.org/10.1007/s10878-014-9765-6

On Pure Space vs Catalytic Space

Sagar Bisoyi[1], Krishnamoorthy Dinesh[2], and Jayalal Sarma[1][✉]

[1] Indian Institute of Technology Madras, Chennai, India
{CS17S020,jayalal}@cse.iitm.ac.in
[2] Chinese University of Hong Kong, Sha Tin, China
krishnamoorthydinesh@cuhk.edu.hk

Abstract. This paper explores the power of catalytic computation when the catalytic space ($c(n)$, the full memory for which the content needs to be restored to original content at the end of the computation) is much more than exponential in the pure space ($s(n)$, the empty memory which does not have any access/restoration constraints). We study the following three regimes of the relation between $s(n)$ and $c(n)$ and explore the class CSPACE($s(n), c(n)$) in each of them.

- **Low-end regime :** $s(n) = O(1)$ We define the classes CR and CNR (nondeterministic variant of CR) where $s(n) = O(1)$ and $c(n) = $ poly(n). Exploring the connection between computational power of one counter machines (OC) and constant pure space catalytic Turing machines, we observe that OC $\subseteq$ CR and show that CR $\subseteq$ OC $\implies$ CR $\neq$ CNR. We prove that L $\not\subseteq$ CSPACE($O(1), o(\sqrt{n})$)
- **Low-end non-constant regime:** $s(n) = o(\log \log n)$: Let M be an oblivious catalytic Turing machine using $s(n)$ pure space and $c(n)$ catalytic space such that $s(n) + \log c(n) = o(\log \log n)$ then $L(M)$ is regular. This strengthens the classical theorem on $s(n) = o(\log \log n)$ to the case of catalytic Turing machines.
- **High-end regime:** $s(n) = O(c(n)^\epsilon)$: We show an implementation of incremental dynamic program using catalytic machines, thus showing that Knapsack problem (with n items, sum of their costs as C and the capacity of the bag as K) can be solved in $O(n \log n \log C + \log(nKC))$ pure space and $O(n^2 K C^3 \log^2 K \log n)$ catalytic space. Hence, catalytic algorithms can lead to a non-trivial saving in the pure space required for computation when K is $\Omega(n)$.

Our techniques include interesting generalizations of crossing sequence arguments and implementations of incremental dynamic programs using catalytic algorithms. These may be of independent interest.

1 Introduction

Space complexity - one of the central subareas of computational complexity strives to understand the power and limitations of space efficient Turing machines - deterministic, non-deterministic and randomized variants. The logarithmic space (in short logspace), is an important regime in this line of study and leads to the famous L vs NL problem, which asks if there is a deterministic logspace algorithm for any problem solvable via a non-deterministic logspace algorithm.

© Springer Nature Switzerland AG 2020
J. Chen et al. (Eds.): TAMC 2020, LNCS 12337, pp. 439–451, 2020.
https://doi.org/10.1007/978-3-030-59267-7_37

A recent innovation in the direction of space complexity is the notion of *catalytic space* introduced by [1] (see [5] for a survey). In this notion, the Turing machine is given larger space (which is full of otherwise useful data) which it is allowed to use for the computation under the promise that the original data that the space contained before the computation will be restored at the end of the computation[1], in addition to the computation being performed correctly. The motivating question then, is whether we will be able to solve more problems using this larger (full) space which we otherwise do not know how to solve in small space.

The class of languages that can be solved in $s(n)$ pure space and $c(n)$ catalytic space (where n is the length of the input) is denoted by $\mathsf{CSPACE}(s(n), c(n))$. Indeed, for the complexity class $\mathsf{CSPACE}(s(n), c(n))$ to be meaningful, it is necessary that $c(n) \geq s(n)$ for otherwise, it will be contained in $\mathsf{SPACE}(s(n))$. Buhrman *et al.* [1] limits their study to the class that $c(n)$ cannot be more than $2^{O(s(n))}$ and restricts the study to the class $\mathsf{CL} = \bigcup_{c>0} \mathsf{CSPACE}(c \log n, n^c)$. They prove that this class is surprisingly powerful and contains the complexity class TC^1 which[2] in turn contains NL too. That is, there is a deterministic algorithm for reachability problem which uses polynomial catalytic space and logarithmic pure-space which can be viewed (1) as an interesting step in both theoretical quest for understanding the power of logspace bounded algorithms, and (2) the practical aspects of designing space bounded algorithms for solving the fundamental problem of reachability testing.

A natural question is whether the restriction that $c(n)$ cannot be more than $2^{O(s(n))}$ is required. Indeed, if this is not the case, the catalytic Turing machine cannot even store the index to the catalytic tape in the pure space. However, it can be easily seen that the model can accept even non-regular languages and hence deserves a careful study.[3]

Low-End Regime: When $s(n)$ is Constant: As a first step towards understanding the computational power of catalytic Turing machines when $c(n)$ is not upper bounded by $2^{O(s(n))}$, we study the extreme case when $s(n)$ is constant and $c(n)$ is poly(n). We call such a class as *catalytic regular* (CR). That is, $\mathsf{CR} = \bigcup_{c \geq 0} \mathsf{CSPACE}(O(1), n^c)$. Let OC denote the set of all languages that can be accepted by a two-way counter machine with counter value restricted to at most a polynomial in the input length. It is easy to observe that $\mathsf{OC} \subseteq \mathsf{CR}$. On the other hand, we explore the nondeterministic version of CR which we will define by $\mathsf{CNR} = \bigcup_{c \geq 0} \mathsf{CNSPACE}(O(1), n^c)$ by using the standard definition of non-deterministic catalytic Turing machines (where the restoration condition is imposed for all non-deterministic paths). Indeed, power of non-determinism in

---

[1] See Sect. 2 for a formal definition.

[2] The class of languages that can be computed by uniform Boolean circuits of depth $O(\log n)$ and size $O(n^k)$ for a constant k, over $\{\mathsf{MAJ}, \neg\}$ gates.

[3] Koucký [5], while justifying the choice that $c(n) = (2^{O(s(n))})$ in [1], remarks that this restricted variant of catalytic Turing machines (with $c(n) = 2^{\omega(s(n))}$) are possibly equivalent to counter machines in terms of the languages that they accept.

this setting is a natural question which amounts to, is CR = CNR? Although we do not answer this question, we prove the following connection.

Theorem 1. *If* CR $\subseteq$ OC, *then* CR $\neq$ CNR.

The heart of the argument is a non-deterministic algorithm for a language L_{dg} (see Sect. 2 for a definition) due to Duris and Galil [3] which is known to be outside OC.

Given the definition of the catalytic regular class, a natural attempt to compare with the standard complexity classes, leads to the question - can we simulate logarithmic pure space with polynomial catalytic space, that is - is L $\subseteq$ CR ? Indeed, both containments are far from clear. To this end, using tools from communication complexity and a variant of the crossing sequence argument for catalytic Turing machines, we show,

Theorem 2. L $\not\subseteq$ CSPACE$(O(1), o(\sqrt{n}))$.

We also strengthen the above theorem to show that L $\not\subseteq$ CNSPACE$(O(1), o(\sqrt{n}))$ thus making progress on even the question is L $\subseteq$ CNR.

Low-End Non-constant Regime: When $s(n)$ **is** $o(\log \log n)$ **:** Continuing the attempt to understand the power catalytic Turing machines when $c(n)$ is not upper bounded by $2^{O(s(n))}$. In the standard Turing machine model, it is known that languages accepted by Turing machines using at most $o(\log \log n)$ space is regular. We ask whether this statement would hold if we augment such Turing machines with a catalytic space at most $o(\log n)$. We prove the following general theorem in this context.

Theorem 3. *Let M be an oblivious catalytic TM that uses $s(n)$ pure space and $c(n)$ catalytic space such that $s(n) + \log c(n) = o(\log \log n)$. Then $L(M)$ is regular.*

The main technical part involved in the above theorem is an adaptation of the crossing sequence argument for catalytic Turing machines. Note that we require obliviousness restriction for the Turing machine in the above theorem, since it is not known in general how to convert a general catalytic Turing machine to an oblivious one accepting the same language without an additive $\Omega(\log n)$ increase in space.

High-end Regime: When $s(n) = O(c(n)^\epsilon)$ **for** $\epsilon < 1$: Buhrman *et al.* [1] shows that even though we have $c(n) \leq 2^{O(s(n))}$, there are nontrivial computational tasks which, uses catalytic space in a non-trivial way, and can be done in much less catalytic space than $2^{O(s(n))}$. They show (Theorem 18 in [1]) that performing iterated multiplication of n matrices (with integer entries) of order $2^{\sqrt{\log n}}$ (sublinear in n) can be done in $O(\log n)$ pure space and $O(2^{\sqrt{\log n}})$ catalytic space.

Our results in this regime, presents an argument that the catalytic space indeed helps in reducing the amount of pure space required to solve a problem. We prove this for the Knapsack problem where, we are given n items with i^{th} item of weight $w_i \geq 0$ of reward $c_i \geq 0$ and a non-negative integer K

(Knapsack capacity). The goal is to find $\max_{I \subseteq [n]} \sum_{i \in I} c_i$ subject to the constraint that $\sum_{i \in I} w_i \leq K$. Let C be the sum of the cost of all items.

Theorem 4. *The Knapsack problem can be solved in $O(n \log n \log C + \log(nKC))$ pure space and $O(n^2 KC^3 \log^2 K \log n)$ catalytic space.*

On the other hand, this problem can be solved by a dynamic programming based algorithm that uses $O(K \log C)$ pure space, where C is the sum of the costs of the different items. Hence, catalytic algorithm can do a significant saving in the pure space required for computation ($O(K \log C)$ versus $O(n \log K \log C)$) when K is $\Omega(n)$.

2   Preliminaries

In this section, we define the complexity theoretic preliminaries required for the rest of the paper. $\mathsf{SPACE}(s(n))$ (resp. $\mathsf{NSPACE}(s(n))$) is the class of all languages that can be decided by a (non)-deterministic Turing machine using $O(s(n))$ space of the work-tape. The class $\mathsf{SPACE}(\log n)$ (resp. $\mathsf{NSPACE}(s(n))$) is denoted by L (resp. NL).

Catalytic Computation: A *Catalytic Turing Machine* (defined by [1]) is a deterministic Turing machine with an input tape, a work-tape and a catalytic tape (both equipped with a left end marker symbol). M uses pure space $s(n)$ and catalytic space $c(n)$ if for all inputs $x \in \{0,1\}^n$ and catalytic tape contents $w \in \{0,1\}^{c(n)}$, $M(x, w)$ on input x halts with w on its catalytic tape and the worktape head does not move beyond $s(n)$ from the left endmarker ($\vdash$) and the catalytic space head does not move beyond $c(n)$ cells from the left endmarker. In this case, we denote, $s(n)$ as the pure space and $c(n)$ as the catalytic space.
    $\mathsf{CSPACE}(s(n), c(n))$ is the class of all languages decided by a catalytic Turing machine using pure space $s(n)$ and auxiliary space $c(n)$. Thus, a language is in $\mathsf{CSPACE}(s(n), c(n))$ if there is a catalytic Turing machine with $s(n)$ pure space and $c(n)$ catalytic space such that $\forall x \in \{0,1\}^n$, $\forall w \in \{0,1\}^{c(n)}$: (1) $M(x, w)$ accepts on input x if and only $x \in L$ and (2) $M(x, w)$ halts with w as the content of the catalytic tape. Let $\mathsf{Time}_M(x, w)$ denote the time taken by catalytic machine M on catalytic content w on the input x.
    A non-deterministic catalytic Turing machine is defined (in [2]) similar to the above, except that the restoration condition is applied to all non-deterministic paths of the Turing machine. That is, a language L is said to be accepted by a non-deterministic catalytic Turing machine M if for all input $x \in \{0,1\}^n$, $\forall w \in \{0,1\}^{c(n)}$ (1) $M(x, w)$ has at least one non-deterministic path that accepts if and only if $x \in L$ and (2) in all non-deterministic paths of $M(x, w)$, $M(x, w)$ halts with w as the content of the catalytic tape. Thus, $\mathsf{CNSPACE}(s(n), c(n))$ is the class of all languages decided by a catalytic Turing machine using pure space $s(n)$ and auxiliary space $c(n)$.
    $\mathsf{CL} = \bigcup_{c>0} \mathsf{CSPACE}(c \log n, n^c)$ $\mathsf{CNL} = \bigcup_{c>0} \mathsf{CNSPACE}(c \log n, n^c)$

Transparent Programs: [1]. The model is a register machine over a ring R working with read-only input registers $x_1, x_2, \ldots x_n$ and working registers $r_1, r_2, \ldots r_m$ where each register holds an element of R. An instruction is of the form $r_i \leftarrow r_i \pm u \times v$, where u and v are fixed elements from R or registers other than r_i. When the $+$ and $-$ are interchanged we can undo the effect of the instruction and hence it is called the inverse of the original instruction and these instructions are called reversible instructions. A program is a list of instructions and a reversible program is a program where each instruction is reversible. A function $f(x)$ is said to be *transparently computed by a reversible program* to a register r_i if the program, when executed on the registers with in initial values $\alpha_1, \alpha_2, \ldots \alpha_i \ldots \alpha_m$ ends with values $\alpha_1, \alpha_2, \ldots \alpha_i + f(x) \ldots \alpha_m$ in the respective registers. When the function is computed by the program in a transparent way, by a slight misuse of terminology, we also call the reversible program to be a transparent program.

Proposition 1 (Arithmetic circuits to Transparent programs). *Suppose f (on n variables) is computed by a layered arithmetic circuit of size s and depth d, with the ring operations being $+$ and $\times$ (bounded fan-in), then f can be computed by a transparent program (over the same ring) of length $O(s2^{d+1})$, $O(s)$ registers and n inputs. If the fan-in of $+$ gate is allowed to be unbounded, then the same holds except that length of the resulting program is $O(s^2 2^{d+1})$.*

The above proposition follows from Corollary 6 [1]. The original claim works for general arithmetic circuits and also allows $+$ fan-in to be unbounded. Additionally, the following proposition implies that we can simulate transparent programs catalytically.

Proposition 2 (Transparent programs to Catalytic algorithms). *Given a transparent program (over ring R) of length ℓ, with m registers and n inputs (each from R), there exists a catalytic TM that can simulate the transparent program in catalytic space $O(m \cdot \log |R|)$ and pure space $O(\log \ell + \log n + \log |R|)$.*

The above proposition follows from Lemma 15 of [1] and improvement on the catalytic space is based on the remark appearing in the proof of Lemma 15.

Counter Machines: A two-way deterministic counter (denoted as 2DC) machine M is a 5-tuple $(Q, \Sigma, q_0, \delta, F)$ where Q is a finite set of states, q_0 a special start state, $F \subseteq Q$ the set of accepting states and Σ is a finite input alphabet. The input is given in the input tape of the machine enclosed in start and end markers. δ is a mapping from $Q \times \Sigma \times \{0, 1\}$ to $Q \times \{L, R\} \times \{-1, 0, +1\}$. The transition function takes three input parameters: the current state, the current symbol being read, and the status of the counter (say, 0 if the counter reads zero and 1 if non-zero) and then it changes the state, moves the head to left or right, and changes the counter value by $-1, 0$ or $+1$. The counter can contain only non-negative integers, decrementing a counter containing 0 is not allowed. Define OC to be the set of languages that can be accepted by 2-way deterministic one counter machines.

Let $L = \{w\#w \mid w \in \Sigma^*\}$ and $L' = \{ww^R \mid w \in \Sigma^*\}$. It is known that (c.f. [3]) both L and L' belongs to OC with counter value being bounded by $O(n)$. Now, consider the language

$$L_{dg} = \left\{ w_0\#w_1\#\dots\#w_k \left| \begin{array}{l} \exists n \geq 1, k \geq 1, \forall i \in \{0,\dots,k\}, w_i \in \Sigma^n \\ \text{and } \exists j \neq 0, w_j = w_0 \end{array} \right. \right\}$$

Duris and Galil [3] showed that this language cannot be recognized by two-way deterministic one-counter machines. That is, $L_{dg} \notin OC$.

Communication Complexity: Given a Boolean function $f : \{0,1\}^n \times \{0,1\}^n \to \{0,1\}$, and n bit strings x, y. Consider the model of computation where there are two parties - Alice (having x) and Bob (having y) and they exchange bits to compute $f(x,y)$. We call the strategy adopted by both the parties as protocol π and the outcome of the protocol is denoted by $\pi(x,y)$. A protocol π is said to compute f if for every $x, y \in \{0,1\}^n \times \{0,1\}^n, \pi(x,y) = f(x,y)$. The cost of a protocol π is $\max_{x,y} |\pi(x,y)|$ and the cost of computing f is the cost of the best cost protocols π computing f. We consider two settings of the model - non-deterministic. If the parties are non-deterministic, then we denote $N^1(f)$ to denote cost of checking if $f(x,y)$ is 1 by an optimal non-deterministic protocol. In the randomized setting, Alice and Bob have access to a shared source of randomness which they can use in the computation. We denote $R_0^{pub}(f)$ as the optimal expected cost of correctly computing f using (public) randomness. For more details on the two settings of the models, see [7].

3   Low-End Regime: Catalytic Regular

As discussed in the introduction, in this section we aim to study the power of catalytic computation with a low-end relationship between the catalytic and pure space. That is, we consider the pure space to be a constant, which is as good as $s(n) = 0$ since the state space is enough to represent the content of the work tape as there are only a constant number of symbols to store. We define, $CR = \bigcup_{c>0} CSPACE(O(1), n^c)$. It is natural in this case to ask if the machine is more powerful than a finite automata. Indeed, it can accept the non-regular language $\{a^n b^n \mid n \geq 0\}$ over the alphabet $\{a, b\}$ by using the head position of the catalytic tape without modifying the catalytic tape at all (hence restoration condition is trivially satisfied by the algorithm). This idea is more general and is captured by the following easy proposition.

Proposition 3. $OC \subseteq CR$.

Indeed, the above proposition follows from the fact that the head position of the catalytic tape can be used as a counter which can do increment and decrement the counter (corresponds to moving the head to right and left respectively) and test for zero (corresponds to testing whether the current symbol read is the left endmarker or not).

The question of is $CR \subseteq OC$? is harder. To address this question, we consider the non-deterministic variant of CR, defined as $CNR = \bigcup_{c>0} CNSPACE(O(1), n^c)$. A natural question is whether non-determinism is more powerful in this resource regime - the CR vs CNR question - if there exists a language which is $CNSPACE(O(1), n^c)$ but not in $CSPACE(O(1), n^{c'})$ for any constant c'.

We now prove Theorem 1 from the introduction. That is, if CR is contained in OC, then it has to be that CR is different from CNR. This follows from the following result about the language L_{dg}. The proof is omitted due to space limitations.

Proposition 4. *The language L_{dg} is in* $CNSPACE(O(1), O(n))$.

Limitations of CR **: The** CR **vs** L **problem:** In this section, we show that there exists a language L in L such that it cannot be computed in $CNSPACE(O(1), o(n))$, thus proving Theorem 2. To this end, we prove a slightly general result on the power of $CSPACE(s(n), c(n))$ for functions $s(n)$ and $c(n)$ (with $s(n) \leq c(n)$). (We do not assume here that $s(n) \geq \log c(n)$).

Lemma 1. *Let $f_n : \{0,1\}^n \times \{0,1\}^n \rightarrow \{0,1\}$ be a family of Boolean functions and M be a catalytic machine using $s(n)$ pure space and $c(n)$ catalytic space accepting the language $L_f = \{xzy \in \{0,1\}^* \mid |x| = |y| = n, z = 0^n$ and $f_n(x,y) = 1, n \geq 1\}$. Then, $R_0^{pub}(f_n) = O(c(n)^2 s(n) \cdot 2^{s(n)})$.*

Proof. The idea is to have Alice and Bob simulate the catalytic machine after fixing the catalytic content at random. Then, a location in z part of the input is chosen at random (using public randomness) as the crossing over point at which the entire configuration of the simulation is exchanged. They output 1 if the machine accepts and 0 otherwise. We bound the expected bits exchanged.

Before starting, Alice and Bob fixes a location in z at random (using public randomness). The protocol is as follows: Alice sets the catalytic content at random and simulates M as long as the input head stays to the left of this location. Once it crosses this location, Alice sends the current configuration of the simulation to Bob who then continues the simulation as long as the input head stays to right of this location. As soon as it crosses, Bob communicates his configuration of the simulation to Alice. This is continued till the simulation is completed. The protocol accepts if and only if the machine M accepts the string xzy.

For the input xzy, we now estimate the average number of times the head of M crosses the boundary. Let $cr(w,i)$ denote the number of times the input head of M crosses the cell i in z on the catalytic content $w \in \{0,1\}^{c(n)}$. The average crossings is given by $\frac{\sum_{i,w} cr(w,i)}{n \cdot 2^c}$ which is same as $\frac{1}{n} \frac{\sum_w \sum_i cr(w,i)}{2^c}$. Using the fact that the total number of crossing cannot exceed the run time (Lemma 2), this is upper bounded by $\frac{1}{n} \frac{\sum_w Time_M(xzy,w)}{2^c}$.

Since M is catalytic, by Lemma 2, the total runtime of M over all the catalytic content is at most the number of configurations which is $O(2^{s+c+\log n+\log s+\log c})$. (for convenience, we drop the parameter n from $c(n)$ and $s(n)$). Hence, the expected number of crossings is $O(c \cdot s \cdot 2^s)$. In each crossing, we send the configuration which is $O(c + \log c + s + \log s) = O(c + s)$ bits. Hence, the expected number of bits exchanged is at most $O((c+s) \cdot c \cdot s \cdot 2^s) = O(c^2 s \cdot 2^s)$ as $s(n) \leq c(n)$. $\qquad \square$

Corollary 1. *For f_n being the equality function on n bits, since[4] $R_0^{pub}(f_n) = \Omega(n)$, any catalytic machine accepting L_f using $s(n) = O(1)$ requires $c(n)$ to be at least $\Omega(n)$. Since $L_f \in \mathsf{L}$, $\mathsf{L} \not\subseteq \mathsf{CSPACE}(O(1), o(\sqrt{n}))$.*

An argument similar to Lemma 1 can be adapted to show the similar result in the case of non-deterministic settings as well thus showing that $\mathsf{L} \not\subseteq \mathsf{CNSPACE}(O(1), o(\sqrt{n}))$. This also shows that the complexity classes AC^0 and $\mathsf{CNSPACE}(O(1), o(n))$ are incomparable as, L_f (defined in Corollary 1) is in AC^0 and not in CR and $\oplus_n \in \mathsf{CNSPACE}(O(1), o(n))$ but not in AC^0. This is in contrast with CL which contains L-uniform TC^1.

4 Low-End Non-constant Regime: $o(\log \log n)$ Pure Space Is Equivalent to No Space

It is known that any $s(n)$ space bounded Turing machine with $s(n) = o(\log \log n)$ can only accepts regular languages. For the setting of catalytic machines, this implication holds when $s(n) + c(n) = o(\log \log n)$. Using standard crossing sequence based arguments we show that the same holds when $s(n) + \log c(n) = o(\log \log n)$ *for oblivious catalytic Turing machines.*

The following result on the number of configurations of a catalytic machine is implicit in Theorem 19 [1] where it is used to conclude that $\mathsf{CL} \subseteq \mathsf{ZPP}$. We state it here for catalytic Turing machines that are also oblivious with an improvement on the count which is crucial for our result.

Lemma 2. *For a catalytic Turing machine M with pure space $s(n)$ and catalytic space $c(n)$, on input string x of length n, there exists a w such that M on initial catalytic tape content w, reaches at most $O(2^{s(n)} \cdot c(n) \cdot s(n) \cdot n)$ many configurations. If M is an oblivious catalytic Turing machine, the bound on the number of configuration can be improved to $O(2^{s(n)} \cdot c(n) \cdot s(n))$.*

We now use the above lemma to prove the main result of this section.

Proof (of Theorem 3). We use a standard recipe for this proof based on crossing sequences and then overcome the technical difficulties in our situation. The argument for showing that $o(\log \log n)$ space bounded Turing machines can compute only regular languages has two parts (1) establish an upper bound on the number of crossing sequences on a given x and, (2) for any c, choose the minimal x which uses more than c cells, and contradict minimality of x by establishing that a shorter string would also have used more than c cells. In our context,

---

[4] Holds since $N^1(f_n) \geq n$ and $R_0^{pub}(f_n) = \Omega(N^1(f_n) - \log n)$. See Chap. 3 of [7] for details.

(1) itself is nontrivial since there are exponentially many configurations in $c(n)$ and $s(n)$. By using the tools from the catalytic computation, we establish that there is a "nice" string $w \in \{0,1\}^{c(n)}$ (depending on x) which when used as the catalytic content for which the number of crossing sequences is still similarly upper bounded. To address (2), by using standard cut-and-paste arguments, we derive that the string w is still a nice string for the shorter string x' we choose as well.

We now execute this plan formally. For an input x of length n and the initial catalytic content $w \in \{0,1\}^{c(n)}$, we first define a semi-configuration of M and crossing sequence of M. Without loss of generality, let the input alphabet set, the catalytic tape alphabet set and the work tape alphabet set be from $\{0,1\}$.

Define the *semi-configuration* of M, during its computation to be a tuple consisting of (1) the current state of M (2) the symbol being scanned by the input head (3) the contents of the catalytic tape and (4) the current position of the catalytic head (5) the contents of the work tape and (6) the current position of the work head. The number of possible *semi-configuration* of M for an input x and initial catalytic content w, denoted by $X = |Q| \times 2 \times 2^{c(n)} \times c(n) \times 2^{s(n)} \times s(n) = O(2^{s(n)+c(n)+\log s(n)+\log c(n)})$.

For an input x such that $|x| = n$, initial catalytic content $w \in \{0,1\}^{c(n)}$ and for any $i \in [n]$, we define the *crossing sequence* at position i, denoted $CS_w^i(x)$, to be the ordered sequence of semi-configuration of M given by $(C_1^i, C_2^i, \ldots, C_t^i)$ (for some $t \geq 1$) whenever the input head is on the ith cell of the input tape.

We first argue that, the constraints on $s(n)$ and $c(n)$ forces the oblivious Turing machine M to have $o(n)$ different crossing sequences possible. Using this we argue that M cannot accept a non-regular language.

By Lemma 2, for a oblivious catalytic machine M, there exists an initial catalytic content w_0 such that for all $i \in [n]$, length of $CS_{w_0}^i(x)$ is, at most the running time of M on (x, w_0) which is, $\ell = O(2^s \cdot c \cdot s)$. Hence, the number of different crossing sequence possible is given by, $\sum_{i=0}^{\ell} X^i = \frac{X^{\ell+1}-1}{X-1} = O(X^\ell)$. This can be bounded by $O(2^{(s+c+\log s+\log c)2^s \cdot c \cdot s})$. Since $s(n) \leq c(n)$ this can in turn be bounded by $O(2^{2(c+\log c)c^2 2^s})$. Now, $2(c+\log c)c^2 2^s = O(2^{3(s+\log c)}) = o(\log n)$ as $s(n) + \log c(n) = o(\log \log n)$. Hence, for any x of length n, there exists a $w_0 \in \{0,1\}^{c(n)}$ for which the number of different crossing sequences is $2^{o(\log n)} = o(n)$.

We now argue that for all x, M never uses space more than some constant c and hence can be simulated by a finite automata thereby implying that $L(M)$ is regular. For the sake of contradiction, assume that for any c, there exists infinitely many x such that M accepts x using more than c cells of catalytic and work tape. For a given c, let x_0 be an input of minimum length using c cells. By earlier argument, there exists a w_0 such that the number of crossing sequences on x_0 is $o(n)$. Hence, there exists a constant n_0 such that for all $n > n_0$ the number of possible crossing sequence of length n is less than $n/3$. This implies there must exist at least three positions $i,j,k \in [n]$ on the input tape such that $CS_{w_0}^i(x_0) = CS_{w_0}^j(x_0) = CS_{w_0}^k(x_0)$. Let x_0 be of the form of $\alpha a \beta a \gamma a \delta$ where $\alpha, \beta, \gamma \in \{0,1\}^*$ and $a \in \{0,1\}$ with the positions of symbol a corresponds to positions i, j, k. By a standard cut-and-paste argument (c.f. Chapter 1, [6]) which can be done even in the presence of a catalytic tape content, we can conclude:

Claim M must accept $x' = \alpha a \gamma a \delta$ and $x'' = \alpha a \beta a \delta$ on catalytic content w_0. In addition, $\mathsf{Time}_M(x', w_0)$ and $\mathsf{Time}_M(x'', w_0)$ are bounded by $\mathsf{Time}_M(x_0, w_0)$.

Let us argue about the amount of catalytic space used up by M. On input x_0, if M used up maximum amount of catalytic space while M's input head is within the sub-string αa of x_0 or γa of x_0 or δ of x_0 then, space used up by M on x' is at least c. Or if M used up maximum amount of catalytic space while M's input head is within the sub-string βa of x_0 then, space used up by M on x'' is at least c. In addition, x' and x'' does not take more time when run on w_0 and hence can only have shorter crossing sequence than x_0. Hence, M accepts a string *shorter* than x_0 and uses space at least c, a contradiction.   □

5   High-End Regime: Implementing Dynamic Programs Using Catalytic Space

In this section, we present our result for a case when $s(n) = O(c(n)^\epsilon)$. We prove that, even in such settings, catalytic algorithms can be provably useful. We achieve this by showing an implementation of a dynamic programming algorithm using less pure space by incorporating the use of catalytic space.

As a classic algorithmic paradigm, dynamic programming has also been studied via the model of incremental dynamic programs by Jukna [4]. We define the model from [4] formally. Every dynamic program naturally induces a sub-problem graph which is a DAG with sub problems as nodes with each of them having a value (denoted by f) and edges connecting them describing how the value $f(v)$ of sub problem v gets computed.

A dynamic program P using max and plus (min and plus resp.) computing a function f is said to be *incremental* if (1) for every input $x = (x_1, \ldots, x_n)$, each edge in the sub-problem graph is determined by at most one data item x_i. (2) for each sub problem v depending on sub problems $u_1, \ldots, u_k$, $f(v) = \max\{f(u_1) + w_1(x), \ldots, f(u_k) + w_k(x)\}$ where $w_i(x)$ is either 0 or some non-zero weight based on x.

In this section, we describe how to solve the Knapsack problem using a catalytic algorithms using small amount of pure space (Sect. 5.2). As a first step, the above framework can be used to view the Knapsack problem as an incremental dynamic program (as observed in [4]). Since such a program is a DAG with max and $+$ as operations, it can naturally be viewed as a (max, $+$) arithmetic circuit. Hence, to obtain a catalytic algorithm, it suffices to convert the associated arithmetic circuit to catalytic algorithm which we achieve in Proposition 5. The part that needs to be handled carefully is obtaining a catalytic algorithm for the max operation (which we explain in Sect. 5.1).

Catalytic Algorithms for Computing Arithmetic Circuits. We describe how to convert an arithmetic circuit (over a finite ring R) to a catalytic algorithm evaluating the polynomial computed by the arithmetic circuit. We achieve this by first converting the arithmetic circuit to a transparent program (Proposition 1) and then running the transparent program using a catalytic Turing machine (Proposition 2). These are implicit in [1] and are stated explicitly in Sect. 2.

In the following proposition, starting from an arithmetic circuit, we describe how to obtain a catalytic algorithm computing them.

Proposition 5. *Given a layered arithmetic circuit over ring R of size s and depth d, there exists a catalytic algorithm using $O(s \log |R|)$ catalytic space and $O(d + \log s + \log |R|)$ pure space evaluating the polynomial corresponding to the arithmetic circuit.*

5.1   Transparent Program to Compute max

In this section, we describe how to obtain a catalytic algorithm computing the max over any finite field. For the Knapsack problem, we choose a large enough field (based on K) and use this algorithm along with Proposition 5 to prove Theorem 4 from Introduction.

Let $(\mathbb{F}_q, <)$ be any lattice. Given two transparent programs P_1 and P_2 computing $r_1 \leftarrow r_1 + f_1$ and $r_2 \leftarrow r_2 + f_2$ over the finite field $\mathbb{F}_q$, we show how to obtain a transparent program computing $\max\{z, y\}$ where maximum is computed according to $<$ relation.

The idea is to obtain an arithmetic circuit computing $\max\{\cdot, \cdot\}$ when viewed as a polynomial over $\mathbb{F}_q[z, y]$. We then use Proposition 1 to obtain a transparent program computing max.

Proposition 6. *Let P_1 and P_2 be transparent programs compute $r_1 \leftarrow \tau_1 + f_1$ and $r_2 \leftarrow \tau_2 + f_2$ over the finite field $\mathbb{F}_q$. Then, there exists transparent programs I, P'' such that P given by I, P_1, P_2, P'' compute $r \leftarrow r + \max\{f_1, f_2\}$. The program P'' is of length $O(q^4)$ and uses and $O(q^3)$ registers. The program I can be of length $O(q^4)$. A similar statement also holds for computing min.*

Proof. Following polynomial computes max over $\mathbb{F}_q$ according to the order $<$. For $a, b \in \mathbb{F}_q$ let $\max\{a, b\}$ denote the maximum of a, b in $\mathbb{F}_q$ according to $<$ relation. $\max\{z, y\} = \sum_{a \in \mathbb{F}_q} \sum_{b \in \mathbb{F}_q} c_{a,b} \cdot \prod_{k \neq a}(z - k) \prod_{k \neq b}(y - b)$ where $c_{a,b} = \frac{\max\{a,b\}}{\prod_{k \neq a}(a - k) \prod_{k \neq b}(b - k)}$. Observe that if $z = a$ and $y = b$, then $c_{a,b} \prod_{k \neq a}(z - k) \prod_{k \neq b}(y - b) = \max(a, b)$ and is zero otherwise. Hence this polynomial correctly computes max. This polynomial can be computed by an arithmetic formula with bounded fan-in $\times$ and unbounded fan-in $+$ of size $q^2(2q + 1) + 2q = O(q^3)$ and depth $2 + \log q = O(\log q)$. We unify all occurrences of z (resp. y) as leaf into a single z (resp. y) to get an arithmetic circuit of the same size and depth as the formula. Applying Proposition 1 gives us a transparent program P' computing $r \leftarrow r + \max\{z, y\}$ of length $O(q^4)$ and $O(q^3)$

registers. Let the registers in P' be denoted by s_i and a_i denote its initial content for $1 \le i \le O(q^3)$.

We modify P' to compute $r \leftarrow r + \max\{f_1, f_2\}$ as follows. In the construction of P', there must be lines which reads z of the form $s_i \leftarrow s_i \pm x$. Let T_z (resp. T_y) be the set of indices of the registers that reads z (resp. y) in the form as mentioned. The modification is as follows: For each $i \in T_z$, we replace the read of z by $s_i \leftarrow s_i \pm r_1$ (with appropriate sign) and for each $j \in T_y$, we replace the read of y by $s_i \leftarrow s_i \pm r_2$ (with the appropriate sign). Let P'' be the resulting program.

Let I denote the following instructions: (1) for each $i \in T_z$; $s_i \leftarrow s_i - r_1$ and (2) for each $j \in T_y$; $s_j \leftarrow s_j - r_2$. The final program P computing $\max\{f_1, f_2\}$ is I, P_1, P_2, P''. We now argue correctness. If we compute $s_i \leftarrow s_i \pm f_1(x)$ for every $i \in T_z$ (by running P_1 or P_1^{-1} appropriately) and similarly for T_y, the resulting program along with P' will compute maximum. The program P'' is also doing the same without computing f_1 or f_2 (or their inverse) repeatedly. We compute f_1 and f_2 once (via P_1 and P_2 in P) and copy the values to all the registers that reads z and y (via lines 1, 2 and the modifications that is done to P').   □

Proposition 7 (($\max, +$) **Arithmetic circuits to Catalytic algorithms**). *Given a layered arithmetic circuit over $\mathbb{F}_q$ of size s and depth d with max gates of fan-in t, there exists a catalytic algorithm using $O(stq^3(\log q))$ catalytic space and $O(d \log q \log t + \log s + \log q + \log t)$ pure space computing the polynomial corresponding to the arithmetic circuit.*

Proof Replacing the max gates by an arithmetic circuit computing max, will increase the size by a factor of q^3 and depth by a factor of $\log q$. The result now follows by the fact that there exist a transparent program for computing $\max\{\cdot, \cdot\}$ of $O(q^4)$ length and $O(q^3)$ registers (Proposition 6) and by applying Proposition 5.   □

5.2   Implementing Incremental Dynamic Programming

We now apply the Proposition 7 in the context of Knapsack problem. Observe that a Knapsack problem on n items with Knapsack capacity K can be expressed as a $(\max, +)$ circuit with size nK, depth n and max fan-in of n. Recall that C is the sum of the cost of all items. We choose a prime q such that $C \le q < 2C$ (such a prime always exists) with the natural associated ordering $\mathbb{F}_q$. By Proposition 7, we get a catalytic algorithm solving the Knapsack problem in $O(n \log n \log C + \log(nKC))$ pure space and $O(n^2 KC^3 \log^2 K \log n)$ catalytic space. This completes the proof of Theorem 4.

On the other hand, this problem can be solved by a DP algorithm that uses $O(K \log C)$ pure space. To see this, let T be an array of size K. Then, for each $i \in \{1, 2, \ldots, n\}$ and each j from $\{0, 1, 2, \ldots K - w_i\}$, do $T_{K-j} \leftarrow \max\{T_{K-j}, T_{K-j-w_i} + c_j\}$ (where c_j and w_j is the reward and the weight of jth item). The optimal value will be available in T_K at the end of the iteration. Hence, our catalytic algorithm can do a significant saving in pure space ($O(K \log C)$ versus $O(n \log K \log C)$) when K is $\Omega(n)$.

Acknowledgments. The authors thank the anonymous reviewers for their constructive comments. Part of the work was done while the second author was at IIT Madras and was supported by the ERP funding CSE1718842RFERMNJA.

References

1. Buhrman, H., Cleve, R., Koucký, M., Loff, B., Speelman, F.: Computing with a full memory: catalytic space. STOC **2014**, 857–866 (2014)
2. Buhrman, H., Koucký, M., Loff, B., Speelman, F.: Catalytic space: non-determinism and hierarchy. Theory Comput. Syst. **62**(1), 116–135 (2018). https://doi.org/10.1007/s00224-017-9784-7
3. Duris, P., Galil, Z.: Fooling a two way automaton or one pushdown store is better than one counter for two way machines. Theor. Comput. Sci. **21**, 39–53 (1982)
4. Jukna, S.: Limitations of incremental dynamic programming. Algorithmica **69**(2), 461–492 (2014). https://doi.org/10.1007/s00453-013-9747-6
5. Koucký, M.: Catalytic computation. Bull. EATCS **118** (2016)
6. Kozen, D.: Theory of Computation. Texts in Computer Science. Springer, London (2006). https://doi.org/10.1007/1-84628-477-5
7. Kushilevitz, E., Nisan, N.: Communication Complexity. Cambridge University Press, Cambridge (1997)

Author Index

Printed in the United States
By Bookmasters